ZINC BIOCHEMISTRY, PHYSIOLOGY, AND HOMEOSTASIS
Recent Insights and Current Trends

Edited by

W. MARET
Center for Biochemical and Biophysical Sciences and Medicine
Harvard Medical School, Boston, MA, USA

Reprinted from *BioMetals*, Volume 14, Numbers 3-4, 2001

KLUWER ACADEMIC PUBLISHERS
DORDRECHT / BOSTON / LONDON

A C.I.P. Catalogue record for this book is available from the Library of Congress.

ISBN 0-7923-6712-X

Published by Kluwer Academic Publishers,
P.O. Box 17, 3300 AA Dordrecht, The Netherlands

Sold and distributed in North, Central and South America
by Kluwer Academic Publishers,
101 Philip Drive, Norwell, MA 02061, U.S.A.

In all other countries, sold and distributed
by Kluwer Academic Publishers,
P.O. Box 322, 3300 AH Dordrecht, The Netherlands

Printed on acid-free paper

Printed in the Netherlands

Table of Contents

Cover illustration

Zinc Homeostasis: Schematic of major pathways in the regulation of cellular zinc, including importers (e.g. Zrt1 and hZip1), exporters (e.g. Znt-1 and Zrc1), metallothionein (MT), and an action of zinc on transcriptional regulators such as MTF-1, an activator induced by zinc, and Zap1, an activator repressed by zinc.

Zinc Function: Schematic of some major functions of cellular zinc.

Dr. David J. Eide is acknowledged for providing the first figure and the template for the second figure.

BioMetals **14:** 187–190, 2001.

Editorial

Zinc biochemistry, physiology, and homeostasis – recent insights and current trends

Wolfgang Maret
Center for Biochemical and Biophysical Sciences and Medicine, Harvard Medical School, Seeley G. Mudd Bldg., 250 Longwood Ave., Boston, MA 02115, USA (Phone: 617-432-5685; Fax: 617-566-3137; E-mail: wolfgang_maret@hms.harvard.edu)

Reviews in this special issue summarize recent discoveries in the biology of zinc in order to draw further attention to the significance of this metal, which is essential for growth and development (Vallee & Falchuk 1993). Recognition of its full potential was much delayed, because its chemical properties and the way it is utilized in biology posed serious challenges to the investigator. Thus, zinc is colorless and diamagnetic, properties that render it invisible to most spectroscopic methods. Therefore, physicochemical approaches brought comparatively less insight into the biology of zinc than into that of e.g. copper or iron. Moreover, unlike iron, where 80% of a total of about 3 g in a human is in the heme group alone, similar total amounts of zinc are spread among thousands of proteins. This dilution effect makes it much more difficult to establish the presence and role of zinc in low abundance proteins. Yet, from 1950 on, an ever-increasing number of zinc enzymes was discovered (Vallee & Galdes 1984; Vallee & Auld 1990). The number of just those with known 3D structures is now already 200.

TFIIIA was the first transcription factor to be identified as a zinc protein (Hanas *et al.* 1983). When the term DNA-binding *finger* was introduced for the nine repetitive domains in this protein (Miller *et al.* 1985), discovery followed a different path. The close spacing of metal ligands in the primary sequences of zinc finger proteins allowed recognition of recurring zinc binding motifs. Consequently, it became common practice to define any new protein with such a motif as a zinc protein, thus assuming the presence of zinc rather than determining it directly. On this basis, hundreds of zinc finger proteins were identified within about 15 years. The domains that zinc organizes in these functionally and structurally diverse proteins are key elements for the molecular recognition of nucleic acids, proteins, or lipids (Laity *et al.* 2001). With blueprints of entire genomes now in hand, we are beginning to grasp the size of the *zinc proteome*, at least with regard to the number of zinc finger proteins (Clarke & Berg 1998). Over one thousand genes in the human genome encode members of three protein families with zinc finger domains alone, i.e., C2H2 zinc fingers, RING fingers and LIM domains (International Human Genome Sequencing Consortium 2001). In other words, the number of genes containing zinc finger domains exceeds 3% of the about 32,000 identified human genes. The set of signatures for other zinc proteins is less complete or even unknown for sites such as those between protein subunits. Therefore, not all sites can be accounted for by homology searches or data base mining, and a final count is still out.

Having established this impressive number of zinc proteins, the question remains how all these proteins acquire their zinc at the right time. The total zinc concentration of a eukaryotic cell is quite high, i.e., about 200 μM (Palmiter & Findley 1995). The bulk of it is bound very tightly in proteins. One corollary of this strong interaction is that the concentration of 'free' or freely available zinc is very low, with estimates given as picomolar for mammalian cells (Peck & Ray 1971; Simons 1991). This seems to rule out this pool as the source for incorporation into proteins. A key issue, and not a trivial one, therefore, is the chemical form in which zinc is made available. Recent discoveries make it clear that an elaborate homeostatic system of proteins regulates cellular zinc distribution and perhaps controls a hierarchy of zinc-dependent functions. In fact, the coordination dynamics of new types of zinc proteins involved in zinc trafficking now open fascinating and unprecedented aspects of an ingenious bioinorganic chemistry. Discoveries of zinc transporters gave the first clues about the participation of specific proteins in a homeostatic system (Palmiter & Findley 1995; Eide 1997). Zinc sensors that induce gene transcription as a result of either too much or too little cellular zinc are the second type of molecules belonging to this system, while a third is metallothionein (MT). The function of MT might serve as an example of the uniquely biological chemistry that has evolved in zinc metabolism to deal with the problem of distributing zinc. MT links zinc distribution to the redox state of the cell (Maret & Vallee 1998). The molecular basis for this coupling is that the bonding to sulfur donor atoms of cysteine ligands confers redox activity on the otherwise redox-inert zinc atom. Thus, a change in the cellular redox potential toward more oxidizing conditions can induce kinetic lability in zinc/sulfur coordination sites and thereby provide a driving force for zinc transfer against thermodynamic gradients, e.g. from its tight binding in MT to sites of lower affinity. MT does not only occur in its zinc-loaded form, but also in its apoform thionein (T) at varying ratios with regard to MT (Yang *et al.* 2001). T is an efficient endogenous chelating agent and an effective thermodynamic sink for zinc. Regulation of MT at the protein level by ligand binding (Jiang *et al.* 1998) and at the gene level by multiple inducers provides a means to control the availability of zinc by adjusting the amount of MT and the MT/T ratio. Moreover, eukaryotic cells compartmentalize zinc and MT/T in their organelles. Questions are now which energy, signals, and mechanisms control their subcellular translocations, and what are the characteristics of the pools of available zinc in different compartments of the cell. Finally, on an organizational level higher than the cell, it remains a major challenge to identify the hormones that regulate body zinc homeostasis.

Another important aspect is that zinc has regulatory functions. This is best illustrated by its role in neurotransmission (Frederickson *et al.* 2000). So-called zinc-containing neurons innervate the forebrain and contain zinc in synaptic vesicles. Nerve stimulation releases this zinc into the synaptic cleft where it has a neuromodulatory function, either outside the postsynaptic cell or inside it, or at both locales. It is intriguing that regulatory functions may not be limited to neurotransmission and may involve control of biological function by transient zinc binding in general.

The chapters in this special issue have been arranged such that there is a progression from single molecules to cells and then to whole organisms. The opening chapters start with the perennial task of how to make the spectroscopically silent zinc visible for investigations. In proteins, metal substitution techniques are the classical approach to achieve this. Thus, cobalt is frequently used to study the function and coordination environment of zinc in proteins *in vitro*. When analyzing zinc or imaging zinc fluxes *in vivo*, substitution, of course, is not an option. Instead, visualization is approached in very much the same manner as for calcium, i.e. with specific fluorophore sensors. **Kimura and Aoki** summarize the structural types of presently available zinc fluorophores and their chemical and physical properties, define the limits of their use for *in vitro* and *in vivo* studies, and examine questions that stipulate development of new fluorophores. **Fierke and Thompson** have developed protein fluorescence sensors based on the very high sensitivity and selectivity of carbonic anhydrase for zinc. They have used the understanding of the coordination chemistry of this protein in combination with protein engineering to optimize affinity for zinc, selectivity and binding kinetics as well as the physical properties of the fluorophore. **Andrews** then describes how cells sense zinc by metal-response element-binding transcription factor-1 (MTF-1), a six zinc-finger protein that is essential for embryonic development. MTF-1 has been a paradigm for metal-dependent gene expression of metallothionein and has now been shown to coordinate the expression of other proteins involved in zinc homeostasis, e.g., zinc-transporter-1. This theme is enlarged in the chapters of **Hantke, Gaither and Eide** who discuss transporters and their regulators. Prokaryotes have at least three types of zinc export systems as well as high- and low-affinity uptake systems, all under control of individual repressors and activators. Gaither and Eide point out the major advances that have been made in the last decade through the discovery of two families of zinc transporters and their regulators in eukaryotes. Many of those belonging to the ZIP (Zrt, Irt-like Protein) family are involved in uptake, including one member that transports zinc out of a cellular compartment. Those of the CDF (cation diffusion facilitator) family mediate zinc efflux or zinc transport into cellular compartments. Neither zinc binding sites nor mechanisms of actions of these proteins are known. **Auld** gives a unique, comprehensive, and up-to-date account of those zinc proteins for which this information is available. He summarizes the number of zinc sites in proteins as characterized by high resolution structural methods, their characteristics and their classification in structural, catalytic, co-catalytic, and protein interface sites. Though zinc finger proteins are not even included, the sheer number of zinc sites in proteins illustrates to what extent zinc proteins impact cell functions. The following chapters then focus on cellular aspects. Zinc deficiency increases apoptosis and halts differentiation and proliferation of cells. **Truong-Tran, Carter, Ruffin and Zalewski** review the evidence suggesting a role of zinc in apoptosis. Zinc is cytoprotective and suppresses apoptotic pathways. Among the multiple targets and mechanisms, regulation of caspase activity is perhaps the critical zinc-dependent event. The chapter by **Beyersmann and Haase** covers the essential role of zinc in cellular proliferation and differentiation. They propose a new role of zinc in cGMP signaling. Again zinc is involved at multiple levels of signal transduction, and the most sensitive zinc-dependent targets in growth and development have not yet been identified. Two chapters then address the role of zinc in brain, where it has a specific function as a neuromodulator in addition to its other typical cellular functions. **Frederickson and Bush** focus on the role of zinc in zinc-containing synaptic terminals in brain. Zinc acts as a signaling substance akin to conventional neurotransmitters in normal physiology. It also exerts pathophysiological effects by participating in plaque deposition in Alzheimer disease and by causing injury to neurons in excitotoxicity. **Takeda** outlines the role of the brain barrier systems for brain zinc homeostasis and discusses learning impairment and olfactory dysfunctions as a result of diet-induced zinc deficiency. **Rink and Gabriel** provide an overview of the extracellular and immunological actions of zinc. The responsiveness of leukocytes and the action of immunostimulants critically depend on the concentration of zinc. Both zinc deficiency and supraphysiological levels of zinc modulate the immune response. Evidence provided in these chapters suggests intracellular, extracellular, and intercellular signaling functions of zinc. **Falchuk and Montorzi** give an account of a complex zinc storage, transport, and distribution system in *Xenopus laevis* oocyte and embryo development. In the frog, the liver protein vitellogenin transports zinc to the egg. A cytosolic pool amounts to about 10% of the total zinc and is responsible for embryogenesis and organogenesis in this closed system. The second pool contains 90% of the total zinc bound to lipovitellin in the yolk sac and is mobilized for later stages of development. **Krebs and Hambidge** discuss tracer kinetic techniques with stable zinc isotopes in humans, which clearly demonstrate the role of the gastrointestinal tract in maintaining whole body zinc homeostasis. Excessive losses, increased requirements, and redistribution are perturbations of homeostasis in disease processes. Such zinc metabolic studies define the regulation of zinc balance to the point where specific questions can be addressed regarding the molecules involved in homeostasis.

Faced with the rich zinc biochemistry outlined in this special issue one wonders why so far zinc has had so little impact on medicine. Perhaps the zinc homeostatic system is so critical and efficient that organisms cannot afford to compromise the supply and distribution of zinc. There is only one known genetic disease of zinc metabolism, acrodermatitis enteropathica, and its pathogenesis is unknown. No other disease has been linked directly to zinc, clearly marginalizing its clinical significance. Yet, imbalances are known. Even mild zinc deficiency, which is quite common, impairs immunological function and defense mechanisms, and increases infectivity. On the other hand, accumulation of zinc in neurons leads to neurodegeneration. Therefore, zinc may be a major etiological factor, and nutritionally a most important metal for human health. Insights into the factors that control, interact with, and perturb cellular zinc and lead to reversible or irreversible changes will fill the large gap between the now known functions of zinc at the molecular level and gross biological observations of stunted growth, developmental abnormalities, and other characteristic signs of zinc deficiency. Therapeutic intervention will eventually follow, not necessarily only with zinc itself, but by characterizing substances that participate in the efficient regulation of zinc availability in the gastrointestinal tract and by targeting specific proteins in the zinc homeostatic system.

References

Clarke ND, Berg JM. 1998 Zinc Fingers in *Caenorhabditis elegans*: Finding Families and Probing Pathways. *Science* **282**, 2018–2022.

Eide D. 1997 Molecular biology of iron and zinc uptake in eukaryotes. *Curr Opin Cell Biol* **9**, 573–577.

Frederickson CJ, Su SW, Silva D, Frederickson CJ, Thompson RB. 2000 Importance of Zinc in the Central Nervous System: The Zinc-Containing Neuron. *J Nutr* **130**, 1471S–1483S.

Hanas JS, Hazuda D, Bogenhagen DF, Wu FY-H, Wu C-W. 1983 *Xenopus* trancription factor A requires zinc for binding to the 5S gene. *J Biol Chem* **258**, 14120–14125.

International Human Genome Sequencing Consortium. 2001 Initial sequencing and analysis of the human genome. *Nature* **409**, 860–921.

Jiang L-J, Maret W, Vallee BL. 1998 The ATP-metallothionein complex. *Proc Natl Acad Sci USA* **95**, 9146–9149.

Laity JH, Lee BM, Wright PE. 2001 Zinc finger proteins: new insights into structural and functional diversity. *Curr Opin Struct Biol* **11**, 39–46.

Maret W, Vallee BL. 1998 Thiolate ligands in metallothionein confer redox activity on zinc clusters. *Proc Natl Acad Sci USA* **95**, 3478–3482.

Miller J, McLachlan AD, Klug A. 1985 Repetitive zinc-binding domains in the protein transcription factor IIIA from *Xenopus* oocytes. *EMBO J* **4**, 1609–1614.

Palmiter RD, Findley SD. 1995 Cloning and functional characterization of a mammalian zinc transporter that confers resistance to zinc. *EMBO J* **14**, 639–649.

Peck Jr. EJ, Ray Jr. WJ. 1971 Metal Complexes of Phosphoglucomutase *in vivo*. *J Biol Chem* **246**, 1160–1167.

Simons TJB. 1991 Intracellular Free Zinc and Zinc Buffering in Human Red Blood Cells. *J Membr Biol* **123**, 63–71.

Vallee BL, Galdes A. 1984 The Metallobiochemistry of Zinc Enzymes. *Adv Enzymol Relat Areas Mol Biol* **56**, 283–430.

Vallee BL, Auld DS. 1990 Zinc Coordination, Function, and Structure of Zinc Enzymes and Other Proteins. *Biochemistry* **29**, 5647–5659.

Vallee BL, Falchuk KH. 1993 The Biochemical Basis of Zinc Physiology. *Physiol Rev* **73**, 79–118.

Yang Y, Maret W, Vallee BL. 2001 Differential fluorescence labeling of cysteinyl clusters uncovers high tissue levels of thionein. *Proc Natl Acad Sci USA* **98**, 5556–5559.

BioMetals 14: 191–204, 2001.

Review

Chemistry of zinc(II) fluorophore sensors

Eiichi Kimura* & Shin Aoki
*Department of Medicinal Chemistry, Faculty of Medicine, Hiroshima University, Minami-ku, Hiroshima, 734-8551, Japan; *Author for correspondence (Tel: +81-82-257-5320; Fax: +81-82-257-5324; E-mail: ekimura@hiroshima-u.ac.jp)*

Received 12 January 2001; accepted 28 January 2001

Key words: carbonic anhydrase, fluorophore, macrocyclic polyamine, sensor, sulfonamide, zinc, zinquin

Abstract

The biological role of the zinc(II) ion has been recognized in DNA and RNA synthesis, apoptosis, gene expression, or protein structure and function. Therefore, development of useful zinc(II) sensors has recently been attracting much interest. Chemistry for selective and efficient detection of trace Zn^{2+} is a central issue. Recently, various types of zinc-fluorophores are emerging, comprising bio-inspired aromatic sulfonamide derivatives, zinc-finger peptides attached to fluorescent dyes, or fluorophore-pendant macrocyclic polyamines. The chemical principles, properties and limitations of these Zn^{2+}-fluorophores are discussed.

Introduction

Qualitative and quantitative analyses of trace metal ions with selective analytical reagents have become extremely important for environmental and biological applications (Czarnik 1995). A remarkable development of fluorescent indicators has already been made for biologically important divalent metal ions, in particular Ca^{2+} and Mg^{2+}, with quite a few practical fluorophores such as Fura-2 (**1**), Quin-2 (**2**) and Mag-indo-1 (**3**) (Grynkiewicz *et al.* 1985; Tsien 1989; Tsien & Pozzan 1989; Haugland 1996).

The criteria for good sensors are (i) stability, (ii) metal selectivity, (iii) metal affinity, (iv) signal transduction, (v) fluorescent signaling, (vi) kinetically rapid sensitization, (vii) ease of delivery to target systems, and (viii) availability. For measurement of dynamic mechanisms of intracellular metal ions, the typical concentrations in resting cells should be known: for instance, $[Ca^{2+}]$ = 50–200 nM. Therefore, for the metal affinity criteria, Ca^{2+}-selective biosensors should possess a K_d (dissociation constant) near the median concentration at physiological pH. When a normal median concentration gives a 50% sensing signal, sensors could most effectively detect both concentration increases and concentration decreases. Fura-2 (K_d = 145 nM) and Quin-2 (K_d = 60 nM), in this regard, are quite appropriate probes for measurement of intracellular Ca^{2+} concentrations (Haugland 1996). As for the desirable fluorescent signaling properties,

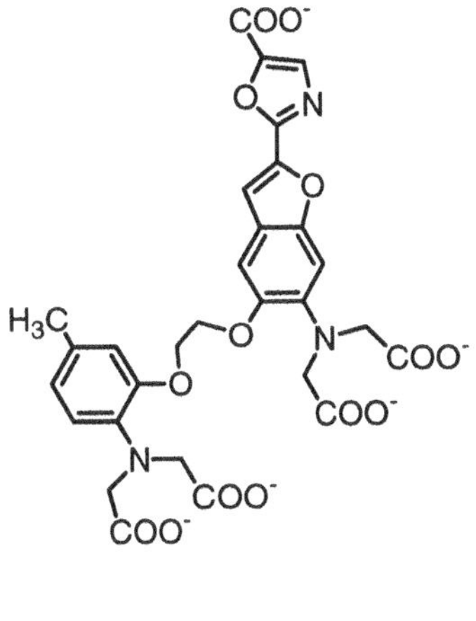

Fura-2
$K_d(Ca^{2+})$: 145 nM
Em: 512 nm (without Ca^{2+})
Em: 505 nm (with Ca^{2+})

Structure 1.

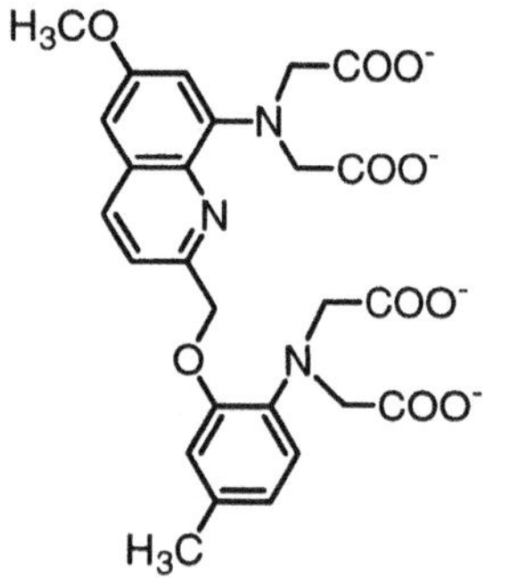

Quin-2
$K_d(Ca^{2+})$: 60 nM
Em: 495 nm (without Ca^{2+})
Em: 495 nm (with Ca^{2+})

Structure 2.

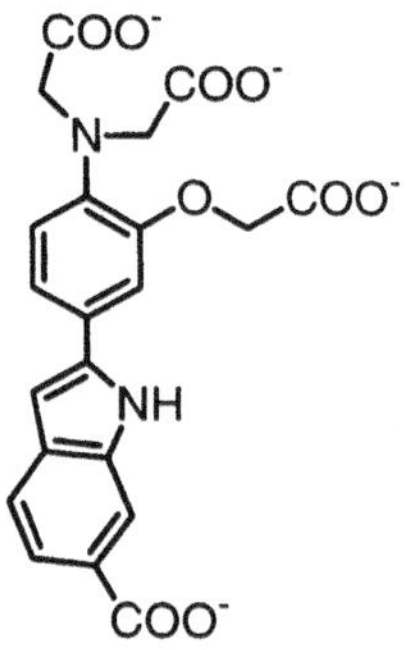

Mag-indo-1
$K_d(Mg^{2+})$: 2.7 mM
Em: 480 nm (without Mg^{2+})
Em: 417 nm (with Mg^{2+})

Structure 3.

(a) intense fluorescence, (b) excitation wavelengths exceeding 340 nm (to pass through glass microscope objectives and minimize UV-inducing cell damage) corresponding to available laser sources, and (c) desirably, emission wavelengths to shift >80 nm before and after metal complexation, so that ratiometric titration can be utilized (for quantification) rather than mere intensity changes.

Development of classical zinc(II)-fluorophores

Zinc(II) is an essential metal ion in the active sites of more than 300 enzymes such as carbonic anhydrase, carboxypeptidase A, class II aldolase, β-lactamase, alkaline phosphatase, phosphotriesterase, and collagenase (Fraústo da Silva & Williams 1991; Kimura 1994; Kimura & Koike 1996; Lipscomb & Sträter 1996; Sträter *et al.* 1996). Moreover, the importance of zinc(II), which is critical for the growth and survival of cells, is becoming recognized in biology, physiology, and pathology (Vallee & Falchuk 1993; Lippard & Berg 1994); e.g., protein structure and function (Cox & McLendon 2000), DNA and RNA synthesis, gene expression (Greisman & Pabo 1997), transcription mediated by NO, apoptosis (Berendji *et al.* 1997), and brain metabolism or diseases (Frederickson *et al.* 1987; Cuajungco & Lees 1997; Choi & Koh 1998). The concentration of free Zn^{2+} within biological cells varies from about 1 nM in the cytoplasm of many cells to about 1 mM in some vesicles. The need for useful zinc-fluorophores to quantify trace Zn^{2+} in these biological mechanisms has become more urgent.

A zinc(II)-fluorophore 6-methoxy-8-*p*-toluenesulfonamido-quinoline (TSQ) **4** was first used as a histochemical stain for Zn^{2+} in various tissue sections of brain, heart, and some other tissues (Frederickson *et al.* 1987). While TSQ was the only available Zn^{2+}-specific fluorophore in the presence of much higher concentrations of Ca^{2+} and Mg^{2+}, the complex structures and stability constants of the TSQ-Zn^{2+} complexes were neither identified nor characterized. The complexation of TSQ with free Zn^{2+} probably occurs not only in a stoichiometry of 2:1 TSQ/Zn^{2+}, but also in a 1:1 complex that may equilibrate with protein-binding. The fluorescence intensity (i.e., quantum yield) of the complex(es) varies with the media. Accordingly, TSQ needed to be carefully studied for quantitative analysis of Zn^{2+}.

Zalewski's group developed Zinquin **5** and extensively used it for cellular physiological studies (Zalewski *et al.* 1993; Zalewski *et al.* 1994a, b). An ester group was incorporated at 6-position of **5**, so that after the neutral lipophilic probe **5** permeates into the cell, the ester would be hydrolyzed to a carboxylate anionic form **6** by intracellular esterases to stay within the cell. Thus, **5** became the first practical zinc-fluorophore to examine the role of Zn^{2+} in regulation of cell growth. Zinquin **5** could monitor loosely bound, labile intracellular Zn^{2+} (but not tightly bound Zn^{2+} in zinc-enzymes or zinc-finger proteins) by fluorescence video image analysis or fluorometric spectroscopy. For instance, the importance of cellular Zn^{2+} distribution in the process of apoptosis was first assayed by **5** (Zalewski *et al.* 1994a) in zinc-rich cells such as hepatocytes and pancreatic islet β-cells where the fluorescence was very intense

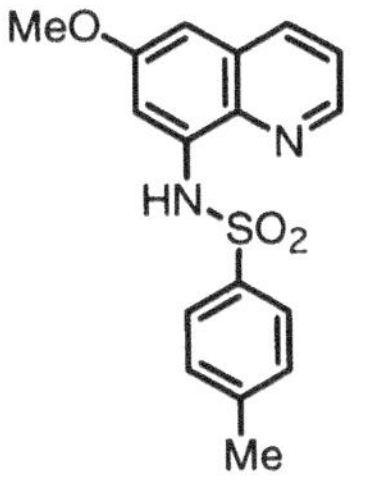

TSQ
Em: 495 nm
Ex: 334 nm
Structure 4.

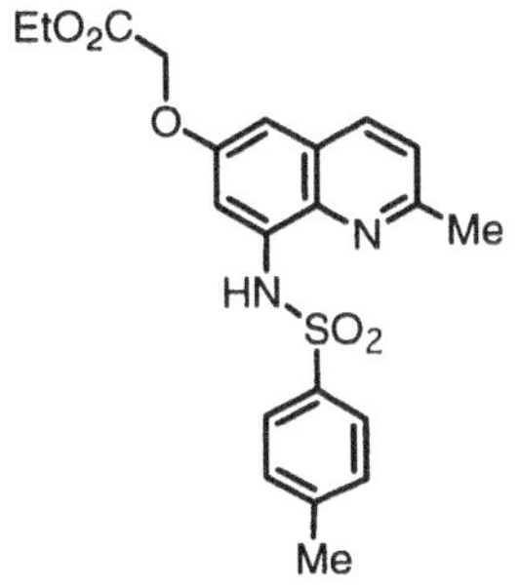

Zinquin
Em: 490 nm
Ex: 370 nm
Structure 5.

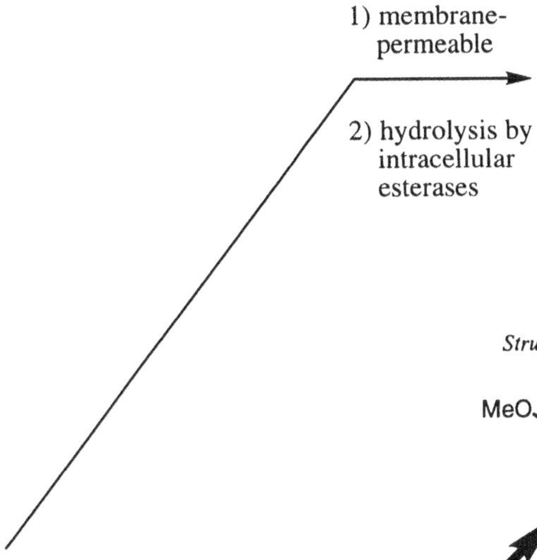

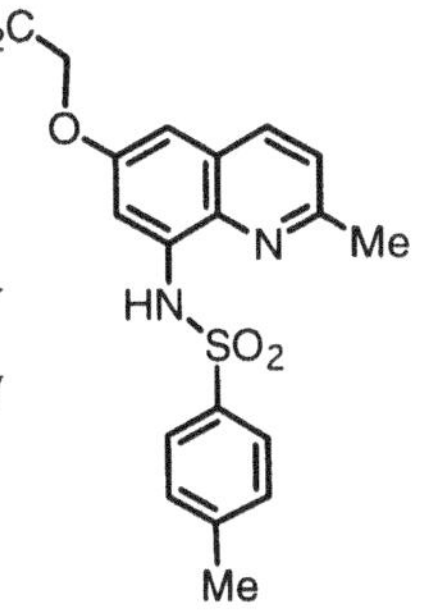

Zinquin acid
Structure 6.

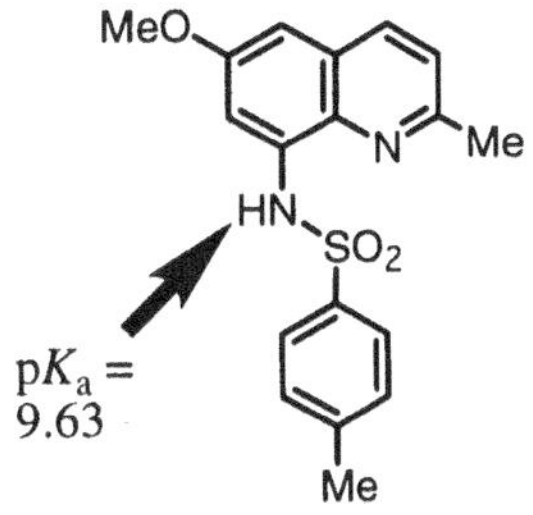

2-Me-TSQ
Em: 485 nm
Ex: 362 nm
Structure 7.

(Coyle *et al.* 1994; Zalewski *et al.* 1994b). Very weak fluorescence (at 490 nm) of a 2 μM solution of **5** at pH 7.4 was increased with subnanomolar free Zn^{2+} and was saturated at 1 μM Zn^{2+}. The fluorescence was enhanced 20-fold by 1 μM Zn^{2+}. Other biologically relevant metal ions (Ca^{2+}, Mg^{2+}, Cu^{2+}, Fe^{2+}, Fe^{3+}, Mn^{2+}, Co^{2+}, etc.) did not affect the Zn^{2+}-dependent fluorescence of **5**, which empirically seemed to be a practical fluorophore for probing Zn^{2+} concentrations ranging 100 pM–10 nM. By fluorometric titration, **5** was shown to form 1:1 and the subsequent 2:1 complexes with Zn^{2+} with binding constants of 7.0×10^6 M^{-1} and 11.7×10^6 M^{-1} at pH 7.4 (Zalewski *et al.* 1993). However, the structures of these complexes were not referred to.

In addressing the basic chemistry of the TSQ fluorophores **4** and **5**, O'Halloran's group recently studied 2-methyl-6-methoxy-8-*p*-toluenesulfonamido-quinoline (2-Me-TSQ) (**7**) (Nasir, *et al.* 1999; Fahrni & O'Halloran 1999). The deprotonation constant of the sulfonamide in **7** was determined to be 9.63 in a 80:20 (v/v) mixture of DMSO/water (I = 0.1 ($KClO_4$)) at 25 °C by potentiometric pH titration. The formation constants, log β_1 of 8.43 ± 0.38 and log β_2 of 18.24 ± 0.24 for the 1:1 and 2:1 complexes of **7** with Zn^{2+} were established $\beta_1 = [(\mathbf{7}^-) - Zn^{2+}]/[\mathbf{7}][Zn^{2+}]$ (M^{-1}) and $\beta_2 = [(\mathbf{7}^-)_2 - Zn^{2+}]/[\mathbf{7}]^2[Zn^{2+}]$ (M^{-2}). It was revealed that the 2:1 $\mathbf{7}^-$-Zn^{2+} tetrahedral complex (**8**) is the dominant species at neutral pH in DMSO/water, in which the deprotonated imide N^- and aromatic N atom of **7** coordinated to zinc(II), as confirmed by the X-ray crystal structure analysis. It was assumed that the adoption of the distorted tetrahedral geometry by the two methyl groups at 2-position of quinoline rings made **7** a Zn^{2+}-selective staining reagent in living cells.

Further, Zinquin **5** (ester form) and its carboxylic acid form **6** (Zinquin acid) have been more elaborately characterized (Hendrickson *et al.* 1997; Fahrni & O'Halloran 1999). Under physiological conditions (pH 7.2), the two forms of Zinquin **5** and **6** bind to Zn^{2+} to form 2:1 complexes with similar overall binding constants, e.g., log K_{app} of 13.5 for $(\mathbf{6}^-)_2$-Zn^{2+} complex (K_{app} =

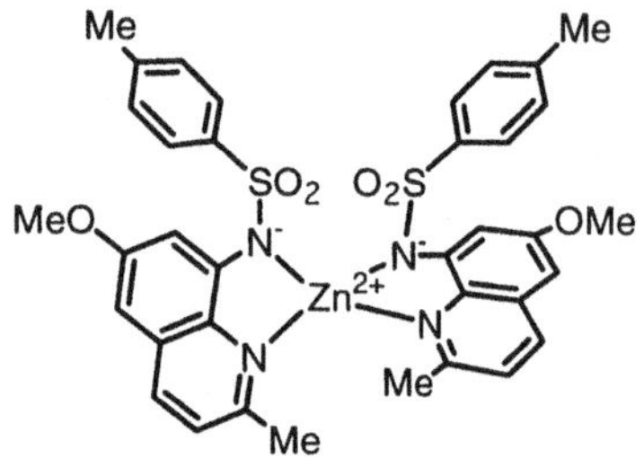

2:1 (7^-)–Zn^{2+} complex
Em: 485 nm
Ex: 362 nm

Structure 8.

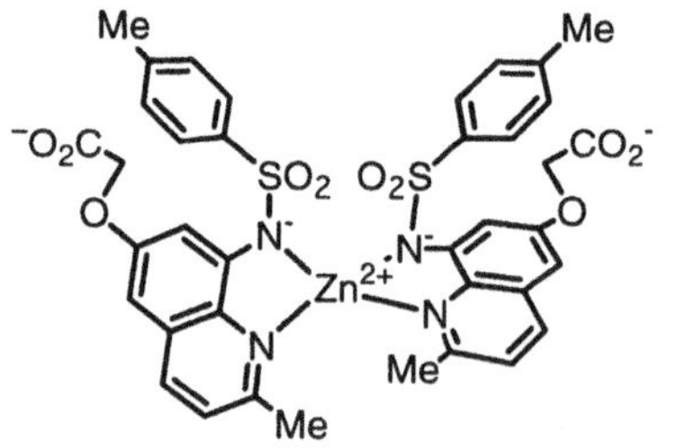

2:1 (6^-)-Zn^{2+} complex
Em: 490 nm
Ex: 370 nm

Structure 9.

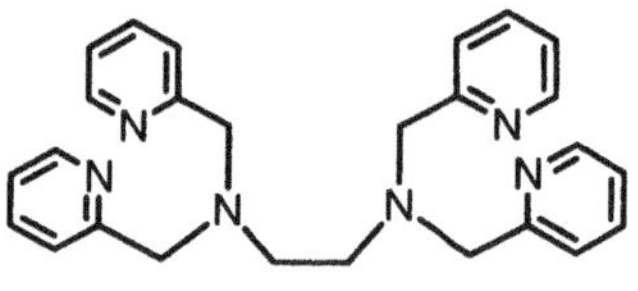

TPEN

Structure 10.

$[(\mathbf{6}^-)_2 - [(\mathbf{6})_2 - Zn^{2+}]/[\mathbf{6}]^2[Zn^{2+}]\,(M^{-2}))$ (Fahrni & O'Halloran 1999). In the presence of 50 μM Zinquin acid **6**, the lower detection of Zn^{2+} was ca. 4 pM and the fluorescence intensity reaches saturation above 100 nM Zn^{2+}. It was evident earlier, however, that the Zn^{2+}-Zinquin stability constants were apparently not large enough to permit interaction of Zinquin with extremely tightly bound (K_d << 1 nM) Zn^{2+} in metalloenzymes or zinc-finger proteins. A typical intracellular Zn^{2+} chelator N,N,N',N'-tetrakis(2-pyridylmethyl)ethylendiamine (TPEN) **10**, which apparently has a much higher affinity for Zn^{2+}, masked the Zn^{2+}-dependent fluorescence of Zinquin in lymphocyte cells (Zalewski *et al.* 1993). O'Halloran's group determined the affinity of various ligands to Zn^{2+} in terms of free $[Zn^{2+}]$ (M), referred to as -log$[Zn^{2+}]$ at [ligand] = 10 mM and [initial Zn^{2+}] = 1 mM at pH 7, 0.1 M ionic strength, and 25 °C: Zinquin acid (**6**) 9.3; TPEN (**10**) 16.0; EDTA (**11**) 14.3; carbonic anhydrase (CA) 12.4 (Fahrni & O'Halloran 1999). Thus, it was quantitatively confirmed that TPEN **10** and CA bind Zn^{2+} with much higher affinities and therefore Zinquins do not mobilize tightly bound Zn^{2+} from enzymes such as carbonic anhydrase (CA). It was suggested that once formed in the cell, the 2:1 complex **8**, like **9**, is not membrane-permeable, since these two fluorophores show similar staining pattern in the presence of Zn^{2+} in living cells (Nasir *et al.* 1999).

Development of zinc-fluorophores from the zinc-containing enzyme, carbonic anhydrase

Mann and Keilin first discovered that sulfonamides inhibit carbonic anhydrase (CA) (Mann & Keilin 1940). Chen and Kernohan presented evidence that bovine erythrocyte CA incorporates equimolar dansylamide **12** to form a highly fluorescent complex **13** with a dissociation constant K_d of 2.5×10^{-7} M at pH 7.4 (Chen & Kernohan 1967). For comparison, the K_d value for Zn^{2+} binding to apoCA is much smaller 4×10^{-12} M (Kiefer *et al.* 1993). The fluorescence of free **12** in water has an emission peak at 580 nm with a quantum yield of only 0.055, but the CA-bound dansylamide **13** dramatically shifted the emission maximum down to 468 nm with much higher quantum yield of 0.84 (excited at 326 nm). The large blue shift of emission was rationalized by the well-shielded and extremely hydrophobic dansyl binding site and in addition by the sulfonamide group losing a proton (to SO_2NH^-) upon binding to CA (Scheme 1). Thus, dansylamide **12** was thought to be a good fluorescent probe of CA or free Zn^{2+} in the presence of apoCA. The pK_a value of the SO_2NH_2 group in **12** was 9.8 either at ground state or excited state. Mere deprotonation of the free **12** (in the absence of CA) in alkaline solution shifted

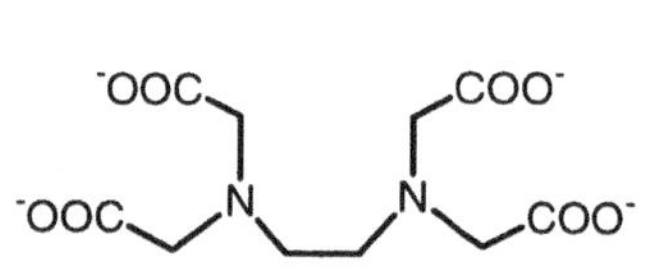

EDTA

Structure 11.

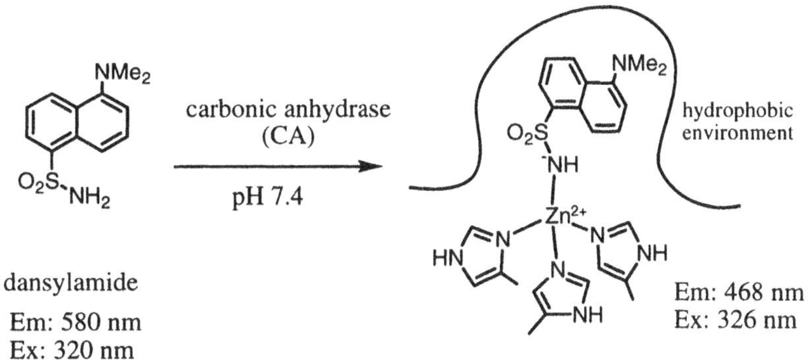

dansylamide
Em: 580 nm
Ex: 320 nm
Structure 12.

Em: 468 nm
Ex: 326 nm
Structure 13.

Scheme 1.

the emission peak less dramatically from 580 nm to 540 nm with the quantum yield from 0.055 to 0.085.

These chemical principles were adopted to a new CA-based fiber optic zinc-biosensor developed by Thompson's group (Thompson & Jones 1993; Thompson & Patchan 1995; Thompson & Maliwal 1998; Thompson *et al.* 1998). The concentration of Zn^{2+} is proportional to the ratio of fluorescence intensities (at 580 nm vs. at 480 nm) at 10–1000 nM (with 1 μM apoCA (metal-free CA) and 10 μM **12** in pH 7.4 HEPES buffer). An advantage with the CA-dansylamide system is that a great wavelength shift in fluorescence with or without Zn^{2+} in CA permits the ratiometric detection at two different wavelengths to be correlated with the analytical level. This linear range interestingly corresponds to the zinc(II) concentration range in the ocean. A fiber optic sensor constructed using this approach showed the zinc-detection limit 10-fold inferior (Thompson & Jones 1993). For practical application of this CA-based sensor to measure environmental Zn^{2+} (e.g., in sea water), a serious problem is reversibility. The off-rate of Zn^{2+} from CA is ca. 10^{-8} s^{-1}, which is too slow to permit taking continuous data. Another problem is fiber attenuation.

Further development from the CA inhibitors are sulfonamide-fluorescein conjugate **14** (Elbaum *et al.* 1996) and dapoxyl sulfonamide **15** (Thompson *et al.* 1999). The design of **14** was achieved by an iterative structure-based procedure in which the X-ray crystal structure of the CAII-arylsulfonamide complex was determined and analyzed for the optimal attachment point of fluorescein. The probe **14** bound tightly only to the Zn^{2+}-bound CAII ($K_d = 2.3$ nM) and exhibited fluorescence anisotropy that is proportional to the concentration of bound Zn^{2+} in the range of 10–100 nM. Dapoxyl sulfonamide **15** exhibited a large increase

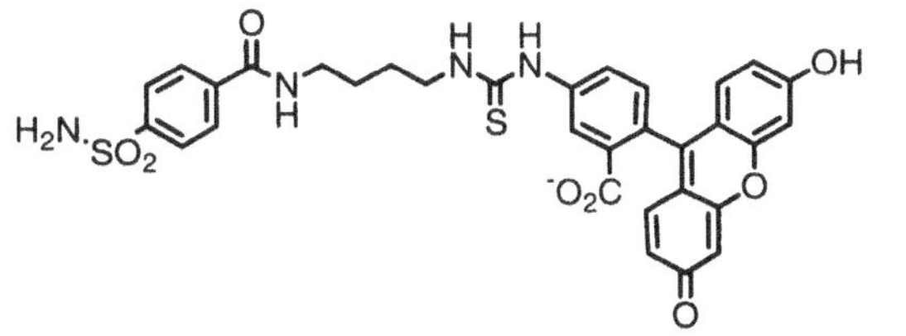

λ_{em} = 520 nm (in CA)
λ_{ex} = 488 nm

Structure 14.

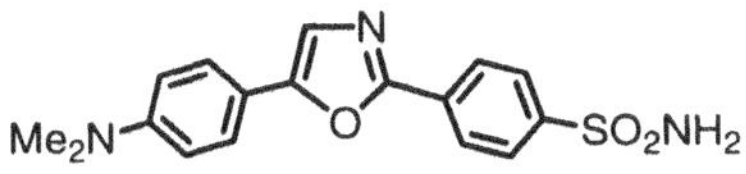

dapoxyl sulfonamide

λ_{em} = 506 nm (in CA)
λ_{ex} = 466 nm

Structure 15.

(90-fold) with blue shift (from 605 nm to 530 nm) in its fluorescence emission upon binding to holoCA (the dissociation constant is 0.3 μM). Ratiometric detection of Zn^{2+} was possible by measuring the emission intensity at 535 nm vs. 685 nm. The increase of emission seems to be due to the 20-fold increase in the fluorescence lifetime of **15** bound to Zn^{2+} in CA (τ changes from 0.22 ns to 3.80 ns upon binding to CA).

Zn^{2+} Chelation-enhanced fluorescence (CHEF)

The chelation-enhanced fluorescence (CHEF) was reported with anthracene derivatives having chelating moieties (e.g., 9,10-bis(2,5-dimethyl-2,5-diazahexyl) anthracene **16**) for Zn^{2+} in CH_3CN (Huston *et al.*

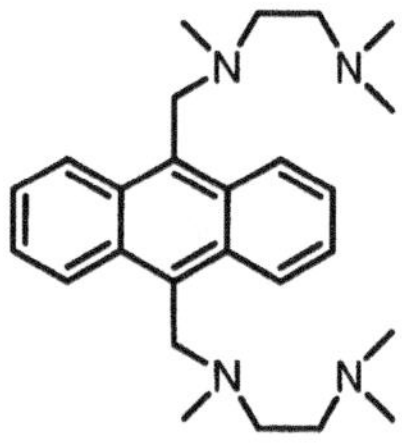

Structure 16.

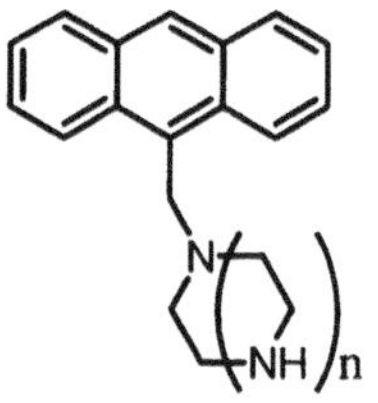

(n = 1 – 5)

Structure 17.

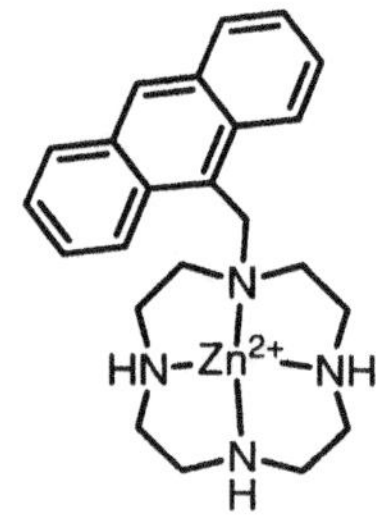

Structure 18.

1988; Fabbrizzi 1997). It was extended to a macrocyclic system **17** (Akkaya *et al.* 1990; Huston *et al.* 1990; Czarnik 1992). A large CHEF effect by Zn^{2+} (14.4-fold) and Cd^{2+} (9-fold) was observed with **17** ($n = 2$) at pH 10 in aqueous solution. However, the protonation(s) and metal complexation at the macrocyclic polyamine moiety commonly inhibit the quenching process by the unprotonated benzyl nitrogen atom. The diprotonated ligand **17** ($n = 2$)$\cdot 2H^+$ at pH 7 (pK_a values for a monosubstituted cyclen are ~12, ~11, <2, and <2; Koike *et al.* 1996b; Kimura 1997) showed almost 120-fold larger fluorescence intensity than that of the unprotonated ligand **17** ($n = 2$) at pH 12. In fluorescence titration of **17** ($n = 2$) (10 μM) with Zn^{2+} (0–20 μM), impractically alkaline pH 12 buffer should be used, where **17** ($n = 2$) is unprotonated and hence competition between H^+ and Zn^{2+} for the macrocycle does not occur. Under these conditions, the emission maximum at 416 nm (excited at 335 nm) increases linearly till nearly 1:1 complexation (to **18**). Thus, **17** ($n = 2$) cannot be a practical zinc-fluorophore in aqueous solution under normal biological pH conditions.

The drawback of **17** ($n = 2$) was the protonation of the macrocyclic polyamine part at neutral pH, which inhibits photoinduced electron transfer (PET) (Prasanna de Silva *et al.* 1997), resulting in the strong fluorescence emission even in the absence of Zn^{2+}. To remedy this monosubstituted macrocyclic tetraamine properties, Nagano's group devised ACF-1 (**19a**) and ACF-2 (**19b**), in which a fluorescein dye and three methyl groups are attached to four nitrogens of the macrocyclic tetraamine (Hirano *et al.* 2000). Fluorescein is advantageous in that it has a high quantum yield of fluorescence in aqueous solution and its excitation wavelength is in the harmless visible range. The tetraalkylations of the macrocyclic tetraamines in **19** lower the amine pK_a values (~8, ~6, <2, and <2) (Maumela *et al.* 1995), so that protonations would be less likely at neutral pH and Zn^{2+} would bind to **19** without competition against protons. As expected, the fluorescence of **19** (5 μM) was quenched at pH 7.5 due to PET. Upon addition of Zn^{2+} (0–5 μM), the fluorescence emission intensity of **19a** and **19b** linearly increased up to 14- and 26-fold, respectively, with little change in the position of excitation and emission maxima ($\lambda_{ex} = 495$ nm, $\lambda_{em} = 515$ nm for **19a** and $\lambda_{ex} = 505$ nm, $\lambda_{em} = 525$ nm for **19b**). The detection limit of **19** was noted as 500 nM of Zn^{2+} under these conditions. ACF-2 **19b** was selective only for Zn^{2+} and Cd^{2+} and the fluorescence of Zn^{2+}–**19** complexes was not interfered by other transition metal ions, Fe^{2+}, Fe^{3+}, Ni^{2+}, Co^{2+}, or Mn^{2+}. It is unfortunate, however, that the complexation constant with Zn^{2+} was not reported. Applicability to biological systems has yet to be determined.

For detection of trace Zn^{2+} in living cells, good permeability of the Zn^{2+}-sensors is crucial, e.g. Zinquin **5**. Recently, Lippard and Tsien *et al.* developed a membrane-permeable Zn^{2+}-fluorophore, Zinpyr-1 (**20**), which is a conjugate of 2′,7′-dichlorofluorescein with bis(2-pyridylmethyl)amine (DPA), a homologue to the membrane-permeable heavy metal chelator, TPEN (**10**) (Walkup *et al.* 2000). Zinpyr-1(**20**) (0.5 μM) has an excitation maximum at 515 nm in 50 mM PIPES (pH 7.0) with 100 mM KCl with a quantum yield of 0.39, and saturated with Zn^{2+} (25 μM), the resulting Zn^{2+}-complex showing λ_{ex} at 507 nm and a quantum yield of 0.89. The apparent dissociation constant, K_d, of the Zn^{2+}-**20** complex was determined fluorometrically to be 0.7 ± 0.1 nM

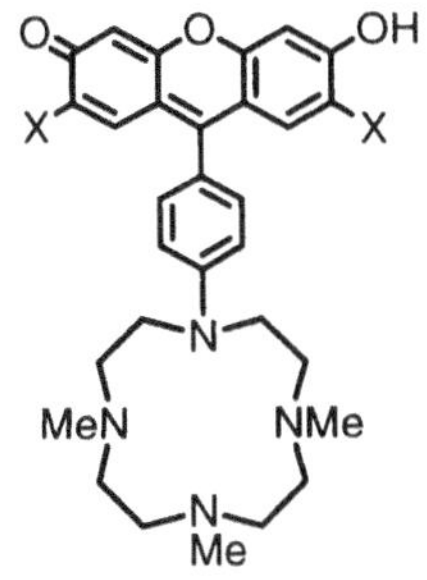

a: X = H **b**: X = Cl

ACF-1 ACF-2

λ_{em} = 515 nm λ_{em} = 525 nm
λ_{ex} = 495 nm λ_{ex} = 505 nm

Structure 19.

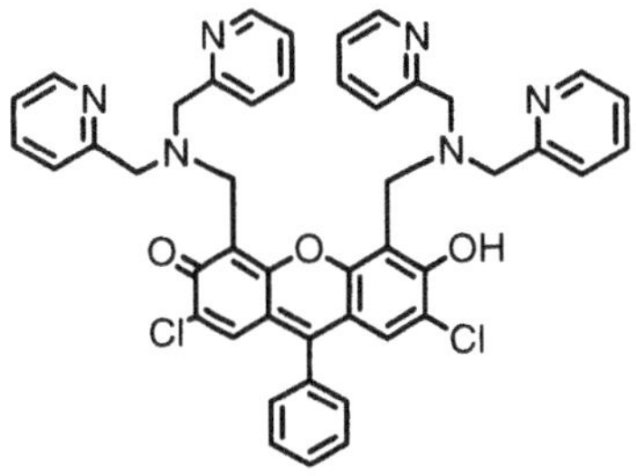

Zinpyr-1

λ_{em} = ca. 525 nm (with Zn^{2+})
λ_{ex} = 515 nm (without Zn^{2+})
λ_{ex} = 507 nm (with Zn^{2+})

Structure 20.

under the given conditions (emission was measured between 509 and 650 nm). The response of **20** to the concentration change of Zn^{2+} in COS-7 cells was demonstrated by influxed Zn^{2+} (exogeneous concentration of 50 μM) with the aid of a zinc ionophore 2-mercaptopyridine *N*-oxide (pyrithione, 20 μM). The enhanced fluorescence due to the complexation with Zn^{2+} was observed, which then was decreased by addition of TPEN **10** having a much higher affinity ($K_d = 10^{-13}$ M at pH 7). Preliminary results showed that Zinpyr-1 stains the Golgi and acidic cellular compartments.

Peptide fluorescent probes for Zn^{2+}

As a bio-inspired zinc-fluorophore, a peptide **21** (25 amino acids) containing a zinc-finger motif (a strong Zn^{2+}-binding site, Cys_2/His_2) covalently attached with a dansylamide residue, was synthesized (i) for selective and efficient Zn^{2+}-binding ($K_d = 1.4 \times 10^{-10}$ M at pH 7) and (ii) to create a hydrophobic environment around the encapsulated dansylamide residue upon its Zn^{2+} complexation (Walkup & Imperiali 1996). The addition of Zn^{2+} (0.1–1 μM) to the peptide **21** (1.4 μM) in pH 7 HEPES buffer resulted in a linearly increasing emission peak at 475 nm (excited at 333 nm) up to 2.4 fold (the emission maximum of the Zn^{2+}-**21** complex was 525 nm). In the absence of Zn^{2+}, the emission maximum was 560 nm. The dissociation constant, K_d, of the Zn^{2+}-**21** complex was $K_d = 1.4 \times 10^{-10}$ M at pH 7. The presence of 0.5 M Na^+, 50 mM Mg^{2+}, and 100 μM Co^{2+} did not interfere with the Zn^{2+} analysis. The covalently attached fluorescent reporter dansylamide is sensitive to metal-induced conformational changes of the supporting peptide framework and yet is remote from the Zn^{2+}-binding site. The enhanced emission is mostly due to the reporter to be placed in a hydrophobic environment. Problems with the zinc-finger peptide, in addition to its synthetic availability, would be vulnerabilities to air oxidation of cysteine residues and to redox active metal ions such as Cu^{2+}, despite the high affinity to Zn^{2+}. In application to the reductive environment of cells, this may not be problematic, but the digestion by proteases is another problem.

A more oxidatively robust peptidyl zinc-fluorophore was later synthesized by substitution of the peptide from Cys_2/His_2 to Cys/His_3 at the Zn^{2+}-binding site (Walkup & Imperiali 1997). However, the zinc affinity dropped by an order of magnitude ($K_d = 3 \times 10^{-9}$ M at pH 7) and the fluorescence response to Zn^{2+} became smaller (fluorescence enhancement was 1.3 fold upon addition of Zn^{2+}) for a sensitive zinc-fluorophore (the emission maxima are 552 and 548 nm in the absence and the presence of Zn^{2+}, respectively). Another modification of the zinc finger motif from Cys_2/His_2 to $Cys/Asp/His_2$ yielded a zinc sensor with enhanced oxidative stability, but with a further weakened affinity ($K_d = 6.5 \times 10^{-8}$ M at pH 7), although this one is responsive to submicromolar to micromolar concentrations of Zn^{2+} in the presence of redox active Cu^{2+} or Fe^{2+}.

In order to minimize the size of the peptidyl component for Zn^{2+} detection, the new heptapeptide **22**, in which unnatural amino acids having a chelating hydroxylquinoline (oxine) unit, were synthesized (Walkup & Imperiali 1998). The chiral amino acids

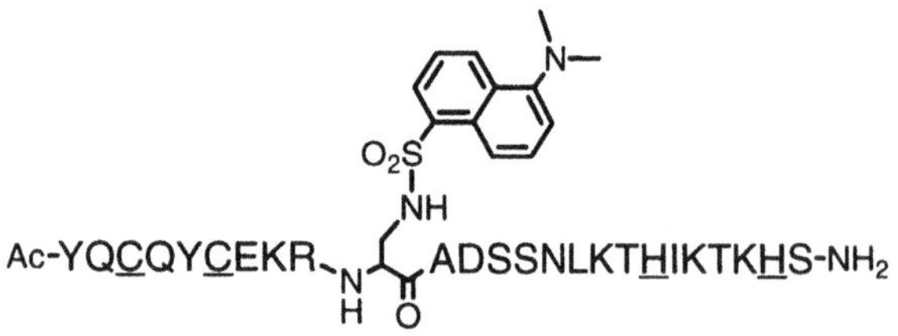

Structure 21.

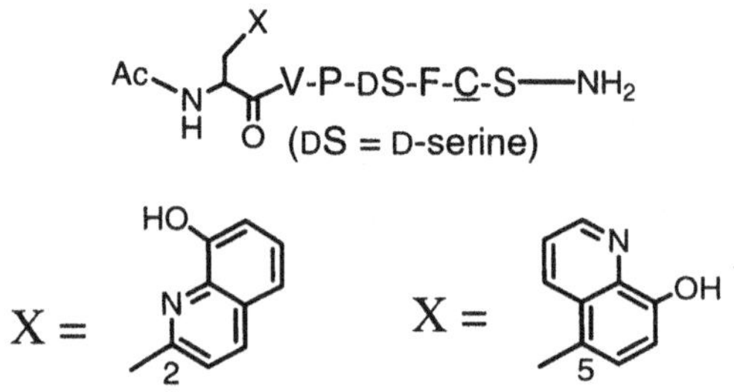

(2Oxn) (5Oxn)

a **b**

Structure 22.

having an oxine unit, **22a** and **22b**, were prepared via diastereoselective alkylation of the pseudoephedrine glycinamide. Cysteine was incorporated for selective binding to Zn^{2+} ion. Although **22b** was found to form complicated metal complexes due to the formation of a disulfide dimer under normal conditions, **22a** showed 1:1 complexation with Zn^{2+}. By UV and fluorescence titrations of **22a** with Zn^{2+} in pH 7.0 buffer (50 mM HEPES with 150 mM NaCl), the dissociation constant, K_d, was determined to be 17 μM. The limit of detection was less than 250 nM.

Another type of a zinc finger consensus peptide **23** (Cys_2/His_2) was designed (Godwin & Berg 1996). The peptide was attached to two fluorescent dyes, fluorescein (F) as the energy donor and lissamine (L) as the acceptor, to visualize zinc binding. In the absence of Zn^{2+}, the peptide is unfolded and the dyes are relatively far apart (i.e., small intramolecular energy transfer occurs between F and L). Upon Zn^{2+} binding to the Cys_2/His_2 site, the peptide folds to bring the two fluorophores closer together, increasing the amount of intramolecular energy transfer. The binding of **23** (3.7 μM) to Zn^{2+} at pH 7.1 was monitored by increasing fluorescence (ca. 2.3-fold) at 596 nm (excitation at 430 nm) with an increase in $[Zn^{2+}]$, which provided the 1:1 and 1:2 Zn^{2+}-**23** complex formation (the estimated K_d for the 1:1 complex is 1×10^{-12} M at pH 7.1). Again, the complexation had to be carried out under a reductive atmosphere to avoid peptide oxidation.

A bio-inspired macrocyclic fluorophore

While engaged in elucidating roles of Zn^{2+} in zinc enzymes (in particular carbonic anhydrase (CA)) by means of macrocyclic polyamine complexes (e.g., Zn^{2+}-1,5,9-triazacyclododecane (Zn^{2+}-[12]aneN_3) **24** and Zn^{2+}-1,4,7,10-tetraazacyclododecane (Zn^{2+}-cyclen) **25**, Kimura's group has discovered intrinsic acid properties of Zn^{2+} (Kimura 1992, 1994, 2001; Kimura *et al.* 1992, 1997a,b, 1999; Koike & Kimura 1991, 1998; Kimura & Shionoya 1994, 1996; Kimura & Kikuta 2000). One of the most outstanding properties of Zn^{2+} revealed was a strong affinity to aromatic sulfonamides, as illustrated by formation of coordination bonds between Zn^{2+} and deprotonated sulfonamide N^- anions at physiological pH. A new chemical model **24c** was presented to account for aromatic sulfonamide anions being good ligands for Zn^{2+} ion at the active center of CA to be strong inhibitors (Scheme 3) (Koike *et al.* 1992). The zinc enzyme models **24** and **25** form stable 1:1 complexes with deprotonated weak acids such as thymine derivatives (**25c**) (Shionoya *et al.* 1993, 1994; Kimura *et al.* 1998, 2000; Aoki *et al.* 1998a,b; Kikuta *et al.* 1999; Aoki & Kimura 2000) and barbital (**25d**) (Koike *et al.* 1996a; Fujioka *et al.* 1996; Aoki *et al.* 2000) in neutral aqueous solution, which results from the Zn^{2+}-bound OH^- species generated with pK_a values of 7.3 (for **24a** $\rightleftarrows$ **24b** + H^+) and 7.9 (for **25a** $\rightleftarrows$ **25b** + H^+) acting as bases to dissociate the acidic protons. The resulting conjugate bases strongly bind to Zn^{2+}, which compensate for the unfavorable deprotonations at neutral pH. It was further demonstrated that tosylamidopropyl[12]aneN_3 **26** yields a very stable four-coordinate, tetrahedral zinc(II) complex **27** under physiological pH (Scheme 2), where the aromatic sulfonamide N^- anion strongly binds to Zn^{2+} ion from the fourth coordination site (Koike *et al.* 1992).

On the basis of these basic studies on the CA-model, a dansylamide-pendant macrocyclic tetraamine (dansylamidoethylcyclen) **28** was designed for a new type of selective and efficient zinc-fluorophore (Scheme 3) (Kimura 1997; Kimura & Koike 1998; Koike *et al.* 1996b). The 12-membered macrocyclic tetraamine, cyclen (L) forms a much more stable Zn^{2+} complex (K=[ZnL]/$[Zn^{2+}]$[L] = $10^{15.3}$ M^{-1}) than [12]aneN_3 (K = $10^{8.4}$ M^{-1}) in H_2O at 25 °C.

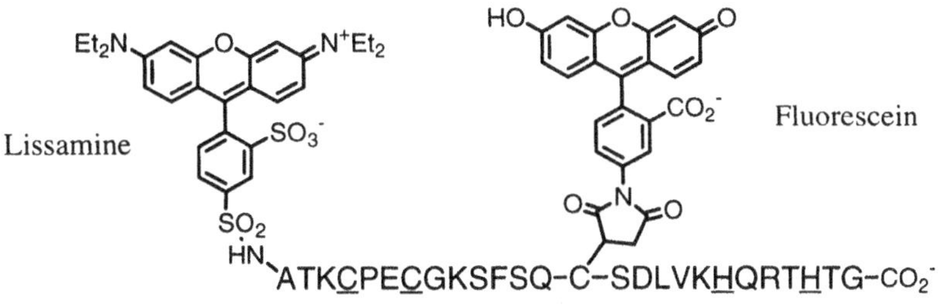

Structure 23.

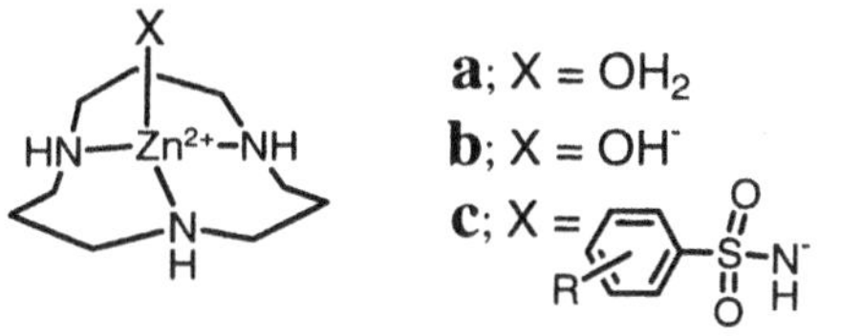

Structure 24.

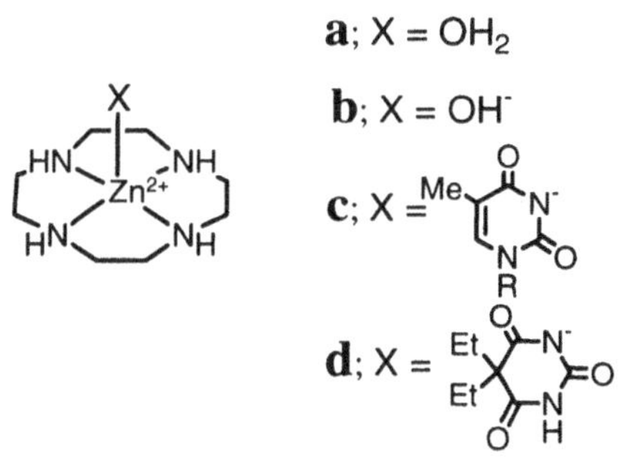

Structure 25.

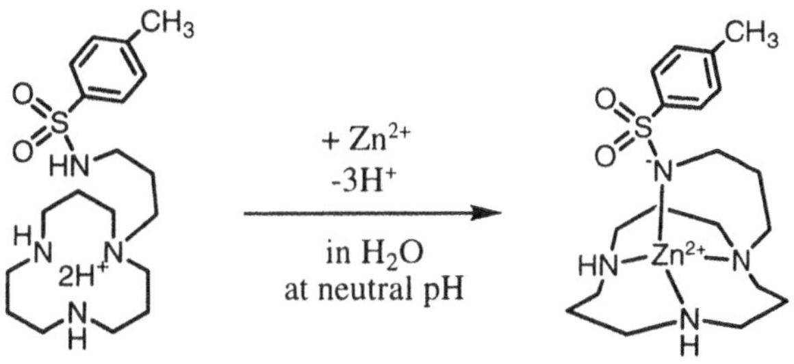

Structure 26. *Structure 27.*

Scheme 2.

The complexation constant for **29** ($K = [Zn^{2+}\text{-}H_{-1}L]/[L \cdot 2H^+][Zn^{2+}]$) was established to be $10^{20.8}$ M^{-1} at 25 °C with $I = 0.10$ ($NaNO_3$) by potentiometric pH titrations. For comparison with other Zn^{2+}-fluorophores earlier described, the dissociation constants, K_d, of **29** are 1.4×10^{-10} M at pH 7.0 to 5.5×10^{-13} M at pH 7.8. The K_d values for **29** appreciably change because of competing protonation. As the pH is raised to 7.8, the K_d value for **29** becomes smaller than those for the zinc finger peptides (e.g., 5.7×10^{-12} M at pH 7.0), which do not so dramatically change at pH 7.8. Most remarkably, 1 μM of **28** (almost in $L \cdot 2H^+$ form) sequesters nearly 100% of Zn^{2+} (1 μM) in the form of stoichiometric Zn^{2+}-$H_{-1}L$ **29** at physiological pH. Such a strong and pH-dependent affinity to Zn^{2+} is one of the most characteristic properties of **28** (in comparison to Zinquin **5** and anthracene-pendant cyclen **17**($n = 2$)) and will be useful for quantifying trace amounts of free Zn^{2+} or bioligand-bound Zn^{2+} in environmental and biological systems.

The dansylamide deprotonation of **28** with Zn^{2+} at pH 7.8 increased the emission intensity by 4.9-fold at 540 nm and 10-fold at 490 nm, while the non-metallated dansylamide deprotonation of L to $H_{-1}L$ without Zn^{2+} at high pH (> 12) brought about only ca. 20% increase in the fluorescence emission intensity. To the contrary, the dansylamide deprotonation with Cu^{2+} completely quenched the fluorescence. The fluorescence maximum of H_2L (582 nm) at neutral pH blue-shifted upon zinc(II) complexation (Zn^{2+}-$H_{-1}L$) to 540 nm. A greater fluorescence blue shift (580 nm to 468 nm) and intensity enhancement (15.3-fold in quantum yield excited at 320 nm) were reported for the dansylamide complexation with CA, which were accounted for by the hydrophobic environment and the deprotonation of dansylamide on Zn^{2+} at the active center of CA.

The fluorescence changes of **28** (5 μM) with various metal ions (5 μM) at pH 7.3 (HEPES buffer) and 25 °C are summarized in Figure 1. The addition of various concentrations of Zn^{2+} (0–10 μM) resulted in increased emission upon excitation at 330 nm as shown in Figure 2. The response (at 528 nm) was linear between 0.1 and 5 μM until it reached a 1:1 $[Zn^{2+}]/[\mathbf{28}]$ ratio, and then became a plateau, indicating that the increase in fluorescence is due to the 1:1 Zn^{2+}-$H_{-1}L$ **29** formation, and moreover, Zn^{2+}-$H_{-1}L$ is stable even at subnanomolar concentrations. On the other hand, Cu^{2+} linearly diminished the fluorescence emission

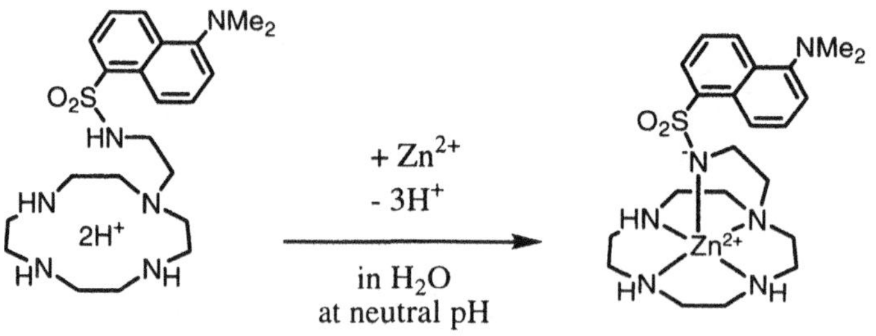

dansylamidoethylcyclen
L·2H⁺
λ_{em} = 555 nm
Structure 28.

Zn–H$_{-1}$L

λ_{em} = 528 nm
Structure 29.

Scheme 3.

until complete quenching at [**28**]/[Cu^{2+}] = 1, although Cu^{2+} forms the most stable five-coordinate complex at 25 °C. Other fluorescence-quenching metal ions (paramagnetic Co^{2+}) that tend to bind fairly strongly with cyclen also caused some quenching, although the effects were not so drastic as Cu^{2+}. Other metals such as Na^{+}, K^{+}, Ca^{2+}, Mg^{2+}, Fe^{2+}, Fe^{3+}, Mn^{2+}, or Mg^{2+} gave negligible effect on the fluorescence of **29**. Preliminary experiments with **28** showed low toxicity against several cell lines and good cell-permeability. Further study of **28** and its homologues for both Zn^{2+} recognition and fluorescence signaling would find useful applications in Zn^{2+} biology.

Other types of zinc-fluorophores

Several other types of zinc-fluorophores have been reported. However, they are less practical than those earlier described. A tripodal ligand, tris[2-(5-dimethylamino-1-naphthalenesulphonamido)ethyl]-amine **30**, was synthesized (Prodi *et al.* 1999). A quantum yield of fluorescence emission of **30** enhanced with a blue shift upon addition of Zn^{2+} and Cd^{2+} in CH_3CN/H_2O (1:1 (v/v)) at relatively high pH 9.5 required to remove all of the sulfonamide protons (excitation at 340 nm), while Cu^{2+} and Co^{2+} quenched the emission. For comparison, an emission of the monomeric control compound **31** increased by Zn^{2+}, but the complexation was prevented by formation of metal hydroxide at the pH employed. Addition of Na^{+}, K^{+}, Ca^{2+}, Sr^{2+}, Ba^{2+}, Eu^{3+}, Ni^{2+}, Mn^{2+}, Pb^{2+}, Fe^{2+}, Fe^{2+}, and Cr^{2+} up to 50 mM caused negligible change of fluorescence spectra of **30** and **31**.

Recently, a lanthanide(III) complex **32** attached with a chelator for Zn^{2+} has been reported (Reany *et al.* 2000). While addition of Zn^{2+} induced negligible UV absorption spectral change of the Eu^{3+} complex **32a**, a small blue shift was observed in absorption of the Tb^{3+} complex **32b** from 255 nm to 250 nm upon complexation with Zn^{2+}. Fluorescence emission of **32a** and **32b** increased by 26% and 42%, respectively, upon complexation with Zn^{2+} (excitation at 262 nm), possibly due to inhibition of PET (from the benzylic nitrogen to the intermediate aryl singlet excited state). In a fluorescence emission spectra of **32b**, two bands were observed at 440 nm and 365 nm. By complexation with Zn^{2+}, the former band shifted from 442 nm to 430 nm with decrease of intensity and the latter band increased without a shift. Affinity constants, log β, of **32a** with Zn^{2+} determined by UV and fluorescence titrations were 5.4–6.0 and those of **32b** with Zn^{2+} are 5.5–6.4. Small changes of UV and emission spectra of **32a** and **32b** by addition of Ca^{2+} and Mg^{2+} were observed, from which log β values for Ca^{2+}-**32a**, Ca^{2+}-**32b**, Mg^{2+}-**32a**, and Mg^{2+}-**32b** were determined to be 3.9–4.0, 3.8–4.0, 1.9–2.1, and 2.0–2.5, respectively, suggesting the possible detection of these metal ions at submillimolar to millimolar order concentration. However, it is seen that the biological application for Zn^{2+} detection is limited.

Concluding remarks

As reviewed here, several useful Zn^{2+}-selective fluorescent probes are now developed. Some sensors are chelators with fluorescent dyes such as fluores-

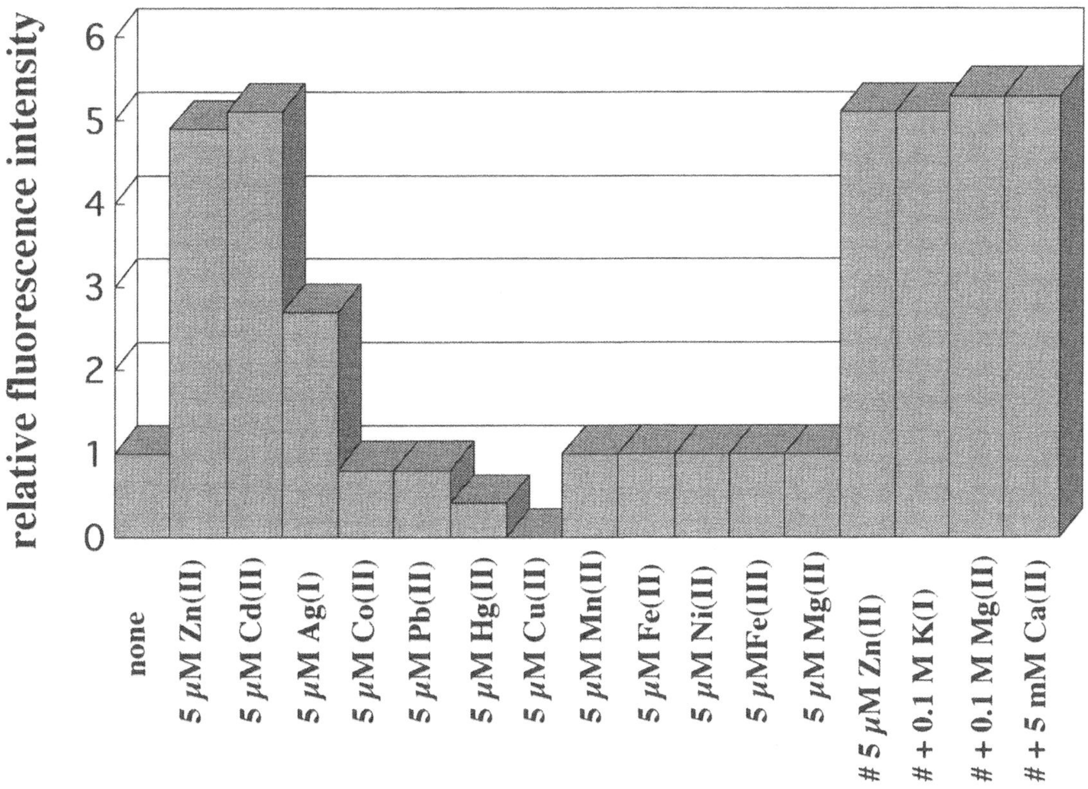

Fig. 1. Relative fluorescence intensity of **28** (5 μM) responding to 1 eq of various metal ions at pH 7.3 (1 mM HEPES with $I = 0.1$ ($NaNO_3$)) and 25 °C. The data marked with # were obtained without the supporting electrolyte.

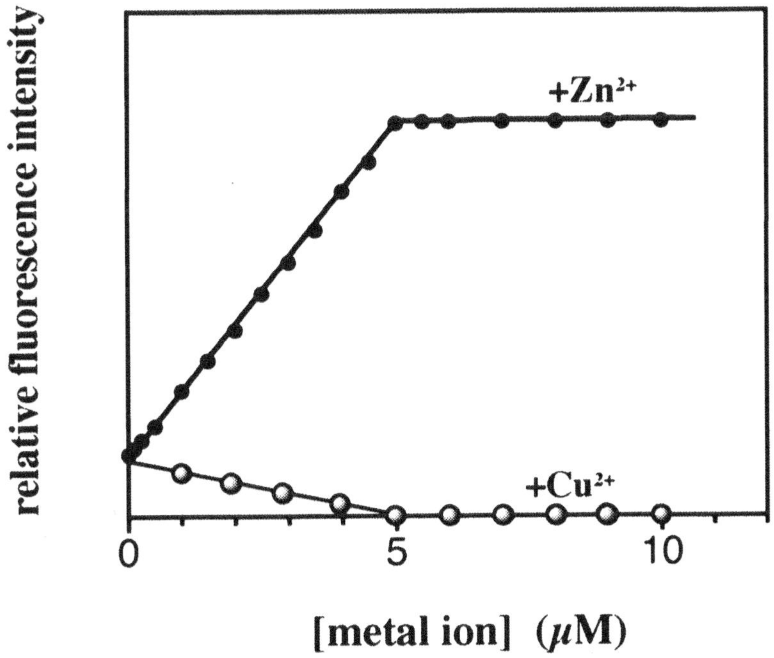

Fig. 2. Fluorescence titration curves of **28** (5 μM) with Zn^{2+} and Cu^{2+} at pH 7.3 (1 mM HEPES with $I = 0.1$ ($NaNO_3$)) and 25 °C.

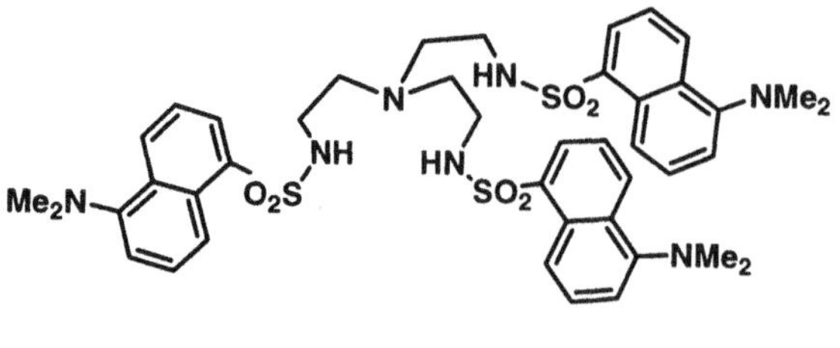

λ_{em} = 540 nm
λ_{ex} = 334 nm

Structure 30.

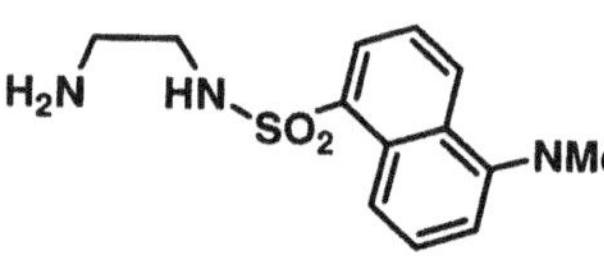

Structure 31.

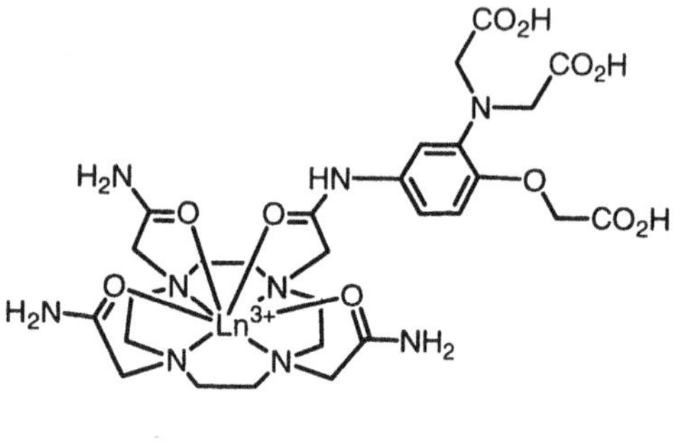

a : Ln = Eu
b : Ln = Tb λ_{em} = 440, 345 nm

Structure 32.

cein, whose microenvironments yields stronger fluorescence upon Zn^{2+}-complexation. Other sensors using dansylamide strongly fluoresce by the deprotonated sulfonamide binding to Zn^{2+} at neutral pH. Chemically ideal Zn^{2+} fluorophores will require the following criteria; (i) The ligands should be easy to make and chemically and biologically robust; (ii) the affinity to Zn^{2+} should be sufficiently high, capable of complexing with trace free Zn^{2+} or protein (e.g., zinc finger)–bound at physiological pH; (iii) the Zn^{2+} selectivity (either by Zn^{2+}-selective chelation or Zn^{2+}-selective fluorescence) should be high and the perturbation by other metal ions should be minimal (e.g., Hg^{2+}, Pb^{2+}). Further, for biological applications, one has to pay attention to (i) kinetic aspects (e.g., how fast is the Zn^{2+} chelation?) , (ii) permeability into cells and how long it stays before diffusion to be responsive to intracellular Zn^{2+}, (iii) excitation of probes with harmless irradiation wavelengths, and (iv) improvement in the fluorescence efficiency. It is evident that we still need to search for new fluorophores and study more biologically suitable Zn^{2+}-fluorophores as an extension of the present basic chemical knowledge.

References

Akkaya EU, Huston MH, Czarnik AW. 1990 Chelation–enhanced fluorescence of anthrylazamacrocycle conjugate probes in aqueous solution. *J Am Chem Soc* **112**, 3590–3593.

Aoki S, Honda Y, Kimura E. 1998a The first selective and efficient transport of imide-containing nucleosides and nucleotides by lipophilic cyclen–zinc(II) complexes (cyclen = 1,4,7,10-tetraazacyclododecane). *J Am Chem Soc* **120**, 10018–10026.

Aoki S, Sugimura C, Kimura E. 1998b Efficient inhibition of photo[2+2]cycloaddition of thymidilyl(3′-5′)thymidine and promotion of photosplitting of the *cis-syn*-cyclobutane thymine dimer by dimeric zinc(II)-cyclen complexes containing *m*- and *p*-xylyl spacers. *J Am Chem Soc* **120**, 10094–10102.

Aoki S, Kimura E. 2000 Highly selective recognition of thymidine mono- and diphosphate nucleotides in aqueous solution by ditopic receptors zinc(II)-bis(cyclen) complexes (cyclen = 1,4,7,10-tetraazacyclododecane). *J Am Chem Soc* **122**, 4542–4548.

Aoki S, Shiro M, Koike T, Kimura E. 2000 Three-dimensional supermolecules assembled from a tris(Zn^{2+}-cyclen) complex and di- and trianionic cyanuric acid in aqueous solution (cyclen = 1,4,7,10-tetraazacyclododecane). *J Am Chem Soc* **122**, 576–584.

Berendji D, Kolb-Bachofen V, Meyer KL, Grapenthin O, Weber H, Wahn V, Kröncke K-D. 1997 Nitric oxide mediates intracytoplasmic and intranuclear zinc release. *FEBS Lett* **405**, 37–41.

Berg JM. 1995 Zinc finger domains: from predictions to design. *Acc Chem Res* **28**, 14–19.

Chen RF, Kernohan JC. 1967 Combination of bovine carbonic anhydrase with a fluorescent sulfonamide. *J Biol Chem* **242**, 5813–5823.

Choi DW, Koh JY. 1998 Zinc and brain injury. *Annu Rev Neurosci* **21**, 347–375.

Cox EH, McLendon GL. 2000 Zinc-dependent protein folding. *Curr Opin Chem Biol* **4**, 162–165.

Coyle P, Zalewski PD, Philcox JC, Forbes IJ, Ward AD, Lincoln SF, Mahadevan I, Rofe AM. 1994 Measurement of zinc in hepatocytes by using a fluorescent probe, Zinquin: relationship to metallothionein and intracellular zinc. *Biochem J* **303**, 781–786.

Cuajungco MP, Lees GJ. 1997 Zinc metabolism in the brain: relevance to human neurodegenerative disorders. *Neurobiol Disease* **4**, 137–169.

Czarnik AW. 1992 *Fluorescent Chemosensor for Ion and Molecule Recognition*. Washington, DC: American Chemical Society.

Czarnik AW. 1994 Chemical communication in water using fluorescent chemosensors. *Acc Chem Res* **27**, 302–308.

Czarnik AW. 1995 Desperately seeking sensors. *Chem Biol* **2**, 423–428.

Elbaum D, Nair SK, Patchan MW, Thompson RB, Christianson DW. 1996 Structure-based design of a sulfonamide probe for fluorescence anisotropy detection of zinc with a carbonic anhydrase-based biosensor. *J Am Chem Soc* **118**, 8381–8387.

Fahrni CJ, O'Halloran TV. 1999 Aqueous coordination chemistry of quinoline-based fluorescence probes for the biological chemistry of zinc. *J Am Chem Soc* **121**, 11448–11458.

Fraústo da Silva J, Williams RJP. 1991 *The Biological Chemistry of the Elements*. Oxford: Clarendon Press.

Fabbrizzi L, Francese G, Licchelli M, Perotti A, Taglietti A. 1997 Fluorescent sensor of imidazole and histidine. *J Chem Soc Chem Comm* 581–582.

Frederickson CJ, Kasarskis EJ, Ringo D, Frederickson RE. 1987 A quinoline fluorescence method for visualizing and assaying the histochemically reactive zinc (bouton zinc) in the brain. *J Neurosci Meth* **20**, 91–103.

Fujioka H, Koike T, Yamada N, Kimura E. 1996 A new bis(zinc(II)-cyclen) complex as a novel chelator for barbiturates and phosphates. *Heterocycles* **42**, 775–787.

Godwin HA, Berg JM. 1996 A fluorescent zinc probe based on metal-induced peptide folding. *J Am Chem Soc* **118**, 6514–6515.

Greisman HA, Pabo CO. 1997 A general strategy for selecting high-affinity zinc finger proteins for diverse DNA target sites. *Science* **275**, 657–661.

Grynkiewicz G, Poenie M, Tsien RY. 1985 A new generation of Ca^{2+} indicators with greatly improved fluorescence properties. *J Biol Chem* **260**, 3440–3450.

Haugland RP. 1996 *Handbook of Fluorescent Probes and Research Chemicals*, 6th edn. Eugene: Molecular Probes.

Hendrickson KM, Rodopoulos T, Pittet P-A, Mahadevan I, Lincoln SF, Ward AD, Kurucsev T, Duckworth PA, Forbes IJ, Zalewski PD, Betts WH. 1997 Complexation of zinc(II) and other divalent metal ions by the fluorophore 2-methyl-8-(toluene-*p*-sulfonamido)-6-quinolyloxyacetic acid in 50% aqueous solution. *J Chem Soc Dalton Trans* 3879–3882.

Hirano T, Kikuchi K, Urano Y, Higuchi T, Nagano T. 2000 Novel zinc fluorescent probes excitable with visible light for biological applications. *Angew Chem Int Ed* **39**, 1052–1054.

Huston MH, Haider KW, Czarnik AW. 1988 Chelation-enhanced fluorescence in 9,10-bis(TMEDA)anthracene. *J Am Chem Soc* **110**, 4460–4462.

Huston MH, Englem NC, Czarnik AW. 1990 Chelatoselective fluorescence perturbation in anthrylazamacrocycle conjugate probes. Electrophilic aromatic cadmiation. *J Am Chem Soc* **110**, 7054–7056.

Kiefer LL, Krebs JF, Paterno SA, Fierke CA. 1993 Engineering a cysteine ligand into the zinc binding site of human anhydrase. *Biochemistry* **32**, 9896–9900.

Kikuta E, Murata M, Katsube N, Koike T, Kimura E. 1999 Novel recognition of thymine base in double-stranded DNA by zinc(II)-macrocyclic tetraamine complexes appended with aromatic groups. *J Am Chem Soc* **121**, 5426–5436.

Kimura E, Shiota T, Koike T, Shiro M, Kodama M. 1990 A zinc(II) complex of 1,5,9-triazacyclododecane ([12]aneN_3) as a model for carbonic anhydrase. *J Am Chem Soc* **112**, 5805–5811.

Kimura, E. 1992 Macrocyclic polyamines with intelligent functions. *Tetrahedron* **48**, 6175–6217.

Kimura E, Shionoya M, Hoshino A, Ikeda T, Yamada Y. 1992 A model for catalytically active zinc(II) ion in liver alcohol dehydrogenase: a novel "hydride transfer" reaction catalyzed by zinc(II)–macrocyclic polyamine complexes. *J Am Chem Soc* **114**, 10134–10137.

Kimura E. 1994 Macrocyclic polyamine zinc(II) complexes as advanced models for zinc(II) enzymes. In: Karlin KD. ed. *Progress in Inorganic Chemistry*. Vol. 41. New York: John Wiley & Sons: 443–490.

Kimura E, Shionoya M. 1994 Macrocyclic polyamine complex beyond metalloenzyme models. In: Fabbrizzi L, Poggi A. eds. *Transition Metals in Supramolecular Chemistry*. Dordrecht, The Netherlands: Kluwer Academic Publishers, 245–259.

Kimura E, Shionoya M. 1996 Zinc complexes as targeting agents for nucleic acids. In: Sigel A, Sigel H. eds. *Metal Ions in Biological Systems*. Vol. 33. New York: Marcel Dekker: 29–52.

Kimura E. 1997 A novel biomimetic zinc(II)-fluorophore, dansylamidoethyl-pendant macrocyclic tetraamine. *South African J Chem* **50**, 240–248.

Kimura E, Aoki S, Koike T, Shiro M. 1997a A tris(Zn^{II}–1,4,7,10-tetraazacyclododecane) complex as a new receptor for phosphate dianions in aqueous solution. *J Am Chem Soc* **119**, 3068–3076.

Kimura E, Koike T, Aoki S. 1997b Why are zinc phosphatases multinuclear? *J Synth Org Chem [Japan]* **55**, 1052–1061.

Kimura E, Koike T. 1998 Recent development of zinc-fluorophores. *Chem Soc Rev* **27**, 179–184.

Kimura E, Ikeda T, Aoki S, Shionoya M. 1998 Macrocyclic zinc(II) complexes for selective recognition of nucleobases in single- and double-stranded polynucleotides. *J Biol Inorg Chem* **3**, 259–267.

Kimura E, Gotoh T, Koike T, Shiro M. 1999 Dynamic enolate recognition in aqueous solution by zinc(II) in a phenacyl-pendant cyclen complex: implications for the role of zinc(II) in class II aldolases. *J Am Chem Soc* **121**, 1267–1274.

Kimura E, Kitamura H, Ohtani K, Koike T. 2000 Elaboration of selective and efficient recognition of thymine base in dinucleotides (TpT, ApT, CpT, and GpT), single-stranded d(GTGACGCC), and double-stranded $d(CGCTAGCG)_2$ by Zn^{2+}-acridinylcyclen (acridinylcyclen = (9-acridinyl)methyl-1,4,7,10-tetraazacyclododecane). *J Am Chem Soc* **122**, 4668–4677.

Kimura E, Kikuta E. 2000 Why zinc in zinc enzymes? From biological roles to DNA base-selective recognition. *J Biol Inorg Chem* **5**, 139–155.

Kimura E. 2001 Model studies for molecular recognition of carbonic anhydrase and carboxypeptidase. *Acc Chem Res* **34**, 171–179.

Koike T, Kimura E. 1991. Roles of zinc(II) ion in phosphatases. A model study with zinc(II)-macrocyclic polyamine complexes. *J Am Chem Soc* **113**, 8935–8941.

Koike T, Kimura E, Nakamura I, Hashimoto Y, Shiro M. 1992 The first anionic sulfonamide-binding zinc(II) complexes with a macrocyclic triamine: chemical verification of the sulfonamide inhibition of carbonic anhydrase. *J Am Chem Soc* **114**, 7338–7345.

Koike T, Takashige M, Kimura E, Fujioka H, Shiro M. 1996a Bis(Zn^{II}-cyclen) complex as a novel receptor of barbiturates in aqueous solution. *Chem Europ J* **2**, 617–623.

Koike T, Watanabe T, Aoki S, Kimura E, Shiro M. 1996b. A novel biomimetic zinc(II)-fluorophore, dansylamidoethyl-pendant macrocyclic tetraamine 1,4,7,10-tetraazacyclododecane (cyclen). *J Am Chem Soc* **118**, 12696–12703.

Lippard SJ, Berg JM. 1994 *Principles of Bioinorganic Chemistry*. Mill Valley: University Science Books.

Lipscomb WN, Sträter N. 1996 Recent advances in zinc enzymology. *Chem Rev* **96**, 2375–2433.

Mann T, Keilin D. 1940 Sulphanilamide as a specific inhibitor of carbonic anhydrase. *Nature* **146** 164–165.

Maumera H, Hancock RD, Carlton L, Reibenspies JH, Wainwright KP. 1995 The amide oxygen as a donor group. Metal ion complexing properties of tetra-*N*-acetamide substituted cyclen: a crystallographic, NMR, molecular mechanics, and thermodynamic study. *J Am Chem Soc* **117**, 6698–6707.

Nasir MS, Fahrni CJ, Suhy DA, Kolodsick KJ, Singe CP, O'Halloran TV. 1999 The chemical cell biology of zinc: structure and intracellular fluorescence of a zinc-quinolinesulfonamide complex. *J Biol Inorg Chem* **4**, 775–783.

Prasanna de Silva A, Gunaratne HQN, Gunnlaugsson T, Huxley AJM, McCoy CP, Rademacher JT, Rice TE. 1997 Signaling recognition events with fluorescent sensors and switches. *Chem Rev* **97**, 1515–1566.

Prodi L, Bolletta F, Montalti M, Zaccheroni N. 1999. Searching for new luminescent sensors: synthesis and photophysical properties of a tripodal ligand incorporating the dansyl chromophore and of its metal complexes. *Eur J Inorg Chem* 455–460.

Reany O, Gunnlaugsson T, Parker D. 2000 Selective signalling of zinc ions by modulation of terbium luminescence. *J Chem Soc Chem Comm* 473–474.

Shionoya M, Kimura E, Shiro M. 1993 A new ternary zinc(II) complex with [12] and N_4 (=1,4,7-10-tetraazacyclododecane) and AZT (=3′-azido-3′-deoxythymidine). Highly selective recognition of thymidine and its related nucleosides by a zinc(II) macrocyclic tetraamine complex with novel complementary associations. *J Am Chem Soc* **115**, 6730–6737.

Shionoya M, Ikeda T, Kimura E, Shiro M. 1994 Novel "multipoint" molecular recognition of nucleobases by a new zinc(II) complex of acridine-pendant cyclen (cyclen = 1,4,7,10-tetraazacyclododecane). *J Am Chem Soc* **116**, 3848–3859.

Sträter N, Lipscomb WN, Klabunde T, Krebs B. 1996 Two-metal ion catalysis in enzymatic acyl-and phosphoryl-transfer reactions *Angew Chem Int Ed* **35**, 2024–2055.

Thompson RB, Jones ER. 1993 Enzyme-based fiber optic zinc biosensor. *Anal Chem* **65**, 730–734.

Thompson RB, Patchan MW. 1995 Lifetime-based fluorescence energy transfer biosensing of zinc. *Anal Biochem* **227**, 123–128.

Thompson RB, Maliwal BP. 1998 Expanded dynamic range of free zinc ion determination by fluorescence anisotropy. *Anal Chem* **70**, 1749–1754.

Thompson RB, Maliwal BP, Feliccia VL, Fierke CA, McCall K. 1998 Determination of picomolar concentrations of metal ions using fluorescence anisotropy: biosensing with a "reagentless" enzyme transducer. *Anal Chem* **70**, 4717–4723.

Thompson RB, Maliwal BP, Zeng H-H. 1999 Improved fluorophores for zinc biosensing using carbonic anhydrase. *Proc SPIE-Int Soc Opt Engin* **3603**, 14–22.

Tsien R. 1989 Fluorescent indicator of ion concentrations. *Meth Cell Biol* **30**, 127–156.

Tsien R, Pozzan T. 1989 Measurement of cytosolic free Ca^{2+} with Quin2. *Methods Enzymol* **172**, 230–262.

Vallee BL, Falchuk KH. 1993 The biochemical basis of zinc physiology. *Physiol Rev* **73**, 79–118.

Walkup GK, Imperiali B. 1996 Design and evaluation of a peptidyl fluorescent chemosensor for divalent zinc. *J Am Chem Soc* **118**, 3053–3054.

Walkup GK, Imperiali B. 1997 Fluorescent chemosensors of divalent zinc based on zinc finger domains. Enhanced oxidative stability, metal binding affinity, and structural and functional characterization. *J Am Chem Soc* **119**, 3443–3450.

Walkup GK, Imperiali B. 1998 Stereoselective synthesis of fluorescent α-amino acids containing oxine (8-hydroxyquinoline) and their peptide incorporation in chemosensors for divalent zinc. *J Org Chem* **63**, 6727–6731.

Walkup GK, Burdette SC, Lippard SJ, Tsien RY. 2000 A new cell-permeable fluorescent probe for Zn^{2+}. *J Am Chem Soc* **122**, 5644–5645.

Zalewski PD, Forbes IJ, Betts WH. 1993 Correlation of apoptosis with change in intracellular labile Zn(II) using Zinquin [(2-methyl-8-p-toluenesulphonamido-6-quinolyloxy)acetic acid], a new specific fluorescent probe for Zn(II). *Biochem J* **296**, 403–408.

Zalewski, PD, Forbes IJ, Seamark RF, Borlinghaus R, Betts WH, Lincoln SF, Ward AD. 1994a Flux of intracellular labile zinc during apoptosis (gene-directed cell death) revealed by a specific chemical probe, Zinquin. *Chem Biol* **3**, 153–161.

Zalewski PD, Millard SH, Forbes IJ, Kapaniris O, Slavotinek A, Betts WH, Ward AD, Lincoln SF, Mahadevan I. 1994b Video image analysis of labile zinc in viable pancreatic islet cells using a specific fluorescent probe for zinc. *J Histochem Cytochem* **42**, 877–884.

BioMetals 14: 205–222, 2001.

Review

Fluorescence-based biosensing of zinc using carbonic anhydrase

Carol A. Fierke[1] & Richard B. Thompson[2*]
[1]*Departments of Chemistry and Biochemistry, University of Michigan, Ann Arbor, Michigan, USA;* [2]*Department of Biochemistry and Molecular Biology, University of Maryland School of Medicine, Baltimore, Maryland, USA;* **Author for correspondence (Tel: (410) 706-7142; Fax: (410) 706-7122; E-mail: rthompso@umaryland.edu)*

Received 16 January 2001; accepted 27 February 2001

Key words: biosensor, carbonic anhydrase, fluorescence, hippocampus, wavelength ratiometric, zinc

Abstract

Measurement of free zinc levels and imaging of zinc fluxes remains technically difficult due to low levels and the presence of interfering cations such as Mg and Ca. We have developed a series of fluorescent zinc indicators based on the superb sensitivity and selectivity of a protein, human apo-carbonic anhydrase II, for Zn(II). These indicators transduce the level of free zinc as changes in intensity, wavelength ratio, lifetime, and/or anisotropy; the latter three approaches permit quantitative imaging of zinc levels in the microscope. A unique attribute of sensors incorporating biological macromolecules as transducers is their capability for modification by site-directed mutagenesis. Thus we have produced variants of carbonic anhydrase with improved affinity for zinc, altered selectivity, and enhanced binding kinetics, all of which are difficult to modify in small molecule indicators.

Introduction

Zinc is an essential trace element for eukaryotes that is important in diverse biological processes, as discussed in detail in several other articles in this issue of *BioMetals*. In order to better understand the *in vivo* roles of zinc, we have been developing fluorescence-based biosensor approaches for measuring and imaging free zinc in biological specimens, in a manner akin to the methods in use for studying calcium signaling. The measurement of zinc ion flux *in vivo* is complicated by the low concentration of free zinc ions compared to both total zinc and other divalent cations, such as calcium and magnesium. In this article we will describe various protein-based sensing approaches that we have developed to circumvent these difficulties. In particular, we will focus on the development of a carbonic anhydrase-based biosensor, including both the origin of the exquisite metal selectivity in this protein and the various means by which binding of zinc may be transduced as changes in fluorescence.

It is convenient to first define our terms, particularly 'biosensor' and 'transducer'. A sensor (in this

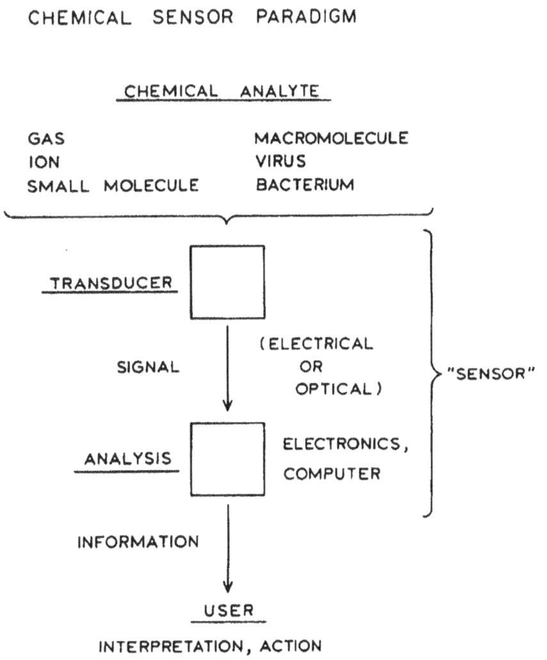

Figure 1. Chemical sensor paradigm

case, a chemical sensor) is a device wherein the presence, amount, or concentration of some chemical analyte interacts with a transducer to produce a signal of some kind (optical and electrical are most common), which can then be converted by some electronic or photonic apparatus into information in a form convenient for interpretation by a human operator (Figure 1) (Thompson & Walt 1994). Although it is not a necessity, in many instances a sensor will also have the capability to measure the analyte continuously for some period, and report the results in real time. While many different approaches are known, a sensor is termed a 'biosensor' if the transducer is a molecule that is biological in origin. In our case the transducer molecule is a variant of human carbonic anhydrase II (CA II), which is found in nature as a metalloenzyme with Zn(II) bound in its active site. The enzymology of CA has been the subject of very thorough study for much of the last century, and the catalytic role of the zinc ion is firmly established (Lindskog *et al.* 1971). It should be emphasized that the interaction of metal ions with the protein structure is both subtle and elegant, and has been the focus of much inquiry (Christianson 1991; Christianson & Fierke 1996; Hakansson *et al.* 1992; Hunt & Fierke 1997; Lindskog & Nyman 1964). Once the zinc is removed, the *apo*-form of CA binds Zn(II) tightly, specifically, and reversibly in the active site, and from a sensing standpoint CA is not so much an enzyme as a talented ligand.

In view of the numbers of colorimetric and fluorescent indicators described in the literature (Fernandez-Gutierrez & Munoz de la Pena 1985; Haugland 1996; White & Argauer 1970), it is reasonable to wonder why one would develop a biosensor for this purpose. The reasons are primarily selectivity and sensitivity. In many biologically relevant circumstances the data indicate that the concentration of free Zn(II) is quite low, in the range of nanomolar or below based mainly on NMR experiments with fluoro-BAPTA (Benters *et al.* 1997; Denny & Atchison 1994). Clearly any biosensor or indicator would require high affinity to be at least partially saturated and generate an adequate signal at such concentrations. We note that some electrochemical sensors are capable of very high sensitivity (and selectivity) for metal ions such as Cu(II), but their response is necessarily slow at low concentrations, and such electrodes have not yet been described for zinc (Belli & Zirino 1993). Moreover, electrochemical approaches would appear to be ill-suited for imaging. Metallofluorescent indicators are known with fairly high affinity, based on modification of multidentate chelating molecules (Grynkiewicz *et al.* 1985; Kuhn *et al.* 1995; Simons 1993). However, for some applications one might wish for better selectivity and/or sensitivity than these probes currently offer. For instance, Fura-2 (part of a family of fluorescent indicators) is about one hundred-fold more sensitive for Zn(II) than for Ca(II), which suggests its use for measuring Zn(II). However, intracellular calcium levels frequently are high enough to induce a signal (indeed, this is the source of Fura's utility), and extracellular Ca in the millimolar range would naturally saturate the indicator, making its use difficult. Even recently developed metallofluorescent indicators do not match the performance of the CA-based systems. For instance, the indicator ZnAF-1 (Hirano *et al.* 2000) has approximately one thousand-fold worse affinity for zinc (0.78 nM) than wild type CA (0.8 pM) (Hunt *et al.* 1999) (and forty thousand-fold worse than high affinity variants, Table 1). Inasmuch as wild type CA-based fluorescence transducers have demonstrated better than two hundred-fold discrimination between Zn(II) and Cd(II) (Thompson *et al.* 1999) and the affinity of ZnAF-1 for Cd^{2+} remains undisclosed, the latter can hardly be said to be "the first Zn^{2+} sensor molecule that can distinguish Cd^{2+} from Zn^{2+}". Metallofluorescent indicators are discussed by Professor Kimura elsewhere in this issue.

By comparison, *apo*carbonic anhydrase has not only very high affinity, but outstanding selectivity as well (Lindskog & Nyman 1964; McCall & Fierke 2000; Thompson *et al.* 1998a). The X-ray crystal structure of human CA isozyme II (Hakansson *et al.* 1992) reveals that the catalytic zinc ion is coordinated at the bottom of a deep active site cleft by three histidine ligands, H94, H96, and H119, and a solvent molecule in tetrahedral fashion (Figures 2 and 3). At physiological pH, the fourth ligand is provided by hydroxide from solvent. These zinc ligands are fully saturated by hydrogen bonds with other protein residues and are surrounded by a shell of hydrophobic residues. This zinc binding site is retained in numerous isozymes of the alpha class of carbonic anhydrases (Christianson & Fierke 1996) as well as the gamma class (Iverson *et al.* 2000). In contrast, the zinc polyhedron in the beta class of carbonic anhydrases is a $HisCys_2H_2O$ site (Kimber & Pai 2000; Mitsuhashi *et al.* 2000). The dissociation constant of human carbonic anhydrase isozymes I and II for zinc at neutral pH is in the pM range (Hunt & Fierke 1997; Lindskog & Nyman 1964), suggesting the possibility of a very sensitive zinc sensor. Moreover, the metal affinity

Table 1. Variants of carbonic anhydrase.

Variant	K_{Zn}, pM	k_{off}, h^{-1}	k_{on}, μM^{-1} sec^{-1}	K_{Cu}, pM	K_{Zn}/K_{Cu}	Reference
Wild type	0.8	0.0003	0.1	0.017	50	(Hunt *et al.* 1999; Hunt & Fierke 1997)
H94A	270,000	>140	0.1			(Kiefer & Fierke 1994)
H94C	33,000	0.5	0.004			(Hunt *et al.* 1999; Kiefer & Fierke 1994)
H94D	15,000	0.7	0.01	5	3,000	(Hunt *et al.* 1999; Kiefer & Fierke 1994)
H119D	25,000	10	0.1	80	300	(Hunt *et al.* 1999; Kiefer & Fierke 1994)
H94E	14,000	30	0.6			(Kiefer & Fierke 1994)
H94N	40,000	5	0.03	10	4,000	(Lesburg *et al.* 1997)
H94Q	8,000	20	0.7			(Lesburg *et al.* 1997)
H119Q	70,000	6	0.02	>70,000	< 1	(Lesburg *et al.* 1997)
T199C	1					(Kiefer *et al.* 1993b)
T199E	0.02					(Ippolito *et al.* 1995a)
T199A	60	0.002	0.01			(Kiefer *et al.* 1995)
Q92A	18	0.001	0.02			(Kiefer *et al.* 1995)
E117A	40	1.5	10			(Kiefer *et al.* 1995)
E117Q	4,000	4,680	300			(Huang *et al.* 1996)
Q92A/E117A	160	1.6	3			(Kiefer *et al.* 1995)
F95M/W97V	1.6	2.1	0.4	0.004	400	(Hunt *et al.* 1999; Hunt & Fierke 1997)
F95I/W97S	6	144	6	0.001	6,000	(Hunt *et al.* 1999; Hunt & Fierke 1997)
F93I/F95M	8	0.4	0.01	0.005	1,600	(Hunt *et al.* 1999; Hunt & Fierke 1997)
F93I/F95M/W97V	11	5.8	0.1	0.003	3,700	(Hunt *et al.* 1999; Hunt & Fierke 1997)
F93S/F95L/W97M	29	96	0.9	0.002	14,000	(Hunt *et al.* 1999; Hunt & Fierke 1997)
F93T/F95S/W97V	92	120	0.4	0.0001	900,000	(Hunt *et al.* 1999; Hunt & Fierke 1997)
F93S/F95T/W97M	76	280	1			(Hunt *et al.* 1999; Hunt & Fierke 1997)

of single zinc sites in metalloenzymes is often in this range or higher indicating that these sites should all be saturated even at nM concentrations of zinc.

Carbonic anhydrase has significantly enhanced specificity for zinc, compared to many small molecule metal chelators as discussed above (McCall & Fierke 2000). Nonetheless, other metals do bind; CA binds Cu(II) about an order of magnitude more tightly than Zn(II), Hg(II) with very high affinity, Cd (II), Co(II) and Ni(II) with nanomolar affinity and Mn(II), Mg(II) and Ca(II) with very low affinity, if at all (Lindskog & Nyman 1964; McCall & Fierke 2000; Thompson *et al.* 1998a). In general, the free concentrations of Cu(II) are estimated to be very low in humans (Rae *et al.* 1999) and it seems likely that this is also true for Hg, Cd, Co and Ni, at least in part due to their toxicity, relative scarcity, and the presence of ligands both extracellularly and intracellularly. Consequently, we anticipate (and have found) little interference with Zn sensing using CA-based sensors. Moreover, some of the sensing schemes described below are even more specific for Zn(II) due to the zinc-dependent binding of the fluorescent aryl sulfonamides, at the active site. In addition, use of a macromolecule transducer, such as CA, confers additional advantages, in that its structure can be subtly altered to 'fine-tune' the selectivity and sensitivity for certain metal ions, as well as the kinetics of binding. Therefore, the tools of genetic engineering can be used to optimize CA as a biosensor using either random mutagenesis and selection techniques ("directed evolution") or structure-based redesign methods (Christianson & Fierke 1996).

Optimization of CA zinc sensor by genetic engineering: Zinc affinity

The main rationale for using carbonic anhydrase as the receptor in a zinc biosensor is that it binds zinc with high affinity and selectivity, as previously described. However, the high zinc affinity of wild-type CA may actually limit the usefulness of a CA-based biosensor. To measure free Zn(II) concentrations in real time, zinc binding must be measured under equilibrium conditions and equilibration must be rapid. Under equilibrium conditions and measuring changes in fluorescence intensity, only zinc concentrations within an order of magnitude of the picomolar Zn(II) dissociation constant of CA (0.1–10 pM) can be accurately measured using a single binding isotherm. While the concentrations of free Zn(II) in biological samples are not yet well-determined in many cases (at least partly due to the lack of a high affinity, selective zinc sensor), the data suggest that the Zn(II) concentration may be higher than 10 pM (Benters *et al.* 1997; Denny & Atchison 1994). Given the uncertainty in the actual Zn(II) concentrations and the possibility of significant fluxes in [Zn(II)], a larger range of detectable zinc concentrations is desirable. One solution to this limitation is to use alternate optical transduction methods that, under certain conditions, can determine zinc concentrations over a larger range (Thompson *et al.* 1998b; Thompson & Patchan 1995a). Alternatively, CA variants with altered zinc affinity could be applied in an array fashion to expand the range of measurable zinc concentrations. To this end, we have successfully used a number of molecular biology techniques, including both structure-based and random mutagenesis, to alter the zinc affinity of carbonic anhydrase by more than 7 orders of magnitude, from 0.02 pM to >1 μM (see below). These mutants, used in an array, could potentially allow quantitation of a wide range of free Zn(II) concentrations ($\sim$2 fM to 10 μM).

We first began altering the zinc affinity of carbonic anhydrase using a structure-based redesign approach. The high resolution structure of CA shows that the single zinc lies at the bottom of a conical cleft where it is coordinated in a tetrahedral fashion by the imidazole side chains of three histidine residues (H94, H96 and H119) and one solvent molecule (Figure 2B) (Eriksson & Jones 1988; Hakansson *et al.* 1992). Studies of both model compounds and proteins suggest that metal affinity and specificity are affected by the geometry of the metal site, including the number and geometric position of the ligands relative to the metal ion and the distances of each ligand from the metal (Alberts & Nadassy 1998; Christianson & Fierke 1996; Glusker 1991; Roe & Pang 1999; Rulisek & Vondrasek 1998). Furthermore, the direct protein ligands of transition metal sites in proteins are typically nested in a hydrogen bond network (Christianson 1991; Christianson & Alexander 1989) and surrounded by a shell of hydrophobic side chains (Yamashita & Wesson 1990). Hence, the entire protein may affect the stability and chemical properties of the protein-metal complex and thus be used to 'tune' the metal affinity. Moreover, these structural features are observable in most metalloenzymes and are likely similarly important for modulating metal affinity, specificity, and reactivity.

The largest alterations in metal affinity in carbonic anhydrase have been obtained by altering the number and chemical nature of the protein side chains that directly coordinate the zinc ion. Removal of one of the ligands, by substitution of histidine with alanine, decreases the zinc affinity by a factor of at least 10^5-fold (Table 1), resulting in a loss of 6–7 kcal/mol of binding energy (Kiefer & Fierke 1994) demonstrating the importance of each of the direct ligands. Surprisingly, replacing the histidine ligands with other residues that could potentially serve as zinc ligands (sulfur of cysteine and the oxygen of glutamate or aspartate) does not enormously increase the binding affinity relative to variants lacking a third protein zinc ligand (Kiefer & Fierke 1994; Kiefer *et al.* 1993a). Variants such as H94C and H94D CA, for example, bind zinc with nanomolar affinity, only approximately 10-fold higher than H94A (Table 1). X-ray crystal structures of these variants (Ippolito and Christianson 1994; Ippolito *et al.* 1995b; Kiefer *et al.* 1993a) reveal that the altered size and shape of the substituted ligand necessitates movement of the surrounding protein structure and/or the zinc ion to maintain tetrahedral geometry. These alterations in the active site structure

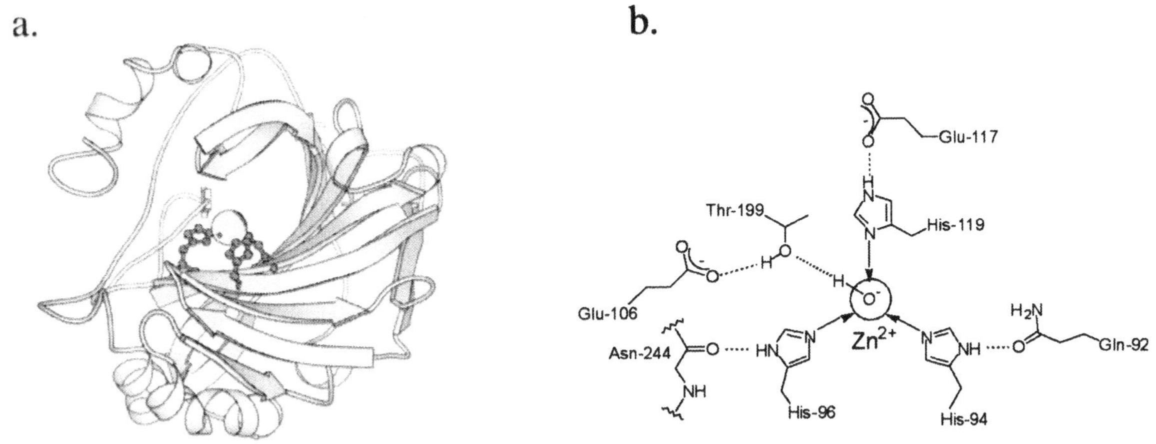

Figure 2. (a) Ribbon diagram of human carbonic anhydrase II; the active site zinc is the sphere coordinated by the three imidazoles. (b) Schematic of active site zinc inner sphere and outer sphere ligands.

are energetically unfavorable, leading to lower binding affinity.

CA metal sites with increased affinity have been created by the addition of a fourth protein ligand to augment the metal polyhedron. Residue T199, which forms a hydrogen bond with the zinc-bound hydroxide of wild type-CA (Figure 2), was substituted with a variety of amino acids capable of directly coordinating zinc, including Cys, Asp, and Glu (Kiefer *et al.* 1993b; Ippolito *et al.* 1995a). Each of these new side chains directly coordinates zinc in tetrahedral geometry, replacing the solvent hydroxide molecule that constitutes the fourth zinc ligand in wild type CA, as demonstrated by high resolution X-ray crystal structure determination (Ippolito *et al.* 1995a; Ippolito & Christianson 1993). The affinity of several of these mutants is comparable to that of wild-type CA (Table 1); in these cases, the favorable energy gained by the additional zinc coordination is virtually offset by the energetically unfavorable movements of the protein. However, the affinity of T199E CA for zinc is greatly enhanced (Ippolito *et al.* 1995a) by addition of this fourth protein ligand, with a K_D of 20 fM, making this the highest affinity zinc site ever engineered.

Additionally, alterations in the hydrogen bond network and hydrophobic shell alter metal affinity. Wild-type CA has four 'second shell' ligands (Figure 2) including hydrogen bonds between the side chains of His 94 and Gln 92, His 119 and Glu 117, and zinc-water and Thr 199. Removal of one of these hydrogen bonds by substitution with alanine decreases metal affinity about 10-fold (Table 1). Furthermore, the decrease in metal affinity is additive

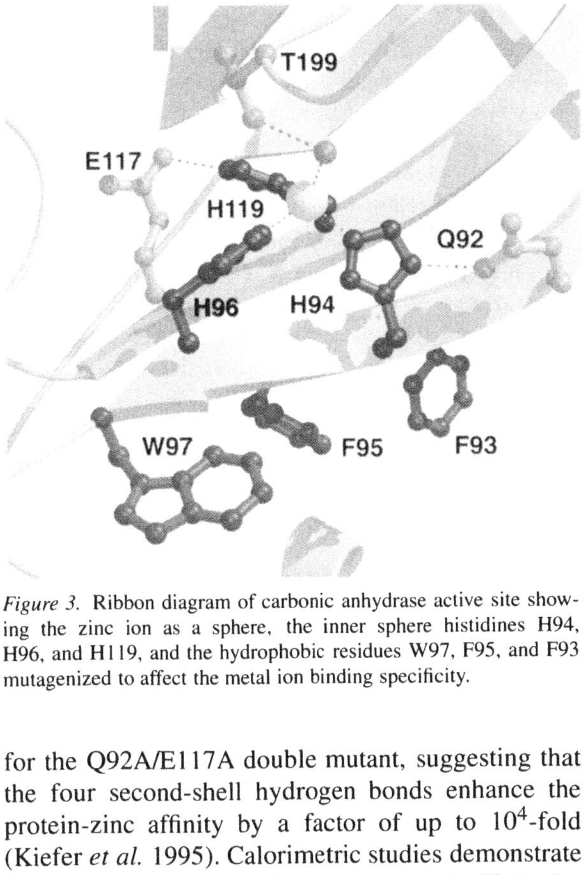

Figure 3. Ribbon diagram of carbonic anhydrase active site showing the zinc ion as a sphere, the inner sphere histidines H94, H96, and H119, and the hydrophobic residues W97, F95, and F93 mutagenized to affect the metal ion binding specificity.

for the Q92A/E117A double mutant, suggesting that the four second-shell hydrogen bonds enhance the protein-zinc affinity by a factor of up to 10^4-fold (Kiefer *et al.* 1995). Calorimetric studies demonstrate that these hydrogen bonds enhance metal affinity by both pre-organizing and desolvating the metal site (DiTusa *et al.* 2001). Alteration of these second-shell ligands allows for subtle changes in the zinc affinity which are essential for creating an array of variants with altered metal affinity.

The shell of hydrophobic side chains surrounding metal binding sites has been proposed to enhance metal affinity by decreasing the mobility of the His ligands or by decreasing the dielectric constant (Yamashita & Wesson 1990). In CA, residues Phe 93, Phe 95, and Trp 97 flank two of the three histidines that coordinate zinc to form a hydrophobic cluster beneath the zinc binding site (Figure 3). To investigate the importance of this hydrophobic shell for determining the metal affinity of CA, we first used cassette mutagenesis to prepare a library of CA variants differing in these three hydrophobic amino acids (Hunt & Fierke 1997). All of these variants were then displayed on filamentous phage as a fusion protein between CA and a minor coat protein (gene 3 protein (g3p)). The phage displaying CA-g3p fusion proteins were then separated on the basis of the zinc affinity of CA using sulfonamide affinity chromatography. Wild-type CA was enriched 20-fold by one round of selection and consensus residues at each position were identified from the enriched variants (I, F, L and M at position 93; I, L and M at position 95; and W and V at position 97). After two rounds of selection, variants that bind to the sulfonamide resin have zinc affinity comparable to that of wild-type CA, indicating that the aromatic residues are not absolutely essential. However, the zinc affinity of mutants containing other substitutions at these positions decreases up to 100-fold (Table 1). Strikingly, the K_{Zn} decreases as the volume of the amino acids at positions 93, 95, and 97 decreases. These experiments demonstrate both that metalloenzyme variants displayed on phage can be selected on the basis of metal affinity and that mutations in the hydrophobic shell can also be used for fine-tuning the zinc affinity of CA.

Zinc specificity

The metal selectivity of WT CA follows the Irving-Williams series (Mn(II)<Co(II)<Ni(II)<Cu(II)> Zn(II)), although the Zn(II) selectivity is increased significantly (several orders of magnitude) compared to most small molecule chelators (Lindskog & Nyman 1964; McCall & Fierke 2000; Thompson *et al.* 1998a). Several features of metal ion binding sites have been hypothesized to alter the transition metal selectivity of chelators. These include: (1) the polarizability of the coordinating atom (Pearson 1966); (2) the relative sizes of the binding site and the metal ion; and (3) the metal ion binding site geometry. Transition metal ions, including Zn(II), Cu(II), and Co(II) are categorized as borderline metals, capable of coordinating O, S, and N with high affinity, but are most often found coordinated to nitrogen (Alberts & Nadassy 1998; Rulisek & Vondrasek 1998). The preferred geometry of metal ions and the optimum ligand distances have been investigated in both model compounds and in proteins (Alberts & Nadassy 1998; Glusker 1991; Roe & Pang 1999; Rulisek & Vondrasek 1998). Zinc is most often observed in tetrahedral geometry in proteins while copper favors square planar and trigonal bipyramidal geometries. In CA, zinc binds with distorted tetrahedral geometry while copper binds with trigonal bipyramidal geometry (Hakansson *et al.* 1992, 1994).

To investigate whether the metal selectivity is tuned by the geometry of the ligands in the metal binding site, we measured the relative affinity and selectivity for various transition metals of mutants of residues Phe93, Phe95 and Trp97 which are located on the same β-strand as two of the histidine ligands (Figure 3) (Cox *et al.* 2000; Hunt *et al.* 1999). Although the zinc and cobalt affinity of these variants decreases as the hydrophobicity of the substituted side chains decreases (Hunt & Fierke 1997), the copper affinity *increases*, resulting in a significant net enhancement in the selectivity of the enzyme for copper (Hunt *et al.* 1999). These data suggest that the hydrophobic shell does not enhance zinc affinity mainly by altering the dielectric constant of the metal binding site. X-ray crystal structures of metal-bound F93I/F95M/W97V and F93S/F95L/W97M CAs (Cox *et al.* 2000) reveal that the coordination geometry of the zinc-bound and copper-bound enzymes remain tetrahedral or trigonal bipyramidal, respectively, as observed in WT CA. However, a conformational change of the direct metal ligand H94 as well as the indirect ligand Q92 occurs in the apo-form of the mutants, thereby eliminating the preorientation of the histidine ligands into a tetrahedral geometry, as observed in the apo-WT enzyme (Hakansson *et al.* 1992). This increased flexibility enhances formation of 5-coordinate trigonal bipyramidal metal coordination geometry relative to 4-coordinate tetrahedral geometry, which in turn increases Cu(II) affinity and decreases Zn(II) affinity. These data demonstrate that aromatic core residues serve mainly a foundational role, as anchors that help to preorient direct and second-shell ligands to optimize the zinc binding geometry and to destabilize alternative geometries. Therefore, the zinc/copper selectivity in CA, and likely other proteins as well, is tuned

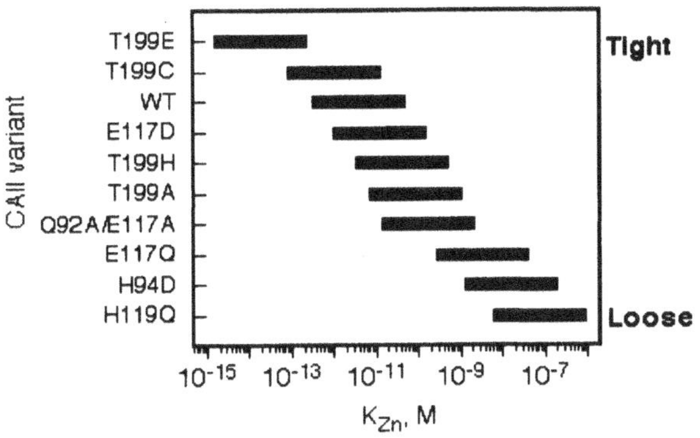

Figure 4. Zinc affinities of variants. Variants are ranked with the highest affinities at the top to lowest at the bottom; the bars give the range of free zinc ion concentration that can be accurately measured by a particular variant.

by stabilizing the preferred tetrahedral geometry of zinc and destabilizing the trigonal bipyramidal copper geometry (Cox *et al.* 2000; Hunt *et al.* 1999).

In addition to enhancing our understanding of zinc affinity and selectivity in proteins, we have prepared a series of variants of CA whose zinc affinity ranges over seven orders of magnitude (Figure 4), and with a range of selectivity as well. The broad range of affinities is of use because an appropriate variant can be chosen which responds well for any given free zinc level from one femtomolar to one micromolar. By comparison, known metallofluorescent zinc indicators respond down to below one nanomolar, but quantitation at levels below this remains troublesome.

Zinc equilibration

A second factor limiting the utility of wild-type CA in a zinc sensor is the extremely slow zinc dissociation rate constant. The $t_{1/2}$ for this dissociation is estimated to be on the order of months (Hunt & Fierke 1997; Lindskog & Nyman 1964), virtually limiting the sensor to a single use rather than continuous, real-time monitoring. In CA, zinc equilibration is limited by both the high zinc affinity (pM) and the slow zinc association rate constant of 10^5 M^{-1} s^{-1} (Henkens & Sturtevant 1968; Kiefer & Fierke 1994), 10^3-fold slower than the diffusion-controlled limit of 10^7–10^8 M^{-1} s^{-1} (Eigen & Hammes 1963). Since the rate constant for zinc equilibration can be approximated by k_{on} [L] + k_{off} and since $K_D = k_{off}/k_{on}$, the observed zinc equilibration rate can be enhanced by increasing k_{on} and/or k_{off}. In mutants with decreased zinc affinity (as described above), the zinc dissociation rate constants often increase (Hunt & Fierke 1997; Kiefer & Fierke 1994; Kiefer *et al.* 1995), roughly paralleling the K_D (Table 1). These mutants therefore further enhance the utility of CA-based zinc sensors.

In some cases, however, the zinc dissociation rate constant increases significantly more than the metal affinity decreases, suggesting that the zinc association rate constant increases significantly (Table 1). In particular, substitution of Glu-117, the hydrogen bond partner of the direct ligand His-119 (Figure 2), with either alanine or aspartate causes the K_D to increase 3- to 10-fold while the dissociation rate constant increases 30- to 70-fold (Kiefer *et al.* 1995; Lesburg & Christianson 1995). Therefore, the calculated zinc association rate constant increases significantly, to $\sim 10^7$ M^{-1} s^{-1} and the half-time for zinc dissociation decreases to about 30 minutes. Additionally, substitution of Glu-117 with Gln causes further increases in K_D and k_{off} (Table 1) (Lesburg *et al.* 1997). In this variant the calculated zinc association rate constant of 3×10^8 M^{-1} s^{-1} is at or near the diffusion-controlled limit and the half-time for zinc dissociation, 0.6 s^{-1}, is certainly fast enough for most biosensor applications. These data indicate that hydrogen bonds between the direct zinc ligands and indirect hydrogen bond acceptors are a primary factor controlling the slow association and dissociation of zinc. These hydrogen bonds may affect zinc equilibration by pre-orienting the metal binding site to reduce conformational energy.

These detailed structure-function studies provide an understanding of the mechanism of metal binding in CA and allow the prediction of additional amino acid substitutions that will affect the properties of the metal binding site. The current CA variants with altered metal affinity, specificity and equilibration kinetics are extremely useful in developing real-time metal ion biosensors. Our demonstrated ability to tune both the metal affinity and metal equilibration kinetics by subtly altering the packing of the zinc binding site has no counterpart in small molecule metallochromic or metallofluorescent indicators, nor in ionophores for electrochemical methods, illustrating the versatility of biological receptors.

Transducing zinc binding as a fluorescence signal

Although *apo*-CA is a very selective, high affinity ligand for Zn(II), it is also necessary that zinc binding creates a signal which can be readily observed and measured. For speed, simplicity, and to avoid the need to consume reagents we do not find transduction employing the enzyme activity itself to be attractive. For use in observing tissues and cells it is desirable that it be an optical signal, preferably fluorescence owing to its sensitivity, flexibility, and the widespread availability of fluorescence microscopes. For a continuous monitor (as opposed to a single determination) the reversible binding of the zinc ion should result in a reversible change in the fluorescence. Thus we would like to arrange matters such that binding of zinc to *apo*-CA results in a change in fluorescence, which we can readily measure, and that dissociation of the metal from the protein reverses that change. Of course, the fractional saturation of the binding site is controlled by the law of mass action, and in particular the concentration (more properly, the activity) of the metal ion.

The vast majority of metallofluorescent indicators transduce metal ion binding as changes in fluorescence intensity: e.g., in most cases interaction with the metal ion results in a change in quantum yield of the fluorophore, leading to a change in the intensity of fluorescence. Attempting to relate a measured fluorescence intensity to an analyte concentration is very prone to artifact, and is seldom done either clinically or in research. Recognizing this problem in attempting to measure Ca(II) concentrations with indicators such as calcein, workers turned to the so-called wavelength-ratiometric probes developed by Roger Tsien and his colleagues. Among these are Fura-2, Indo-1, and a host of successors (Grynkiewicz *et al.* 1985; Haugland 1996). These probes exhibit a shift in excitation and/or emission spectra upon binding the metal, and consequently the ratio of intensities at the two peaks is proportional to the ratio of indicator with metal bound and free. Measuring a ratio of fluorescence intensities instead of a simple intensity eliminates many of the artifacts present in such measurements, and makes calibration more straightforward. Transducing the binding of the metal as a change in fluorescence anisotropy (polarization) confers similar advantages (Thompson *et al.* 1998a; Weber 1956), because the anisotropy can also be measured as a ratio of two intensities. Fluorescence anisotropy microscopy has been known for some time (Dix & Verkman 1990; Fushimi *et al.* 1990). Finally, one can also transduce the binding of the metal as a change in fluorescence lifetime, which similarly avoids the artifacts associated with simple intensity measurements. Moreover, lifetime measurements (as with anisotropy measurements) can under certain conditions exhibit a much broader dynamic range of analyte concentration than is achievable with either simple intensity or intensity ratio measurements (Szmacinski & Lakowicz 1993; Thompson *et al.* 1998b; Thompson & Patchan 1995a). However, lifetime measurements remain complex and relatively expensive, despite advances in instrumentation for their measurement and imaging.

Finally, lifetime measurements are very well suited to configurations of the biosensor wherein the CA is immobilized on quartz, glass, or some other suitable substrate (Clark *et al.* 1999) which is immersed into a solution (such as the cytoplasm of a cell), permitting the Zn level to be measured continuously. A particularly useful example is immobilization on the distal tip of a fiber optic, which can then be inserted into the bloodstream or interstitial spaces of an experimental animal to measure the analyte *in situ* (Thompson 1991). This is a potentially powerful research and diagnostic tool which we hope to exploit for zinc studies.

Zinc sensing using fluorescent aryl sulfonamides and CA

The first approach to fluorometrically determining zinc using carbonic anhydrase built on the pioneering work of Chen & Kernohan (1967). They discovered that dansylamide, one of the class of aryl sulfonamide inhibitors of the enzyme, exhibited dramatic changes in its fluorescence upon binding to *holo*carbonic anhydrase. Aryl sulfonamides had long been known to inhibit the enzyme (Maren 1977), and due to their therapeutic importance in treating glaucoma, hundreds of such compounds had been identified. It was well understood that the inhibition was based upon the active site zinc promoting the ionization of the (weakly acidic) sulfonamide when it is bound, in a manner precisely analogous to the water ordinarily bound as a fourth ligand. The fluorescence of dansyl is well known to be environment-sensitive, and consequently its emission is substantially blue shifted and enhanced upon binding to the enzyme (Figure 5). Thompson and Jones (1993) showed in the absence of the metal that dansylamide's affinity was much reduced, and consequently the fluorescence of dansylamide could be used as an indicator of zinc's presence in the active site. Since the occupancy of the zinc binding site is controlled by the free zinc ion concentration through the law of mass action, one could relate the change in

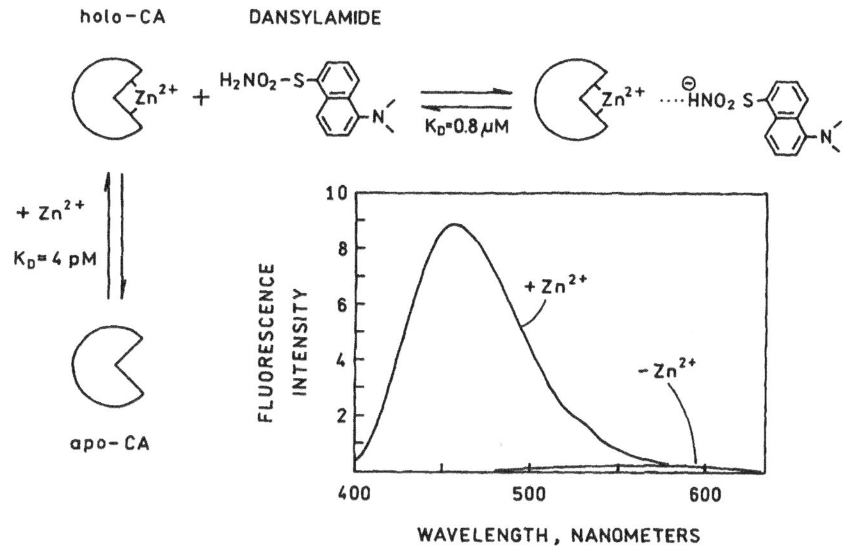

Figure 5. Dansylamide sensing scheme. Free dansylamide emits weak fluorescence (*inset*) in the presence of apo-CA, but when Zn(II) binds to CA, dansylamide binds to holo-CA, emitting strong blue fluorescence.

fluorescence observed to the concentration of zinc. By itself, dansylamide has essentially no affinity for the metal.

In fact, dansylamide's fluorescence changes in several ways upon binding to the protein, each of which may be used to determine zinc concentration. Perhaps most important is the dramatic 100+ nm blue shift and seven-fold increase in quantum yield, because it permits a ratiometric measurement of intensities at emission wavelengths corresponding to the free (560 nm) and bound (450 nm) forms of dansylamide. While this approach enjoys some of the well-known advantages of the ratiometric measurements it should be remembered that it is not entirely insensitive to the concentrations of dansylamide and apoprotein in use. The large change in fluorescence lifetime of dansylamide may also be used in the same way to determine zinc. Although measuring changes in fluorescence lifetime requires more sophisticated instrumentation than measuring intensities, this approach has particular advantages for measurements through fiber optics (Thompson 1991; Thompson 1993). Moreover, the instrumentation is becoming simpler and cheaper (Levy *et al.* 1997; Lippitsch *et al.* 1988; Thompson *et al.* 1992). Finally, under certain circumstances lifetime measurements can exhibit an enhanced dynamic range (Szmacinski & Lakowicz 1993; Thompson & Patchan 1995a); we subsequently showed anisotropy measurements offer the same advantage (Thompson *et al.* 1998b).

Dansylamide suffers from two key drawbacks: first, its extinction coefficient is rather low (ε_{330} = 3300 M^{-1} cm^{-1}), limiting sensitivity. Second, the short wavelength UV necessary for excitation is a problem owing to the expensive microscope optics required, the inconvenience of exciting at these wavelengths, and the high autofluorescence background observed in natural specimens. Clearly it was desirable to have a longer wavelength probe; consequently, we developed ABD-N and ABD-M (Figure 6). These compounds are excitable in the visible and have 2–3 fold larger extinction coefficients than dansylamide, but also exhibit shifts in excitation and emission upon binding to holo-CA, as well as changes in quantum yield, lifetime, and anisotropy (Table 2) (Thompson *et al.* 2000a, 1998b). ABD-N is the more useful due to its higher quantum yield when bound compared with ABD-M. ABD-N can be used to quantitate zinc by changes in intensity, lifetime, anisotropy, or wavelength ratio; the last is illustrated in Figure 7 (Thompson *et al.* 2000a). Both molecules are adequately water-soluble.

In looking for additional fluorescent aryl sulfonamides which might be useful for this approach, we found two others which are noteworthy. The first is Dapoxyl sulfonamide (Figure 6), which in a sense is complementary to ABD-N in that it exhibits a very

Table 2. Aryl sulfonamides used in zinc sensing.

Probe	Free/Bound	λ_{exc}, nm	λ_{em}, nm	QY, relative	$< \tau >$, nsec	K_D, μM	log ε	aniso	ratio	λ_1/λ_2	References
DNSA	Free	330	560	0.08	3		3.52		0.039	450/560	(Chen & Kernohan 1967; Thompson & Jones 1993)
	Bound		450	0.55	22.1	0.8			1.6		
ABD-N	Free	435	602	0.086	0.34		3.94	0.09	3	560/680	(Thompson *et al.* 2000a)
	Bound	420	558	1.0	4.98	0.9		0.23	7.5		
ABD-M	Free	390	528	0.33	0.8		3.9	0.03			(Thompson *et al.* 1998b)
	Bound	380	492	1.0	1.53	0.3		0.32			
Azosulfamide	Free	501				0.01	4.5				(Thompson & Patchan 1995b)
	Bound										
Dapoxyl sulfonamide	Free	365	615	0.01	0.22		4.3		3	535/685	(Thompson *et al.* 2000b)
	Bound		535	1.00	3.60	0.13			38		
BTCS	Free	466	504	0.88	2.53		4.56	0.05			(Thompson *et al.* 2000b)
	Bound			1.00	2.71			0.25			
Elbaum's	Free	495	505	1.0	4.0		4.9	0.01			(Elbaum *et al.* 1996)
	Bound			0.96		0.002		0.10			

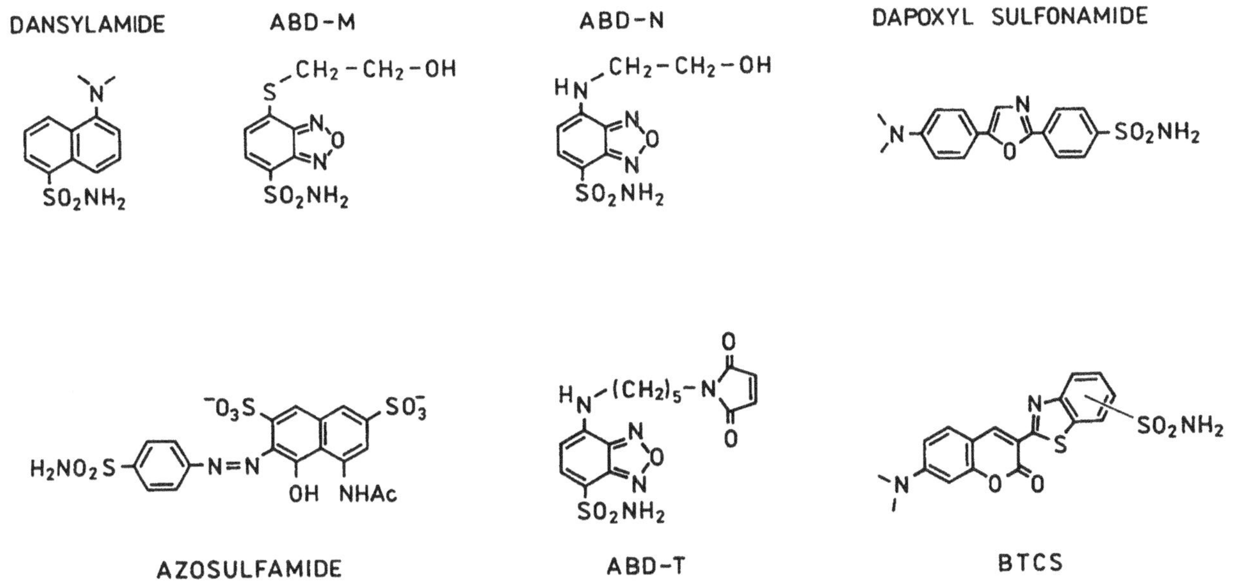

Figure 6. Aryl sulfonamides which bind to carbonic anhydrase.

large fluorescence enhancement upon binding to holo-CA, together with a large blue shift in its fluorescence. However, it has twice as large an extinction coefficient as ABD-N (Table 2). Unlike ABD-N, however, it penetrates cells readily, and stains membrane bilayers. The fluorescence of the bilayer-bound form differs from the CA-bound form, but not dramatically so (Thompson *et al.* 2000b), making intracellular zinc measurements problematic at the present time. It has a significant two-photon excitation cross section, but is not a good anisotropy probe. By comparison, the benzothiazolyl coumarin sulfonamide BTCS (Figure 6) exhibits little change in lifetime or quantum yield upon binding to CA, which makes it an excellent anisotropy probe (Thompson *et al.* 2000b). Spectroscopically, it is similar to the fluorescein moiety in our original anisotropy probe (Elbaum *et al.* 1996), and has a high quantum yield and extinction coefficient.

Fluorescence energy transfer-based zinc biosensing

The environment-sensitive fluorophores most useful in the aryl sulfonamide approach depicted above are in general poor absorbers which often must be excited at relatively short wavelengths. Partly to avoid this problem a more flexible approach was sought. Among the hundreds of aryl sulfonamide inhibitors known, several are colored but otherwise nonfluorescent, and therefore useless in the approach outlined above. However, the propensity of these colored aryl sulfonamides to bind to the holoprotein at the active site when zinc was present (and not to bind in its absence) suggested a somewhat different approach. In particular, it was straightforward to fluorescently label the protein with a covalent derivative, chosen for overlap of its emission with the absorbance of the colored aryl sulfonamide. If the colored sulfonamide binds at the active site it is perforce brought within 25 Å of the label, which thus partly quenches the fluorescence of the label due to Förster resonance energy transfer (Figure 8). In the absence of zinc the sulfonamide is free to diffuse away, so no energy transfer occurs and there is no diminution of the label's emission. Importantly, the lifetime of the label declines commensurately with the intensity, permitting the zinc's quantitation by time-resolved fluorescence measurements, which are more robust than intensity measurements. The response of the system can be optimized by careful selection of the label and colored arylsulfonamide based on their known spectral properties. Use of site-directed CA mutants permitted the label to be positioned in a defined and predetermined fashion to optimize the response (Thompson *et al.* 1996a; Thompson & Patchan 1995b).

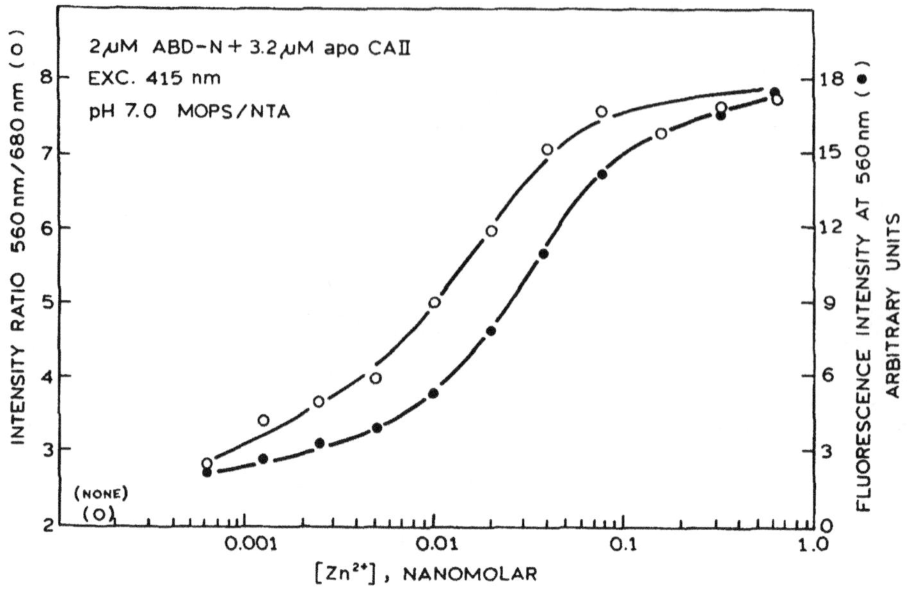

Figure 7. Fluorescence emission intensity at 560 nm (filled circles) and emission intensity ratio at 560 nm/680 nm (open circles) for 3.2 μM apo CA + 2.0 μM ABD-N; excitation at 415 nm.

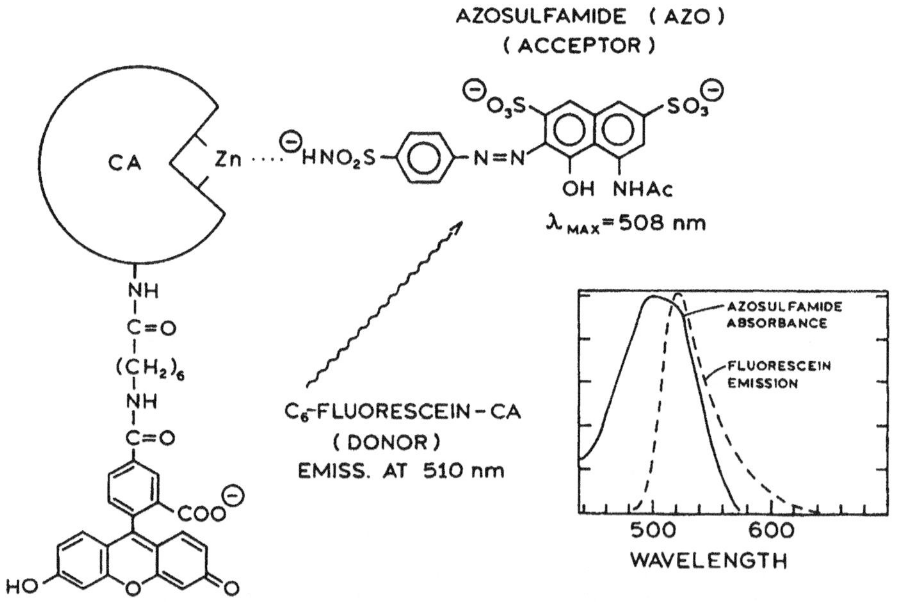

Figure 8. Fluorescence energy transfer zinc sensing approach. In the presence of zinc, azosulfamide binds to the holo-CA and accepts energy from the donor, quenching it; in the absence of zinc, azosulfamide does not bind to the apo-CA and no energy transfer occurs, maintaining the fluorescence intensity and lifetime of the donor. The *inset* shows the spectral overlap of the fluorescein donor emission (dashed line) with the azosulfamide absorbance (solid line).

Zinc sensing using fluorescent-labeled CA

The use of a separate, small molecule such as the aryl sulfonamides described above together with the protein CA has certain drawbacks. First, while CA is a macromolecule that does not ordinarily penetrate lipid bilayers, the aryl sulfonamides are small molecules which can (in some cases), so the two may partition differently in cells. Also, in the case of Dapoxyl sulfonamide the fluorophore binds to bilayers, emitting fluorescence which potentially interferes with that from the fluorophore bound to holo-CA. At some fundamental level, one would like the fluorescent moiety to go anywhere the protein went; e.g., that they be covalently attached. This is termed a 'reagentless' transducer, in that no separate reagent is necessary.

While we have described several fluorescently-labeled CA variants which exhibit changes in fluorescence upon binding metal ions such as Cu(II), Co(II), and Ni(II) (Thompson *et al.* 1996b, 1998a, 1999), it has been somewhat more difficult to design transducers for Zn(II). This is because (unlike the other metals) Zn(II) is not a very good quencher of fluorescence owing to its low atomic number, diamagnetic character, and difficulty of reduction. By comparison ions such as Cu(II), Ni(II), Hg(II), or Co(II) are typically good quenchers because they are paramagnetic, of high atomic number, and/or readily reduced or oxidized. Thus two new approaches were followed with a view to having the presence of zinc in the CA binding site alter the fluorescence of a nearby fluorophore. The rationale for the first was that an environment-sensitive fluorophore in the immediate vicinity of the binding site might well detect a difference between when the binding site was occupied with the zinc ion and when it was unoccupied. The fluorophore-variant combination that showed the greatest differences in fluorescence was N67C-ABD, but these changes (see below) were not attributable to the changes in solvent polarity that manifest themselves as shifts in excitation and emission, such as occur with dansylamide or ABD-N. We also designed another covalently attachable derivative of the ABD fluorophore with an n-hexyl tether (ABD-T) (Thompson *et al.* 1998a), with the objective of attaching it to the protein through a long, flexible link so as to permit it to bind as a fourth ligand to the active site zinc, and exhibit the overt fluorescence changes typical of ABD-N and dansylamide. While this did not occur, the response was still useful, especially with regard to anisotropy and lifetime (see below).

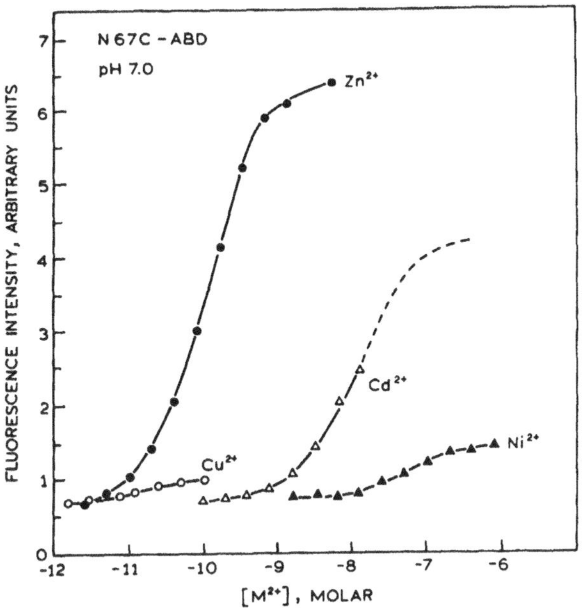

Figure 9. Fluorescence intensity of N67C-ABD as a function of Zn (filled circles), Cu(II) (open circles), Cd(II) (open triangles), and Ni(II) (filled triangles).

N67C-ABD did not show the response expected if there were a change in the polarity of its environment, in that there is little or no shift in the emission or excitation upon zinc binding (Thompson *et al.* 1999). However, there is a substantial increase in intensity and lifetime upon binding zinc. More unusual are the lesser increases in intensity and lifetime upon binding other metal ions (Figure 9). This may be rationalized by considering the fluorescent aryl sulfonamide derivative of β-mercaptoethanol, ABD-M (Table 2). ABD-M shows a short lifetime and modest quantum yield free in solution like its amine cousin, ABD-N; however, upon binding to the holoenzyme it exhibits a more modest increase in quantum yield and lifetime than ABD-N, which makes it a better anisotropy probe. In view of the longer average lifetime and higher apparent quantum yield of ABD-N, it is evident that ABD-M is still fairly well quenched even when bound. It may be that binding offers a degree of protection from quenching by, for instance, water. In the case of N67C-ABD the presence of zinc may limit the accessibility of the fluorophore to water, effectively 'dequenching' it. The other metal ions have the same effect, only they themselves (notably copper and cobalt) also quench the fluorophore by other mechanisms, making the intensity and lifetime enhancements less in their cases. Although it produces

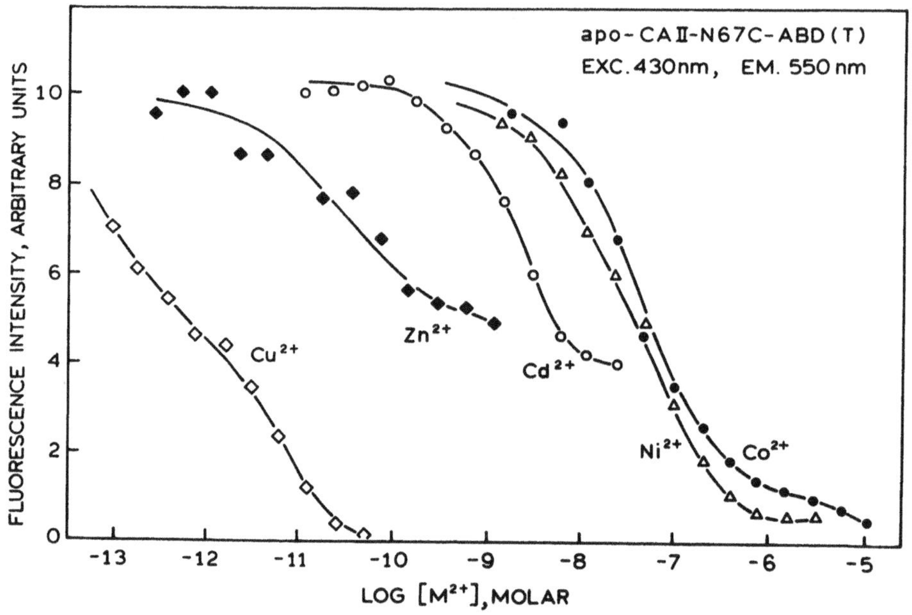

Figure 10. Fluorescence intensity of N67C-ABD-T as a function of Cu(II) (open diamonds), Zn(II) (filled diamonds), Cd(II) (open circles), Ni(II) (triangles), or Co(II) (filled circles) concentration.

a fairly good fluorescence enhancement, one cannot increase the signal by looking selectively at the bound form's emission because the shift is negligible.

The tethered form of ABD, ABD-T, also did not respond as expected, in that when coupled to the introduced thiol at position 67 (N67C) it exhibited no shift in emission or excitation upon zinc binding, effectively ruling out interaction of the sulfonamide moiety with the zinc. In its case zinc quenches the fluorescence (as do the other metals), resulting in reduced intensity (Figure 10) and lifetime, with increased anisotropy (Figure 11). The responses are quite satisfactory, but for intensity imaging purposes the decline in intensity is less desirable than an increase.

Application to studies of brain zinc

Zinc is an ion of substantial biological interest in many respects, as is discussed elsewhere in this issue. One important question is the role of the weakly bound zinc found in axonal boutons in the cortex, especially the hippocampus. Understanding the biological and pathological role(s) of zinc in the brain had been hampered by the difficulty of measuring zinc levels even extracellularly, owing to the high levels of interfering calcium and magnesium, and the transient localized nature of the release upon stimulation.

Recently, ABD-N has been used together with apo-CA to image zinc release from rat hippocampus in organotypic cell cultures (Thompson *et al.* 2000a) and classical slice preparations. For these experiments the apo-CA and ABD-N are incorporated into the artificial cerebrospinal fluid (ACSF) bathing the neural tissue. Because of the high affinity of CA for zinc, the ACSF must be specially prepared to avoid contamination. In particular, the ACSF (minus Ca and Mg) is passed over a chelating resin to remove zinc and other potential interferents, whereupon high purity Ca and Mg salts (electronic or similar grade) are added; ordinary reagent grade salts have unacceptable levels (several parts per million) of metallic impurities. Release is stimulated either electrically, with a brief pulse train, or by other insults (Frederickson *et al.* 2000). The concentration of free zinc may be quantitated ratiometrically (Figure 7) and the kinetics of zinc release quantitated through the microscope (Figure 12) (Suh *et al.*, submitted). We observed these slow release kinetics in our initial experiments (Thompson *et al.* 2000a), but believed that the apparent slow release (over tens of seconds) was due to the slow association rate constant of the wild type enzyme used (Thompson *et al.* 2000c), and that only nanomolar zinc concentrations were present. In fact micromolar levels are present, as may be seen in Figure 12, which also underscores the importance of quantitative methods.

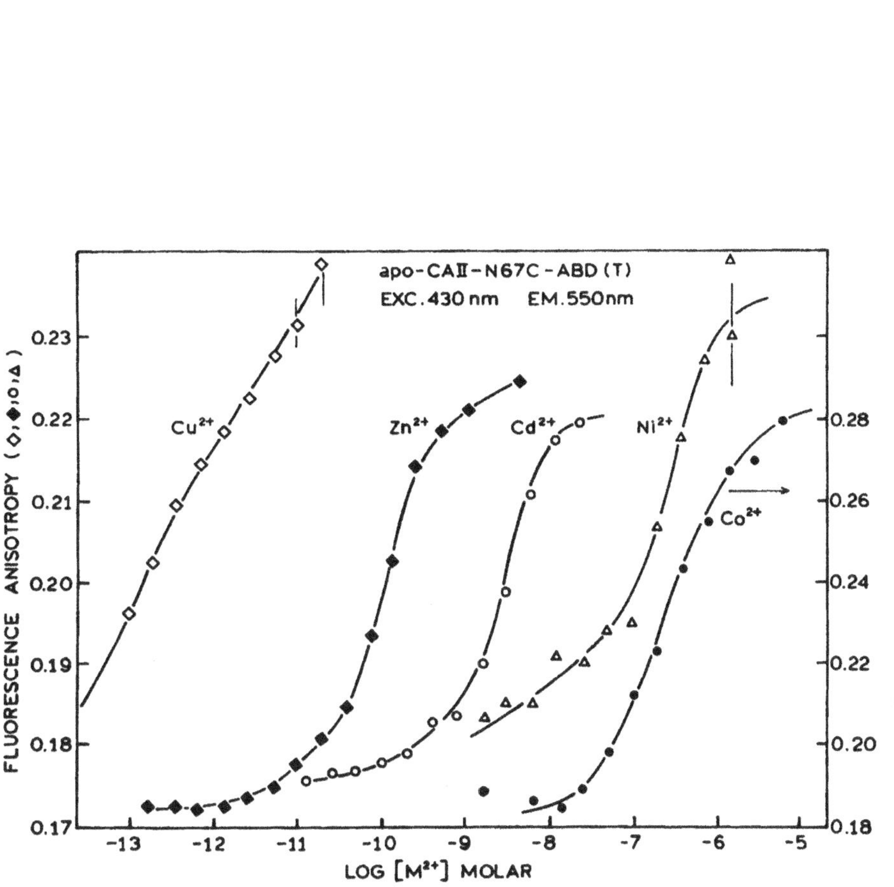

Figure 11. Fluorescence anisotropy of N67C-ABD-T as a function of Cu(II) (open diamonds), Zn(II) (filled diamonds), Cd(II) (open circles), Ni(II) (open circles), or Co(II) (filled circles) concentrations.

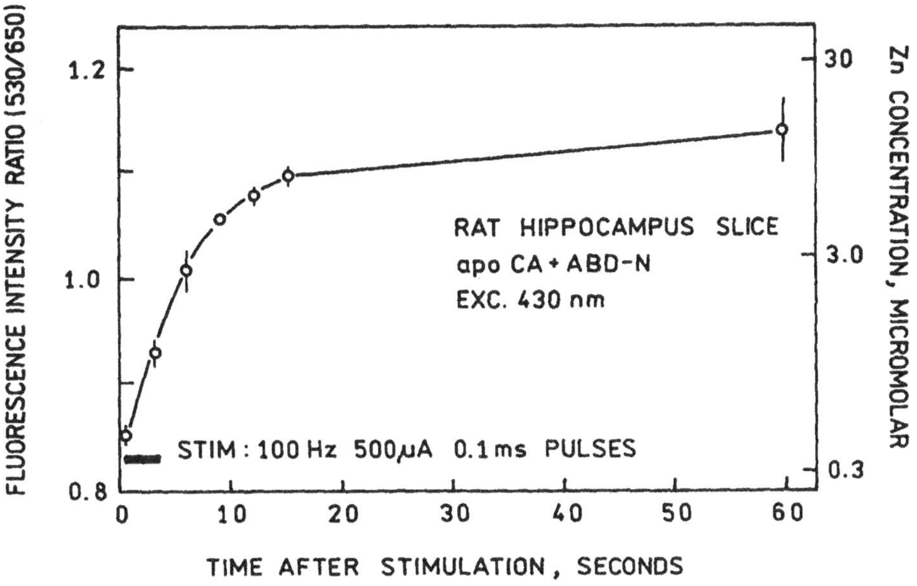

Figure 12. Kinetics of Zn(II) release from rat hippocampal slice preparation following electrical stimulation at the dentate gyrus (unpublished results of Suh, Frederickson, and Thompson). Stimulation was 0.1 msec 500 uA pulses at 100 Hz for 5 seconds. Images were acquired in an Olympus inverted microscope with a 4X objective; the results represent a small area of the field near the stimulating electrode.

The same approach may be used with *in vivo* dialysis collection approaches.

Future prospects: An expressable Zn indicator?

The issues associated with introducing the carbonic anhydrase molecule into the cell are by now widely appreciated, and several expedients have been described for doing this. Among the potential means are injection (Suh *et al.*, unpublished results) electroporation, and introduction of labeled particles (PEBBLES) by gas gun (Clark *et al.* 1999). One potential approach to calcium ion sensing described by Tsien's group is to express the transducer molecule inside the cell by recombinant DNA techniques, using variants of the Green Fluorescent Protein (GFP) from *Aequorea* (Miyawaki *et al.* 1997) as covalently attached label(s). The issue is how to get the binding of zinc to a carbonic anhydrase molecule to perturb (hopefully increase) the intensity of a GFP attached to the CA molecule. This prospect is daunting inasmuch as the fluorophore moiety in GFP is well shielded from outside influences, and typically has a very good quantum yield; indeed, in this context it might be said that GFP suffers from the defects of its virtues. A GFP sufficiently modified to permit zinc binding to perturb its fluorescence might also no longer be very fluorescent. However, the value of this approach, particularly for *in vitro* studies, suggests that it should be pursued if possible. Two approaches which appear promising have been recently described. Pearce and her colleagues (Pearce *et al.* 2000) described a fusion protein consisting of a metallothionein sandwiched between GFP variants capable of energy transfer. While the authors did not report a metal titration, treatment with EDTA to remove metal from the metallothionein resulted in modest changes in energy transfer, suggesting a conformational change had occurred. Jensen *et al.*, 2001 engineered a multidentate zinc binding site with ligands on two different tethered GFP variants capable of energy transfer. In the presence of zinc the variants are brought closer together by mutually binding a zinc ion, with concomitant energy transfer. This approach also succeeded, but the zinc sensitivity was modest at just under millimolar. Both these efforts underscore the power of Tsien's approach, but clearly more work is needed.

References

Alberts IL, Nadassy K. 1998 Analysis of zinc binding sites in protein crystal structures. *Protein Sci* **7**, 1700–1716.

Belli SL, Zirino A. 1993 Behavior and calibration of the copper(II) ion-selective electrode in high chloride media and marine waters. *Anal Chem* **65**, 2583–2589.

Benters A, Flogel U, Schafer T, Leibfritz D, Hechtenberg S, Beyersmann D. 1997 Study of the interactions of cadmium and zinc ions with cellular calcium homeostasis using $_{19}$F-NMR spectroscopy. *Biochem J* **322**, 793–799.

Chen RF, Kernohan J. 1967 Combination of bovine carbonic anhydrase with a fluorescent sulfonamide. *J Biol Chem* **242**, 5813–5823.

Christianson DW. 1991 Structural biology of zinc. *Adv Prot Chem* **42**, 281–355.

Christianson DW, Alexander RS. 1989 Carboxylate histidine zinc interactions in protein – structure and function. *J Am Chem Soc* **111**, 6412–6419.

Christianson DW, Fierke CA. 1996 Carbonic anhydrase – Evolution of the zinc binding site by nature and by design. *Acc Chem Res* **29**, 331–339.

Clark HA, Hoyer M, Philbert MA, Kopelman R. 1999 Optical nanosensors for chemical analysis inside single living cells. 1. Fabrication, characterization, and methods for intracellular delivery of PEBBLE sensors. *Anal Chem* **71**, 4831–4836.

Cox JD, Hunt JA, Compher KM, Fierke CA, Christianson DW. 2000 Structural influence of hydrophobic core residues on metal binding and specificity in carbonic anhydrase II. *Biochemistry* **39**, 13687–13694.

Denny MF, Atchison WD. 1994 Methylmercury-induced elevations in intrasynaptosomal zinc concentrations: an $_{19}$F-NMR study. *J Neurochem* **63**, 383–386.

DiTusa CA, McCall KA, Christensen T, Mahapatro M, Fierke CA, Toone EJ. 2001 Thermodynamics of metal ion binding. II. Metal ion binding by carbonic anhydrase variants. *Biochemistry*, **40**, 5345–5351.

Dix JA, Verkman AS. 1990 Mapping of fluorescence anisotropy in living cells by ratio imaging: Application to cytoplasmic viscosity. *Biophys J* **57**, 231–240.

Eigen M, Hammes GG. 1963 Elementary steps in enzyme reactions as studied by relaxation spectrometry. *Adv Enzymol Relat Areas Mol Biol* **25**, 1–38.

Elbaum D, Nair SK, Patchan MW, Thompson RB, Christianson DW. 1996 Structure-based design of a sulfonamide probe for fluorescence anisotropy detection of zinc with a carbonic anhydrase-based biosensor. *J Am Chem Soc* **118**, 8381–8387.

Eriksson AE, Jones TA. 1988 Refined structure of human carbonic anhydrase II at 2.0 A resolution. *Proteins* **4**, 274–282.

Fernandez-Gutierrez A, Munoz de la Pena A. 1985 Determinations of inorganic substances by luminescence methods. In: Schulman S.G. ed. *Molecular Luminescence Spectroscopy, Part I: Methods and Applications;* New York: Wiley-Interscience, 371–546.

Frederickson CJ, Suh SW, Silva D, Frederickson CJ, Thompson RB. 2000 Importance of zinc in the central nervous system: the zinc-containing neuron. *J Nutr (Suppl.)* **130**, 1471S–1483S.

Fushimi K, Dix JA, Verkman AS. 1990 Cell membrane fluidity in the intact kidney proximal tubule measured by orientation-independent fluorescence anisotropy imaging. *Biophys J* **57**, 241–254.

Glusker JP. 1991 Structural aspects of metal liganding. *Adv Prot Chem* **42**, 1–76.

Grynkiewicz G, Poenie M, Tsien RY. 1985 A new generation of calcium indicators with greatly improved fluorescence properties. *J Biol Chem* **260**, 3440–3450.

Hakansson K, Carlsson M, Svensson LA, Liljas A. 1992 Structure of native and apo carbonic anhydrase II and structure of some of its anion-ligand complexes. *J Mol Biol* **227**, 1192–1204.

Hakansson K, Wehnert A, Liljas A. 1994 X-ray analysis of metal-substituted human carbonic anhydrase II derivatives. *Acta Crys* **D50**, 93–100.

Haugland RP. 1996 *Handbook of Fluorescent Probes and Research Chemicals*. Oregon: Molecular Probes, Inc., Eugene.

Henkens RW, Sturtevant JM. 1968 The kinetics of the binding of Zn(II) by apocarbonic anhydrase. *J Am Chem Soc* **90**, 2669–2676.

Hirano T, Kikuchi K, Urano Y, Higuchi T, Nagano T. 2000 Highly zinc-selective fluorescent sensor molecules suitable for biological applications. *J Am Chem Soc* **122**, 12399–12400.

Huang C-C, Lesburg CA, Kiefer LL, Fierke CA, Christianson DW. 1996 Reversal of the hydrogen bond to zinc ligand histidine-119 dramatically diminishes catalysis and enhances metal equilibration kinetics in carbonic anhydrase II. *Biochemistry 35*, 3439–3446.

Hunt JA, Ahmed M, Fierke CA. 1999 Metal binding specificity in carbonic anhydrase is influenced by conserved hydrophobic amino acids. *Biochemistry* **38**, 9054–9060.

Hunt JA, Fierke CA. 1997 Selection of carbonic anhydrase variants displayed on phage: aromatic residues in zinc binding site enhance metal affinity and equilibration kinetics. *J Biol Chem* **272**, 20364–20372.

Ippolito JA, Baird TT, McGee SA, Christianson DW, Fierke CA. 1995a Structure-assisted redesign of a protein-zinc binding site with femtomolar affinity. *Proc Natl Acad Sci USA* **92**, 5017–5021.

Ippolito JA, Christianson DW. 1993 Structure of a His^3Cys zinc binding site in human carbonic anhydrase II. *Biochemistry* **32**, 9901–9905.

Ippolito JA, Christianson DW. 1994 Structural consequences of redesigning a protein-zinc binding site. *Biochemistry* **33**, 15241–15249.

Ippolito JA, Nair SK, Fierke CA, Christianson DW. 1995b Structure of His94Asp carbonic anhydrase II in a new crystalline form reveals a partially occupied zinc binding site. *Prot Engin* **8**, 975–980.

Iverson TM, Alber BE, Kisker C, Ferry JG, Rees DC. 2000 A closer look at gamma-class carbonic anhydrases: high resolution crystallographic studies of the carbonic anhydrase from *Methanosarcina thermophila*. *Biochemistry* **39**, 9222–9231.

Jensen KK, Martini L, Schwartz TW. 2001 Enhanced fluorescence resonance energy transfer between spectral variants of green fluorescent protein through zinc-site engineering. *Biochemistry* **40**, 938–945.

Kiefer LL, Fierke CA. 1994 Functional characterization of human carbonic anhydrase II variants with altered zinc binding sites. *Biochemistry* **33**, 15233–15240.

Kiefer LL, Ippolito JA, Fierke CA, Christianson DW. 1993a Redesigning the zinc binding site of human carbonic anhydrase II: Structure of a $His^2Asp\text{-}Zn_{2+}$ metal coordination polyhedron. *J Am Chem Soc* **115**, 12581–12582.

Kiefer LL, Krebs JF, Fierke CA. 1993b Engineering a cysteine residue into the zinc binding site of carbonic anhydrase II. *Biochemistry* **32**, 9896–9900.

Kiefer LL, Paterno SA, Fierke CA. 1995 Hydrogen bond network in the metal binding site of carbonic anhydrase enhances zinc affinity and catalytic efficiency. *J Am Chem Soc* **117**, 6831–6837.

Kimber MS, Pai EF. 2000 The active site architecture of *Pisum sativum* beta-carbonic anhydrase is a mirror image of that of alpha-carbonic anhydrases. *EMBO J* **19**, 1407–1418.

Kuhn MA, Hoyland B, Carter S, Zhang C, Haugland RP. 1995 Fluorescent ion indicators for detecting heavy metals. *SPIE Conference on Adv Fluor Sens Tech II (San Jose, California)*, Vol. 2388, 238–244.

Lesburg CA, Christianson DW. 1995 X-ray crystallographic studies of engineered hydrogen bond networks in a protein-zinc binding site. *J Am Chem Soc* **117**, 6838–6844.

Lesburg CA, Huang C-C, Christianson DW, Fierke CA. 1997 Histidine to carboxamide ligand substitutions in the zinc binding site of carbonic anhydrase II alter metal coordination geometry but retain catalytic activity. *Biochemistry* **36**, 15780–15791.

Levy R, Guignon EF, Cobane S, St. Louis E, Fernandez S. 1997 Compact, rugged, and inexpensive frequency domain fluorometer. *SPIE Conference on Advances in Fluorescence Sensing Technology III*, San Jose, CA vol. 2980, 81–89.

Lindskog S, Henderson LE, Kannan KK, Liljas A, Nyman PO, Strandberg B. 1971 Carbonic anhydrase. In: Boyer PD, ed. *The Enzymes*. New York: Academic Press: 587–665.

Lindskog S, Nyman PO. 1964 Metal-binding properties of human erythrocyte carbonic anhydrases. *Biochim Biophys Acta* **85**, 462–474.

Lippitsch ME, Pusterhofer J, Leiner MJP, Wolfbeis OS. 1988 Fiber-optic oxygen sensor with the fluorescence decay time as the information carrier. *Anal Chim Acta* **205**, 1–6.

Maren TH. 1977 Use of inhibitors in physiological studies of carbonic anhydrase. *Am J Physiol* **232**, F291–F297.

McCall KA, Fierke CA. 2000 Colorimetric and fluorimetric assays to quantitate micromolar concentrations of transition metals. *Anal Biochemistry* **284**, 307–315.

Mitsuhashi S, Mizushima T, Yamashita E, Yamamoto M, Kumasaka T, Moriyama H, Ueki T, Miyachi S, Tsukihara T. 2000 X-ray structure of beta carbonic anhydrase from the red alga, *Porphyridium purpureum*, reveals a novel catalytic site for CO_2 hydration. *J Biol Chem* **275**, 5521–5526.

Miyawaki A, Llopis J, Heim R, McCaffery JM, Adams JA, Ikura M, Tsien RY. 1997 Fluorescent indicators for Ca_{2+} based on green fluorescent proteins and calmodulin. *Nature* **388**, 882–887.

Pearce LL, Gandley RE, Han W, Wasserloos K, Stitt M, Kanai AJ, McLaughlin MK, Pitt BR, Levitan ES. 2000 Role of metallothionein in nitric oxide signaling as revealed by a green fluorescent fusion protein. *Proc Natl Acad Sci USA* **97**, 477–482.

Pearson RG. 1966 Acids and bases. *Science* **151**, 172–177.

Rae TD, Schmidt PJ, Pufahl RA, Culotta VC, O'Halloran TV. 1999 Undetectable intracellular free copper: the requirement of a copper chaperone for superoxide dismutase. *Science* **284**, 805–808.

Roe RR, Pang YP. 1999 Zinc's exclusive tetrahedral coordination governed by its electronic structure. *J Mol Model* **5**, 134–140.

Rulisek L, Vondrasek J. 1998 Coordination geometries of selected transition metal ions (Co_{2+}, Ni_{2+}, Cu_{2+}, Zn_{2+}, Cd2+, and Hg_{2+}) in metalloproteins. *J Inorg Biochem* **71**, 115–127.

Simons TJB. 1993 Measurement of free zinc ion concentration with the fluorescent probe mag-fura-2 (furaptra). *J Biochem Biophys Meth* **27**, 25–37.

Szmacinski H, Lakowicz JR. 1993 Optical measurements of pH using fluorescence lifetimes and phase-modulation fluorometry. *Anal Chem* **65**, 1668–1674.

Thompson RB. 1991 Fluorescence-based fiber optic sensors. In: Lakowicz JR, ed. *Topics in Fluorescence Spectroscopy. Vol. 2: Principles*. New York: Plenum Press: 345–365.

Thompson RB. 1993 Fiber optic ion sensors based on phase fluorescence lifetime measurements. *SPIE Conference on Advances in Fluorescence Sensing Technology*, Los Angeles, CA, vol. 1885, 290–299.

Thompson RB, Frisoli JK, Lakowicz JR. 1992 Phase fluorometry using a continuously modulated laser diode. *Anal Chem* **64**, 2075–2078.

Thompson RB, Ge Z, Patchan MW, Fierke CA. 1996a Performance enhancement of fluorescence energy transfer-based biosensors by site-directed mutagenesis of the transducer. *J Biomed Optics* **1**, 131–137.

Thompson RB, Ge Z, Patchan MW, Huang C-C, Fierke CA. 1996b Fiber optic biosensor for Co(II) and Cu(II) based on fluorescence energy transfer with an enzyme transducer. *Biosensors Bioelectron* **11**, 557–564.

Thompson RB, Jones ER. 1993 Enzyme-based fiber optic zinc biosensor. *Anal Chem* **65**, 730–734.

Thompson RB, Whetsell WO Jr., Maliwal BP, Fierke CA, Frederickson CJ. 2000a Fluorescence microscopy of stimulated Zn(II) release from organotypic cultures of mammalian hippocampus using a carbonic anhydrase-based biosensor system. *J Neurosci Meth* **96**, 35–45.

Thompson RB, Maliwal BP, Feliccia VL, Fierke CA, McCall K. 1998a Determination of picomolar concentrations of metal ions using fluorescence anisotropy: biosensing with a 'reagentless' enzyme transducer. *Anal Chem* **70**, 4717–4723.

Thompson RB, Maliwal BP, Fierke CA. 1998b Expanded dynamic range of free zinc ion determination by fluorescence anisotropy. *Anal Chem* **70**, 1749–1754.

Thompson RB, Maliwal BP, Fierke CA. 1999 Selectivity and sensitivity of fluorescence lifetime-based metal ion biosensing using a carbonic anhydrase transducer. *Anal Biochem* **267**, 185–195.

Thompson RB, Maliwal BP, Zeng HH. 2000b Zinc biosensing with multiphoton excitation using carbonic anhydrase and improved fluorophores. *J Biomed Optics* **5**, 17–22.

Thompson RB, Patchan MW. 1995a Fluorescence lifetime-based biosensing of zinc: origin of the broad dynamic range. *J Fluoresc* **5**, 123–130.

Thompson RB, Patchan MW. 1995b Lifetime-based fluorescence energy transfer biosensing of zinc. *Anal Biochem* **227**, 123–128.

Thompson RB, Walt DR. 1994 Emerging strategies for molecular biosensors. *Naval Res Rev* **46**, 19–29.

Thompson RB, Zeng HH, Loetz M, Fierke C. 2000 Issues in enzyme-based metal ion biosensing in complex media. *In-vitro Diagnostic Instrumentation (San Jose, CA)*, vol. 3913, 120-127.

Weber G. 1956 Photoelectric method for the measurement of polarization of fluorescence of solutions. *J Opt Soc Am* **46**, 962.

White CE, Argauer RJ. 1970 *Fluorescence Analysis: A Practical Approach*. New York: Marcel Dekker, Inc.

Yamashita MM, Wesson L. 1990 Where metal ions bind in proteins. *Proc Natl Acad Sci USA* **87**, 5648–5652.

BioMetals 14: 223–237, 2001.

Review

Cellular zinc sensors: MTF-1 regulation of gene expression

Glen K. Andrews
Department of Biochemistry and Molecular Biology, University of Kansas Medical Center, Kansas City, KS 66160-7421, USA (Tel: (913)588-6935; Fax: (913)588-7035; E-mail: gandrews@kumc.edu)

Received 15 January 2001; accepted 15 March 2001

Key words: metal-response element, MTF-1, metalloregulatory, metallothionein, transcription, zinc, zinc-transporter-1, γ-glutamylcysteine synthetase

Abstract

Zinc metabolism in higher eukaryotes is complex, being controlled by uptake, efflux, and storage in individual cells, as well as in peripheral tissues and organs. Recently there have been advances in the understanding of the genes involved in these processes and their regulation. Metal-response element-binding transcription factor-1 (MTF-1) functions as a cellular zinc sensor which coordinates the expression of genes involved in zinc homeostasis, as well as protection against metal toxicity and oxidative stresses. In mice, these are known to include the metallothionein (MT), the zinc-transporter-1 (ZnT1) and the γ-glutamylcysteine synthetase heavy chain (γGCS_{hc}) genes. The cysteine-rich MTs function as an intracellular metal-chelators that bind zinc with high affinity, whereas the transmembrane protein ZnT1 exports zinc from the cell. γ-Glutamylcysteine synthetase controls the rate limiting step in glutathione (GSH) biosynthesis. GSH, which is present in mM concentrations in cells, effectively chelates large amounts of zinc *in vitro*. Both MT and GSH also function as antioxidants. The current model suggests that the zinc-finger domain of MTF-1 directly (and reversibly) binds to zinc. This metalloregulatory protein then adopts a DNA-binding conformation and translocates to the nucleus, where it binds to metal-response elements in these gene promoters leading to increased transcription. The six zinc-finger domain of this factor is highly conserved from insects to mammals, and biochemical studies confirm that the zinc-fingers are heterogeneous in function and in zinc-binding. Furthermore, the mouse MTF-1 gene is essential for development of the embryo, thus underscoring the importance of this transcription factor.

Abbreviations: γGCS_{hc}, γ-glutamylcysteine synthetase heavy chain; GSH – glutathione; MRE – metal-response element; MTF-1 – metal-response element-binding transcription factor-1; MT – metallothionein; TnT lysate – coupled transcription-translation lysate; USF – upstream stimulatory factor; ZIP – zinc-iron related transport protein; ZnT1 – zinc-transporter-1

Introduction

Regulation of gene expression by transition metals has been demonstrated in organisms ranging from bacteria to mammals (O'Halloran 1993). Metals regulate genes involved in protection against metal toxicity, as well as those involved in the homeostasis of essential metals, which themselves can be toxic. Transcription factors (activators and repressors), which directly interact with metal ions, and subsequently signal changes in gene expression are known as metalloregulatory proteins (reviewed in O'Halloran 1993; DeMoor & Koropatnick 2000). These cellular metal sensors can regulate gene transcription, and/or mRNA stability and translation. Transition metals are ubiquitous in our diet and environment and have a great impact on gene expression and, therefore, organismal function. Of particular focus here is the essential metal zinc.

The transcription factor MTF-1 coordinates the expression of genes which are important in the homeostasis of zinc and in protection against metal-toxicity and oxidative stress. In mice, these are known to include the MT-I/II, ZnT1 and γGCS_{hc} genes although it seems likely that other genes are also regulated by MTF-1. This article provides a brief overview of the functions of these proteins and then describes our current understanding of the mechanisms by which MTF-1 senses zinc and regulates their expression.

Overview of higher eukaryotic MTs

In prokaryotes, lower eukaryotes, plants and throughout the animal kingdom, one of the most intensely-studied examples of metal-regulation of gene transcription is that of the MT genes (Andrews 1990; Thiele 1992; Klaassen *et al.* 1999; DeMoor & Koropatnick 2000; Miles *et al.* 2000). In mammals, birds and fishes, the MT genes are remarkably responsive to zinc. Induction of these after exposure to zinc has been documented in many cell-types in culture and in intact animals. Furthermore, these genes are dramatically depressed in specific tissues (e.g., intestine, pancreas) taken from animals exposed to dietary zinc deficiency.

The metal-inducible MT genes from higher eukaryotes encode proteins of 60 to 68 amino acids in length which contain 20 cysteine residues and no aromatic amino acids (Kägi & Schäffer 1988). The placement of cysteine residues is absolutely conserved, and the majority of other changes in amino acid sequence are conservative. The cysteine-rich metal-binding clusters are distributed in α and β domains of the protein (Nordberg & Nordberg 2000). MTs adopt their specific, biologically unique tertiary structure only upon metal binding. These proteins can generally bind seven zinc or cadmium ions or up to 12 copper ions. Although MT is isolated as a $zinc_7$-complex from most mammalian tissues, it is a cadmium/zinc-complex under toxicological conditions (Kägi & Schäffer 1988; Kägi 1991). MT is also isolated as a copper/zinc-complex from animals with inherited disorders of copper metabolism, such as the Menkes and Wilson diseases in humans (Vulpe & Packman 1995) or the LEC rat (Sugawara *et al.* 1991). Furthermore, in *Drosophila melanogaster*, MT is naturally found complexed with copper (Maroni *et al.* 1995).

MTs are the most-abundant intracellular zinc-binding proteins in higher eukaryotes (Kägi & Schäffer 1988), and a significant percentage (5 to 20%) of the total cellular zinc is found complexed with MT under normal physiological conditions. Multiple isoforms of the protein are often present and the complexity of the MT gene family vary among organisms (1 in bacteria, 4 in mice, 16 in humans) and levels of expression of individual genes varies among tissues in higher eukaryotes (Miles *et al.* 2000). None of the MT genes that have been genetically inactivated in any species is an essential gene for that organism (Michalska & Choo 1993; Masters *et al.* 1994; Jensen *et al.* 1996). However, neither the effects of loss-of-function mutation in the mouse MT-IV gene, nor those of a complete loss of all four mouse MT genes have been examined. Functions for MTs in protection against metal toxicity, zinc-deficiency and oxidative stress have been demonstrated. For example, mouse MT-I and -II can provide a biologically important reservoir of zinc under zinc-limiting conditions (Dalton *et al.* 1996a; Andrews & Geiser 1999), and protect the animal against cadmium toxicity and oxidative stress (Lazo *et al.* 1995; Palmiter 1998; Lazo *et al.* 1998; Klaassen *et al.* 1999). In contrast, MT predominantly functions to protect against copper toxicity in yeast and the fruit fly (Karin *et al.* 1984; Mehra & Winge 1991; Jensen *et al.* 1996; Zhang *et al.* 2000). MT can sequester reactive oxygen and hydroxyl radicals, and provide for zinc, copper, or cadmium exchange with other proteins (Roesijadi 2000). Although primarily a metal-binding cytoplasmic protein, MT can translocate to the nucleus and may protect DNA from oxidative damage and participate in zinc exchange with zinc-dependent transcription factors (Cherian & Apostolova 2000; Roesijadi 2000).

Overview of a mouse zinc-transporter (ZnT) family

Recently, four mammalian genes involved in efflux or vesicular transport of zinc have been identified (McMahon & Cousins 1998a). Genes involved in the uptake of zinc (ZIP genes) are discussed elsewhere in this journal issue. Mouse ZnT1 through 4 are peptides of 359 to 503 amino acids, with six membrane-spanning domains, a histidine-rich intracellular loop, and a long intracellular carboxyl-terminal tail (Palmiter & Findley 1995; Palmiter *et al.* 1996a,b). The functional domains of the ZnT proteins are not well defined, but similar histidine-rich regions in other metal-transporters suggest a role of this domain in zinc chelation and transport. Transport function of these proteins is not energy dependent, and it is thought that they function as multimers (Palmiter & Findley 1995). ZnT1 is homologous to zinc and cobalt resistance genes of yeast (Palmiter & Findley 1995). It functions to efflux zinc from cells, is localized to the plasma membrane, and is apparently expressed in most cell- and tissue-types (Palmiter & Findley 1995; Palmiter *et al.* 1996a). Exceptionally high level expression of the ZnT1 gene occurs in the visceral endoderm of the early mouse embryo and in the placenta (Langmade *et al.* 2000). These cells surround the developing mouse embryo and play a key role in nutrient transport and protection. Cultured cells which actively express ZnT1 are more resistant to zinc-toxicity (Palmiter & Findley 1995; Langmade, Ravindra & Andrews, unpublished observation). Mouse ZnT2 causes the vesicular accumulation of zinc in endosomal vesicles (Palmiter *et al.* 1996a), and is most similar in structure to ZnT3 which is responsible for the accumulation of zinc in synaptic vesicles in the brain (Wenzel *et al.* 1997; Cole *et al.* 1999). Targeted deletion of ZnT3 is not lethal (Cole *et al.* 1999). ZnT4 was discovered to be the Lethal Milk locus in the mouse (Huang & Gitschier 1997). This zinc-efflux protein is highly expressed in the mammary gland. Aberrant expression of ZnT4 causes severe zinc-deficiency to develop in the pups of mutant mothers. ZnT4 may also be involved in more general zinc homeostasis in the adult (Huang & Gitschier 1997).

Except for the finding of cell-specific expression patterns of ZnT genes, little else is known about their regulation. Among the ZnT genes, zinc-induction of ZnT1 has been documented in cultured neurons, and fibroblasts (Palmiter & Findley 1995; Tsuda *et al.* 1997; Langmade *et al.* 2000), and in the rat intestine after oral gavage with zinc (McMahon & Cousins 1998b; Davis *et al.* 1998). Furthermore, ZnT1 expression in enterocytes and the visceral endoderm of the embryo is responsive to changes in dietary zinc levels (McMahon & Cousins 1998b; Langmade *et al.* 2000). Furthermore, ZnT1 is an essential gene and homozygous knockout of the ZnT1 gene is lethal to the developing embryo (R.D. Palmiter, personal communications). Thus, ZnT1 appears to play a key role in zinc homeostasis.

Overview of γGCS

The enzyme γGCS is a key regulatory enzyme in the synthesis of GSH (Anderson 1998; Griffith & Mulcahy 1999; Wild & Mulcahy 2000; Soltaninassab *et al.* 2000). GSH is an important intracellular tripeptide (γ-glutamylcysteinylglycine) with multiple functions ranging from antioxidant defense to cell proliferation (Lu 1999). GSH is present in higher eukaryotic cells in millimolar concentrations (Griffith 1999), and it interacts with hydroxyl radicals, peroxinitrite, and hydroperoxides, as well as reactive electrophiles (Griffith & Mulcahy 1999). GSH also chelates metal ions with relatively high affinity. It binds zinc with an equilibrium constant of 2×10^{-8} M (Chaberek & Martell 1959; Ballatori 1994), and might compete for or facilitate metal interactions with proteins. When oxidized, GSH facilitates release of zinc from MT (Maret 1994), and when reduced it facilitates transfer of copper to MT (Da Costa Ferreira *et al.* 1993). Thus, GSH plays a role in zinc metabolism.

The synthesis of GSH is tightly regulated at the key step which is the ATP-dependent synthesis of γ-glutamylcysteine by the enzyme γGCS (Griffith & Mulcahy 1999). It is a heterodimeric zinc metalloprotein that belongs to a unique class of enzymes that gain activity due to the formation of a reversible disulfide bond (Soltaninassab *et al.* 2000). In the rat, a ≈28 kDa light chain and 73 kDa heavy chain form the holoenzyme. Details of the catalytic mechanism and structure of this enzyme have been recently reviewed elsewhere (Griffith & Mulcahy 1999). The enzyme's two subunits are encoded by separate genes which display both differential and coordinate regulation (Wild & Mulcahy 2000; Soltaninassab *et al.* 2000). In the mouse, the γGCS_{hc} (heavy chain) gene is essential for development of the embryo past d8.5 of gestation, but not for cell growth in culture (Shi *et al.* 2000). Expression of these genes is up-regulated in response to oxidants and metals (Griffith 1999; Wild & Mulcahy 2000; Soltaninassab *et al.* 2000). Recent studies suggest that the transcription factor Nrf2 in combination with other bZIP proteins mediates gene induction in response to oxidants, but AP-1 and NF-κB may also be involved (Wild & Mulcahy 2000). The mouse γGCS_{hc} gene is a target for MTF-1 (Günes *et al.* 1998), as is discussed below.

MTF-1: A zinc-dependent, positive transcriptional regulator in higher eukaryotes

Metal response elements mediate zinc-induction

All of the zinc-activated MT genes have promoter elements termed metal response elements (MRE), which

Species	Gene		MRE sequence	Orientation
mouse	MT-I	a	CTT**TGCGCCC**GGACT	→
		b	GTT**TGCACCC**AGCAG	←
		c	AAG**TGCGCTC**GGCTC	→
		d	CTC**TGCACTC**CGCCC	→
		e	CTG**TGCACAC**TGGCG	→
mouse	ZnT-1	a	CTT**TGCAGAC**GGTTT	←
		b	CTT**TGCACTC**GGAAC	←
human	γGCS_{hc}	a	CCT**TGCACAC**GCCTC	→
		b	GAC**TGCGCCC**GAGAG	→
chicken	MT	a	CGC**TGCGCGC**AGCAC	→
		b	TGC**TGCGCGC**AGCGC	←
		c	CTC**TGCGCTC**GGTTG	←
		d	CTG**TGCGCAC**CGCCT	→
		e	CGG**TGCGCAC**AGCGT	←
trout	MT-A	a	TTC**TGCACAC**GGCAC	→
		b	GCT**TGCACAC**GGTTT	←
		c	CAC**TGCGCAC**AATAA	→
		d	CAG**TGCACAC**GGTAC	←
		e	ATT**TGCACAC**GGGCA	←
		f	CTT**TGCGCTC**GTCGA	←
Drosophila	MTn	a	AGA**TGCTCTC**GGTTT	←
		b	CTT**TACACAC**GGGTC	←
		c	TTT**TGCACAC**GCCGG	←
		d	ATT**TGGAGCC**GGCCG	→
		e	TTC**TGCACAC**GTCTC	→
Consensus MRE			CTN**TGCRCNC**GGCCC	

Position

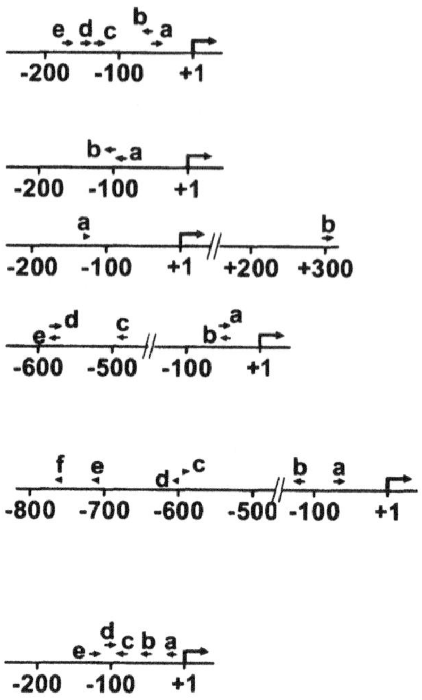

Fig. 1. Metal response elements are found in the proximal promoters of MTF-1 regulated genes. The proximal promoter region (+1 designates the transcription start site) of MT genes from higher eukaryotes (insects to mammals) contains multiple copies of metal response elements (MRE_{a-f}) which represent binding sites for the metalloregulatory protein MTF-1. The mouse ZnT1 and γGCS_{hc} genes contain two MREs each. Arrows indicate the sense or antisense orientation of each MRE, and their positions in each promoter. The 12 bp MRE sequence is well conserved and the core bases are essential for MTF-1- binding.

are present in multiple copies in the proximal promoters. Two MREs are present in the mouse ZnT1 gene promoter (Palmiter & Findley 1995) and two MREs are present in the mouse γGCS_{hc} gene (Günes *et al.* 1998). Multiple MREs function synergistically to confer response to zinc, cadmium and oxidative stress in transfected mammalian cells (Stuart *et al.* 1984, 1985; Koizumi *et al.* 1999), and it is thought that two or more MREs are required for a promoter to exhibit significant metal responsiveness. Neither the spacing nor orientation of MREs in these promoters appears to be critical for function. Metal responsiveness is dependent on the MRE core consensus sequence **TGCRCNC** (Stuart *et al.* 1984, 1985; Cizewski Culotta & Hamer 1989) which is found within an extended consensus sequence of 12 bp (Figure 1).

MTF-1 is an essential protein that binds to MREs in a zinc-dependent manner

The MRE is a binding site for the transcription factor MTF-1. MTF-1 was first cloned from the mouse by screening an expression library for MRE-binding activity (Radtke *et al.* 1993). The native protein was also purified from human cells by binding-site affinity chromatography, and the human MTF-1 gene was cloned (Brugnera *et al.* 1994; Otsuka *et al.* 1994; Koizumi *et al.* 1999). MTF-1 genes were subsequently cloned in *Drosophila* (Zhang *et al.* 2000), pufferfish (Auf der Maur *et al.* 1999), and the chicken (Jiang & Andrews, unpublished data) (Figure 2), which revealed that MTF-1 is a conserved protein in higher eukaryotes.

Binding of MTF-1 to the MRE is dependent on zinc and is easily disrupted when zinc is depleted (Andrews 2000). Furthermore, MTF-1 binds tightly to functional MREs, but not to nonfunctional MRE-like

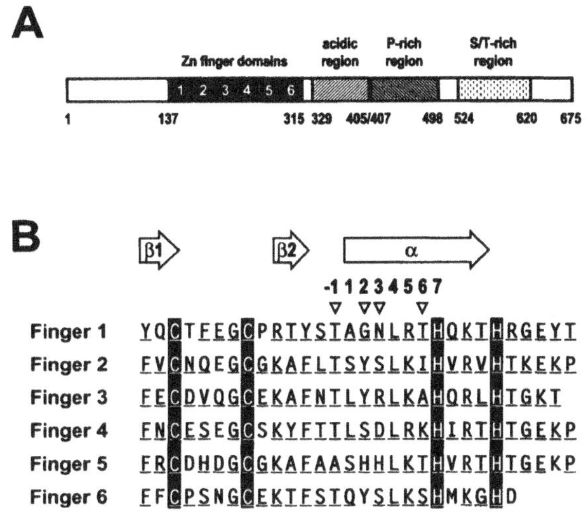

Sequence identity between species:

	Fugu	Chicken	Mouse	Human
Drosophila	68%	68%	67%	67%
Fugu		92%	92%	92%
Chicken			97%	97%
Mouse				99%

Fig. 2. The transcription factor MTF-1 contains several functional domains, and the zinc-finger domain is highly conserved among higher eukaryotes. **A**: Diagrammatic representation of the functional domains of mouse MTF-1. An amino terminal domain precedes the six zinc-finger domain, which is followed by three transactivation domains located in the carboxyl-terminal region of the protein. These domains are acidic, proline-rich and serine-threonine-rich, respectively (Radtke *et al.* 1995). Except for the zinc-fingers, other regions of the protein are not highly conserved. **B**: The highly conserved six zinc-finger domain of MTF-1 is shown here as a consensus sequence which was derived by comparisons of the amino acid sequence of the fingers from the five indicated species of mammals, birds, fish and insects. The cysteine and histidine residues (boxed) coordinate a zinc atom in similar C_2H_2-type zinc-fingers. The underlined amino acids are present at that position in 4 out of the 5 species examined, and the amino acid sequence identity in this domain is over 90% between fish and man. The *Drosophila* sequence is more divergent (67% identity). The reversible binding of zinc to specific zinc-fingers of MTF-1 apparently modulates DNA-binding activity of this transcription factor.

sequences (Koizumi *et al.* 1999). High affinity binding is dependent on the MRE core bases (Figure 1). MTF-1 binds with high affinity to the MREs from the mouse ZnT1 and γGCS_{hc} promoters (Günes *et al.* 1998; Langmade *et al.* 2000).

Homozygous knockout of the mouse MTF-1 gene in cultured cells eliminates heavy metal-induced MT and ZnT1 gene expression, as well as basal expression of these genes (Heuchel *et al.* 1994; Langmade *et al.* 2000). Homozygous knockout of this gene in mice abolishes expression of the MT-I gene, and significantly attenuates the expression of the ZnT1 and γGCS_{hc} genes in the embryonic liver and visceral endoderm of the yolk sac (Günes *et al.* 1998; Langmade *et al.* 2000; Andrews *et al.* 2001). Furthermore, the mouse MTF-1 gene is essential for fetal development (Günes *et al.* 1998). Embryos homozygous for MTF-1 null mutations die on d14 of gestation, which is after development of the yolk sac and initial formation of the liver. However, the fetal liver fails to develop properly (or degenerates) in these mice, while development of the nervous system and visceral yolk sac is not impaired (Günes *et al.* 1998; Lichtlen *et al.* 1999; Andrews *et al.* 2001).

MTF-1 structure

MTF-1 is a zinc-finger transcription factor in the Cys_2His_2 family (Figure 2A). The six zinc-finger domain has been highly conserved during evolution (Figure 2), while significant divergence has occurred in the remainder of the protein (Auf der Maur *et al.* 1999). Although, the precise mechanisms by which MTF-1 activates MT gene expression in response to metals remain unknown, this observation is consistent with the concept that the zinc-finger domain of MTF-1 is critical for both its metalloregulatory and DNA-binding functions in response to zinc (Müller *et al.* 1995; Dalton *et al.* 1997). The transactivation domains of MTF-1 are less well understood than the zinc-finger domain, but intra-molecular interactions are important for optimal MTF-1 function (Radtke *et al.* 1995; Müller *et al.* 1995). The carboxyl-termini of human and mouse MTF-1 contain three transactivation domains which are acidic, proline-rich and serine-threonine-rich, respectively (Radtke *et al.* 1995; Müller *et al.* 1995) (Figure 2A). The transactivation domain from the VP16 transcription factor can replace the function of these MTF-1 transactivation domains to produce a metal-responsive factor in transfected cells (Palmiter 1994; Radtke *et al.* 1995), but the transactivation domain of the zinc-finger protein Sp1 cannot (Bittel *et al.* 2000).

MTF-1 binding activity, nuclear translocation and occupancy of MREs in vivo is responsive to zinc

Treatment of cells with zinc *in vivo* causes a rapid, dramatic increase in DNA-binding activity of MTF-1 measured *in vitro* (Dalton *et al.* 1997; Koizumi *et al.* 1999), and this is accompanied by the nuclear translocation of MTF-1 (Smirnova *et al.* 2000; Otsuka *et al.* 2000). Western blot analysis revealed that zinc-treatment causes the rapid accumulation of immunoreactive MTF-I in the nuclear fraction (Figure 3A). In the untreated cells, 80% of the MTF-1 was detected in the cytosolic fraction and it was not active in DNA-binding assays (Smirnova *et al.* 2000). Within 30 min of treatment of cells with zinc, essentially all of this protein was found in the nucleus and electrophoretic mobility shift assays using the MRE binding site detected a parallel increase in DNA-binding activity (Smirnova *et al.* 2000) (Figure 3B).

The activation and translocation of MTF-1 in response to zinc (or oxidative stress) is accompanied by the occupancy of MREs in the mouse MT-I promoter *in vivo* (Palmiter 1987; Dalton *et al.* 1996b). *In vivo* genomic footprinting using ligation-mediated PCR provided evidence for increased protein-DNA interactions with bases in the MRE core sequences and surrounding bases in cells treated with zinc for 1 h (Figure 4). Shown here are footprints for MRE-c and MRE-d of the mouse MT-I promoter. In these same samples there was no apparent change in protein-DNA interactions with Sp1 or USF1 binding sites in this promoter (Dalton *et al.* 1996b). These transcription factors are constitutively active to bind DNA. The activation and nuclear translocation of MTF-1 in response to zinc parallels increases in the relative rate of transcription of the MT-I gene.

Studies of the *in vivo* occupancy of the MREs in the mouse ZnT1 and γGCS_{hc} genes in zinc-treated cells have not been reported. However, the γGCS_{hc} proximal promoter drives MTF-1-dependent and zinc-dependent expression of a reporter gene in a transient transfection assay (Günes *et al.* 1998). In contrast, the ZnT1 promoter has been reported to be unresponsive to zinc under similar assay conditions (Palmiter & Findley 1995). However, MTF-1 binds *in vitro* to the MREs in the ZnT1 promoter, and the expression and metal regulation of the endogenous gene is MTF-1-dependent and is lost in cultured cells from MTF-1 knockout mice. In addition, zinc elicits a rapid transcriptional response of the ZnT1 gene which parallels that of the MT-I gene in cultured cells (Langmade *et al.* 2000). Thus, it is assumed here that MTF-1 directly modulates expression of the ZnT1 gene by interacting with MREs in that promoter.

MTF-1 binding activity is reversibly induced in vitro *by zinc*

MTF-1 in the cytosol of untreated cells is not active to bind DNA, but it can be fully activated *in vitro* by low micromolar (5 to 15 μM) concentrations of exogenous zinc at temperatures above 4 °C (Dalton *et al.* 1997; Bittel *et al.* 1998; Smirnova *et al.* 2000; Otsuka *et al.* 2000). Similarly, human and mouse MTF-1 synthesized *in vitro* in a coupled transcription-translation system (TnT lysate) is not active to bind DNA, but can be activated by exogenous zinc (1 to 5 μM) at elevated temperature (Bittel *et al.* 1998) (Figure 5A). This activation does not occur at 37 °C in the absence of exogenous zinc, nor does it occur at 4 °C in the presence of exogenous zinc. The three zinc-finger (Cys_2His_2) protein Sp1 synthesized under these conditions is largely constitutively active to bind DNA without the addition of exogenous zinc. Once activated by zinc to bind DNA, MTF-1 is more sensitive to metal chelators than is Sp1 (Radtke *et al.* 1993; Dalton *et al.* 1997). Furthermore, the DNA-binding activity of native and recombinant MTF-1 can be reversibly modulated by zinc (Okajima *et al.* 1993; Heuchel *et al.* 1994; Otsuka *et al.* 1994; Dalton *et al.* 1997). Diluting active MTF-1 into buffer in which the zinc concentration is below 0.6 μM, leads to a time – and temperature – dependent loss of binding, which can be completely restored by readjusting the zinc concentration in the reaction to 30 μM (Figure 5B).

In contrast to these effects of zinc, other transition metals (e.g., cadmium) which are potent inducers of MT, ZnT1 and γGCS_{hc} gene expression, do not activate the DNA-binding activity of mammalian MTF-1 *in vitro* and exert only modest effects *in vivo* on this activity (Bittel *et al.* 1998; Koizumi *et al.* 1999; Smirnova *et al.* 2000). These metal ions may cause the redistribution of zinc which in turn activates some MTF-1 to bind DNA (Palmiter 1994), but other mechanisms may also be involved. It should also be emphasized that increased DNA-binding activity alone seems unlikely to completely explain the metalloregulatory functions of MTF-1. Mouse MTF-1 can activate MRE-driven reporter gene expression in yeast only in response to zinc, and not to cadmium or oxidative stress (Andrews *et al.* unpublished results). Mouse MTF-1 functions as a zinc-sensor to activate MRE-driven gene expression in mammalian, yeast and *Drosophila* cells (Bittel *et al.* 2000; Andrews *et al.* unpublished results), and *Drosophila* MTF-1 can activate gene expression in response to zinc in transfected mammalian cells (Zhang *et al.* 2000). Thus, this metalloregulatory function of MTF-1 has been highly conserved during evolution.

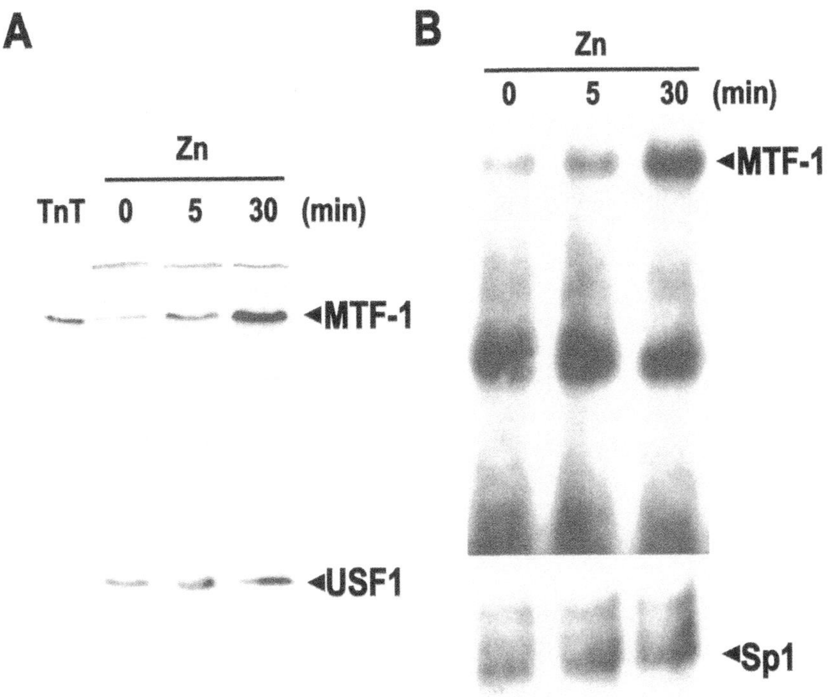

Fig. 3. Western blot and mobility shift assay detection of MTF-1 in nuclear extracts from Hepa cells at different times after zinc-treatment. Hepa cells were treated with 100 μM $ZnSO_4$ for 5 or 30 min and then separated into cytosolic and nuclear fractions (shown), as described (Smirnova *et al.* 2000). **A**: Nuclear extracts were analyzed by Western blotting for MTF-1. *Upper panel*: Recombinant mouse MTF-1 synthesized *in vitro* in a TnT lysate was used as a positive control for MTF-1. *Lower panel*: the extracts were Western blotted for USF1. **B**: *Upper panel*: nuclear extracts were analyzed for DNA-binding activity using a labeled MRE oligonucleotide. *Lower panel*: the same extracts were assayed using an Sp1 specific oligonucleotide. Arrows point to specific MTF-1 and Sp1 complexes with their respective oligonucleotides. Reproduced with permission from (Smirnova *et al.* 2000).

MTF-1 activation by zinc reflects functional heterogeneity of its zinc-fingers

Zinc-activation of the DNA-binding activity of MTF-1 involves reversible interactions with zinc (Radtke *et al.* 1993), and these interactions occur with the zinc-finger domain (Dalton *et al.* 1997). This was first demonstrated by deletion mutagenesis of the protein (Dalton *et al.* 1997), and subsequently by finger swapping experiments (Bittel *et al.* 2000), finger mutation experiments (Koizumi *et al.* 2000), and analyses of the purified recombinant MTF-1 finger domain (Chen *et al.* 1998, 1999). Replacing the zinc-finger domain of Sp1 with that of MTF-1 results in a chimeric protein that requires exogenous zinc for activation of MRE-binding activity, similar to native MTF1 (Bittel et al. 2000) (Figure 6). In contrast, the three zinc-fingers of Sp1, in the context of the MTF-1 peptide backbone, are constitutively active to bind DNA and do not require exogenous zinc.

The precise mechanisms by which zinc reversibly interacts with the zinc-fingers of MTF-1 and facilitates a DNA-binding conformation remain to be determined. Although, contradictory evidence has been reported with regard to the functions of the individual zinc-fingers of MTF-1, it is clear in each of these studies that the six zinc-fingers exhibit functional heterogeneity (Chen *et al.* 1999; Koizumi *et al.* 2000; Bittel *et al.* 2000). Some are involved in DNA-binding whereas others appear to be important for sensing zinc.

Analyses of the purified recombinant zinc-finger domain of human MTF-1 suggests that three or four zinc-fingers play a structural role in folding, DNA-binding and DNA-bending (Chen *et al.* 1998). The reduced peptide which bound enough zinc to fold three or four fingers was able to bind an MRE with high affinity (K_{app} 3.8 × 10^8 M^{-1}) and specificity. This finding is consistent with the fact that a single zinc-finger of this type can interact tightly with three or four bases in DNA (Rebar *et al.* 1996). Thus, effective MRE-binding could be accomplished with three or

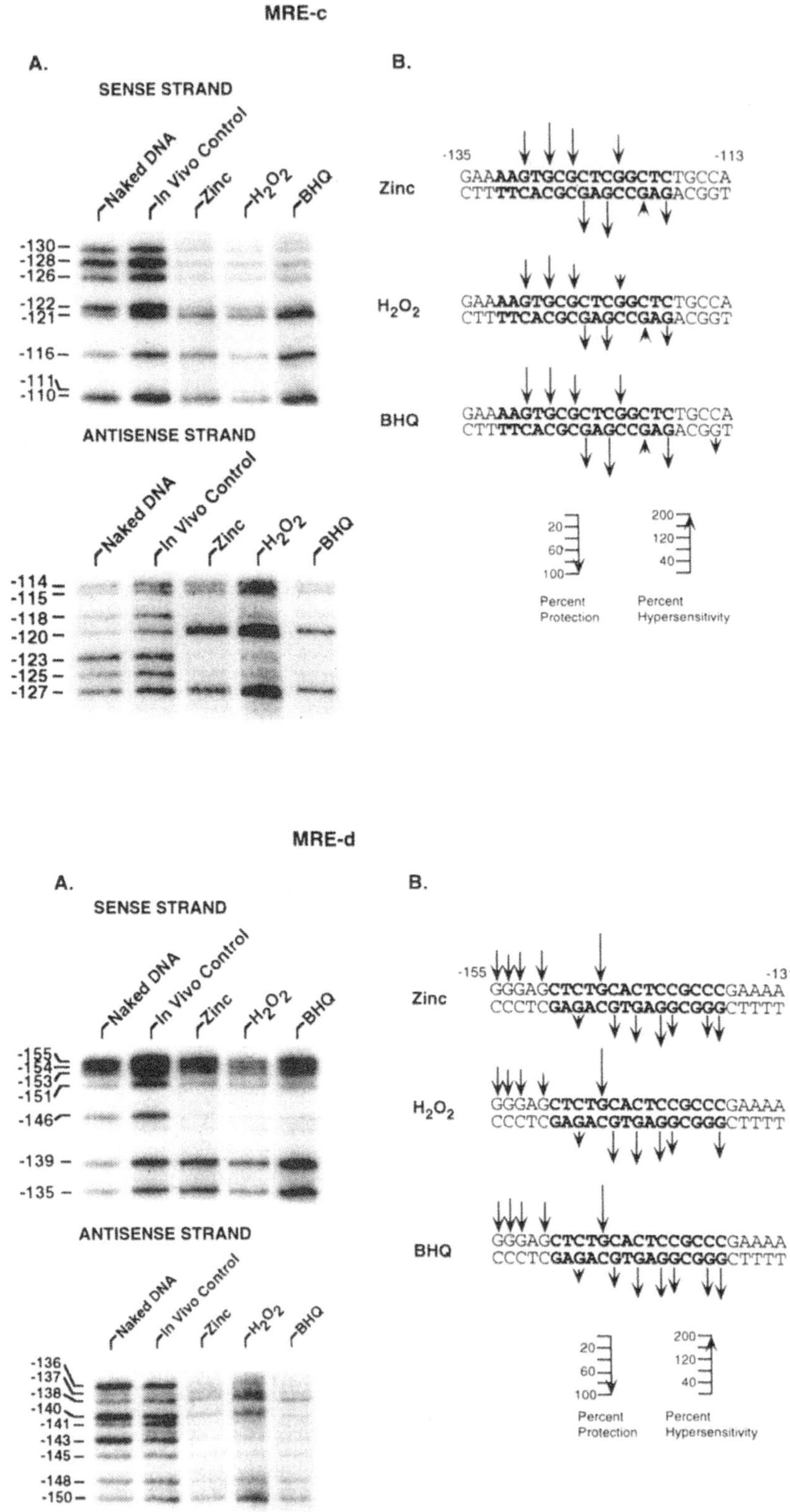

Fig. 4. In vivo genomic footprints over MRE-c and MRE-d in cells treated with zinc or oxidative stress-inducing agents. Mouse Hepa cells were treated with 2.5 mM H_2O_2 for 0.5 h, or 400 μM tBHQ (tert-butyl hydroquinone) or 100 μM $ZnCl_2$ for 1 h, and *in vivo* genomic footprints in the MT-I promoter were determined using ligation-mediated PCR of DNA from cells treated with dimethylsulfate. **A**: Shown are the results of the regions of the sequencing gel corresponding to MRE-c (Upper panel) and MRE-d (Lower panel). **B**: Band intensities were quantitated by phosphoimage analysis of dried gels and protection of $\geq$20% and hypersensitivity of $\geq$40% of individual guanine residues in treated cells compared with those in control cells were calculated for both the sense and antisense strand. Reproduced with permission from (Dalton *et al.* 1996b).

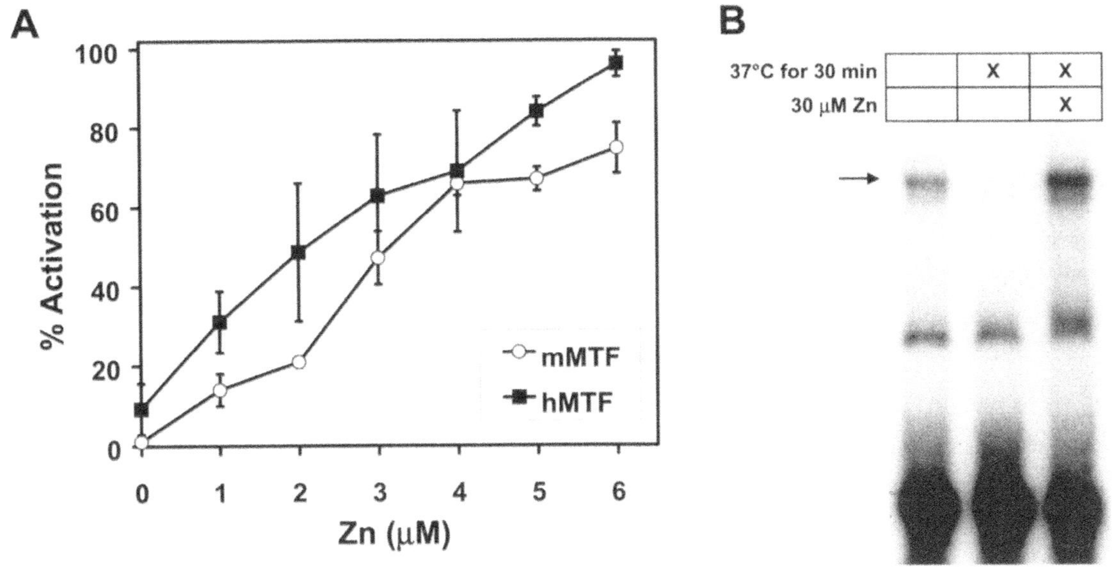

Fig. 5. Zinc reversibly activates MTF-1 DNA-binding activity. 'Recombinant' mouse and human MTF-1 were synthesized *in vitro* in a TnT lysate, as described (Dalton *et al.* 1997), and its DNA-binding activity was detected by a mobility shift assay using labeled MRE oligonucleotide. **A**: The effects of zinc concentration on the DNA-binding activity of mouse MTF-1 (mMTF) or human MTF-1 (hMTF) were examined (Bittel *et al.* 1998). Binding reactions were assembled with the indicated concentrations of exogenous zinc and incubated at 37 °C for 15 min before labeled MRE-s oligonucleotide was added and the reactions were subjected to electrophoresis. The amount of MTF-1/MRE complex was quantitated by phosphorimage analysis. Reactions containing 30 μM Zn served as the 100% activation standards and values shown represent the average ± SEM of three determinations. **B**: Mobility shift assay was performed using recombinant mouse MTF-1 that was activated (37 °C, 15 min) after addition of zinc (60 μM) directly to the TNT lysate. Binding reactions, minus labeled MRE-s, were then assembled in which the activated recombinant MTF-1 and the exogenous zinc were diluted over 100-fold (<0.6 μM). Binding reactions were then incubated at 4 °C (Lane 1) or at 37 °C for 1 h, as indicated. After incubation at 37 °C for 1 h, one binding reaction was placed at 4 °C (Lane 2), whereas the other binding reaction was readjusted to 30 μM zinc (Lane 3) and incubated further at 37 °C for 15 min. Labeled MRE-s was then added and all binding reactions were analyzed by electrophoresis. The gel was dried and labeled MRE-s detected by autoradiography. The arrow indicates the specific MTF-1/MRE-s complex. Reproduced with permission from (Dalton *et al.* 1997; Bittel *et al.* 1998).

four zinc-fingers. These studies also demonstrated that MTF-1 binding alters the DNA structure of the MRE. (Chen *et al.* 1998). The additional, and lower affinity binding of two or three zinc atoms did not cause a further conformational change in the finger domain. This finding is unexpected because zinc-finger domains adopt their conformation upon zinc-binding. This heterogeneity of zinc-binding is, however, consistent with a zinc-sensing function within the zinc-finger domain. Subsequent analysis of deletion mutants of the purified MTF-1 zinc-finger domain suggested that fingers 5 and 6 correspond to the weak zinc-binding fingers which stabilize the constitutive tight DNA-binding activity of fingers 1 to 4 (Chen *et al.* 1999).

However, the studies reviewed earlier reveal that native MTF-1 exists in a latent DNA-binding form, not a constitutively active DNA-binding form. The temperature requirement for zinc-activation of MTF-1 suggests that zinc-binding induces a conformation change in the native protein (Dalton *et al.* 1997, 2000). Although much evidence supports this concept, a genetic study using transfected cells suggested that the function of MTF-1 may be inhibited in the cell by a zinc-sensitive inhibitor (Palmiter 1994). Inactivation of the putative inhibitor could lead to heightened expression of the MT-I gene. No such inhibitor has been identified, and mouse MTF-1 functions in insect cells and yeast cells (Bittel *et al.* 2000), which suggests that a specific zinc-sensitive inhibitor is not required for MTF-1 to sense zinc. It is conceivable that the peptide domains of MTF-1 which surround the zinc-fingers, also influence their folding and DNA-binding, as was recently suggested (Koizumi *et al.* 2000).

No one has reported the successful purification of full-length recombinant MTF-1, and this problem has impeded detailed physical analyses of the native and mutant proteins. However, the effects of finger deletions in the mouse MTF-1 zinc-finger domain in the context of the native peptide backbone have been examined using mobility shift assays with

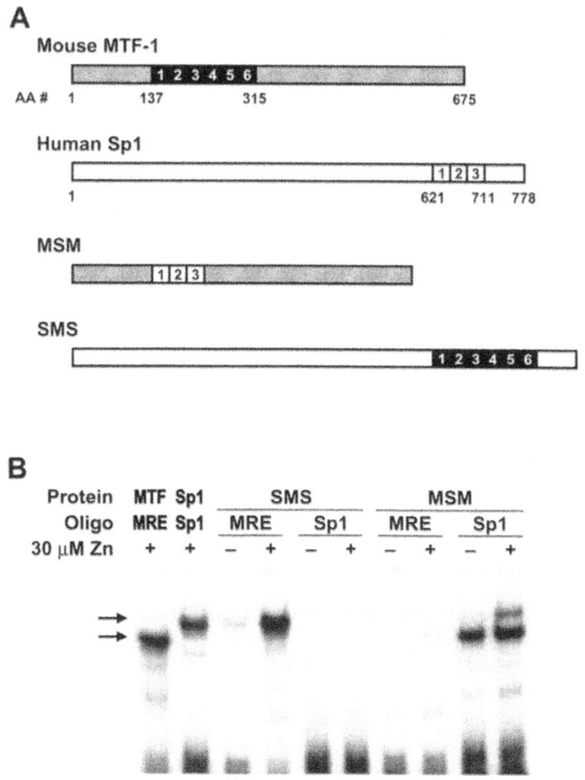

Fig. 6. The zinc-finger domain of MTF-1 mediates reversible activation by zinc: Zinc-finger swapping between MTF-1 and Sp1. **A**: Diagram of the chimeric proteins created by swapping the zinc-finger domains of MTF-1 and Sp1. The zinc-fingers of MTF-1 and Sp1 were switched without gain or loss of amino acids in the backbone of each. The domains which were exchanged encompass the first cysteine of finger 1 to the last histidine of the final finger. The Sp1 peptide backbone with the MTF-1 finger cassette is termed SMS, whereas the MTF-1 peptide backbone with the Sp1 finger cassette is called MSM. **B**: Full-length MTF-1 and Sp1 and chimeric SMS and MSM constructs were used to program TnT lysates, and zinc-activated DNA-binding (Dalton *et al.* 1997) was assessed by mobility shift assay using the labeled MRE or Sp1 oligonucleotides, as indicated. Arrows indicate the specific protein-DNA complexes. Reproduced with permission from Bittel *et al.* (2000).

proteins synthesized in a TnT lysate system (Figure 7A). This approach precludes measurements of binding constants, and can only provide qualitative information on relative DNA-binding activity. However, functional activity of the expressed proteins was also monitored using transient transfection assays in yeast (Figure 7B), MTF-1 knockout cells or *Drosophila* cells (Bittel *et al.* 2000).

Consistent with studies of the isolated finger domain, these studies also demonstrated functional heterogeneity of the zinc-fingers of MTF-1 and mapped core DNA-binding activity to fingers 2, 3 and 4 (Bittel *et al.* 2000). However, zinc-dependent activation of both DNA-binding and of reporter gene expression was mapped to zinc-finger 1 (Bittel *et al.* 2000). Deletion of finger 1 resulted in a protein which bound DNA constitutively, and zinc-response, but not basal expression, was lost in all three transfection systems (Figure 7). In contrast, these functions were unaffected by deletion of fingers 5 and 6 (Figure 7B). This observation was confirmed in a recent study of human MTF-1 which demonstrated that mutation of the second cysteine residue (cys to tyr) in finger 5 or 6, which would preclude zinc-binding and folding of the finger, had little affect on DNA-binding or transcriptional activation in transfected cells (Koizumi *et al.* 2000). These results are inconsistent with a major role of fingers 5 or 6 in the DNA-binding or gene activation processes of MTF-1, yet these fingers are highly conserved during evolution. Thus, it seems very unlikely that these zinc-fingers have no function.

Surprisingly, finger 1 of MTF-1 may constitute a unique zinc-sensing domain which is important for its metalloregulatory function (Bittel *et al.* 2000). Consistent with this concept, transfer of MTF-1 finger 1 to a position immediately preceding the three zinc-fingers of Sp1 resulted in a chimeric protein which requires exogenous zinc to activate DNA-binding *in vitro*, unlike native Sp1 which binds DNA constitutively (Figure 8). This suggests that in the absence of sufficient zinc, zinc-finger 1 of MTF-1 adopts a conformation which impedes the DNA-binding activity of adjacent zinc-fingers. Zinc binding apparently relieves that inhibition, which suggests that a conformational change in this finger occurs upon binding zinc. Consistent with these concepts, mutation of the second cysteine residue in finger 1 created a protein which did not bind DNA or activate transcription in response to zinc (Koizumi *et al.* 2000). Thus, zinc-finger 1 of MTF-1 is apparently the zinc-sensitive inhibitor of MTF-1 activity.

Future perspectives

Most studies of MTF-1 structure and function suggest that this protein serves to sense zinc levels in the cell by a direct and reversible interaction of zinc with a subset of zinc-fingers; the precise mechanism of this process remains to be determined. Clearly, structural studies of purified recombinant MTF-1 are needed. Unfortunately, purification of sufficient, full-length recombinant MTF-1 from bacteria has proven difficult.

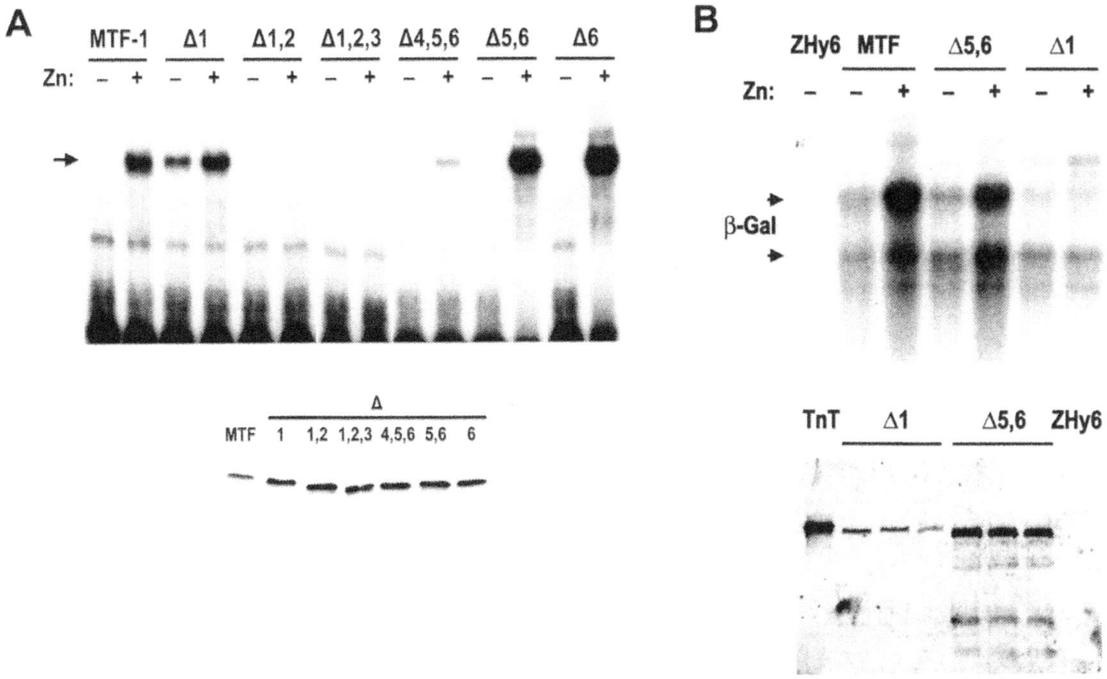

Fig. 7. Analysis of the DNA-binding of mouse MTF-1 zinc-finger deletion constructs, and their function in yeast. **A**: Upper panel: fingers were deleted (Δ) from the first cysteine of the deleted finger to the amino acid preceding the first cysteine of the next finger, as indicated (Bittel *et al.* 2000). Proteins were synthesized *in vitro* in a TnT lysate and analyzed by mobility shift assay for binding to labeled MRE-s. DNA-binding activity was assessed before (−) and after (+) the addition of 30 μM zinc, and incubation at 37 °C. The arrow indicates the specific protein-DNA complex. Lower panel: the synthesis of MTF-1 and its finger deletion mutants was confirmed by Western blotting (Smirnova *et al.* 2000). **B**: Function of MTF-1, Δ1 or Δ5,6 finger deletion mutants in yeast. The *Saccharomyces cerevisiae* strain ZHy6 has a mutation in the *ZAP1* gene which results in severely attenuated zinc transporter gene expression, and increased dependence of these cells on zinc levels in the culture medium (Zhao & Eide 1997; Davis *et al.* 1996). Top panel: Northern blot probed for β-galactosidase expression in ZHy6 cells cotransfected with an MTF-1, Δ1 or Δ5,6 expression vector plus an MRE-d5-β-galactosidase reporter. Cells were grown overnight in medium containing 2 μM zinc and brought to 60 μM zinc as indicated for 1 h. Arrows indicate the presence of two β-galactosidase transcripts. Bottom panel: proteins extracted from three independent colonies of yeast strain ZHy6 transformed with the Δ1 or Δ5,6 expression vector were analyzed by Western blotting. Recombinant mouse MTF-1 synthesized in a TnT lysate was used as a positive control in lane 1. The lane labeled ZHy6 contained protein from the non-transfected parental yeast strain. Reproduced with permission from Bittel *et al.* (2000).

One of the complications of this experimental system is the fact that zinc is a ubiquitous and essential metal ion. Defining assay conditions *in vitro* which control zinc availability is relatively simple, but defining such conditions *in vivo* in transfection assays is difficult. Cells adapt to zinc availability to maintain homeostasis. Thus, transfection studies may not accurately reveal some details of MTF-1 function (e.g., fingers 5 and 6 function?). Furthermore, the redox environment in the cell may affect zinc-activation of MTF-1 (Bittel *et al.* 1998; Koizumi *et al.* 2000). GSH alters the sensitivity of MTF-1 to zinc-activation *in vitro* (Bittel *et al.* 1998; Koizumi *et al.* 2000). This effect probably involves competition for zinc binding, although the redox state of the cysteine residues in the zinc-finger of MTF-1 also influences zinc binding (Chen *et al.* 1998).

It also remains to be determined how cadmium, and other metal ions, affect MTF-1 activity leading to the activation of gene expression. Cadmium treatment, using concentrations that are maximally effective to induce MT and ZnT1 gene expression in mammalian cells (Langmade *et al.* 2000), causes only a modest increase in MTF-1 binding activity *in vivo* and this metal has no effect *in vitro* (Koizumi *et al.* 1992; Bittel *et al.* 1998; Koizumi *et al.* 1999; Smirnova *et al.* 2000). It is conceivable that these metals may utilize specific co-activators of MTF-1 and/or activate signal transduction cascades that impinge on MTF-1 to affect MT gene transcription.

The transactivation domains of MTF-1 contain several potential sites for phosphorylation by known kinases, but functional phosphorylation, or other modifications (acetylation, methylation) of MTF-1 has not

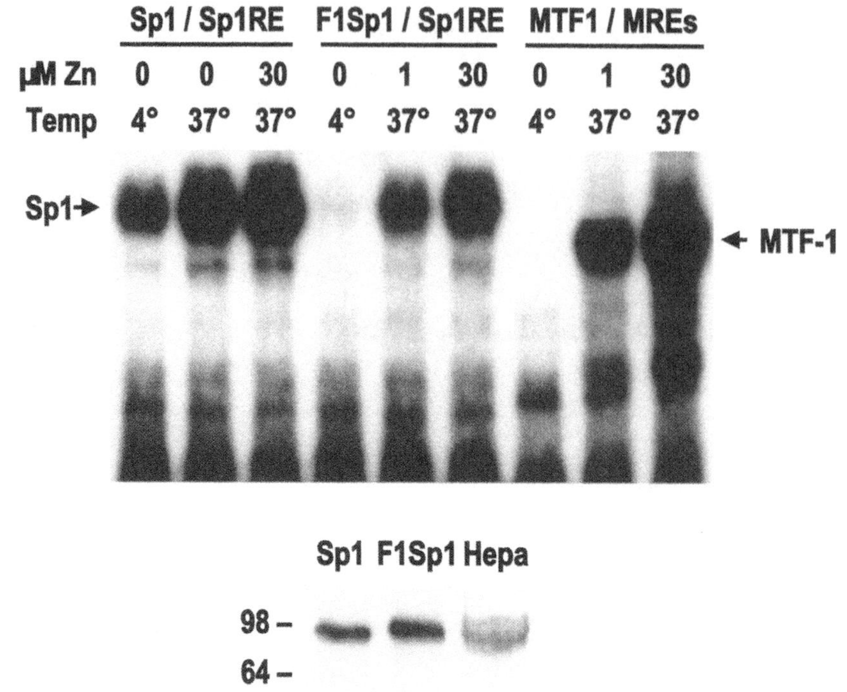

Fig. 8. Comparison of zinc-activation of DNA-binding of MTF-1, Sp1 and a chimeric F1Sp1 which contains the first zinc-finger from mouse MTF-1. The F1Sp1 fusion construct contains the Sp1 amino acid sequence to lysine 632 which is followed by 29 residues encompassing finger 1 of MTF-1 including the 3 amino acids immediately preceding the first cysteine and the 3 amino acids following the last histidine. This is followed by the three zinc-fingers of Sp1 and the remainder of the Sp1 peptide backbone (Bittel *et al.* 2000). Upper panel: TnT lysates were programmed with MTF-1, Sp1 or chimeric F1Sp1. Mobility shift assays containing an aliquot from the indicated TnT lysates were adjusted to the indicated concentrations of zinc and incubated at 4 °C or 37 °C for 15 min before electrophoresis. Binding to the Sp1 oligonucleotide (Sp1RE) or MREs was monitored. Arrows indicate the specific protein-DNA complexes. Lower panel: Western blot detection of Sp1 or F1Sp1 in these TnT lysates or in a whole cell extract from Hepa cells using an Sp1 antibody. The arrow indicates the position of the 98 kDa molecular weight marker. Reproduced with permission from Bittel *et al.* (2000).

been demonstrated. A recent study reported that MTF-1 from cadmium treated cells displays altered mobility during native gel electrophoresis, consistent with a possible posttranslational modification or conformation change in the protein (Otsuka *et al.* 2000). The nature and function (if any) of that potential modification are unknown. Inhibition of histone deacetylase activity renders cultured cells hypersensitive to metal-induction of MT gene expression (Andrews & Adamson 1987), which suggests that the threshold for sensitivity to metals may have different set points in different cells based on nucleosome structures. That MTF-1 molecules may interact is suggested by the findings that two or more MREs cooperate to confer metal-responsiveness (Stuart *et al.* 1984; Koizumi *et al.* 1999), and a single palindromic MRE directs metal-regulation in avian MT promoters (Shartzer *et al.* 1993). However, dimerization of MTF-1 has not been demonstrated.

MTF-1 may also cooperate with other transcription factors to regulate gene expression. The bHLH protein upstream stimulatory factor-1 (USF1) may play a role in the cadmium activation of MT-I gene expression in cultured cells (Li *et al.* 1998), and this protein apparently cooperates with MTF-1 to regulate the high level expression of MT genes in the visceral endoderm cells of the early mouse embryo (Andrews *et al.* 2001). Whether or not these proteins directly interact with each other is under investigation. Finally, methylation status of the mouse MT-I gene represses its activity and inducibility by metal ions (Lieberman *et al.* 1983; Lu *et al.* 1999), although methylation of the MRE does not appear to inhibit the *in vitro* binding of MTF-1 (Radtke *et al.* 1996).

In summary, current models of the mechanisms by which metals regulate MT gene expression suggest that the reversible interaction of zinc with the metalloregulatory protein MTF-1 modulates its DNA-binding structure. Reversible occupancy of perhaps a single zinc-binding site in a zinc-finger may stabilize the DNA-binding form of the protein, leading to nuclear translocation, binding to MRE and increased transcription of MT, ZnT1 and γGCS_{hc} genes. These genes protect the cell from metal deficiency, metal toxicity and oxidative stresses. Further validation of these models for MTF-1 function awaits the crystal structures of active DNA-bound MTF-1 and inactive MTF-1.

Acknowledgements

This work was supported, in part, by NIH grants (ES05704; CA61262) to GKA. We are indebted to Jim Geiser and Steve Eklund for excellent technical assistance. I also want to mention the scientists who contributed significantly to the studies performed in my laboratory. Specifically, Drs Tim Dalton, Qingwen Li, Doug Bittel, Irina Smirnova, Rudy Ravindra, Pat Daniels, Luchuan Liang and students Josh Langmade and Huimin Jiang who each worked in my laboratory. In addition, Drs Lashitew Gedamu, Susan Samson, Peter Lichtlen, Walter Schaffner, and Dennis Winge have been invaluable collaborators in these studies.

References

Anderson ME. 1998 Glutathione: an overview of biosynthesis and modulation. *Chem Biol Interact* **111–112**, 1–14.

Andrews GK. 1990 Regulation of metallothionein gene expression. *Prog Food Nutr Sci* **14**, 193–258.

Andrews GK. 2000 Regulation of metallothionein gene expression by oxidative stress and metal ions. *Biochem Pharm* **59**, 95–104.

Andrews GK, Adamson ED. 1987 Butyrate selectively activates the metallothionein gene in teratocarcinoma cells and induces hypersensitivity to metal induction. *Nucl Acids Res* **15**, 5461–5475.

Andrews GK, Geiser J. 1999 Expression of metallothionein-I and -II genes provides a reproductive advantage during maternal dietary zinc deficiency. *J Nutr* **129**, 1643–1648.

Andrews GK, Lee DK, Ravindra R, Lichtlen P, Sirito M, Sawadogo M, Schaffner W. 2001 The transcription factors MTF-1 and USF1 regulate mouse metallothionein-I gene expression in visceral endoderm cells during early development. *EMBO J.* **20**, 1114–1122.

Auf der Maur A, Belser T, Elgar G, Georgiev O, Schaffner W. 1999 Characterization of the transcription factor MTF-1 from the Japanese pufferfish (*Fugu rubripes*) reveals evolutionary conservation of heavy metal stress response. *Biol Chem* **380**, 175–185.

Ballatori N. 1994 Glutathione mercaptides as transport forms of metals. *Adv Pharmacol* **27**, 271–298.

Bittel D, Dalton T, Samson S, Gedamu L, Andrews GK. 1998 The DNA-binding activity of metal response element-binding transcription factor-1 is activated *in vivo* and *in vitro* by zinc, but not by other transition metals. *J Biol Chem* **273**, 7127–7133.

Bittel D, Smirnova I, Andrews GK. 2000 Functional heterogeneity in the zinc fingers of the metalloregulatory transcription factor, MTF-1. *J Biol Chem* **275**, 37194–37201.

Brugnera E, Georgiev O, Radtke F, Heuchel R, Baker E, Sutherland GR, Schaffner W. 1994 Cloning, chromosomal mapping and characterization of the human metal-regulatory transcription factor MTF-1. *Nucl Acids Res* **22**, 3167–3173.

Chaberek S, Martell AE. 1959 *Organic Sequestering Agents*. New York: Wiley and Sons Inc.

Chen XH, Agarwal A, Giedroc DP. 1998 Structural and functional heterogeneity among the zinc fingers of human MRE-binding transcription factor-1. *Biochemistry* **37**, 11152–11161.

Chen XH, Chu MH, Giedroc DP. 1999 MRE-binding transcription factor-1: Weak zinc-binding finger domains 5 and 6 modulate the structure, affinity, and specificity of the metal-response element complex. *Biochemistry* **38**, 12915–12925.

Cherian MG, Apostolova MD. 2000 Nuclear localization of metallothionein during cell proliferation and differentiation. *Cell Mol Biol* **46**, 347–356.

Cizewski Culotta V, Hamer DH. 1989 Fine mapping of a mouse metallothionein gene metal response element. *Mol Cell Biol* **9**, 1376–1380.

Cole TB, Wenzel HJ, Kafer KE, Schwartzkroin PA, Palmiter RD. 1999 Elimination of zinc from synaptic vesicles in the intact mouse brain by disruption of the *ZnT3* gene. *Proc Natl Acad Sci USA* **96**, 1716–1721.

Da Costa Ferreira AM, Ciriolo MR, Marcocci L, Rotilio G. 1993 Copper(I) transfer into metallothionein mediated by glutathione. *Biochem J* **292**, 673–676.

Dalton TD, Bittel D, Andrews GK. 1997 Reversible activation of the mouse metal response element-binding transcription factor-1 DNA binding involves zinc interactions with the zinc-finger domain. *Mol Cell Biol* **17**, 2781–2789.

Dalton TP, Fu K, Palmiter RD, Andrews GK. 1996a Transgenic mice that over-express metallothionein-I resist dietary zinc deficiency. *J Nutr* **126**, 825–833.

Dalton TP, Li QW, Bittel D, Liang LC, Andrews GK. 1996b Oxidative stress activates metal-responsive transcription factor-1 binding activity – Occupancy *in vivo* of metal response elements in the metallothionein-I gene promoter. *J Biol Chem* **271**, 26233–26241.

Dalton TP, Solis WA, Nebert DW, Carvan MJ, III. 2000 Characterization of the MTF-1 transcription factor from zebrafish and trout cells. *Comp Biochem Physiol [B]* **126**, 325–335.

Davis SR, McMahon RJ, Cousins RJ. 1998 Metallothionein knockout and transgenic mice exhibit altered intestinal processing of zinc with uniform zinc-dependent zinc transporter-1 expression. *J Nutr* **128**, 825–831.

Davis W, Jr., De Sousa PA, Schultz RM. 1996 Transient expression of translation initiation factor eIF- 4C during the 2-cell stage of the preimplantation mouse embryo: Identification by mRNA differential display and the role of DNA replication in zygotic gene activation. *Dev Biol* **174**, 190–201.

DeMoor JM, Koropatnick DJ. 2000 Metals and cellular signaling in mammalian cells. *Cell Mol Biol* **46**, 367–381.

Griffith OW. 1999 Biologic and pharmacologic regulation of mammalian glutathione synthesis. *Free Radic Biol Med* **27**, 922–935.

Griffith OW, Mulcahy RT. 1999 The enzymes of glutathione synthesis: gamma-glutamylcysteine synthetase. *Adv Enzymol Relat Areas Mol Biol* **73**, 209–267.

Günes Ç, Heuchel R, Georgiev O, Müller KH, Lichtlen P, Blüthmann H, Marino S, Aguzzi A, Schaffner W. 1998 Embryonic lethality and liver degeneration in mice lacking the metal-responsive transcriptional activator MTF-1. *EMBO J* **17**, 2846–2854.

Heuchel R, Radtke F, Georgiev O, Stark G, Aguet M, Schaffner W. 1994 The transcription factor MTF-I is essential for basal and heavy metal-induced metallothionein gene expression. *EMBO J* **13**, 2870–2875.

Huang LP, Gitschier J. 1997 A novel gene involved in zinc transport is deficient in the lethal milk mouse. *Nature Genet* **17**, 292–297.

Jensen LT, Howard WR, Strain JJ, Winge DR, Culotta VC. 1996 Enhanced effectiveness of copper ion buffering by *CUP1* metallothionein compared with *CRS5* metallothionein in *Saccharomyces cerevisiae*. *J Biol Chem* **271**, 18514–18519.

Kägi JHR, Schäffer A. 1988 Biochemistry of metallothionein. *Biochemistry* **27**, 8509–8515.

Kägi JHR. 1991 Overview of metallothionein. *Methods Enzymol* **205**, 613–626.

Karin M, Najarian R, Haslinger A, Valenzuela P, Welch J, Fogel S. 1984 Primary structure and transcription of an amplified genetic locus: the CUP1 locus of yeast. *Proc Natl Acad Sci USA* **81**, 337–341.

Klaassen CD, Liu J, Choudhuri S. 1999 Metallothionein: An intracellular protein to protect against cadmium toxicity. *Annu Rev Pharmacol Toxicol* **39**, 267–294.

Koizumi S, Suzuki K, Ogra Y, Gong P, Otsuka F. 2000 Roles of zinc fingers and other regions of the transcription factor human MTF-1 in zinc-regulated DNA binding. *J Cell Physiol* **185**, 464–472.

Koizumi S, Suzuki K, Ogra Y, Yamada H, Otsuka F. 1999 Transcriptional activity and regulatory protein binding of metal-responsive elements of the human metallothionein-IIA gene. *Eur J Biochem* **259**, 635–642.

Koizumi S, Yamada H, Suzuki K, Otsuka F. 1992 Zinc-specific activation of a HeLa cell nuclear protein which interacts with a metal responsive element of the human metallothionein-II_A gene. *Eur J Biochem* **210**, 555–560.

Langmade SJ, Ravindra R, Daniels PJ, Andrews GK. 2000 The transcription factor MTF-1 mediates metal regulation of the mouse ZnT1 gene. *J Biol Chem* **275**, 34803–34809.

Lazo JS, Kondo Y, Dellapiazza D, Michalska AE, Choo KHA, Pitt BR. 1995 Enhanced sensitivity to oxidative stress in cultured embryonic cells from transgenic mice deficient in metallothionein I and II genes. *J Biol Chem* **270**, 5506–5510.

Lazo JS, Kuo SM, Woo ES, Pitt BR. 1998 The protein thiol metallothionein as an antioxidant and protectant against antineoplastic drugs. *Chem Biol Interact* **112**, 255–262.

Li QW, Hu NM, Daggett MAF, Chu WA, Bittel D, Johnson JA, Andrews GK. 1998 Participation of upstream stimulatory factor (USF) in cadmium- induction of the mouse metallothionein-I gene. *Nucl Acids Res* **26**, 5182–5189.

Lichtlen P, Georgiev O, Schaffner W, Aguzzi A, Brandner S. 1999 The heavy metal-responsive transcription factor-1 (MTF-1) is not required for neural differentiation. *Biol Chem* **380**, 711–715.

Lieberman MW, Beach LR, Palmiter RD. 1983 Ultraviolet radiation-induced metallothionein-I gene activation is associated with extensive DNA demethylation. *Cell* **35**, 207–214.

Lu SC. 1999 Regulation of hepatic glutathione synthesis: current concepts and controversies. *FASEB J* **13**, 1169–1183.

Lu ZH, Cobine P, Dameron CT, Solioz M. 1999 How cells handle copper: A view from microbes. *J Trace Elem Exp Med* **12**, 347–360.

Maret W. 1994 Oxidative metal release from metallothionein via zinc-thiol/disulfide interchange. *Proc Natl Acad Sci USA* **91**, 237–241.

Maroni G, Ho AS, Laurent T. 1995 Genetic control of cadmium tolerance in *Drosophila melanogaster*. *Environ Health Perspect* **103**, 1116–1118.

Masters BA, Kelly EJ, Quaife CJ, Brinster RL, Palmiter, RD. 1994 Targeted disruption of metallothionein I and II genes increases sensitivity to cadmium. *Proc Natl Acad Sci USA* **91**, 584–588.

McMahon RJ, Cousins RJ. 1998a Mammalian zinc transporters. *J Nutr* **128**, 667–670.

McMahon RJ, Cousins RJ. 1998b Regulation of the zinc transporter ZnT-1 by dietary zinc. *Proc Natl Acad Sci USA* **95**, 4841–4846.

Mehra RK, Winge DR. 1991 Metal ion resistance in fungi: Molecular mechanisms and their regulated expression. *J Cell Biochem* **45**, 30–40.

Michalska AE, Choo KHA. 1993 Targeting and germ-line transmission of a null mutation at the metallothionein I and II loci in mouse. *Proc Natl Acad Sci USA* **90**, 8088–8092.

Miles AT, Hawksworth GM, Beattie JH, Rodilla V. 2000 Induction, regulation, degradation, and biological significance of mammalian metallothioneins. *Crit Rev Biochem Mol Biol* **35**, 35–70.

Müller HP, Brugnera E, Georgiev O, Badzong M, Müller KH, Schaffner W. 1995 Analysis of the heavy metal-responsive transcription factor MTF-1 from human and mouse. *Somat Cell Mol Genet* **21**, 289–297.

Nordberg M, Nordberg GF. 2000 Toxicological aspects of metallothionein. *Cell Mol Biol* **46**, 451–463.

O'Halloran TV. 1993 Transition metals in control of gene expression [see comments]. *Science* **261**, 715–725.

Okajima A, Miyazawa K, Kitamura N. 1993 Characterization of the promoter region of the rat hepatocyte-growth-factor/scatter-factor gene. *Eur J Biochem* **213**, 113–119.

Otsuka F, Iwamatsu A, Suzuki K, Ohsawa M, Hamer DH, Koizumi S. 1994 Purification and characterization of a protein that binds to metal responsive elements of the human metallothionein IIA gene. *J Biol Chem* **269**, 23700–23707.

Otsuka F, Okugaito I, Ohsawa M, Iwamatsu A, Suzuki K, Koizumi S. 2000 Novel responses of ZRF, a variant of human MTF-1, to *in vivo* treatment with heavy metals. *Biochim Biophys Acta* **1492**, 330–340.

Palmiter RD. 1987 Molecular biology of metallothionein gene expression. *Experientia Suppl* **52**, 63–80.

Palmiter RD. 1994 Regulation of metallothionein genes by heavy metals appears to be mediated by a zinc-sensitive inhibitor that interacts with a constitutively active transcription factor, MTF-1. *Proc Natl Acad Sci USA* **91**, 1219–1223.

Palmiter RD. 1998 The elusive function of metallothioneins. *Proc Natl Acad Sci USA* **95**, 8428–8430.

Palmiter RD, Cole TB, Findley SD. 1996a ZnT-2, a mammalian protein that confers resistance to zinc by facilitating vesicular sequestration. *EMBO J* **15**, 1784–1791.

Palmiter RD, Cole TB, Quaife CJ, Findley SD. 1996b ZnT-3, a putative transporter of zinc into synaptic vesicles. *Proc Natl Acad Sci USA* **93**, 14934–14939.

Palmiter RD, Findley SD. 1995 Cloning and functional characterization of a mammalian zinc transporter that confers resistance to zinc. *EMBO J* **14**, 639–649.

Radtke F, Georgiev O, Müller H-P, Brugnera E, Schaffner W. 1995 Functional domains of the heavy metal-responsive transcription regulator MTF-1. *Nucl Acids Res* **23**, 2277–2286.

Radtke F, Heuchel R, Georgiev O, Hergersberg M, Gariglio M, Dembic Z, Schaffner W. 1993 Cloned transcription factor MTF-1 activates the mouse metallothionein I promoter. *EMBO J* **12**, 1355–1362.

Radtke F, Hug M, Georgiev O, Matsuo K, Schaffner, W. 1996. Differential sensitivity of zinc finger transcription factors MTF-1, Sp1 and Krox-20 to CpG methylation of their binding sites. *Biol Chem* **377**, 47–56.

Rebar EJ, Greisman HA, Pabo CO. 1996 Phage display methods for selecting zinc finger proteins with novel DNA-binding specificities. *Methods Enzymol* **267**, 129–149.

Roesijadi G. 2000 Metal transfer as a mechanism for metallothionein-mediated metal detoxification. *Cell Mol Biol* **46**, 393–405.

Shartzer KL, Kage K, Sobieski RJ, Andrews GK. 1993 Evolution of avian metallothionein: DNA sequence analyses of the turkey metallothionein gene and metallothionein cDNAs from pheasant and quail. *J Mol Evol* **36**, 255–262.

Shi ZZ, Osei FJ, Kala G, Kala SV, Barrios RJ, Habib GM, Lukin DJ, Danney CM, Matzuk, MM, Lieberman MW. 2000 Glutathione synthesis is essential for mouse development but not for cell growth in culture. *Proc Natl Acad Sci USA* **97**, 5101–5106.

Smirnova IV, Bittel DC, Ravindra R, Jiang H, Andrews, GK. 2000 Zinc and cadmium can promote the rapid nuclear translocation of MTF-1. *J Biol Chem* **275**, 9377–9384.

Soltaninassab SR, Sekhar KR, Meredith MJ, Freeman ML. 2000 Multi-faceted regulation of gamma-glutamylcysteine synthetase. *J Cell Physiol* **182**, 163–170.

Stuart GW, Searle PF, Chen HY, Brinster RL, Palmiter RD. 1984 A 12-base-pair DNA motif that is repeated several times in metallothionein gene promoters confers metal regulation to a heterologous gene. *Proc Natl Acad Sci USA* **81**, 7318–7322.

Stuart GW, Searle PF, Palmiter RD. 1985 Identification of multiple metal regulatory elements in mouse metallothionein-I promoter by assaying synthetic sequences. *Nature* **317**, 828–831.

Sugawara N, Sugawara C, Katakura M, Takahashi H, Mori M. 1991 Copper metabolism in the LEC rat: Involvement of induction of metallothionein and disposition of zinc and iron. *Experientia* **47**, 1060–1063.

Thiele DJ. 1992 Metal-regulated transcription in eukaryotes. *Nucl Acids Res* **20**, 1183–1191.

Tsuda M, Imaizumi K, Katayama T, Kitagawa K, Wanaka A, Tohyama M, Takagi T. 1997 Expression of zinc transporter gene, ZnT-1, is induced after transient forebrain ischemia in the gerbil. *J Neurosci* **17**, 6678–6684.

Vulpe CD, Packman S. 1995 Cellular copper transport. *Annu Rev Nutr* **15**, 293–322.

Wenzel HJ, Cole TB, Born DE, Schwartzkroin PA, Palmiter RD. 1997 Ultrastructral localization of zinc transporter-3 (ZnT-3) to synaptic vesicle membranes within mossy fiber boutons in the hippocampus of mouse and monkey. *Proc Natl Acad Sci USA* **94**, 12676–12681.

Wild AC, Mulcahy RT. 2000 Regulation of gamma-glutamylcysteine synthetase subunit gene expression: insights into transcriptional control of antioxidant defenses. *Free Radic Res* **32**, 281–301.

Zhang B, Egli D, Georgiev O, Schaffner W. 2001 The Drosophila homolog of mammalian zinc-finger factor MTF-1 activates transcription in response to heavy metals. *Mol Call Biol* **21**, 4505–4514.

Zhao H, Eide DJ. 1997 Zap1p, a metalloregulatory protein involved in zinc-responsive transcriptional regulation in *Saccharomyces cerevisiae*. *Mol Cell Biol* **17**, 5044–5052.

BioMetals 14: 239–249, 2001.

Review

Bacterial zinc transporters and regulators

Klaus Hantke
Mikrobiologie/Membranphysiologie, Universität Tübingen, Auf der Morgenstelle 28, D-72076 Tübingen, Germany (Tel: +49-7071 2974645; Fax; +49-7071 295843; E-mail: hantke@uni-tuebingen.de)

Received 25 January 2001; accepted 16 March 2001

Key words: zinc, transporter, regulator, RND type exporter, P-type ATPase, ABC transporter

Abstract

Zn^{2+} homeostasis in bacteria is achieved by export systems and uptake systems which are separately regulated by their own regulators. Three types of Zn^{2+} export systems that protect cells from high toxic concentrations of Zn^{2+} have been identified: RND multi-drug efflux transporters, P-type ATPases, and cation-diffusion facilitators. The RND type exporters for Zn^{2+} are only found in a few gram-negative bacteria; they allow a very efficient export across the cytoplasmic membrane and the outer membrane of the cell. P-type ATPases and cation-diffusion facilitators belong to protein families that are also found in eukaryotes. The exporters are regulated in bacteria by MerR-like repressor/activators or by ArsR-like repressors. For the high-affinity uptake of Zn^{2+}, several binding-protein-dependent ABC transporters belonging to one class have been identified in different bacteria. Zn^{2+} ABC transporters are regulated by Zur repressors, which belong to the Fur protein family of iron regulators. Little is known about low-affinity Zn^{2+} uptake under zinc-replete conditions. One known example is the phosphate uptake system Pit, which may cotransport Zn^{2+} in *Escherichia coli*. Similarly, the citrate–metal cotransporter CitM in *Bacillus subtilis* may help to supply Zn^{2+}.

Introduction

Zinc is, for most if not all bacteria, an essential trace element. Many bacterial enzymes contain zinc in the active center or in a structurally important site; the large group of DNA-binding proteins with zinc-finger motifs, however, are mainly found in eukaryotes and rarely in bacteria (Clarke & Berg 1998). In 1974, Bucheder & Broda showed in a careful study that the uptake of $^{65}Zn^{2+}$ in *Escherichia coli* is energy-dependent; in starved cells under anoxic conditions, uptake is stimulated by glucose and is more strongly stimulated by the addition of oxygen.

Only recently, with the availability of molecular genetic methods, have further details of bacterial Zn^{2+} transport and Zn^{2+} homeostasis been revealed. The first transporters identified were shown to confer resistance to high Zn^{2+} concentrations. In many cases, these transporters export, in addition to Zn^{2+}, also other toxic ions such as Cd^{2+}, Co^{2+}, and Pb^{2+}. In eukaryotes, resistance to high concentrations of Zn^{2+} is not only achieved by export, but also by binding to metallothionein. Among bacteria, this type of protein has to date only been observed in cyanobacteria (Robinson *et al.* 1998). Also, high-affinity zinc-uptake systems have been detected that help bacteria to live at extremely low Zn^{2+} concentrations, which are found, for example, in the serum of animals and humans. In addition, there are less well characterized low-affinity uptake systems that contribute to Zn^{2+} supply in a Zn^{2+}-rich, but non-toxic environment.

Figure 1 shows an overview of the different systems found in gram-negative bacteria grown under high, medium, or low Zn^{2+} concentrations. Examples from various bacteria are combined in this figure and will be addressed in the following paragraphs. What is noteworthy in the figure is the lack of uptake systems. Specific transporters active under Zn^{2+}-replete conditions have not been identified. Only a few unspecific systems are known to transport Zn^{2+}among

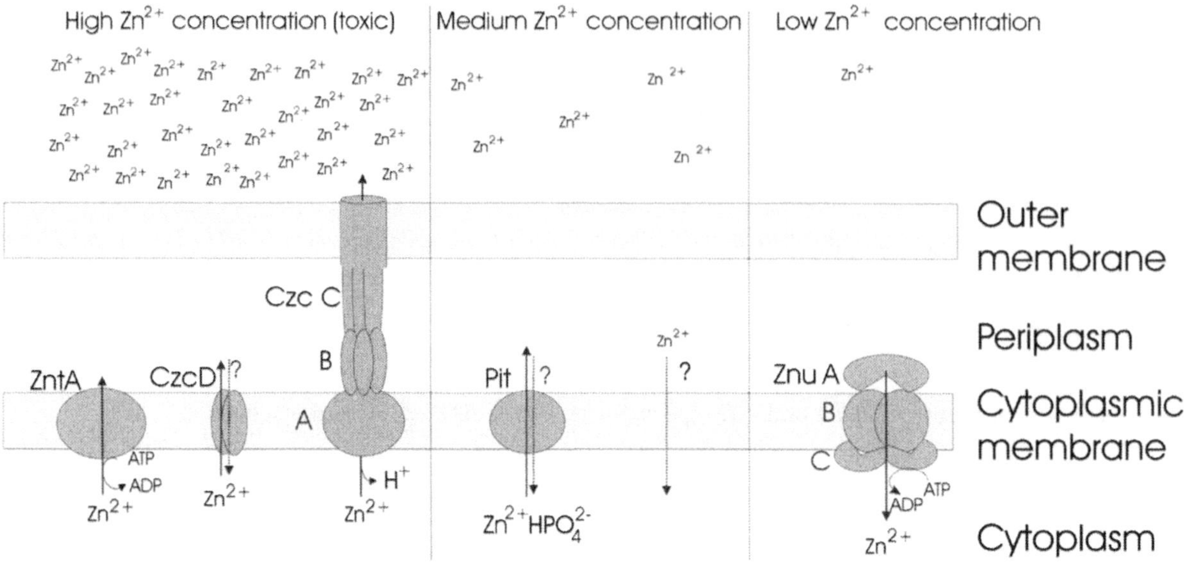

Fig. 1. Pathways for uptake and efflux of Zn^{2+} in gram-negative bacteria. Depending on the Zn^{2+} concentration in the medium, different types of Zn^{2+} transporters are synthesized. At limiting Zn^{2+} concentrations, binding-protein-dependent ABC transporters are induced (right). The Pit-like proteins might act as co-transporters to help satisfy the Zn^{2+} demands of the cell under Zn^{2+}-replete conditions (middle). Exporters of the CzcABC-like RND transporters seem to be very efficient in protecting the cells against toxic Zn^{2+} concentrations. Also, CzcD-like cation facilitators and P-type ATPases such as ZntA protect the cells against high Zn^{2+} concentrations (left). Obvious gaps in our knowledge are apparent: little is known about the passage of divalent ions across the outer membrane and about which transport systems supply the cells under Zn^{2+}-replete conditions.

other divalent cations. Although the outer membrane has many binding sites for divalent cations, little is known how Zn^{2+} passes through this membrane. Gram-positive bacteria, which do not have an outer membrane (Figure 1), have many binding sites for divalent cations in the teichoic acids, the polymeric ribitol phosphates bound to the thick peptidoglycan layer. These systems seem to act as cation exchangers on the surface of bacteria.

Zn^{2+} export systems (1): Cation diffusion facilitators

The highly Zn^{2+}-resistant gram-negative bacterium *Ralstonia metallidurans* (formerly Alcaligenes eutrophus) CH34, isolated from a decantation tank of a zinc factory, has a minimal inhibitory concentration for Zn^{2+} of 12 mM in a Tris-based medium. The *czcNICBADRS* gene cluster found on a large plasmid determines the high metal resistance by encoding two systems for the export of Zn^{2+}, Co^{2+}, and Cd^{2+} (Anton *et al.* 1999; Grosse *et al.* 1999). CzcD is a member of the cation diffusion facilitator family and exports these metals across the cytoplasmic membrane (Anton *et al.* 1999).

The *Staphylococcus aureus* Zn^{2+}-resistance determinant ZntA shows 38% identity to CzcD from *Ralstonia eutropha* (Xiong & Jayaswal 1998). A *zntA* mutant is sensitive to 0.5 mM Zn^{2+}compared to 5 mM for the parent strain. ZntR, a member of the ArsR family, regulates the expression of *zntA*. The nomenclature is very unfortunate since these ZntA and ZntR proteins are not related to the ZntA and ZntR proteins of *E. coli* and other bacteria, which are treated later in this review. Several other members of this protein family characterized in eukaryotes have important functions in the loading of vesicles with Zn^{2+} and in exporting Zn^{2+} from the cytoplasm (Palmiter *et al.* 1996); in bacteria sequence similarities indicate a wider distribution (Paulsen & Saier 1997). One homologue in *E. coli* encoded by *ybgR* is also active in Zn^{2+} transport (Patzer & Hantke, unpublished).

Zn^{2+} export systems (2): RND type exporters

The CzcD protein seems to be the first line of defense of *Ralstonia metallidurans* against high Zn^{2+} concentrations, since in a czcD mutant expression of the *czcCBA* genes is induced and the *czcD* mutant is slightly more sensitive to Zn^{2+} than the parent strain

(Anton *et al.* 1999). The proteins CzcA, CzcB, and CzcC form a sophisticated transport system that exports Zn^{2+}, Co^{2+}, and Cd^{2+} across two membranes, the cytoplasmic membrane and the outer membrane, thereby possibly also protecting the periplasmic space of *R. metallidurans* from the toxic metals. CzcA is a cation-proton antiporter located in the cytoplasmic membrane. CzcB seems to be a connector protein that is distantly related to AcrA (22% amino acid identity), a protein located in the periplasm as part of an acriflavin export pump. CzcC has a distant relationship to TolC and may connect CzcB to the outer membrane, forming a CzcABC protein complex (Rensing *et al.* 1997a). This arrangement allows extrusion of the toxic metals from the cytoplasm into the medium. The whole export system belongs to the widely distributed RND (resistance, nodulation, division) protein family (Tseng *et al.* 1999). The protein CzcA alone without CzcB and CzcC allows only a low level of metal ion resistance.

CzcS and CzcR belong to the large family of the two-component histidine sensor kinase regulators. Studies with *czcR* and *czcS* deletions, LacZ fusions and the analysis of *czcCBA* mRNA synthesis under inducing and non-inducing conditions has shown that CzcR/S regulate the *czc* genes (Grosse *et al.* 1999). The system was induced with 0.3 mM Zn^{2+} and less well with 0.3 mM Cd^{2+}, indicating a preference for Zn^{2+}.

Zn^{2+} export systems (3): P-type ATPases

P-type ATPases, named for the phosphorylated aspartate enzyme intermediate, form a large family of cation-transporting membrane proteins found in eukaryotes and bacteria. Two related subgroups transport Cu^{+} and Ag^{+}, and Zn^{2+}, Cd^{2+}, and Pb^{2+} (reviewed by Gatti *et al.* 2000).

A Cd^{2+}- and Zn^{2+}-resistance determinant encoded by the genes *cadA* and *cadC* has been found on the staphylococcal plasmid pI258. In a *Staphylococcus aureus* strain, the cloned genes raised the minimal inhibitory concentration of Cd^{2+} from 0.005 mM to 2.56 mM, while the minimal inhibitory concentration of Zn^{2+} was raised from 0.6 mM to 1.8 mM (Yoon & Silver 1991). These values illustrate the high toxicity of Cd^{2+} for the unprotected cell.

CadA is a P-type ATPase (Tsai *et al.* 1992), and CadC, important for full resistance, is a regulatory protein with similarities to ArsR proteins, which regulate arsenite/antimonite resistance ATPases (Shi *et al.* 1994; Rosenstein *et al.* 1994). The CadA/C transport system is also found in the gram-negative bacterium *Stenotrophomonas maltophila* and has 96% sequence identity to the staphylococcal CadA/C (Alonso *et al.* 2000). This high level of sequence identity of the two proteins indicates a recent horizontal gene transfer of *cadA* and *cadC* between a gram-positive and a gram-negative bacterium. In *Listeria monocytogenes*, the related CadA and CadC proteins (66% and 48% sequence identity to *S. aureus* CadA and CadC, respectively), are encoded by a transposon on a plasmid and confer only Cd^{2+} resistance and no Zn^{2+} resistance (Lebrun *et al.* 1994). The lack of Zn^{2+} resistance may be explained by the high intrinsic resistance of the *L. monocytogenes* and *B. subtilis* strains used (to 7 and 3.5 mM Zn^{2+}, respectively), which indicates the presence of additional Zn^{2+} exporters encoded on the chromosome. Also in *S. aureus*, the pI258 encoded CadA/C proteins increased the Zn^{2+} resistance level only by a factor of 3 which indicates that protection from Cd^{2+}may be the main function.

The *E. coli* ZntA protein, another member of the P-type ATPases, confers Cd^{2+} and Zn^{2+} resistance (Beard *et al.* 1997; Rensing *et al.* 1997b). The Zn^{2+} content of *E. coli* has been estimated as a nominal concentration of 0.6 mM if it were all free (Kung *et al.* 1976). In mammalian cells this value has been estimated at 0.2 mM (Palmiter & Findley 1995). The concentration of free Zn^{2+} is difficult to determine, the results depend on the methods used. *In vitro* studies suggest that thiolate-bound Cd^{2+} or Zn^{2+} are the best substrates for ZntA (Sharma *et al.* 2000). In the cell, most of these metals might be bound as thiolates of cysteine or glutathione. The main function of CadA- and ZntA-like proteins seems to be the extrusion of Cd^{2+} and Pb^{2+}, whereas Zn^{2+} export seems to be only an additional function.

In *Proteus mirabilis*, a mutation in a ZntA homologue leads to a defective swarming behavior with slower migration than the parent strain (Lai *et al.* 1998; Rensing *et al.* 1998). This unusual phenotype might arise through the disturbed Zn^{2+} homeostasis having an effect on swarm-cell differentiation.

In humans, two homologous P-type ATPases transport copper. Mutations in these proteins cause disorders of copper metabolism known as Wilson and Menkes diseases. To study the effects of Wilson disease mutations in *E. coli*, two mutants with site-directed mutations in ZntA were constructed, His475Gln and Glu470Ala, the human counterparts of which cause Wilson disease. Both mutant proteins show a reduced metal-ion-stimulated ATPase activity (about 30–40% of the wild-type activity) and are phosphorylated much less efficiently than the wild-type proteins. These results suggest that the mutations affect major stages in the transport process of both P-type ATPases (Okkeri & Haltia 1999).

Members of the MerR/ZntR family or of the ArsR/SmtB family regulate Pb^{2+}, Cd^{2+}, and Zn^{2+} extrusion

The regulator of *zntA* in *E. coli*, ZntR, induces *zntA* transcription at 19 μM Cd^{2+} or 100 μM Zn^{2+} (Noll & Lutsenko 2000). Also Pb^{2+} is an efficient inducer of the system (Binet & Poole 2000). ZntR is a MerR-like regulator that binds as a repressor to the *zntA* promoter; in the presence of Cd^{2+} or Zn^{2+}, ZntR is converted into a transcriptional activator that changes the conformation of the promoter region and makes it a better substrate for the RNA polymerase (Outten *et al.* 1999).

MerR is a well-studied regulator of a Hg^{2+} detoxification system. Hg^{2+} is recognized by amino acid residue C82 from one monomer of the MerR dimer and by C117 and C126 from the second monomer of MerR. In a systematic study, mutated MerR was screened for other metal specificities. Mutations in 11 positions changed the responsiveness from Hg^{2+} to Hg^{2+} and Cd^{2+} (Caguiat *et al.* 1999), making the regulator less specific. Interestingly, in these mutants, the response to Zn^{2+} changed only slightly.

Other P-type ATPases like CadA are regulated by members of the ArsR/SmtB superfamily of repressors. The structure of SmtB has been solved. It is a dimeric repressor protein with a typical helix-turn-helix motif, the arrangement of the three core helices and the beta hairpin is similar to the HNF-3/forkhead, CAP and diphtheria toxin repressor proteins (Cook *et al.* 1998). However, the Zn^{2+} binding sites have not been characterized well. Both types of regulators, ArsR/SmtB and MerR/ZntR, are also found in connection with other metal transporters. Each of these regulators regulates very specifically only one system.

Binding-protein-dependent ABC transporters drive high-affinity metal uptake

High-affinity Zn^{2+} uptake systems in gram-negative and gram-positive bacteria have been characterized. One of the first described was in *E. coli* (Patzer & Hantke 1998). During the selection of recombinants with iron-regulated *lacZ* fusions using the transposing phage Mud1, fusions regulated by the availability of Zn^{2+} were obtained. Cells carrying these fusions grow as red colonies on MacConkey agar with Zn^{2+}-complexing chelators, indicating derepression of the *lacZ* fusion, while addition of $ZnCl_2$ leads to repression and growth of white colonies. On complex nutrient agar plates, growth of these mutants is inhibited by 5 mM EGTA or 0.4 mM EDTA. When $ZnCl_2$ is spotted on filter paper discs, a zone of growth around the disc is observed. Other metals, such as Ni^{2+}, Cu^{2+}, Mn^{2+}, and Fe^{2+}, do not stimulate growth. A much smaller zone of growth is observed with Co^{2+}, which might substitute for Zn^{2+} in some proteins and might lower the need for Zn^{2+}.

The Mud1 phage had inserted into one of three genes encoding a binding-protein-dependent ABC transporter. The gene *znuA* (zinc uptake) encodes a periplasmic binding protein, *znuB* encodes an integral membrane protein, and *znuC* encodes the ATPase component of the transporter. To prove the *in vivo* function of these genes, $^{65}Zn^{2+}$ uptake was measured in *znu* mutants and in the parent strain. In a HEPES-buffered medium, the uptake of $^{65}Zn^{2+}$ is the same for the mutants and the parent strain. Only when the cells were pre-grown in the presence of 5 mM EGTA did the mutant unexpectedly take up more $^{65}Zn^{2+}$ than the parent strain; this could be interpreted as the induction of another Zn^{2+} transporter in the mutant. Addition of 0.5 mM EGTA to the transport medium lowers the uptake by a factor of 10 in the induced parent strain, whereas no uptake is observed in *znu* mutants regardless of the growth conditions. Unfortunately the Zn^{2+} concentration in the uptake medium was not exactly defined. According to Zn^{2+} determinations, it was below 150 nM, and the free Zn^{2+} concentration was even lower through the addition of 0.5 mM EGTA (in the presence of 3 mM Mg^{2+}).

Sequence similarity searches revealed that ZnuA belongs to a large family of binding proteins that recognize either Zn^{2+}, Mn^{2+}, or $Fe^{?}$ as their substrate (Figure 2). Binding proteins of ABC transporters had been grouped into eight clusters (Tam & Saier 1993), but since these metal binding proteins had new characteristics, they were defined as the cluster 9 family of binding proteins by Dintilhac *et al.* (1997). These authors observed that competence of *adcABC* mutants of *Streptococcus pneumoniae* is Zn^{2+}-dependent. Mutants with mutations in the *adc*

```
PZP_HAEIN     M--K-----KLLKISAISAALL----------------SAPMMANAD--------  24Zn2+
ZNUA_HAEDU    M----------FKKTVLTLAML----------------GVTTVANAD--------  21Zn2+
ZNUA_ECOLI    M--KCYNITLLIFITIIGRIMLHKKTLLFAALSAALWGGATQAADAA--------  45Zn2+
YCDH_BACSU    MFKKWSG----LFVIAACFLLVAACG--------------NSSTKGSADSKGDKL  37Zn2+
ADCA_STRPN    M--KKIS----LLLASLCALFLVACS--------------N------QKQADGKL  29Zn2+
MTSA_STRPN    MGKRMS----LILGAFLSVFLLVAC----------------SSTGAKT-AESDKL  34Me2+
SCAA_STRGO    M-KKCR----FLVLLLLAFVGLAAC----------------SSQKSSTDSSSSKL  34Mn2+
PSAA_STRPN    M-KKLG----TLLVLFLSAIILVAC----------------ASGKKDT-TSGQKL  33Mn2+?
YFEA_YERPE    MLIKKKSPYLKMIERLNSPFLRAAALFTIVAFSSLIST--AALAENNPSDTAKKF  53Mn2+/Fe?
MNTC_SYNSP    MATSFASRGGLLASGLAIAFWLTGCGTAEVTTSNAPSEEVTAVTTEVQGETEEKK  55Mn2+
MNTA_BACSU    MRQG------LMAAVLFATFALTGC---------------GTDSAGK--SADQQL  32Mn2+
TROA_TREPA    MIRE------RICACVLALGMLTGF---------------THAFGSKDAAADGKP  34Mn2+?
              *          ^

PZP_HAEIN     -VLASVKPLGFIVSSIADGVTGTQVLVPAGASPHDYNLKLSDIQKVKSADLVVWI  78Zn2+
ZNUA_HAEDU    -VLTSIKPLGFIANAITDGVTETKVLLPVTASPHDYSLKPSDIEKLKSAQLVVWV  75Zn2+
ZNUA_ECOLI    -VVASLKPVGFIASAIADGVTETEVLLPDGASEHDYSLRPSDVKRLQNADLVVWV  99Zn2+
YCDH_BACSU    HVVTTFYPMYEFTKQIVKDKGDVDLLIPSSVEPHDWEPTPKDIANIQDADLFVYN  92Zn2+
ADCA_STRPN    NIVTTFYPVYEFTKQVAGDTANVELLIGAGTEPHEYEPSAKAVAKIQDADTFVYE  84Zn2+
MTSA_STRPN    KVVATNSIIGDMTKVMAGDKIDLHSIVPIGQDPHEYEPLPEDVEKTSNADVIFYN  89Me2+
SCAA_STRGO    NVVATNSIIADITKNIAGDKINLHSIVPVGQDPHKYEPLPEDVKKTSKADLIFYN  89Mn2+
PSAA_STRPN    KVVATNSIIADITKNIAGDKIDLHSIVPIGQDPHEYEPLPEDVKKTSEADLIFYN  88Mn2+?
YFEA_YERPE    KVVTTFTIIQDIAQNIAGDVAVVESITKPGAEIHDYQPTPRDIVKAQSADLILWN 108Mn2+/Fe?
MNTC_SYNSP    KVLTTFTVLADMVQNVAGDKLVVESITRIGAEIHGYEPTPSDIVKAQDADLILYN 110Mn2+
MNTA_BACSU    QVTATTSQIADAAENIGGKHVKVTSLMGPGVDPHLYKASQGDTKKLMSADVVLYS  87Mn2+
TROA_TREPA    LVVTTIGMIADAVKNIAQGDVHLKGLMGPGVDPHLYTATAGDVEWLGNADLILYN  89Mn2+?
               ^ ^^   ^   ^  ^         ^   ^ ^ # ^     ^^    ^*^   ^

PZP_HAEIN     GEDIDSFLDKPISQIERKKVITIADLADVKPLLSKAHHEHFHEDGDHDHDHKHEH 133Zn2+
ZNUA_HAEDU    GDGLEAFLEKSIDKLPKEKVLRLEDVPGIKMIV--------------DATKKKDH 116Zn2+
ZNUA_ECOLI    GPEMEAFMQKPVSKLPGAKQVTIAQLEDVKPLLMKSIHG---DDDDHDHAEKSDE 151Zn2+
YCDH_BACSU    SEYMETWVPSAEKSMGQGHAVFVNASKGIDLMEGSEEEHEEHDHGEHEHSHAMDP 147Zn2+
ADCA_STRPN    NENMETWVPKLLDTLDKKKVKTIKATGDMLLLPGGEEEEGDHDHGEEGHHHEFDP 139Zn2+
MTSA_STRPN    GINLEDGGQAWFTKLVKNA----QKTKNKDYFAVSDGIDVIYLEGASEKGKE-DP 139Me2+
SCAA_STRGO    GINLETGGNAWFTKLVENA----QKKENKDYYAVSEGVDVIYLEGQNEKGKE-DP 139Mn2+
PSAA_STRPN    GINLETGGNAWFTKLVENA----KKTENKDYFAVSDGVDVIYLEGQNEKGKE-DP 138Mn2+?
YFEA_YERPE    GMNLER----WFEKFFESI----KDVPSA---VVTAGITPLPIREGPYSGIA-NP 151Mn2+/Fe?
MNTC_SYNSP    GMNLER----WFEQFLGNV----KDVPSV---VLTEGIEPIPIADGPYTDKP-NP 153Mn2+
MNTA_BACSU    GLHLEGKMEDVLQKIGEQK---QSA------AVAEAIPKNKLIPAGEGKTF-DP 131Mn2+
TROA_TREPA    GLHLETKMGEVFSKLRGSR----LVV------AVSETIPVSQRLSLEEAE-F-DP 132Mn2+?
              ^  ^^         ^                                      ^

PZP_HAEIN     KHDHKHDHDHDHDHKHEHKHDHEHHDHDHHEGLTTNWHVWYSPAISKIVAQKVAD 188Zn2+
ZNUA_HAEDU    DH-HDHDHDHDHDHDHEHIHGHHHDK---------DWHIWFSPEASQLAAEQIAE 161Zn2+
ZNUA_ECOLI    DH--------------------------HHGDFNM--HLWLSPEIARATAVAIHG 178Zn2+
YCDH_BACSU    -------------------------------------HVWLSPVLAQKEVKNITA 165Zn2+
ADCA_STRPN    -------------------------------------HVWLSPVRAIKLVEHIRD 157Zn2+
MTSA_STRPN    -------------------------------------HAWLNLENGIIYSKNIAK 157Me2+
SCAA_STRGO    -------------------------------------HAWLNLENGIIYAQNIAK 157Mn2+
PSAA_STRPN    -------------------------------------HAWLNLENGIIFAKNIAK 156Mn2+?
YFEA_YERPE    -------------------------------------HAWMSPSNALIYIENIRK 169Mn2+/Fe?
MNTC_SYNSP    -------------------------------------HAWMSPRNALVYVENIRQ 171Mn2+
MNTA_BACSU    -------------------------------------HVWFSIPLWIYAVDEIEA 149Mn2+
TROA_TREPA    -------------------------------------HVWFDVKLWSYSVKAVYE 150Mn2+?
                                                   # * ^          ^
```

Fig. 2. Sequence comparison of selected metal binding proteins (including the signal sequence, which may be the reason for the low similarity at the N-termini) of cluster 9 ABC permeases. The metal specificity of the binding proteins is not always clear; the existence of conflicting reports is indicated by ?. Identical residues are marked by an *, similar residues by ∧, and the positions involved in metal binding in the two known crystal structures of PsaA and TroA by #. Note the H-, D-, and E-rich domain in the Zn^{2+} binding proteins. Abbreviations of the organisms: HAEIN–*Haemophilus influenzae*; HAEDU–*Haemophilus ducreyi*; BACSU–*Bacillus subtilis*; STRPN–*Streptococcus pneumoniae*; STRGO–*Streptococcus gordonii*; YERPE–*Yersinia pestis*; SYNSP–*Synechocystis* sp.; TREPA–*Treponema pallidum*.

```
PZP_HAEIN    KLTAQFPDKKALIAQNLSDFNRTLAEQSEKITAQLANV--KDKGFYVFHDAYGYF 241Zn2+
ZNUA_HAEDU   RLTAQLPEKKAKIAENLAAFKANLADKSNEITQQLQAV--KDKGYYTFHDAYGYF 214Zn2+
ZNUA_ECOLI   KLVELMPQSRAKLDANLKDFEAQLASTETQVGNELAPL--KGKGYFVFHDAYGYF 231Zn2+
YCDH_BACSU   QIVKQDPDNKEYYEKNSKEYIAKLQDLDKLYRTTAKK--AEKKEFITQHTAFGYL 218Zn2+
ADCA_STRPN   TLSADYPDKKETFEKNAAAYIEKLQSLDKAYAEGLSQ--AKEKSFVTQHAAFNYL 210Zn2+
MTSA_STRPN   QLIAKDPKNKETYEKNLKAYVAKLEKLDKEAKSKFDAIAENKKLIVTSEGCFKYF 212Me2+
SCAA_STRGO   RLIEKDPDNKATYEKNLKAYIEKLTALDKEAKEKFNNIPEEKKMIVTSEGCPKYF 212Mn2+
PSAA_STRPN   QLSAKDPNNKEFYEKNLKEYTDKLDKLDKESKDKFNKIPAEKKLIVTSEGAFKYF 211Mn2+?
YFEA_YERPE   ALVEHDPAHAETYNRNAQAYAEKIKALDAPLRERLSRIPAEQRWLVTSEGAFSYL 224Mn2+/Fe?
MNTC_SYNSP   AFVELDPDNAKYYNANAAVYSEQLKAIDRQLGADLEQVPANQRFLVSCEGAFSYL 226Mn2+
MNTA_BACSU   QFSKAMPQHADAFRKNAKEYKEDLQYLDKWSRKEIAHIPEKSRVLVTAHDAFAYF 204Mn2+
TROA_TREPA   SLCKLLPGKTREFTQRYQAYQQQLDKLDAYVRRKAQSLPAERRVLVTAHDAFGYF 205Mn2+?
              ^   *       ^   ^   ^  ^           ^ ^    #^   *^

PZP_HAEIN    NDAYGLKQTGYFTINPLVAPGAKTLAHIKEEIDEHKVNCLFAEPQFTPKVIESLA 296Zn2+
ZNUA_HAEDU   ERAYGLNSLGSFTINPTIAPGAKTLNAIKENIAAHKAQCLFAEPQFTPKVIDSLS 269Zn2+
ZNUA_ECOLI   EKQFGLTPLGHFTVNPEIQPGAQRLHEIRTQLVEQKATCVFAEPQFRPAVVESVA 286Zn2+
YCDH_BACSU   AKEYGLKQVPIAGLSPDQEPSAASLAKLKTYAKEHNVKVIYFEEIASSKVADTLA 273Zn2+
ADCA_STRPN   ALDYGLKQVAISGLSPDAEPSAARLAELTEYVKKNKIAYIYFEENASQALANTLS 265Zn2+
MTSA_STRPN   SKAYGVPSAYIWEINTEEEGTPDQISSLIEKLKVIKPSALFVESSVDRRPMETV- 266Me2+
SCAA_STRGO   SKAYNVPSAYIWEINTEEEGTPDQIKSLVEKLRKTKVPSLFVESSVDDRPMKTV- 266Mn2+
PSAA_STRPN   SKAYGVPSAYIWEINTEEEGTPEQIKTLVEKLRQTKVPSLFVESSVDDRPMKTV- 265Mn2+?
YFEA_YERPE   AKDYGFKEVYLWPINAEQQGIPQQVRHVIDIIRENKIPVVFSESTISDKPAKQV- 278Mn2+/Fe?
MNTC_SYNSP   ARDYGMEEIYMWPINAEQQFTPKQVQTVIEEVKTNNVPTIFCESTVSDKGQKQV- 280Mn2+
MNTA_BACSU   GNEYGFKVKGLQGLSTDSDYGLRDVQELVDLLTEKQIKAVFVESSVSEKSINAVV 259Mn2+
TROA_TREPA   SRAYGFEVKGLQGVSTASEASAHDMQELAAFIAQRKLPAIFIESSIPHKNVEALR 260Mn2+?
             ^ ^^^      ^^^  ^    ^  ^ ^    ^   ^^ *       ^ ^

PZP_HAEIN    KNTKVNVGQLDPIGD-------KVTLGKNSYATFLQSTADSYMECL-AK------ 337Zn2+
ZNUA_HAEDU   KSTAVKVGQLDPLGA-------KVKLSKTAYPQFLQAIADEFSQCL-TQ------ 310Zn2+
ZNUA_ECOLI   RGTSVRMGTLDPLGT-------NIKLGKTSYSEFLSQLANQYASCLKGD------ 328Zn2+
YCDH_BACSU   SEIGAKTEVLNTLEGL----SKEEQDKGLGYIDIMKQNLDALKDS---------- 314Zn2+
ADCA_STRPN   KEAGVKTDVLNPLESL----TEEDTKAGENYISVMEKNLKALKQTTDQEGPAIEP 316Zn2+
MTSA_STRPN   ----SKDSGIPIYSEIFTDSIAKKGKPGDSYYAMMKWNLD--------------- 302Me2+
SCAA_STRGO   ----SKDTNIPIYAKIFTDSIAEKGEDGDSYYSMMKYNLD--------------- 302Mn2+
PSAA_STRPN   ----SQDTNIPIYAQIFTDSIAEQGKEGDSYYSMMKYNLD--------------- 301Mn2+?
YFEA_YERPE   ----SKETGAQYGGVLYVDSLSGEKGPVPTYISLINMTVD--------------- 314Mn2+/Fe?
MNTC_SYNSP   ----AQATGARFGGNLYVDSLSTEEGPVPTFLDLLEYDAR--------------- 316Mn2+
MNTA_BACSU   EGAKEKGHTVTIGGQLYSDAMGEKGTKEGTYEGMFRHNIN--------------- 299Mn2+
TROA_TREPA   DAVQARGHVVQIGGELFSDAMGDAGTSEGTYVGMVTHNID--------------- 300Mn2+?
                 ^      ^    #         ^^

PZP_HAEIN    ---------------------------   337Zn2+
ZNUA_HAEDU   ---------------------------   310Zn2+
ZNUA_ECOLI   ---------------------------   328Zn2+
YCDH_BACSU   ----------------------LLVKS   319Zn2+
ADCA_STRPN   EKAEDTKTVQNGYFEDAAVKDRTLSDYA  344Zn2+
MTSA_STRPN   -------KISEGL------AK-------  310Me2+
SCAA_STRGO   -------KISEGL------AK-------  310Mn2+
PSAA_STRPN   -------KIAEGL------AK-------  309Mn2+?
YFEA_YERPE   -------TIAKGF------GQ-------  322Mn2+/Fe?
MNTC_SYNSP   -------VITNGLLAGTNAQQ-------  330Mn2+
MNTA_BACSU   -------TITKAL-------K-------  306Mn2+
TROA_TREPA   -------TIVAAL------AR-------  308Mn2+?
```

Fig. 2. Continued.

operon have a lower growth rate and are competence-defective during a stage after the interaction with the competence-stimulating peptide. Competence is restored after addition of Zn^{2+} to the chemically defined medium. The predicted AdcA sequence is similar to that of binding proteins of ABC transporters (22% identity to ZnuA) and has the characteristic N-terminal lipid anchor found in gram-positive bacteria. AdcB is similar to membrane components of ABC transporters (29% identity to ZnuB), and AdcC is an ATP binding protein (36% identity to ZnuC).

The PsaA protein (pneumococcal surface adhesin A) was identified as a second member of the cluster 9 proteins in *S. pneumoniae* (Dintilhac *et al.* 1997). The virulence and competence of *psaA* mutants are Mn^{2+}-dependent. Proteins closely related to PsaA have been detected in various caries-associated streptococci. The proteins FimA (fimbrial adhesin precursor) and ScaA (co-aggregation-mediating adhesin precursor) had been implicated in adhesion to fibrin monolayers, adhesion to saliva-coated hydroxyapatite, or aggregation with a *Streptomyces* strain, since mutants with mutations in the respective genes show a reduced adhesion and the proteins generated protective antibodies. A double function of these proteins as metal binding proteins and adhesins seems unlikely. Mutants with mutations in these transporters probably lose their fitness necessary for virulence and adhesion. Unfortunately, many homologues of these binding proteins are annotated in the databases as adhesins, which might be incorrect.

A crystal structure of the Mn^{2+}-specific binding protein PsaA from *S. pneumoniae* is available (Lawrence *et al.* 1998). The four amino acids H67, H139, E205, and D280 bind the metal. It has been noted by Lawrence *et al.* (1998) that in this structure, this site is occupied by Zn^{2+} in a tetrahedral coordination. Mn^{2+} is usually found in an octahedral geometry, which is difficult to reconcile with PsaA being a Mn^{2+} binding protein.

TroA is a metal binding protein in *Treponema pallidum* and also belongs to the cluster 9 family of ABC metal binding proteins. In the crystal structure, the four amino acids H68, H133, H199, and D279 bind Zn^{2+} (Lee *et al.* 1999). In Figure 2, it is obvious that these four amino acids are positioned in the places equivalent to those of the amino acids involved in binding in the PsaA structure. Lee *et al.* (1999) argue that the difference — E205 in PsaA and H199 in TroA — might be responsible for the substrate specificity of TroA for Zn^{2+} and PsaA for Mn^{2+} (Lee *et al.* 1999; Deka *et al.* 1999). However, the *troABC* genes encoding the metal ABC transporter are regulated by the DtxR homologue TroR, which only accepts Mn^{2+} and not Zn^{2+} (Posey *et al.* 1999), and the high similarity to binding proteins with Mn^{2+} specificity argue for a Mn^{2+} specificity of this transport system (Figure 2). Further research is needed to clarify this point.

A recent report on the closely related protein MtsA from *Streptococcus pyogenes* further challenges the situation. The binding protein MtsA shows more than 70% identity to PsaA, ScaA, and FimA, which have been shown to be manganese transporters. In contrast, recombinant MtsA does not show a specific interaction with Mn^{2+}, but does with Cu^{2+}, Zn^{2+}, and Fe^{3+} *in vitro*, and an *mtsA* mutant has a 50% lower iron content and a 30% lower Zn^{2+} content than the parent strain (Janulczyk *et al.* 1999). From these results, it has been postulated that the mtsABC genes encode a binding-protein-dependent transport system with a very broad metal specificity for Cu^{2+}, Fe^{3+}, and Zn^{2+}. Transport experiments with the mtsA mutant and the parent strain should verify the postulates.

The Yfe transport system (Figure 2) transports iron and manganese and also possesses an unusually low specificity (Bearden & Perry 1999). The cluster 9 family of binding-protein-dependent transport systems and the controversy surrounding their metal specificity has been extensively discussed by Claverys (2001).

Related Zn^{2+} binding-protein-dependent uptake systems in other bacteria have also been characterized. In *B. subtilis*, the *ycdHlyceA* genes are regulated by Zn^{2+} and the repressor Zur. Two lines of evidence suggest that the *ycdH* operon encodes a high-affinity Zn^{2+} transporter. First, a *ycdH* mutant is impaired in growth in low- Zn^{2+} medium. Second, mutation of *ycdH* alters the regulation of both the *yciC* (a gene regulated by Zn^{2+} and the repressor Zur) and *ycdH* operons such that much higher levels of exogenous Zn^{2+} are required for repression (Gaballa & Helmann 1998). In addition, the encoded proteins belong to the cluster 9 family of binding-protein-dependent ABC metal transporters.

The Zn^{2+} binding protein Pzp1 of *Hemophilus influenzae* was isolated because it is a homologue of the streptococcal FimA protein. Pzp1 was expected to be an adhesin (see above), but later studies indicated a periplasmic location. The recombinant protein purified from *E. coli* contains two Zn^{2+} ions per monomer. A pzp1 mutant was shown to be defective in growth. Suppression was achieved by addition of 100 μM Zn^{2+}, while no suppression was obtained with 100 μM Ca^{2+}, Mg^{2+}, Cu^{2+}, Ni^{2+}, Cd^{2+}, Mn^{2+}, or Fe^{3+} (Lu *et al.* 1997). Co^{2+} was not tested. A *znuA* mutant of *Haemophilus ducreyi* shows a strongly decreased virulence (Lewis *et al.* 1999).

The Pzp1 protein contains an extended region rich in histidine, aspartate, and glutamate residues, HHEHFHEDGDHDHDHK HEHKHDHKHD HDHDHDHKHE HKHDHEHHDH DHHEGLTTNW HVW, which may have a function in binding of Zn^{2+} or delivering Zn^{2+} to other proteins. It is interesting to note that only unambiguous Zn^{2+} binding proteins contain this H-, D-, and E-rich structure in front of the conserved HXW motif (position 171/173 of Pzp1 in Figure 2). H171 is in a Zn^{2+} binding position in the crystal structures of PsaA and TroA. Structural studies of Pzp1 in comparison to other ZnuA proteins with much smaller histidine-rich regions would be interesting.

Zur is a regulator of several high-affinity Zn^{2+} uptake systems

In order to identify the Zn^{2+}-dependent regulator of the znu genes in *E. coli*, constitutive mutants were isolated and tested for complementation by a gene bank of *E. coli*. A complementing gene, *yjbK* of the *E. coli* genome, was identified and named *zur* (for zinc uptake regulation). The Zur protein shows 27% sequence identity with the iron regulator Fur. High-affinity $^{65}Zn^{2+}$ transport of the constitutive zur mutant is tenfold higher than that of the uninduced parental strain (Patzer & Hantke 1998).

By inactivation of the *zur* gene, it has been demonstrated that Zur acts as a repressor and not as an activator. Some chromosomal mutant *zur* alleles have been sequenced to correlate the loss of Zur function with individual mutations. Wild-type and mutant Zur proteins have been purified to electrophoretic homogeneity. A considerable portion of the Zur mutant proteins, except ZurΔ46–91, accumulate in inclusion bodies. Wild-type Zur and ZurΔ46–91 form homo- and heterodimers. Dimerization is independent of metal ions since it also occurs in the presence of metal chelators (Patzer & Hantke 2000). Using an *in vivo* titration assay, the site affording Zur regulation was narrowed down to a 31-bp region in the promoter region of *znuA* and *znuCB*. This location was confirmed by DNase I footprinting assays. A single region comprising a nearly perfect palindrome is protected, which indicates direct binding of Zur. Zinc chelators completely inhibit DNA binding of Zur, and addition of Zn^{2+} in low concentrations enhances binding. Zur occupies its binding site only in the presence of Zn^{2+} or other divalent metal cations at low concentrations, as shown by DNase I footprinting. Zur protects a 29-nt approximate palindrome on each strand of the znu operator with a 3′ stagger of 4 nt (Patzer & Hantke 2000). This footprint resembles that of typical DNA binding dimers, such as classical helix-turn-helix proteins, e.g., the CI repressor from bacteriophage λ (Jordan & Pabo 1988). The observed 3′ stagger is indicative for coverage of the minor groove at the ends, but provides no information about the protein–DNA recognition contacts. Analysis of the mutant Zur proteins suggested an amino-terminal DNA contact domain around residue 65 and a carboxy-terminal dimerization and Zn^{2+}-binding domain. Footprinting experiments have indicated that although most of the mutant Zur proteins bind to the *znu* promoter *in vitro*, no protection is observed *in vivo* (Patzer & Hantke 2000).

Zur is active only in the reduced form. As a cytoplasmic protein, it has predominantly reduced thiols rather than oxidized disulfides due to the reducing conditions in the cytoplasm (Gilbert 1990). *In vitro*, the cysteine residues of Zur are easily oxidized to disulfides, as judged by the slower migration in SDS-polyacrylamide gels under reducing conditions compared to non-reducing conditions. Oxidized Zur does not bind DNA or considerable amounts of Zn^{2+}. This might indicate that Zn^{2+} is mainly bound by some of the nine cysteines found in Zur. In *E. coli* Fur, there are only four cysteines, all of which are conserved in Zur. At least two to three Zn^{2+} ions per dimer bind specifically to Zur (Patzer & Hantke 2000). It remains to be seen whether one of these Zn^{2+} ions is bound to the same site in Zur as was found in Fur at the conserved cysteines (positions 92 and 95; Jacquamet *et al.* 1998). These two cysteines are found in all Fur-like proteins except those from *Pseudomonas* and related organisms. The repressing activity of the Zur protein is Zn^{2+} specific since addition of Cd^{2+}, Hg^{2+}, Pb^{2+}, Mn^{2+}, Fe^{2+}, and Cu^{2+} led to a derepression of the Znu transport system *in vivo* (Patzer & Hantke 2000).

Zur, like Fur, seems to be widespread among bacteria, even in gram-positive bacteria and cyanobacteria, as indicated by sequence similarity searches. However, only in some cases has a functional Zn^{2+}-dependent regulator been demonstrated. Since small changes in the sequence might change the metal specificity, predictions always have to be examined. In *B. subtilis*, apart from Fur and PerR (regulator of the peroxide stress response) (Bsat *et al.* 1998), the third Fur-like homologue YqfV may act as Zur (Gaballa & Helmann 1998). The sequence similarity to *E. coli* Zur is not strong. A *zur* gene has also been found in *L. monocytogenes* (Dalet *et al.* 1999). Based on sequence similarity, the proposed Fur protein in *Staphylococcus epidermidis* (Heidrich *et al.* 1996) may be Zur rather than Fur. It is possible that other proteins designated as Fur homologues will turn out to be Zur proteins. In many of the partially sequenced genomes of various bacterial species, a Zur equivalent is found, e.g., in *Salmonella* strains, *Klebsiella pneumoniae, Yersinia pestis, Vibrio cholerae, Bordetella pertussis, Caulobacter crescentus, Pseudomonas aeruginosa*, and *Neisseria* strains. Since these organisms also possess a system homologous to Znu, it has been proposed that these proteins likewise are regulators of Zn^{2+} uptake. An alignment reveals two groups of Zur proteins — one is found in gram-negative bacteria, the second in gram-positive bacteria. Zur from gram-positive bacteria is more similar than Zur from gram-negative bacteria to the *E. coli* Fur (Patzer & Hantke 2000). No evidence for autoregulation of Zur or for the influence of other regulators on Zur has been found. Zur has the very limited function of regulating Zn^{2+} uptake and metabolism in an environment poor in Zn^{2+}. To date, besides the *znuA* promoter, only two other Zur binding sites have been identified on the *E. coli* chromosome. No similarities to genes with known function have been found.

Low-affinity Zn^{2+} uptake systems

Little is known about low-affinity uptake systems for Zn^{2+} in *E. coli* or other bacteria. These systems are active at rich non-toxic concentrations of Zn^{2+} when *znuABC* is repressed by Zur. Cotransport of metals (van Veen *et al.* 1994) and Zn^{2+}(Beard *et al.* 2000) with phosphate via the Pit inorganic phosphate transport system has been observed. This system seems to be mainly responsible for the uptake of Zn^{2+}, but possibly also under certain conditions the efflux by metal exchange is catalyzed.

In *B. subtilis*, two systems have been found to be regulated by Zur and Zn^{2+}, the above-mentioned ABC transporter YcdHI, YceA, and the membrane protein YciC, which was assumed to be part of a low-affinity transport system for Zn^{2+} (Gaballa & Helmann 1998). YciC has similarity to nitrile hydratase activating proteins and may have a function in Zn^{2+} homeostasis and not in transport, but has not been studied further.

In *B. subtilis*, the metal–citrate uptake protein CitM has been shown to have a broad specificity for cotransport of Mg^{2+}, Ni^{2+}, Mn^{2+}, Co^{2+}, and Zn^{2+}; however, Ca^{2+}, Ba^{2+}, and Sr^{2+} are recognized by the different but homologous transporter CitH (Krom *et al.* 2000). To which extent CitM contributes to the Zn^{2+} supply in the presence of citrate is not known.

ZraG and ZraH regulate the Zn^{2+} binding protein ZraP

In *E. coli* the periplasmic Zn^{2+} binding protein ZraP (former designation YjaI) had been postulated to be induced by the ZntR homologue PmtR from *Proteus mirabilis* (Noll *et al.* 1998). *E. coli* transformed with a PmtR-encoding plasmid grows better in the presence of 1 mM Zn^{2+} than the same strain transformed with only the vector. In addition, more ZraP is produced by the *pmtR*-containing *E. coli* strain. Recent studies by Leonhartsberger *et al.* (2001) indicate that the observed effects may be indirect. Their work has revealed that *zraP* is induced by the autoregulated two-component histidine kinase regulatory system ZraS/R encoded at a distance of 237 bp from *zraP*. ZraS, with the sequence signature of a sensor kinase, is found in the membrane fraction, and specific binding of the activator ZraR to the promoter region of *zraP* and *zraSR* has been demonstrated. Besides being a periplasmic Zn^{2+} binding protein, the function of ZraP is unknown. It might be a Zn^{2+} binding protein that protects periplasmic enzymes from high Zn^{2+} concentrations or it might be a sensor for the Zn^{2+} concentration in the periplasm which is reported to ZraS. ZraS then would act as a transmitter of the signal, and ZraR could induce an unknown protein in addition to ZraP and ZraS/R. The question remains whether other genes are regulated by ZraS/R.

Is EDDS a zincophore?

[S,S]Ethylenediamine disuccinic acid (EDDS) is produced by *Amycolatopsis orientalis* under Zn^{2+}-limiting conditions (Zwicker *et al.* 1998) and complexes Zn^{2+} and other divalent cations. EDDS is an isomer of the well-known chelator EDTA and is structurally related to certain siderophores (iron chelators). Siderophores are produced by bacteria and fungi under iron-limiting conditions. The siderophores bind Fe^{3+} specifically; these complexes are taken up by specific transport systems found in nearly every type of bacterium (Braun *et al.* 1998). Similarly, it is possible that *A. orientalis* uses EDDS as a zincophore to satisfy its Zn^{2+} demands. Experimental proof for this hypothesis is lacking. *E. coli* is unable to utilize Zn^{2+} bound to EDDS (Patzer & Hantke 1998).

Conclusions

Zinc transporters have been found in the following protein families: RND multi-drug efflux transporters, P-type ATPases, cation diffusion facilitators for export of the toxic metal Zn^{2+}, binding-protein-dependent ABC transporter, and phosphate or citrate metal co-transporters for the uptake of Zn^{2+} necessary for growth. It is remarkable that nearly every system is regulated by its own regulator. In *E. coli*, three regulators — ZntR, ZraS/R, and Zur — are known to respond to Zn^{2+}. This is clearly different than the iron regulation where Fur (or DtxR in certain gram-positive bacteria) acts as a global regulator of many genes related to iron uptake and the oxidative stress response (Hantke & Braun 2000). Zn^{2+}-dependent regulators of export systems derepress or activate above 0.1 mM Zn^{2+}, while the Zur-like proteins derepress the uptake system below 10 μM Zn^{2+}.

It is astonishing that the reports on the metal specificity of different binding-protein-dependent ABC transporters found in pathogenic bacteria are so contradictory. In the crystal structures of PsaA and TroA, Zn^{2+} is found in the metal binding site, although *in vivo* observations and regulation indicate that Mn^{2+} is the main substrate (Claverys 2001). It is possible that these transporters are relatively unspecific because the host regulates its Me^{2+} concentrations within certain limits. This constant environment may allow an 'uncontrolled' uptake of different divalent cations by the pathogen. Since these systems have an important impact on the virulence of pathogens, this field will remain a subject of research and the results are eagerly awaited.

Acknowledgements

I thank Volkmar Braun (this institute) for critically reading the manuscript and for discussions, and Karen A. Brune for editing the manuscript. The research of the author is supported by the Deutsche Forschungsgemeinschaft (HA 1186/3-1) and by the Fonds der Chemischen Industrie.

References

Alonso A, Sanchez P, Martinez JL. 2000 *Stenotrophomonas maltophila* D457R contains a cluster of genes from gram-positive bacteria involved in antibiotic and heavy metal resistance. *Antimicrob Agents Chemother* **44**, 1778–1782.

Anton A, Grosse C, Reissmann J, Pribyl T, Nies DH. 1999 CzcD is a heavy metal ion transporter involved in regulation of heavy metal resistance in *Ralstonia* sp. strain CH34. *J Bacteriol* **181**, 6876–6881.

Beard SJ, Hashim R, Wu G, Binet MR, Hughes MN, Poole RK. 2000 Evidence for the transport of zinc(II) ions via the *pit* inorganic phosphate transport system in *Escherichia coli*. *FEMS Microbiol Lett* **184**, 231–235.

Bearden SW, Perry RD. 1999 The Yfe system of *Yersinia pestis* transports iron and manganese and is required for full virulence of plague. *Mol Microbiol* **32**, 403–414.

Binet MR, Poole RK. 2000 Cd(II), Pb(II) and Zn(II) ions regulate expression of the metal-transporting P-type ATPase ZntA in *Escherichia coli*. *FEBS Lett* **473**, 67–70.

Braun V, Hantke K, Köster W. 1998 Bacterial iron transport: mechanisms, genetics, and regulation. *Met Ions Biol Syst* **35**, 67–145.

Bsat N, Herbig A, Casillas-Martinez L, Setlow P, Helmann JD. 1998 *Bacillus subtilis* contains multiple Fur homologues: identification of the iron uptake (Fur) and peroxide regulon (PerR) repressors. *Mol Microbiol* **29**, 189–198.

Bucheder F, Broda E. 1974 Energy-dependent zinc transport by *Escherichia coli*. *Eur J Biochem* **45**, 555–559.

Caguiat JJ, Watson AL, Summers AO. 1999 Cd(II)-responsive and constitutive mutants implicate a novel domain in MerR. *J Bacteriol* **181**, 3462–3471.

Clarke ND, Berg JM. 1998 Zinc fingers in *Caenorhabditis elegans*: finding families and probing pathways. *Science* **282**, 2018–2022.

Claverys J-P. 2001 A new family of high affinity ABC manganese and zinc permeases. *Res Microbiol* **152**, 231–243.

Cook WJ, Kar SR, Taylor KB, Hall LM. 1998 Crystal structure of the cyanobacterial metallothionein repressor SmtB: a model for metalloregulatory proteins. *J Mol Biol* **275**, 337–346.

Dalet K, Gouin E, Cenatiempo Y, Cossart P, Hechard Y. 1999 Characterisation of a new operon encoding a Zur-like protein and an associated ABC zinc permease in *Listeria monocytogenes*. *FEMS Microbiol Lett* **174**, 111–116.

Deka RK, Lee YH, Hagman KE, Shevchenko D, Lingwood CA, Hasemann CA, Norgard MV, Radolf JD. 1999 Physicochemical evidence that *Treponema pallidum* TroA is a zinc-containing metalloprotein that lacks porin-like structure. *J Bacteriol* **181**, 4420–4423.

Dinthilhac A, Alloing G, Granadel C, Claverys J-P. 1997 Competence and virulence of *Streptococcus pneumoniae*: AdcA and PsaA mutants exhibit a requirement for Zn and Mn resulting from inactivation of putative ABC metal permeases. *Mol Microbiol* **25**, 727–739.

Gaballa A, Helmann JD. 1998 Identification of a zinc-specific metalloregulatory protein, Zur, controlling zinc transport operons in *Bacillus subtilis*. *J Bacteriol* **180**, 5815–5821.

Gatti D, Mitra B, Rosen BP. 2000 *Escherichia coli* soft metal ion-translocating ATPases. *J Biol Chem* **275**, 34009–34012.

Gilbert HF. 1990 Molecular and cellular aspects of thiol-disulfide exchange. *Adv Enzymol Relat Areas Mol Biol* **63**, 69–172.

Grosse C, Grass G, Anton A, Franke S, Santos AN, Lawley B, Brown NL, Nies DH. 1999 Transcriptional organization of the *czc* heavy-metal homeostasis determinant from *Alcaligenes eutrophus*. *J Bacteriol* **181**, 2385–2393.

Hantke K, Braun V. 2000 The art of keeping low and high iron concentrations in balance. In: Storz G, Hengge-Aronis R. eds. *Bacterial stress responses*. Washington, D.C.: ASM Press: 275–288.

Heidrich C, Hantke K, Bierbaum G, Sahl HG. 1996 Identification and analysis of a gene encoding a Fur-like protein of *Staphylococcus epidermidis*. *FEMS Microbiol Lett* **140**, 253–259.

Jacquamet L, Aberdam D, Adrait A, Hazemann JL, Latour JM, Michaud-Soret I. 1998 X-ray absorption spectroscopy of a new zinc site in the Fur protein from *Escherichia coli*. *Biochemistry* **37**, 2564–2571.

Janulczyk R, Pallon J, Bjorck L. 1999 Identification and characterization of a *Streptococcus pyogenes* ABC transporter with multiple specificity for metal cations. *Mol Microbiol* **34**, 596–606.

Jordan SR, Pabo CO. 1988 Structure of the lambda complex at 2.5 Å resolution: details of the repressor-operator interactions. *Science* **242**, 893–899.

Krom BP, Warner JB, Konings WN, Lolkema JS. 2000 Complementary metal ion specificity of the metal-citrate transporters CitM and CitH of *Bacillus subtilis*. *J Bacteriol* **182**, 6374–6381.

Lai HC, Gygi D, Fraser GM, Hughes C. 1998 A swarming-defective mutant of *Proteus mirabilis* lacking a putative cation-transporting membrane P-type ATPase. *Microbiology* **144**, 1957–1961.

Lawrence MC, Pilling PA, Epa VC, Berry AM, Ogunniyi AD, Paton JC. 1998 The crystal structure of pneumococcal surface antigen PsaA reveals a metal-binding site and a novel structure for a putative ABC-type binding protein. *Structure* **6**, 1553–1561.

Lebrun M, Audurier A, Cossart P. 1994 Plasmid-borne cadmium resistance genes in *Listeria monocytogenes* are similar to *cadA* and *cadC* of *Staphylococcus aureus* and are induced by cadmium. *J Bacteriol* **176**, 3040–3048.

Lee YH, Deka RK, Norgard MV, Radolf JD, Hasemann CA. 1999 *Treponema pallidum* TroA is a periplasmic zinc-binding protein with a helical backbone. *Nat Struct Biol* **6**, 628–633.

Leonhartsberger S, Huber A, Lottspeich F, Böck A. 2001 The *hydH/G* genes from *Escherichia coli* code for a zinc and lead responsive two-component regulatory system. *J Mol Biol* **307**, 93–105.

Lewis DA, Klesney-Tait J, Lumbley SR, Ward CK, Latimer JL, Ison CA, Hansen EJ. 1999 Identification of the *znuA*-encoded periplasmic zinc transport protein of *Haemophilus ducreyi*. *Infect Immun* **67**, 5060–5068.

Lu D, Boyd B, Lingwood CA. 1997 Identification of the key protein for zinc uptake in *Haemophilus influenzae*. *J Biol Chem* **272**, 29033–29038.

Noll M, Lutsenko S. 2000 Expression of ZntA, a zinc-transporting P1-type ATPase, is specifically regulated by zinc and cadmium. *IUBMB Life* **49**, 297–302.

Okkeri J, Haltia T. 1999 Expression and mutagenesis of ZntA, a zinc-transporting P-type ATPase from *Escherichia coli*. *Biochemistry* **38**, 14109–14116.

Outten CE, Outten FW, O'Halloran TV. 1999 DNA distortion mechanism for transcriptional activation by ZntR, a Zn(II)-responsive MerR homologue in *Escherichia coli*. *J Biol Chem* **274**, 37517–37524.

Palmiter RD, Findley SD. 1995 Cloning and functional characterization of a mammalian zinc transporter that confers resistance to zinc. *EMBO J* **14**, 639–649.

Palmiter RD, Cole TB, Findley SD. 1996 ZnT-2, a mammalian protein that confers resistance to zinc by facilitating vesicular sequestration. *EMBO J* **15**, 1784–1791.

Patzer SI, Hantke K. 1998 The ZnuABC high-affinity zinc uptake system and its regulator Zur in *Escherichia coli*. *Mol Microbiol* **28**, 1199–1210.

Patzer SI, Hantke K. 2000 The zinc-responsive regulator Zur and its control of the *znu* gene cluster encoding the ZnuABC zinc uptake system in *Escherichia coli*. *J Biol Chem* **275**, 24321–24332.

Paulsen IT, Saier MHJ. 1997 A novel family of ubiquitous heavy metal ion transport proteins. *J Membr Biol* **156**, 99–103.

Posey JE, Hardham JM, Norris SJ, Gherardini FC. 1999 Characterization of a manganese-dependent regulatory protein, TroR, from *Treponema pallidum*. *Proc Natl Acad Sci USA* **96**, 10887–10892.

Rensing C, Pribyl T, Nies DH. 1997a New functions for the three subunits of the CzcCBA cation-proton antiporter. *J Bacteriol* **179**, 6871–6879.

Rensing C, Mitra B, Rosen BP. 1997b The zntA gene of *Escherichia coli* encodes a Zn(II)-translocating P-type ATPase. *Proc Natl Acad Sci USA* **94**, 14326–14331.

Rensing C, Mitra B, Rosen BP. 1998 A Zn(II)-translocating P-type ATPase from *Proteus mirabilis*. *Biochem Cell Biol* **76**, 787–790.

Robinson NJ, Bird AJ, Turner JS. 1998 Metallothionein gene regulation in cyanobacteria. In: Silver S, Walden W. eds. *Metal ions in gene regulation*. New York: ITP: 372–397.

Rosenstein R, Nikoleit K, Götz F. 1994 Binding of ArsR, the repressor of the *Staphylococcus xylosus* (pSX267) arsenic resistance operon to a sequence with dyad symmetry within the *ars* promoter. *Mol Gen Genet* **242**, 566–572.

Sharma R, Rensing C, Rosen BP, Mitra B. 2000 The ATP hydrolytic activity of purified ZntA, a Pb(II)/Cd(II)/Zn(II)-translocating ATPase from *Escherichia coli*. *J Biol Chem* **275**, 3873–3878.

Shi W, Wu J, Rosen BP. 1994 Identification of a putative metal binding site in a new family of metalloregulatory proteins. *J Biol Chem* **269**, 19826–19829.

Tam R, Saier Jr. MH. 1993 Structural, functional, evolutionary relationships among extracellular solute-binding receptors of bacteria. *Microbiol Rev* **57**, 320–346.

Tsai KJ, Yoon KP, Lynn AR. 1992 ATP-dependent cadmium transport by the *cadA* cadmium resistance determinant in everted membrane vesicles of *Bacillus subtilis*. *J Bacteriol* **174**, 116–121.

Tseng TT, Gratwick KS, Kollman J, Park D, Nies DH, Goffeau A, Saier MHJ. 1999 The RND permease superfamily: an ancient, ubiquitous and diverse family that includes human disease and development proteins. *J Mol Microbiol Biotechnol* **1**, 107–125.

van Veen HW, Abee T, Kortstee GJ, Konings WN, Zehnder AJ. 1994 Translocation of metal phosphate via the phosphate inorganic transport system of *Escherichia coli*. *Biochemistry* **33**, 1766–1770.

Xiong A, Jayaswal RK. 1998 Molecular characterization of a chromosomal determinant conferring resistance to zinc and cobalt ions in *Staphylococcus aureus*. *J Bacteriol* **180**, 4024–4029.

Yoon KP, Silver S. 1991 A second gene in the *Staphylococcus aureus cadA* cadmium resistance determinant of plasmid pI258. *J Bacteriol* **173**, 7636–7642.

Zwicker N, Theobald U, Zähner H, Fiedler H-P. 1997 Optimization of fermentation conditions for the production of ethylene-diamine-disuccinic acid by *Amycolatopsis orientalis*. *J Ind Microbiol Biotechnol* **19**, 280–285.

BioMetals 14: 251–270, 2001.

Review

Eukaryotic zinc transporters and their regulation

L. Alex Gaither & David J. Eide*
*Department of Nutritional Sciences, University of Missouri-Columbia, Columbia, MO 65211, USA; *Author for correspondence (Tel: 573-882-9686; Fax: 573-882-0185; E-mail: eided@missouri.edu)*

Received 2 January 2001; accepted 30 April 2001

Key words: CDF, efflux, regulation, storage, transport, uptake, zinc, ZIP

Abstract

The last ten years have witnessed major advances in our understanding of zinc transporters and their regulation in eukaryotic organisms. Two families of transporters, the ZIP (*Z*rt-, *I*rt-like *P*rotein) and CDF (*C*ation *D*iffusion *F*acilitator) families, have been found to play a number of important roles in zinc transport. These are ancient gene families that span all phylogenetic levels. The characterized members of each group have been implicated in the transport of metal ions, frequently zinc, across lipid bilayer membranes. This remarkable conservation of function suggests that other, as yet uncharacterized members of the family, will also be involved in metal ion transport. Many of the ZIP family transporters are involved in cellular zinc uptake and at least one member, the Zrt3 transporter of *S. cerevisiae*, transports stored zinc out of an intracellular compartment during adaptation to zinc deficiency. In contrast, CDF family members mediate zinc efflux out of cells or facilitate zinc transport into intracellular compartments for detoxification and/or storage. The activity of many of these transporters is regulated in response to zinc through transcriptional and post-transcriptional mechanisms to maintain zinc homeostasis at both the cellular and organismal levels.

Introduction

Zinc is an essential nutrient for all organisms on earth. This metal ion plays critical roles in a wide variety of biochemical processes. For example, zinc is a cofactor required for the function of over 300 different enzymes including representatives from all six major functional enzyme classes (Vallee & Auld 1990). Zinc is also an important structural cofactor for many proteins including the ubiquitous zinc finger DNA binding proteins (Rhodes & Klug 1993). Because of these essential roles, organisms must maintain adequate intracellular zinc concentrations to support cell growth even when extracellular or dietary levels are low. To accomplish this feat, cells have evolved with efficient uptake systems to allow accumulation of zinc even when it is scarce. These uptake systems use integral membrane transport proteins to move zinc across the lipid bilayer of the plasma membrane.

Once inside a eukaryotic cell, a portion of the zinc must be transported into intracellular organelles to serve as a cofactor for various zinc-dependent enzymes and processes present in those compartments. For example, the mitochondrial isozyme of alcohol dehydrogenase is a zinc-dependent enzyme (Sytkowski 1977). Therefore, transporter proteins must be present in organelle membranes to facilitate this flux of zinc. Zinc can also be stored in certain intracellular compartments when supplies are high and used later if zinc deficiency ensues. Again, zinc transporters are required to facilitate this transport in and out of organelles.

Although zinc is essential, excess zinc can be toxic to cells (for example, see Koh *et al.* 1996). The mechanism of zinc toxicity is not known but the metal may bind to inappropriate intracellular ligands or compete with other metal ions for enzyme active sites, transporter proteins, etc. Therefore, while maintaining adequate levels of zinc for growth, cells must also control intracellular levels when exposed to excessive zinc concentrations. Several mechanisms exist to detoxify excess zinc including the binding of the metal to cytoplasmic macromolecules. Metallothionein proteins may play such a detoxification role (Hamer 1986). Zinc transporters can also aid in detoxification by facilitating intracellular sequestration within organelles, or efflux of zinc across the plasma membrane. Finally, in multicellular organisms, cellular zinc efflux systems are required for the distribution of dietary zinc to other tissues. For example, in the enterocyte of the mammalian intestine, zinc transporters must take up zinc from the intestinal lumen and then efflux that zinc across the basal lateral membrane into the portal blood.

From this discussion, it should be clear that zinc transporters play a variety of essential roles in zinc metabolism. This review considers our current knowledge of zinc transporters in eukaryotic organisms. Because the activity of these transporters, and hence zinc homeostasis, is often controlled by zinc status, the mechanisms of regulation will also be described. Zinc transporter activity is known to be regulated at both transcriptional and post-translational levels.

Families of zinc transporters in eukaryotes

Many types of transporters have been implicated in zinc transport processes. In bacteria, transporters belonging to the ABC family have been shown to function in zinc uptake. For example, the ZnuABC proteins of *E. coli* are a major source of zinc accumulation for these cells (Patzer & Hantke 1998). Zinc efflux transporter proteins belonging to the family of P-type ATPases have also been identified. The ZntA transporter of *E. coli* is one such protein. ZntA plays an important role in zinc detoxification by pumping the metal ion out of the cell when intracellular zinc levels get too high (Rensing *et al.* 1997).

In eukaryotes, neither ABC transporters nor P-type ATPases have yet been found to play physiological roles in zinc transport. Rather, two other families of transporters have been implicated in this process. The ZIP family plays prominent roles in zinc uptake, transporting zinc from outside the cell into the cytoplasm. ZIP transporters have also been found to mobilize stored zinc by transporting the metal from within an intracellular compartment into the cytoplasm. A second group of transporters, the CDF family, transports zinc in the direction opposite to that of the ZIP proteins, promoting zinc efflux or compartmentalization by pumping zinc from the cytoplasm out of the cell or into the lumen of an organelle. The next two sections will provide an overview of these two families of transporters. Specific examples of the roles they play in eukaryotic zinc transport and their regulation will be considered in later sections of this review. A key point to be made is that all of the characterized members of these families play roles in metal ion transport with zinc often being the substrate. This observation suggests that many other members function in zinc transport. The roles of the many characterized proteins implicated in zinc transport have been assessed based on mutant phenotypes (e.g., sensitivity to zinc excess or deficiency), gene or protein regulation in response to zinc, and direct assays of zinc transport using $^{65}Zn^{2+}$ as a tracer.

The ZIP family of metal ion transporters

The ZIP transporters, so-called for being 'Zrt-, *I*rt-like *P*roteins' are named after two of the first members of the family to be identified, Zrt1 of *Saccharomyces cerevisiae* and Irt1 of *Arabidopsis thaliana* (Eide *et al.* 1996; Zhao & Eide 1996a). Zrt1 is a zinc uptake transporter in yeast and Irt1 is an iron transporter in plants. At the time of this writing, 86 ZIP members are found in the non-redundant protein sequence database at the National Center for Biotechnology Information (NCBI) (Figure 1). This list includes proteins from eubacteria, archeae, fungi, protozoa, insects, plants, and mammals. Recent reviews described the ZIP family as containing only 20–30 members (Eng *et al.* 1998; Guerinot 2000). The 3-fold increase in the number of member genes reported here is largely due to our use of more sensitive database analysis tools. Previous comparisons were performed using BLAST (Basic Local Alignment Search Tool) (Altschul *et al.* 1990) while our analysis used PSI-BLAST (Position-Specific Iterated BLAST), a more sensitive method of database comparisons that relies on a position-specific scoring matrix (Altschul *et al.* 1997). While previous BLAST analyses suggested that the family was limited to eukaryotes, members of the ZIP family have now been found in archeae and eubacteria as well. Thus, contrary to our previous view (Eng *et al.* 1998), we now conclude that the ZIP family has a very ancient origin.

The ZIP family can be split into several subfamilies based on a higher degree of sequence conservation within these groups. Guerinot previously divided the

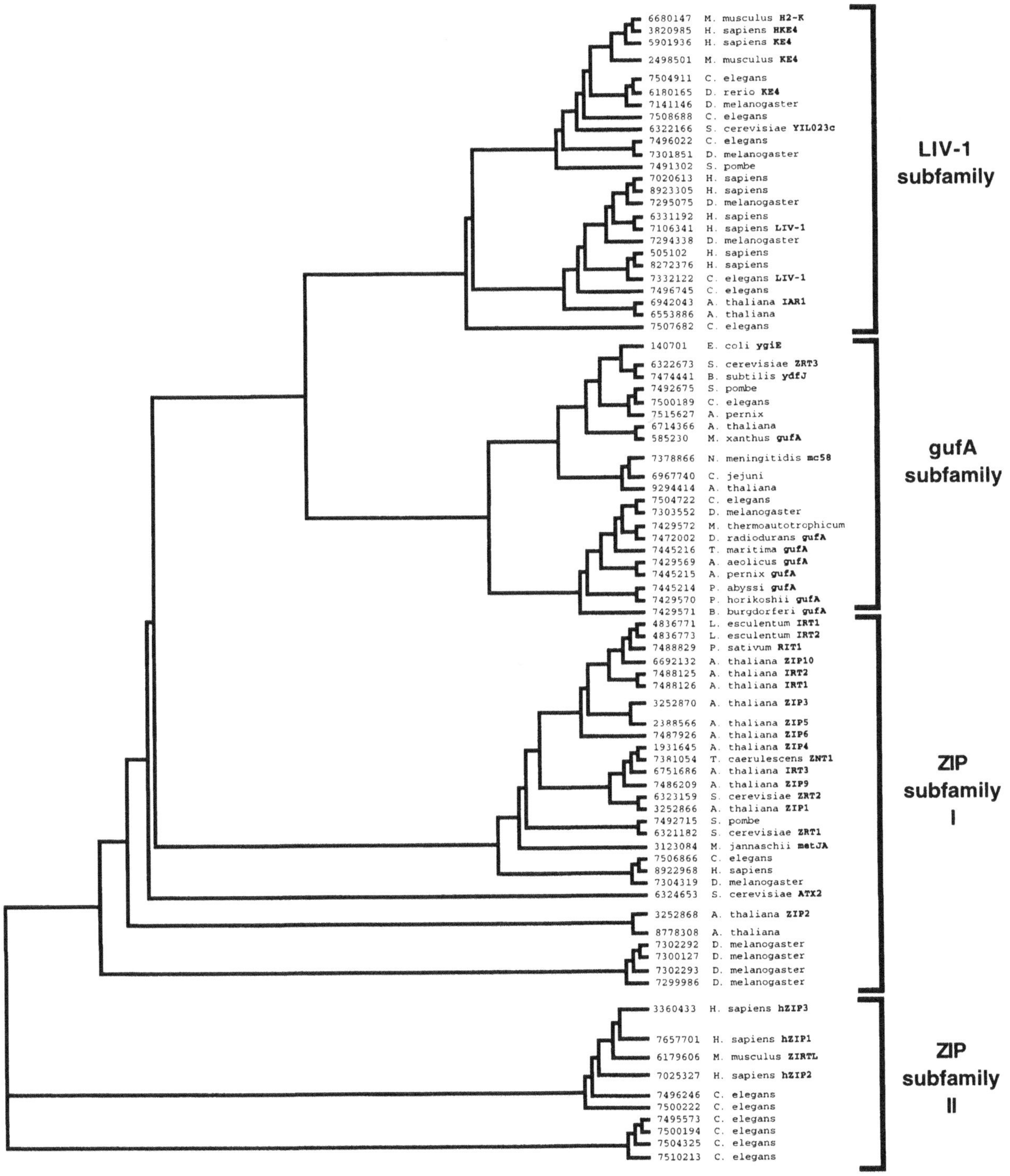

Fig. 1. The ZIP family of metal ion transporters. A dendrogram is shown describing the sequence relationships of ZIP members identified in the NCBI nonredundant protein database (12/00). Related sequences were identified using a PSI-BLAST-generated Position-Specific Scoring Matrix and the resulting dendrogram is a neighbor-joined tree generated using CLUSTALX (Thompson *et al.* 1997). The corresponding accession numbers, source organism, and gene name (*bold*), if any, are given. The different subfamilies within the ZIP family are indicated with *brackets*.

known eukaryotic ZIP proteins into subfamilies I and II (Guerinot 2000) and we have retained that classification here. Subfamily I consists largely of fungal and plant members whereas subfamily II is a smaller group of mammalian and nematode proteins. As described below, many members of ZIP subfamilies I and II are involved in zinc uptake. Our PSI-BLAST analysis identified two additional subfamilies. One new group is designated the gufA subfamily after the gufA protein of *Myxococcus xanthus*. GufA is a protein of unknown function first identified by genomic sequencing (McGowan *et al.* 1993). Several related genes, also of unknown function, have been identified in other prokaryotes and have been named gufA as well. Among the gufA subfamily are proteins from eukaryotes. This includes Zrt3, which is now known to be a zinc transporter in *S. cerevisiae*. The presence of Zrt3 in the gufA cluster of proteins clearly links this subfamily to roles in metal ion transport. The fourth subfamily, the LIV-1 subfamily, is restricted to eukaryotes. LIV-1, the gene after which this subfamily is named, encodes a protein of unknown function that is expressed at higher levels in certain metastatic breast cancer tissues (Manning *et al.* 1995). No members of the LIV-1 subfamily have been functionally characterized. However, their relationship to other ZIP proteins suggests a role in metal ion transport. This role was first proposed by Eng *et al.* (1998). This hypothesis was also proposed by Taylor (2000) based on the high abundance of histidines in various regions of the LIV-1 proteins, a common feature of both ZIP and CDF families. We find here that the sequence similarity between LIV-1 and other ZIP proteins is much greater than was recognized in previous studies.

Most ZIP proteins are predicted to have eight transmembrane domains but some members may have as few as five. Because many of the protein sequences have been determined by computer-assisted editing of probable introns from genomic sequences, the variability in the number of transmembrane domains may be due in large part to errors in this conceptual splicing process. All of the functionally characterized members of the family are predicted to have eight transmembrane domains so this seems the most accurate domain structure for most of the family. The majority of ZIP proteins share a similar predicted topology where the amino and carboxy termini are extracytoplasmic (Figure 2A). Elements of this topology have been confirmed for some members of the family, e.g., the amino terminus of Zrt1 and the carboxy terminus of hZip2 have been shown to be on the extracellular surface of the plasma membrane (Gaither & Eide 2000; Gitan *et al.* 1998). Another feature shared by many of the ZIP proteins is a long loop region located between transmembrane domains III and IV. This region is referred to as the 'variable region' because both its length and sequence shows little conservation among the family members (Guerinot 2000). One feature of the variable region that is shared by several of the ZIP proteins is the presence of many histidine residues. For example, in Zrt1, this sequence is ... HDHTHDE... and in Irt1, the sequence is ... HGHGHGH.... While the function of this motif is unknown, we recognize it as a potential metal binding domain and its conservation in many of the ZIP proteins suggests a role in metal ion transport or its regulation. Finally, the variable region is predicted to be cytoplasmic and this location has been confirmed for Zrt1 (Gitan & Eide 2000).

The greatest degree of conservation among ZIP proteins is found in transmembrane domains IV-VIII (Figure 3). Transmembrane domains IV and V are particularly amphipathic and contain conserved histidine residues frequently with adjacent polar or charged amino acids. Some examples of these motifs are shown in Figure 3. Given their amphipathic nature and sequence conservation, transmembrane domains IV and V are predicted to line an aqueous cavity in the transporter through which the cationic substrate passes (Eng *et al.* 1998). Consistent with this model, mutation of the conserved histidines or adjacent polar/charged residues in transmembrane domains IV and V of Irt1 eliminated its transport function (Rogers *et al.* 2000). Finally, residues important in determining the substrate specificity of Irt1 were mapped to the loop region between domains II and III (Rogers *et al.* 2000). This region is predicted to lie on the outer surface of the membrane and could be the site of initial substrate binding during the transport process.

The mechanism of transport used by the ZIP proteins is still unclear. Zinc uptake by human hZip2 zinc transporter was found to be energy independent (Gaither & Eide 2000). This observation is in conflict with studies of the yeast zinc transporters Zrt1 and Zrt2 which showed strict energy dependence (Zhao & Eide 1996a). Therefore, fungal and human ZIPs may work by different mechanisms. Zinc uptake by hZip2 was not dependent on K^+ or Na^+ gradients but was stimulated by HCO_3^- (Gaither & Eide 2000). It was suggested that hZip2 functions *in vivo* by a Zn^{2+}-HCO_3^- symport mechanism. Alternatively, zinc uptake by these proteins may simply be driven by the concen-

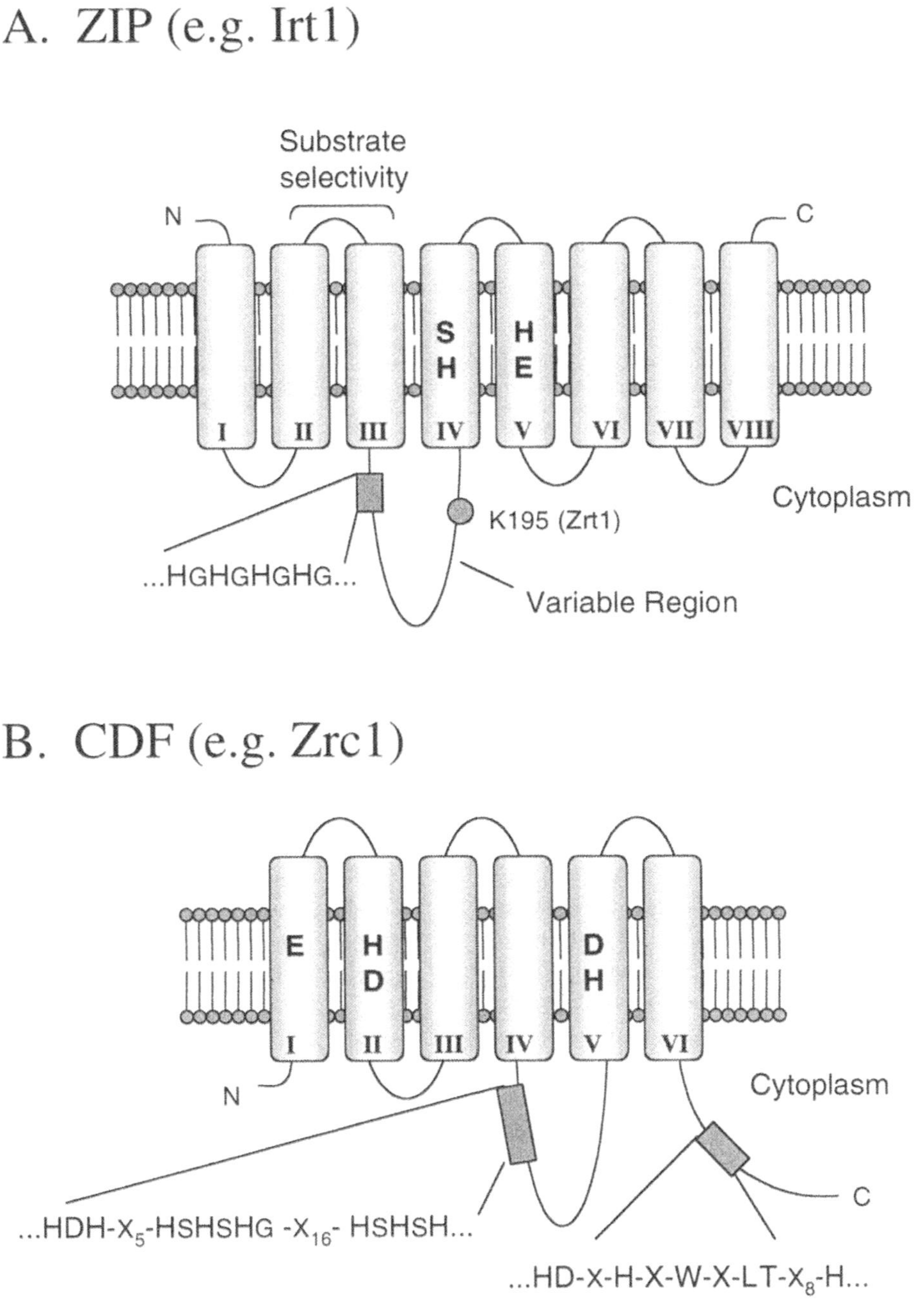

Fig. 2. Predicted membrane topology of ZIP and CDF proteins. A. Irt1, like most other ZIP transporters is predicted to have eight transmembrane domains (I–VIII). The conserved and functionally important residues within domains VI and V are indicated as is the variable region, the ubiquitinated K195 in Zrt1, and the extracellular loop region that affects Irt1 substrate specificity. B. Zrc1, like most other CDF transporters, is predicted to have six transmembrane domains (I–VI). Conserved polar or charged residues within transmembrane domains I, II, and V are indicated. The locations of histidine-rich regions of Irt1 and Zrc1 that are potentially involved in metal binding are also indicated.

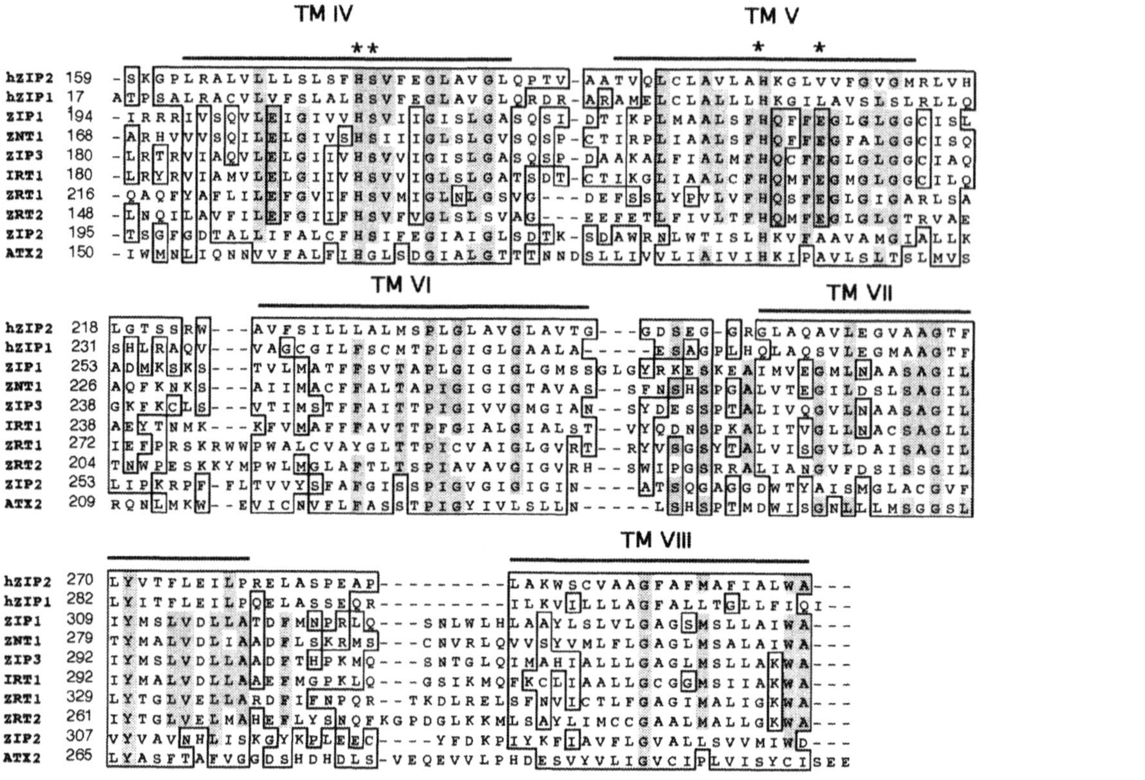

Fig. 3. Sequence conservation in transmembrane domains IV-VIII of characterized ZIP family members. The amino acid sequences of several ZIP transporters are shown and the approximate locations of transmembrane domains (TM) are indicated by a *line* over the corresponding sequences. Similar amino acids (≥ 85% of the sequences shown) are *boxed* and residues that are identical in ≥ 65% of the sequences are *shaded*. The conserved histidines and nearby polar or charged residues in the amphipathic TM IV and V that are required for function of Irt1 (Rogers *et al.* 2000) are indicated with *asterisks*.

tration gradient of the metal ion substrate. Although the total level of zinc in a cell is high (e.g., 200 μM) (Palmiter & Findley 1995), very little of that zinc is present in a 'free' or labile form. Estimates of the labile pool of zinc in cells are in the nanomolar range (Suhy & O'Halloran 1995). Therefore, a concentration gradient of labile zinc across the plasma membrane may be an important driving force for Zn^{2+} uptake. The negative-inside membrane potential found in cells may also be a driving force for uptake of this cation (Stein 1990; Zhang & Allen 1995).

The CDF family of metal ion transporters

Like the ZIPs, members of the CDF family of proteins are found in organisms at all phylogenetic levels. The name CDF stands for 'cation diffusion facilitator' and is based on the early recognition that these proteins commonly play roles in metal ion transport (Nies & Silver 1995). As discussed in later sections of this review, many members of this family have been implicated specifically in the transport of zinc from the cytoplasm out of the cell or into organellar compartments. The CDF family was recently reviewed by Paulsen and Saier (1997). This article described a family of thirteen members. Thanks to more sequence data and better database analysis tools, that number has now grown to 101 (Figure 4).

Also like the ZIPs, the CDF family can be divided into different sub-groups based on clusters of proteins with greater sequence similarities. Based on this clustering, we have divided the CDF family into three subfamilies, I, II, and III. CDF subfamily I contains mostly prokaryotic members from both eubacterial and archeael sources while subfamilies II and III contain approximately equal representation of eukaryotic and prokaryotic members. The CDF proteins listed in Figure 4 range in size from 199–1677 amino acids with most members in the 300–550 residue range. Again, some of the variability, especially for the smaller sequences, may be due to errors in the conceptual assignment of splice junctions. The member

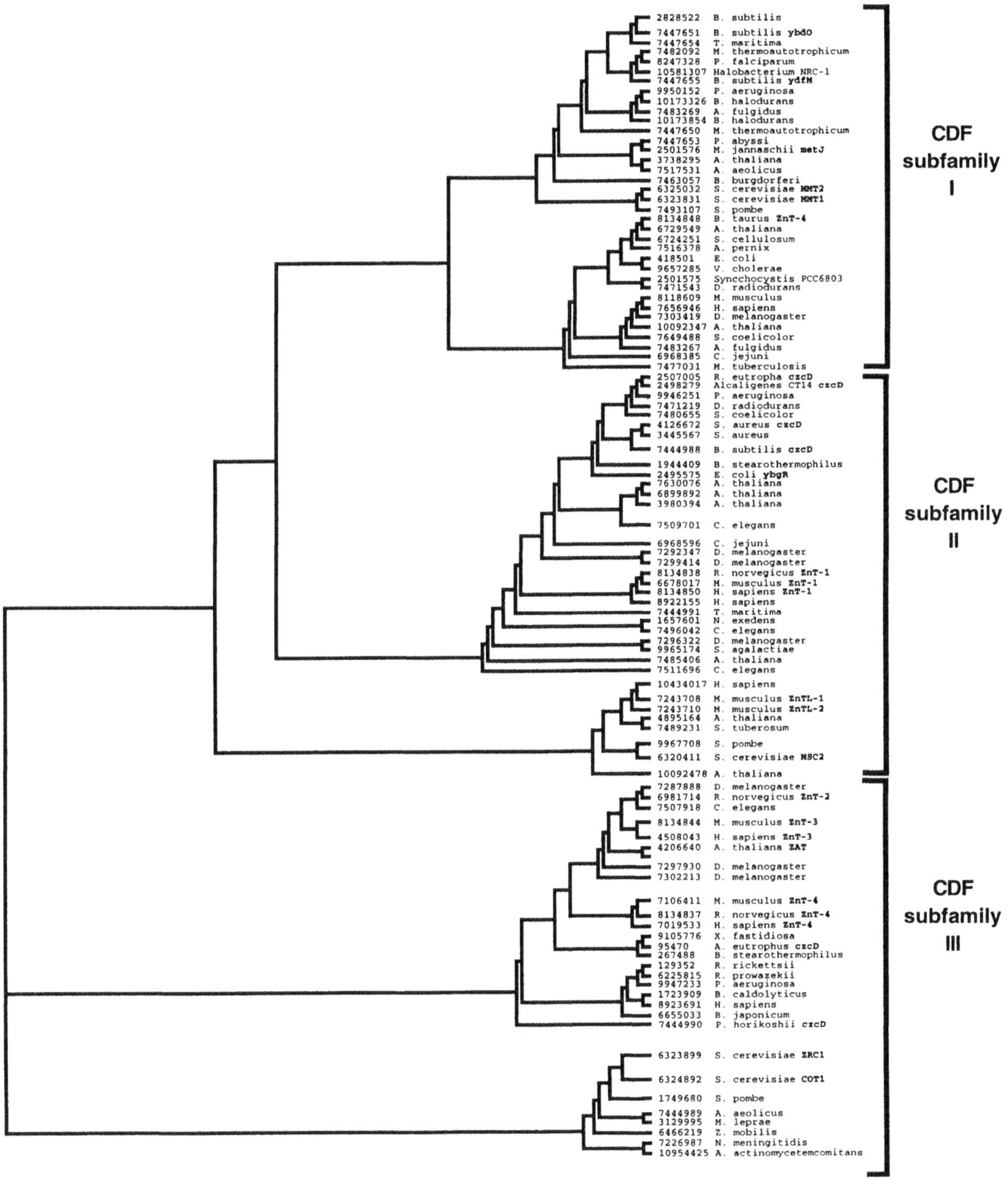

Fig. 4. The CDF family of metal ion transporters. A dendrogram is shown describing the sequence relationships of CDF members identified in the NCBI nonredundant protein database (12/00). Related sequences were identified using a PSI-BLAST-generated Position-Specific Scoring Matrix and the resulting dendrogram is a neighbor-joined tree generated using CLUSTALX (Thompson *et al.* 1997). The corresponding accession numbers, source organism, and gene name (*bold*), if any, are given. The different subfamilies within the CDF family are indicated with *brackets*.

proteins from bacteria average about 300 amino acids while the eukaryotic members are usually larger averaging approximately 450 amino acids. Most members of the CDF family are predicted to have six transmembrane domains. Their predicted membrane topology is similar to that shown for one such protein, Zrc1 from *S. cerevisiae*, depicted in Figure 2B. Notable exceptions to this rule are the Msc2 protein of *S. cerevisiae* and a closely related protein from *Schizosaccharomyces pombe* (accession number 9967708) which are predicted to have 12 transmembrane domains. Comparing the amino acid sequence of these novel proteins to other members of the CDF family suggests that the amino-terminal six transmembrane domains may have been fused onto an archetypical six-domain CDF member some time during their evolution.

Most of the CDF proteins share a similar predicted topology where both the amino and carboxy termini are cytoplasmic (Figure 2B). Some elements of this topology have been confirmed for the czcD protein from *Ralstonia* sp. Strain CH34 (Anton *et al.* 1999). Another feature common among CDF proteins is a long loop region located between transmembrane domains IV and V. This loop frequently contains a histidine-rich motif, $(HX)_n$ where n ranges from 3–6 and X is often S or G, that could function as a potential metal binding domain (Figure 5A). In addition, a second potential metal-binding motif found in the cytoplasmic C-terminal tail has the sequence ... H-D/E-X-H-X-W-X-L-T–X_8-H... (Figure 5B). The function of these domains has not been determined but their sequence conservation and potential for metal binding is intriguing.

The greatest degree of conservation among CDF proteins is found within transmembrane domains I, II, and V (Figure 5C, D). These three transmembrane domains are also highly amphipathic suggesting an important role in substrate transport. This hypothesis is supported by the observation that certain polar or charged amino acids within these domains are among the most highly conserved. To date, the importance of these residues to the metal ion transport function of these proteins has not been investigated. The mechanism of transport used by the CDF proteins has not been closely examined. Analysis of a mammalian CDF, the ZnT-1 protein, suggested that its zinc transport activity was not dependent on ATP (Palmiter & Findley 1995). However, given the negative-inside membrane potential of the plasma membrane and the likelihood that this efflux is also occurring against a zinc concentration gradient (see above), secondary active transport by some mechanism seems probable.

Zinc transport and its regulation in yeast

Much of our understanding of zinc transport and its regulation comes from studies of the yeast *S. cerevisiae*. Many elements of the following discussion are summarized in Figure 6. Zinc uptake in *S. cerevisiae* is time-, temperature-, concentration-dependent and saturable (Fuhrmann & Rothstein 1968; Mowll & Gadd 1983; White & Gadd 1987). Kinetic studies of zinc uptake by cells grown with different amounts of zinc in the medium suggested the presence of at least two uptake systems. One system has a high affinity for zinc with an apparent K_m of 1 μM Zn^{2+} and is active in zinc-limited cells (Zhao & Eide 1996a). The second system has a lower affinity for zinc (apparent K_m of 10 μM Zn^{2+}) and its activity is detectable in zinc-replete cells (Zhao & Eide 1996b). These apparent K_m values are overestimates of the true K_m values because they do not consider the chelation properties of the uptake assay media. Equilibrium calculations suggest that the actual K_m values are approximately 10 and 100 nM for the high and low affinity uptake systems, respectively. Both high and low affinity uptake systems are specific for zinc and probably do not contribute to the accumulation of other metals (Zhao & Eide 1996b).

The *ZRT1* and *ZRT2* genes encode the transporter proteins of the high and low affinity systems, respectively (Zhao & Eide 1996a, b). Zrt1 and Zrt2 are closely related to each other sharing 44% amino acid sequence identity and 67% similarity. Both are members of the ZIP family of metal ion transporters. A *zrt1 zrt2* mutant is viable (Zhao & Eide 1996b) suggesting that additional, as yet uncharacterized zinc uptake systems are also present in this yeast. These other systems are unlikely to be major sources of zinc under any but the most zinc-replete conditions given that *zrt1 zrt2* mutant cells require 105-fold more zinc to grow than wild type cells (Zhao & Eide 1996b). Zinc uptake through other metal transporters or channels could contribute to these uncharacterized zinc accumulation mechanisms. Alternatively, zinc uptake in the *zrt1 zrt2* mutant could be occurring by fluid-phase endocytosis and subsequent mobilization of the accumulated zinc from the vacuole (see below).

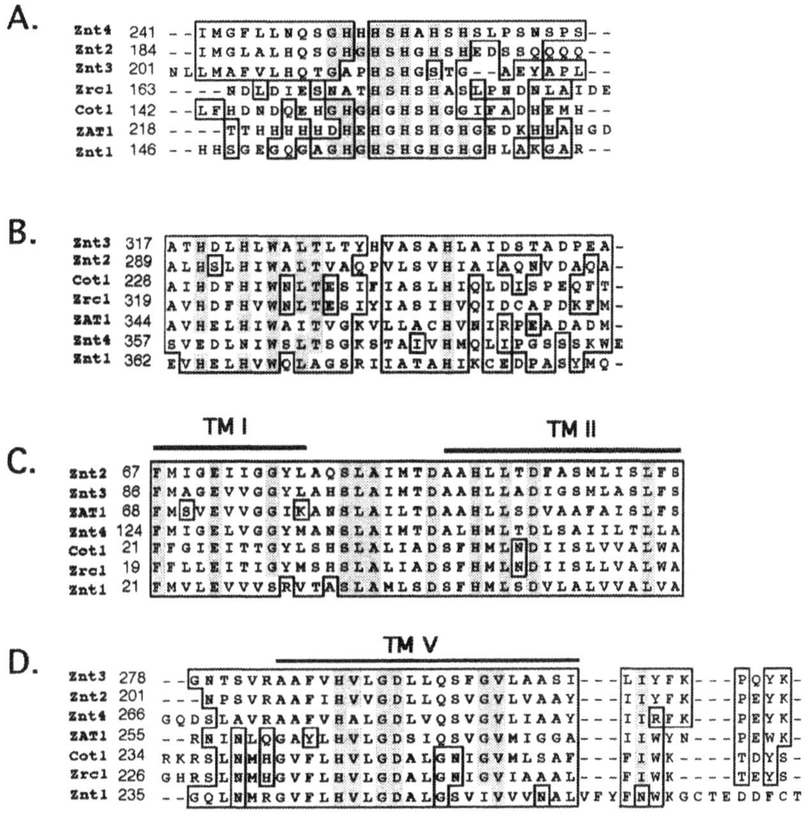

Fig. 5. Sequence conservation in various domains of representative CDF family members. The amino acid sequences of several CDF transporters are shown and the approximate locations of transmembrane domains (TM) are indicated by a *line* over the corresponding sequence. Similar amino acids ($\geq$ 85% of the sequences shown) are *boxed* and residues that are identical in $\geq$ 65% of the sequences are *shaded*. A. Histidine-rich sequences in the predicted cytoplasmic loop between TM IV and V; B. A potential metal binding domain in the carboxy-terminal tail regions; C. Transmembrane domains I and II; D. Transmembrane domain V.

Transcriptional control of zinc uptake in yeast

Zinc uptake in *S. cerevisiae* is controlled at the transcriptional level in response to intracellular zinc levels. The high affinity system is induced more than 30-fold in zinc-limited cells resulting from increased transcription of the *ZRT1* gene (Zhao & Eide 1996a). The low affinity system is also regulated through the control of *ZRT2* transcription (Zhao & Eide 1997). Regulation of these genes in response to zinc is mediated by the product of the *ZAP1* gene (Zhao & Eide 1997). *ZAP1* encodes a transcriptional activator with seven carboxy-terminal C_2H_2 zinc finger domains and two amino terminal activation domains (Figure 7). Zap1 was also found to regulate its own transcription through a positive autoregulatory mechanism. This type of regulatory circuitry would allow for an amplified response to changes in zinc levels and Zap1 activity under progressively zinc-limiting conditions.

Zap1 binds to zinc-responsive elements (ZREs) in the promoters of the *ZRT1*, *ZRT2*, and *ZAP1* genes (Zhao *et al.* 1998). A ZRE consensus sequence, 5′-ACCYYNAAGGT-3′, was identified and found to be both necessary and sufficient for zinc-responsive transcriptional regulation. Despite each gene having one or more consensus ZRE in their promoters, there is differential zinc responsiveness among the *ZRT1*, *ZRT2*, and *ZAP1* genes. Significantly more zinc is required to repress Zap1-dependent expression of the *ZRT2* promoter than is required to repress either the *ZRT1* or *ZAP1* promoters. These data suggest that Zap1 activity on the *ZRT2* promoter may be altered by accessory factors (e.g., other transcription factors) that modulate Zap1's response to zinc on this promoter. If zinc controls the affinity of Zap1 for its ZRE binding sites, one possible model is that other proteins bind to the *ZRT2* promoter and help stabilize binding of Zap1 to

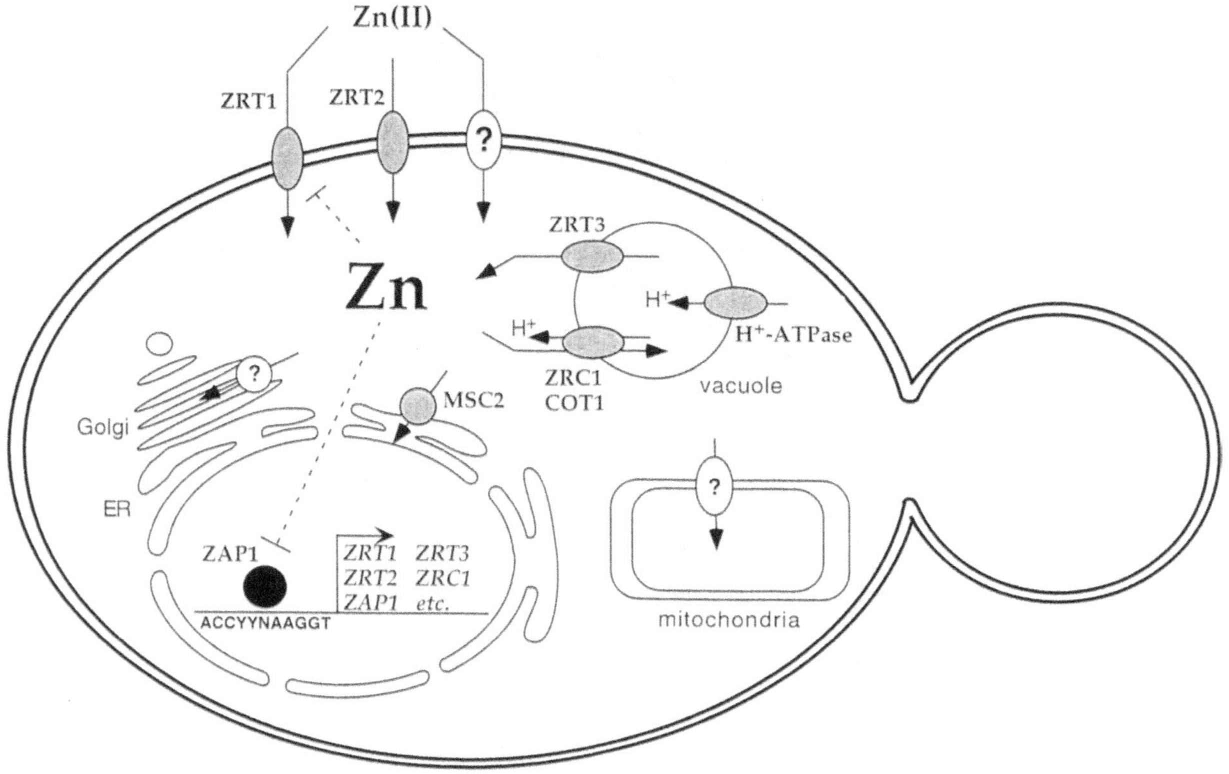

Fig. 6. Model of zinc transport and its regulation in *S. cerevisiae*. Proteins involved in zinc transport are shown in *gray* and the Zap1 regulatory protein is depicted in *black*. *Solid arrows* indicate transport steps and *dashed lines* indicate regulatory interactions. Transporters that are anticipated to exist but have not been identified are indicated by the *question marks*.

the ZREs, thus increasing the affinity of Zap1 for these sites.

The differential sensitivity of the *ZRT1*, *ZRT2*, and *ZAP1* promoters to zinc is also consistent with the different functions of these proteins and suggests the following scenario: basal (i.e., Zap1-independent) expression of the Zrt2 low affinity transporter is sufficient to supply zinc to cells under zinc-replete conditions (Zhao & Eide 1997). As cells enter the initial phases of zinc limitation, their first response is to increase the activity of the Zrt2 transporter. As zinc limitation becomes more severe, the Zrt1 transporter is induced to provide high affinity uptake activity for zinc acquisition. Finally, increased expression of the *ZAP1* gene, allowing for maximum expression of its target genes, would only be needed under conditions of extreme zinc-limitation.

Another intriguing question that remains to be answered is precisely how zinc regulates Zap1 activity. Recent studies that dissect the functional domains of Zap1 have assisted greatly in answering this question. Two activation domains, designated AD I and AD II, were mapped within the protein and both function *in vivo* (Bird *et al.* 2000b) (Figure 7). The complete DNA binding domain was mapped to the carboxy-terminal five zinc fingers (Bird *et al.* 2000a; Zhao *et al.* 1998). Mutations that disrupt the formation of each of these fingers were tested for their ability to complement a *zap1* mutant *in vivo* and for DNA binding affinity *in vitro* (Bird *et al.* 2000a). Each finger was found to be required for high affinity DNA binding. Consistent with this hypothesis, purified Zap1 protein was found to have a stoichiometry of five zinc atoms per monomer of protein (Bird *et al.* 2000a). Mutation of zinc fingers 1 and 2 in AD II had no effect on this stoichiometry suggesting these fingers bind zinc with low affinity. The function of fingers 1 and 2 remains unclear.

The portion of the Zap1 protein required for zinc responsiveness co-localizes to the DNA binding domain and the five C-terminal zinc fingers (Bird *et al.* 2000b). For example, fusion of the Zap1 DNA binding domain (amino acids 687–880) to the Gal4 activation domain resulted in a functional protein *in vivo*.

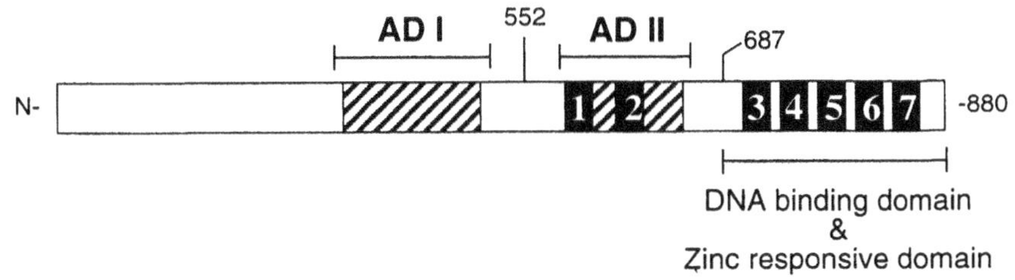


```
       β1 ──►          β2 ──►       α ──────────────►
       579                     -1 1 2 3 4 5 6 7
ZnF1   L K C K W K E C P E S C S S L F D L Q R H L L K D H V S Q D F K H P M E P
ZnF2   L A C N W E D C D F L G D D T C S I V N H I N C Q H G I N F D I Q F A -650
       705
ZnF3   V I C Q W D G C N K S F S S A Q E L N D H L E A V H L T R G K S E
ZnF4   Y Q C L W H D C H R T F P Q R Q K L I R H L K V - H S K Y K P
ZnF5   Y K C K T - - C K R C F S S E E T L V Q H T R T - H S G E K P
ZnF6   Y K C H I - - C N K K F A I S S S L K I H I R T - H T G E K P
ZnF7   L Q C K I - - C G K R F N E S S N L S K H I K T - H Q K K Y K -851
```

Fig. 7. Potential domain structure of Zap1. The Zap1 activation domains AD I and AD II (*hatched boxes*) and zinc fingers (numbered 1–7, *filled boxes*) are shown. The location of the DNA binding and zinc-responsive domains are also indicated and the sequences of the Zap1 zinc fingers are shown below. The conserved Zn^{2+} ligands are *boxed* and the locations of the presumed $\beta 1$, $\beta 2$ and α-helix structures are indicated. The amino acids relative to the start site of the α-helix are *numbered* -1 to 7.

This fusion was zinc regulated to a similar degree as the wild type Zap1 protein. The co-localization of the DNA binding domain and zinc-responsive domain suggested that DNA binding activity of Zap1 may be controlled by zinc binding to this region of the protein. However, the ability of Zap1 to confer zinc-responsive gene expression on a heterologous DNA binding domain, i.e. the DNA binding domain from the Gal4 activator (Bird *et al.* 2000b), suggested that zinc impairs activation domain function. While others are possible, we propose the following models of how Zap1 activity is regulated by zinc. First, we hypothesize that Zap1 is the direct zinc sensor and contains one or more low affinity regulatory zinc binding sites in the DNA binding domain in addition to the five high affinity C_2H_2 zinc fingers. Binding of zinc to these regulatory sites may stabilize a conformation (e.g. a multimeric complex) that sterically impairs both the DNA binding interface and the accessibility of the activation domain to general transcription factors. Alternatively, binding of zinc to the regulatory sites might recruit another protein that represses Zap1 function. In this model, either Zap1 or the accessory protein may contain the regulatory zinc binding site(s). Future studies will determine which, if any, of these models is correct.

Post-translational control of zinc uptake in yeast

A second mechanism in *S. cerevisiae* regulates zinc transporter activity at a post-translational level. In zinc-limited cells, Zrt1 is a stable plasma membrane protein. Exposure to high levels of extracellular zinc triggers a rapid loss of Zrt1 uptake activity and protein. This inactivation occurs through zinc-induced endocytosis of the protein and its subsequent degradation in the vacuole (Gitan *et al.* 1998). Our molecular understanding of these events are summarized in Figure 8. Mutations that inhibit the internalization step of endocytosis also inhibited zinc-induced Zrt1 inactivation and the major vacuolar proteases were required to degrade Zrt1 in response to zinc. Furthermore, immunofluorescence microscopy showed that Zrt1 is in the plasma membrane of zinc-limited cells and is transferred to the vacuole via an endosome-like compartment upon exposure to zinc. Zrt1 inactivation is a relatively specific response to zinc; Cd^{2+} and Co^{2+} trigger the response but less effectively than zinc. Excess zinc does not alter the stability of several other

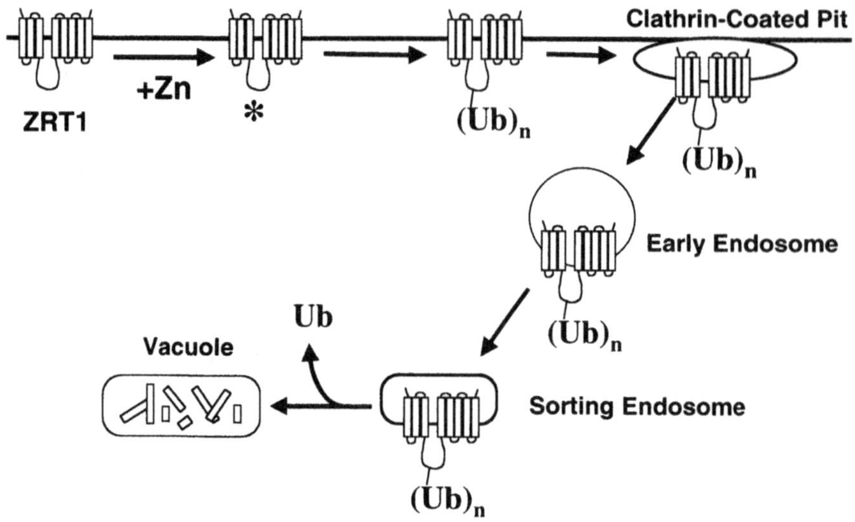

Fig. 8. A model of zinc-responsive post-translational inactivation of Zrt1. High zinc causes the ubiquitination of Zrt1 which results in the protein's migration into clathrin-coated pits and subsequent endocytosis. Zrt1 then passes through the endocytic pathway to the vacuole where it is degraded by vacuolar proteases. The model proposes that zinc alters the conformation of Zrt1 (indicated by the *asterisk*) to make it a better substrate for ubiquitination.

plasma membrane proteins. Therefore, zinc-induced Zrt1 inactivation is a specific regulatory mechanism to shut off zinc uptake activity in cells exposed to high extracellular zinc levels. This system thereby prevents the overaccumulation of this potentially toxic metal.

The mechanism of zinc-induced endocytosis raises a number of exciting new questions. First, while it is clear that zinc induces endocytosis of Zrt1, it is unknown if this response is induced by a mechanism that senses intracellular or extracellular metal ion levels. Second, it is unclear if the signal being monitored is Zn^{2+} ions *per se*, the activity of a zinc-dependent or zinc-inhibited enzyme, or a more indirect consequence of high metal accumulation. The observation that Co^{2+} and Cd^{2+} also induce endocytosis of Zrt1 (Gitan *et al.* 1998) is potentially instructive. Both Co^{2+} and Cd^{2+} have similar coordination chemistries to Zn^{2+} and will bind to protein ligands in a similar fashion. Therefore, the simplest hypothesis is that Zn^{2+} ions trigger endocytosis directly and that Co^{2+} and Cd^{2+} mimic that signal. The lower activity of Co^{2+} and Cd^{2+} in triggering the response may be due to a greater specificity of the sensing mechanism for Zn^{2+} or different uptake efficiencies for different metal ions. A third unanswered question is how the zinc signal is transmitted to Zrt1. This could occur through the metal binding directly to the transporter or through an indirect signal transduction pathway. Recent studies demonstrate that Zrt1 is ubiquitinated prior to endocytosis suggesting that this modification serves as a signal for recruitment of the protein into clathrin-coated pits (Gitan & Eide 2000). A similar role for ubiquitin has also been found for other *S. cerevisiae* and some mammalian plasma membrane proteins (Hicke 1997). Zinc-induced ubiquitination of Zrt1 occurs on a lysine residue (K195) located in the variable region of the protein (Figure 2A). Therefore, the principle question is currently how does zinc control ubiquitination of Zrt1?

The post-translational regulatory system is clearly separate from the transcriptional control system given that inactivation of Zrt1 activity occurs normally in a *zap1* deletion mutant (Gitan *et al.* 1998). However, these two systems undoubtedly work together to maintain the homeostatic control of intracellular zinc levels. It is interesting to note that the transcriptional control system exerts its greatest effect on ZRT1 expression when cell-associated zinc levels vary between 0.01 and 0.07 nmol Zn/10^6 cells (i.e., $\sim$0.5 - 4 $\times$ 10^7 atoms zinc/cell) (Gitan *et al.* 1998). Approximately 90% repression of a *ZRT1* promoter-*lacZ* fusion was observed when cell-associated zinc levels rose to 0.07 nmol/10^6 cells (Zhao *et al.* 1998). In contrast, the post-translational response is triggered only at cell-associated zinc levels of greater than 0.07 nmol Zn/10^6 cells. Thus, we envision a two-tiered regulatory system in which the transcriptional control can respond to moderate changes in zinc availability and the post-translational control responds to more extreme zinc excess. A likely scenario in which the post-translational control would be important for maintaining zinc homeostasis is when zinc-limited cells are suddenly exposed to high levels of zinc. The rapid down-regulation of zinc uptake by Zrt1 endocytosis helps to prevent overaccumulation of zinc. This fast response would not be possible solely through the transcriptional control of a stable plasma membrane protein. During inactivation of zinc uptake activity, other systems may be induced to facilitate storage of the excess zinc or mediate its efflux from the cell.

Intracellular zinc transport in yeast

Once zinc is taken up across the plasma membrane, some of the metal must be transported into organelles such as the mitochondria and compartments of the secretory system to serve as a cofactor of zinc-dependent proteins found within those compartments. Furthermore, the vacuole has been implicated in the storage and detoxification of zinc (Ramsay & Gadd 1997). Very little is known about the specific transporters involved in intracellular zinc trafficking. Three potential intracellular zinc transporters have been identified in *S. cerevisiae*. These transporters are three members of the CDF family, Zrc1, Cot1, and Msc2. *ZRC1* was isolated as a determinant of zinc resistance, i.e. overexpression of *ZRC1* results in increased ability of these cells to tolerate high zinc levels (Kamizono *et al.* 1989). A *zrc1* mutation was later found to increase sensitivity to lipid hydroperoxides and decrease glutathione levels by approximately 40% (Kobayashi *et al.* 1996). The relationship between these phenotypes and zinc, if any, is unknown. The *COT1* gene was isolated in a similar fashion to *ZRC1*, i.e., as a suppressor of cobalt toxicity, but was later found to confer zinc resistance as well (Conklin *et al.* 1994; Conklin *et al.* 1992). Disruption of either *ZRC1* or *COT1* resulted in greater sensitivity to excess zinc further supporting the role of these genes in zinc compartmentalization.

Zrc1 and Cot1 are closely related proteins (60% identity) with approximately 400 amino acids and 6 potential transmembrane domains. The physiological roles of these transporters remain unclear. Neither *ZRC1* nor *COT1* are essential genes, and a *zrc1 cot1* mutant is also viable. Thus, these two genes do not together provide a function essential for growth. Neither protein appears to catalyze zinc efflux from the cell. While Cot1 was originally proposed to be a mitochondrial protein (Conklin *et al.* 1992), the subcellular location of both Zrc1 and Cot1 has recently been identified as the vacuole (Li & Kaplan 1998). This localization was determined with overexpressed proteins and so must be viewed with caution. However, with this caveat aside, these results suggest that these transporters are responsible for zinc sequestration into the vacuole (Li & Kaplan 1998). Because zinc transport into the vacuole has been attributed to a Zn^{2+}/H^+ antiport system (Bode *et al.* 1995; Nishimura *et al.* 1998), this leads to the conclusion that this is the transport mechanism used by Zrc1 and Cot1. This mechanism would then provide a simple explanation for the zinc sensitivity observed in vacuolar H^+-ATPase mutants (Eide *et al.* 1993; Ramsay & Gadd 1997); i.e., mutants defective for vacuolar acidification lack the H^+ gradient necessary to drive zinc sequestration.

More recently, we have identified a gene in yeast, *ZRT3*, that also plays a role in vacuolar zinc transport (MacDiarmid *et al.* 2000). Although distantly related to Zrt1 and Zrt2, Zrt3 is a potential transport protein that is a member of the ZIP family. Like the *ZRT1* and *ZRT2* genes, *ZRT3* is a ZAP1 target gene and is up-regulated in zinc-limited cells. Our analysis of Zrc1, Cot1, and Zrt3 has generated the following scenario of zinc storage in yeast. Zinc-replete wild type cells generate a vacuolar zinc storage pool through the action of the Zrc1 and Cot1 transporters. This pool of stored zinc is in a labile form that can be mobilized when cells are deprived of extracellular zinc. Mobilization of the vacuolar store is the role of Zrt3 whose expression is induced under zinc-limiting conditions. Several aspects of this model have already been confirmed (MacDiarmid *et al.* 2000).

Finally, a third member of the CDF family has recently been implicated in zinc transport in *S. cerevisiae*. This transporter is encoded by the *MSC2* gene. While Msc2 is a member of the CDF family, it differs from most other members by having twelve rather than six transmembrane domains and two rather than one histidine-rich region. Paradoxically, *MSC2* was first identified by a transposon insertion allele that caused an increased frequency of meiotic sister chromatid (MSC) recombination events (Thompson & Stahl 1999). This effect was found to be allele-specific and did not occur when the *MSC2* gene was deleted. The connection between *MSC2* and recombination is still a mystery but a subsequent analysis has suggested a role of Msc2 in zinc transport (Li & Kaplan 2000). An *msc2* deletion mutation caused decreased viability on respired carbon sources and an abnormal cellular morphology when cells were grown at an elevated temperature. Both of these phenotypes were suppressible by zinc supplementation suggesting some defect in zinc metabolism in this strain. The *msc2* mutant also had alterations in intracellular zinc content and an apparent increase in the regulatory zinc pool sensed by Zap1. The Msc2 protein was localized to the nuclear envelope. An attractive hypothesis is that Msc2 mediates zinc transport into the intermembrane space of this compartment. This intriguing model awaits further testing.

Zinc transport and its regulation in plants

Our understanding of zinc transport and its regulation in plants is increasing rapidly with the identification of both ZIP and CDF family genes in many plant species. The number of ZIP family members in plants is remarkable. The *Arabidopsis* genome alone contains 18 such genes representing members from three of the four subclasses of ZIP proteins; only ZIP subfamily II does not contain any plant members. The high number of potential metal ion transport proteins in plants and animals (see below) no doubt stems from the greater diversity of tissue-specific roles to be played by these proteins in multicellular organisms. Irt1 (Iron-regulated transporter 1) was the first ZIP protein to be identified in any organism (Eide *et al.* 1996). The *IRT1* gene was cloned because its expression in a yeast mutant defective for iron uptake suppressed the growth defect of this strain in low iron media. Biochemical studies confirmed that Irt1 expression increased iron uptake in this yeast strain (Eide *et al.* 1996) and later studies demonstrated that Irt1 can also transport Zn^{2+}, Mn^{2+}, and Cd^{2+} (Korshunova *et al.* 1999). The function of Irt1 in plants, however, appears to be largely iron accumulation. Irt1 is expressed solely in roots and only in roots of iron-limited plants. Therefore, if Irt1 participates in the accumulation of metals other than iron, e.g., zinc, it is likely to do so only under iron-limiting conditions. This prediction is consistent with the observation that iron-limited plants accumulate higher levels of other metal ions such as zinc, manganese, and cadmium (Cohen *et al.* 1998; Welch *et al.* 1993).

Four other *Arabidopsis* ZIP transporters, Zip1-4, may play roles in zinc transport (Grotz *et al.* 1998). Expressing Zip1, Zip2, or Zip3 in *S. cerevisiae* confers increased zinc uptake each with distinct biochemical properties. These results indicate that these proteins are zinc transporters. Zip4 expressed in yeast failed to increase zinc uptake perhaps due to poor expression or mislocalization of the protein in the yeast cell. *ZIP1* is expressed predominantly in roots while *ZIP3* and *ZIP4* mRNA could be detected in both roots and shoots. Furthermore, *ZIP1*, *ZIP3*, and *ZIP4* mRNA are induced under zinc-limiting conditions. These results further suggest a role for these proteins in zinc transport. The tissue-specific expression and subcellular localization of these proteins is not known so their precise roles can not yet be assessed. The zinc-responsive regulation of mRNA levels in response to zinc availability demonstrates that some mechanism of regulation exists in plants. This regulation may occur through a transcriptional control mechanism, like that found in yeast, or alternatively by a mechanism that regulates mRNA stability.

Incorrect regulation of zinc transporter expression would likely have a great impact on zinc accumulation in plants. This premise may in part explain the physiology of a unusual group of plants, the metal ion hyperaccumulators. Metal ion hyperaccumulators are plants that take up large quantities of metal ions from the soil. They are of great research interest because of their potential to remove metal pollutants from surface soils in a process called phytoremediation (Raskin 1995). Among the best known hyperaccumulators is *Thlaspi caerulescens*, a member of the Brassicaceae family that also includes *Arabidopsis*. Certain ecotypes of *T. caerulescens* are capable of accumulating up to 30,000 ppm Zn in their shoots without apparent toxic consequences (Brown *et al.* 1995). By comparison, non-hyperaccumulators normally accumulate only 0.1% of that level. Thus, hyperaccumulators must have remarkable ability to accumulate and detoxify metal ions. Biochemical analysis of Zn^{2+} uptake by *T. caerulescens* found that the V_{max} was elevated almost 5-fold compared to a non-hyperaccumulating ecotype, *T. arvense*, with no difference in K_m (Lasat *et al.* 2000). These results suggested that expression of zinc uptake transporters is higher in *T. caerulescens*. A ZIP family member was recently cloned from *T. caerulescens* and *T. arvense* and called *ZNT1* (Pence *et al.* 2000). Znt1 is expressed at a low level and regulated by zinc status in *T. arvense*. In striking contrast, this gene is expressed at a much higher level in *T. caerulescens* and is unaffected by zinc availability. Znt1 expression can explain the increased zinc accumulation in this and perhaps other metal hyperaccumulating plant species.

The genomes of plant species also contain many members of the CDF family; *Arabidopsis* alone encodes ten CDF member genes. These proteins are likely to function in subcellular zinc compartmentalization, as was the case for Zrc1 and Cot1 in yeast, as well as in zinc efflux. To date, only one plant CDF member has been studied, the Zat protein of *Arabidopsis* (van der Zaal *et al.* 1999). *ZAT* mRNA expression is not zinc regulated but the protein does appear to be a zinc transporter. Transgenic plants overexpressing the *ZAT* gene show increased zinc resistance. Zn content in roots of these transgenic plants was also found to increase suggesting that Zat transports zinc into an intracellular compartment, e.g., the vacuole, of root cells. As is the case with any multicellular organism, zinc transporters are required for both cellular zinc uptake as well as efflux to allow the utilization of the metal. In plants, for example, a zinc efflux transporter is required to pass zinc from the root tissue into the xylem for distribution to aerial portions of the plant. CDF proteins such as Zat probably perform this function as well.

Zinc transporters and their regulation in mammals

Several zinc transporters from both the ZIP and CDF families are found in mammalian organisms and many of these have been implicated in zinc transport. To date, twelve ZIP genes have been identified in humans and three have been found in the mouse (Figure 1). Functional data are only available for two of the human genes, *hZIP1* and *hZIP2*, and none of the mouse genes have been characterized. hZip1 and hZip2 appear to play roles in zinc uptake across the plasma membrane. *hZIP2* mRNA expression has only been detected in prostate and uterine tissue indicating restricted tissue-specificity. Functional assays indicated that the hZip2 protein is a functional zinc transporter (Gaither & Eide 2000). When *hZIP2* was overexpressed in K562 erythroleukemia cells grown in culture, these cells accumulated more zinc than control cells due to an increased zinc uptake activity. Moreover, hZip2 protein was localized to the plasma membrane of these cells. These results indicated that *hZIP2* may serve in zinc uptake in the few tissues where it is expressed.

Zinc uptake mediated by hZip2 was biochemically distinct from the endogenous activity of the K562 cell line in a number of respects. For example, zinc uptake mediated by hZip2 was stimulated by HCO_3^- treatment whereas the endogenous system was not affected. Furthermore, several other metal ions [e.g., Co^{2+}, Fe^{2+}, and Mn^{2+}] severely inhibited zinc uptake by hZip2 but the endogenous activity was far less sensitive. We have recently determined that another ZIP transporter, hZip1, is the endogenous zinc uptake system in K562 cells (Gaither & Eide 2001). First, K562 cells express *hZIP1* mRNA and the functional hZip1 protein is localized to the plasma membrane of these cells. Second, overexpression of *hZIP1* mRNA by approximately 2-fold increased zinc uptake activity by 2-fold as well. This increased uptake activity in hZip1 overexpressing cells was biochemically indistinguishable from the endogenous system. Finally, antisense oligonucleotides targeted to inhibit *hZIP1* expression also inhibited the endogenous zinc uptake activity. These results strongly suggest that hZip1 is the endogenous transporter in K562 cells. The antisense *hZIP1* oligonucleotide treatment reduced zinc uptake to 10–20% of control levels suggesting that hZip1 is the major pathway of zinc uptake in these cells.

In marked contrast to the *hZIP2* gene, *hZIP1* is expressed in a wide variety of different cell types. Thus, our results suggest that hZip1 may be the primary component of zinc uptake in many human tissues. This conclusion was supported by a recent study in which a correlation was found between *hZIP1* expression levels and zinc uptake in human malignant cell lines derived from the prostate. Prostate cell lines LNCap and PC-3 possess high levels of zinc uptake activity that is stimulated by prolactin and testosterone. Costello *et al.* (1999) found that *hZIP1* is expressed in LNCap and PC-3 cells and this expression is increased by prolactin and testosterone treatment. Expression of hZIP1 was also repressed by adding zinc to the medium suggesting some regulation of zinc uptake occurs in response to cellular zinc status. A closely related ortholog of hZip1 from the mouse was recently reported (Lioumi *et al.* 1999). This protein was named Zirtl for 'zinc-iron regulated transporter-like' protein. Like *hZIP1*, the *ZIRTL* gene is expressed in a wide variety of tissues. Zirtl fused to GFP was reported to localize to intracellular organelles. However, these studies were preformed without confirmation that the tagged protein retained wild type function. Therefore, it is possible that localization of the GFP-tagged Zirtl protein does not reflect the normal location of the native protein.

One paradox that arises from our studies of hZip1 and hZip2 is that these transporters have a surprisingly low affinity for their substrate. Both transporters have K_m values of approximately 3 μM for free Zn^{2+} ion. Similar K_m values have been reported for zinc transporters in a large number of mammalian cell types (Reyes 1996). The paradox arises when we consider the free Zn^{2+} concentration in mammalian serum. While the total zinc concentration of serum is approximately 20 μM, very little metal is present in an unbound form (Magneson *et al.* 1987). In serum, ~75% Zn^{2+} is bound to albumin and 20% is bound to α2-macroglobulin. Much of the remaining zinc is complexed with amino acids such as histidine and cysteine. Because of the high chelation capacity of serum, the free Zn^{2+} concentration in serum is calculated to be in the low nM range. Given this extremely low concentration of substrate, it was initially unclear how these transporters could contribute to zinc accumulation by mammalian cells under physiological conditions. The solution to this paradox comes from considering the capacity of these transporters relative to the zinc requirements of the cell. Our studies demonstrated that the capacity (i.e., V_{max}) for uptake is so high relative to the cellular demand for zinc that sufficient levels can be obtained despite the apparent low affinity of the transporters.

The potential role of the DCT1/DMT1/Nramp2 Fe^{2+} transporter in zinc uptake should be included in this discussion. DCT1/DMT1/Nramp2 is a member of the Nramp family of transporters and is unrelated to either ZIP or CDF proteins. Gunshin *et al.* (1997) provided evidence that DCT1/DMT1/Nramp2 was capable of Zn^{2+} uptake; *Xenopus* oocytes expressing DCT1/DMT1/Nramp2 displayed cation influx currents indicative of Zn^{2+} movement across the membrane. However, more recent results have indicated that the currents recorded in these oocytes result from Zn^{2+}-induced proton fluxes rather than transport of the metal ion (Sacher *et al.* 2001).

Several members of the CDF family have been implicated in zinc transport processes in mammals. The isolation and characterization of these transporters has been extensively reviewed by McMahon & Cousins (1998a) and will be considered only briefly here. Mammalian CDF family members are involved in the efflux of zinc from the cell or the transport of zinc into intracellular organelles. The analysis in Figure 4 lists seven CDF genes in humans, six in the mouse genome plus a small number of others from the rat and other mammals. Four of the mammalian genes, *ZnT-1*, *ZnT-2*, *ZnT-3*, and *ZnT-4*, have been functionally characterized to some extent so that their roles in zinc metabolism seem clear. ZnT-1 is a zinc efflux transporter in the plasma membrane of mammalian cells (Palmiter & Findley 1995). Given this localization, ZnT-1 may play a role in the cellular detoxification of zinc by exporting unneeded metal ion out of the cell. This role is consistent with the observation that cells that overexpress this transporter show higher zinc resistance than control cells. ZnT-1 may also play a role in the dietary absorption of zinc in the intestine and the recovery of zinc from urine in the renal tubules of the kidney. In the intestine, ZnT-1 is expressed in the enterocytes of the duodenum and the jejunem, i.e. the primary sites of zinc absorption (McMahon & Cousins 1998b). ZnT-1 protein is found localized to the basolateral membrane of enterocytes suggesting a role in transporting zinc out of the enterocyte and into the blood stream. ZnT-1 is also found on the basolateral surface of renal tubule cells (McMahon & Cousins 1998a), a position that would be expected of a protein involved in transporting zinc absorbed from urine back into the circulation. It has been well established that the loss of zinc in urine is very low due to efficient renal reabsorption (Victery *et al.* 1981).

ZnT-2 may play a role in intracellular zinc sequestration and storage similar to that proposed for Zrc1 and Cot1 in yeast. This protein is located in the membrane of an acidic endosomal/lysosomal compartment that accumulates zinc when cells are grown under high zinc conditions (Palmiter *et al.* 1996a). This compartment has been recently identified as the late endosome (Kobayashi *et al.* 1999). ZnT-3 also transports zinc into an intracellular compartment where the metal may play a role in neuronal signaling. *ZnT-3* mRNA has been detected only in the brain and testis and is most abundant in the neurons of the hippocampus and the cerebral cortex (Palmiter *et al.* 1996b). ZnT-3 protein is localized to membranes of synaptic vesicles in these neurons, suggesting that the protein transports zinc into this compartment. Consistent with this hypothesis, a subset of glutamatergic neurons contain histochemically reactive zinc in their synaptic vesicles. The localization of ZnT-3 protein was coincident with these zinc-containing vesicles (Palmiter *et al.* 1996b). Moreover, a *ZnT-3* null mouse line generated by targeted gene disruption failed to accumulate zinc in these vesicles (Cole *et al.* 1999). Thus, ZnT-3 is required for transport of zinc into synaptic vesicles in some types of neurons where it may play a neuromodulatory role (Fredrickson *et al.* 2000).

The fourth characterized mammalian CDF protein is ZnT-4. ZnT-4 is expressed in the mammary gland and is responsible for zinc transport into milk (Huang & Gitschier 1997). In fact, mutations in the *ZnT-4* gene are responsible for the *lethal milk* (*lm*) mutant mouse. The *lm* gene is so-named because pups of any genotype suckled on *lm/lm* dams die before weaning. Death is due to zinc deficiency from an insufficient supply of zinc in the milk (Piletz & Ganschow 1978). ZnT-4 has also been found to be expressed in the intestinal enterocytes where it is localized in endosomal vesicles concentrated at the basolateral membrane (Murgia *et al.* 1999). Thus, like ZnT-1, the function of ZnT-4 in the intestine may be to facilitate transport of zinc into the portal blood.

Our knowledge of how mammalian zinc transporters are regulated is still rudimentary. One point that is becoming increasingly clear is that zinc efflux in many cell types is regulated by zinc. One of the first indications of this effect came from studies of transient forebrain ischemia in gerbils (Tsuda *et al.* 1997). Differential display analysis demonstrated that ZnT-1 mRNA is up-regulated during ischemia, a condition which is known to cause zinc influx into neurons (Koh *et al.* 1996). Consistent with the *ZnT-1* regulation being the result of zinc influx, cultured neurons transiently increased *ZnT-1* mRNA when exposed to zinc. Thus, transcriptional control of *ZnT-1* may contribute to zinc detoxification by promoting efflux. *ZnT-1* is expressed in many cell types so this may be a general mechanism of cellular zinc homeostasis. The transcriptional control of *ZnT-1* may also play a role in zinc absorption. *ZnT-1* mRNA levels were increased in enterocytes following an oral dose of zinc (McMahon & Cousins 1998b). Given the location of the ZnT-1 protein on the basolateral membrane of these cells, a likely hypothesis is that up-regulation of ZnT-1 promotes zinc absorption by facilitating transport into the portal blood. A recent study by Andrews and colleagues (Langmade *et al.* 2000) has shown that increased expression of *ZnT-1* by zinc is mediated by the zinc-responsive MTF-1 transcription factor. This intriguing protein is considered elsewhere in this issue (Andrews, this issue). Cousins and coworkers (Liuzzi, *et al.* 2001) have shown that *ZnT-2* mRNA is up-regulated in response to increased dietary zinc levels suggesting that this gene is also a target of MTF-1 regulation.

Regulation of zinc uptake transporters in mammals is far less well understood. Some biochemical evidence suggests that regulation of these transporters in response to zinc status does occur. For example, zinc uptake in brush border membrane vesicles was found to increase in zinc-deficient rats (Menard & Cousins 1983). Similarly, cultured endothelial cells grown under low zinc conditions had a higher rate of zinc uptake than zinc-replete cells (McClung & Bobilya 1999). These data suggest that zinc deficiency can increase the expression or activity of zinc uptake transporters in some cell types. The mechanism of this regulation is unknown but a potential clue comes from the study of *hZIP1* expression in cultured malignant prostate cell (Costello *et al.* 1999). *hZIP1* mRNA levels were reduced in zinc-treated cells suggesting that a transcriptional control mechanism similar to that described in yeast may be present in mammalian cells. Should it exist, such a mechanism would play a critical role in mammalian zinc homeostasis.

Concluding remarks

Research over the past ten years has produced major advances in our knowledge of zinc transporters and their regulation in eukaryotic organisms. The identification of the ZIP and CDF families of metal ion transporters represents a major step forward. Study of these transporters has identified their various roles in zinc uptake, efflux, compartmentalization, storage, and detoxification. Moreover, the regulatory mechanisms that control the activity of these transporters in response to zinc status are becoming increasingly clear. These include both transcriptional and post-transcriptional mechanisms of regulation and they play important roles in zinc homeostasis and metabolism. Despite this progress, however, we are still very far from a complete picture of these processes in any organism. This is even true of yeast where the current model is the most complete among eukaryotes. Perhaps the greatest challenge that lies ahead will be determining the different roles of the ZIP and CDF family members found in plants and animals. The high number of such proteins in mice, for example, suggests that they play diverse functions in metal ion transport. Functional analyses of these transporters will identify their substrates and determine their biochemical mechanisms of action. Localization studies will determine the tissue- and cell-specific expression patterns of these proteins and define their subcellular locations. These data will tell us much about the potential roles these proteins play in zinc transport. Genetic studies, e.g., targeted gene disruption in mice, will assist in determining transporter function through the phenotypic analysis of the resulting mutants. Finally, the mechanisms that regulate the activity of these transporters need to be analyzed to place these proteins into the context of cellular and organismal zinc physiology and homeostasis. The mechanisms of zinc sensing used in these regulatory circuits will be an exciting new area of research. Clearly, the next ten years of research into zinc transporters and their regulation promises to be as exciting as the last decade.

References

Altschul S, Madden T, Schaffer A, Zhang J, Zhang Z, Miller W, Lipman D. 1997 Gapped BLAST and PSI-BLAST: a new generation of protein database search programs. *Nucl Acids Res* **25**, 3389–3402.

Altschul SF, Gish W, Miller W, Myers EW, Lipman DJ. 1990 Basic local alignment search tool. *J Mol Biol* **215**, 403–410.

Anton A, Grosse C, Reismann J, Pribyl T, Nies DH. 1999 CzcD is a heavy metal transporter involved in regulation of heavy metal resistance in *Ralstonia* sp. Strain CH34. *J Bacteriol* **181**, 6876–6881.

Bird AJ, Evans-Galea M, Blankman E, Zhao H, Luo H, Winge DR, Eide DJ. 2000a Mapping the DNA binding domain of the Zap1 zinc-responsive transcriptional activator. *J Biol Chem* **275**, 16160–16166.

Bird AJ, Zhui H, Luo H, Jensen LT, Srinivasan C, Evans-Galea M, Winge DR, Eide DJ. 2000b A dual role for zinc fingers in both DNA binding and zinc sensing by the Zap1 transcriptional activator. *EMBO J* **19**, 3704–3713.

Bode HP, Dumschat M, Garotti S, Fuhrmann GF. 1995 Iron sequestration by the yeast vacuole. *Eur J Biochem* **228** 337–342.

Brown SL, Chaney RL, Angle JS, Baker AJM. 1995 Zinc and cadmium uptake by hyperaccumulator *Thlaspi caerulescens* grown in nutrient solution. *Soil Sci Soc Am J* **59**, 125–132.

Cohen CK, Fox TC, Garvin DF, Kochian LV. 1998 The role of iron-deficiency stress responses in stimulating heavy-metal transport in plants. *Plant Physiol* **116**, 1063–1072.

Cole TB, Wenzel HJ, Kafer KE, Schwartzkroin PA, Palmiter RD. 1999 Elimination of zinc from synaptic vesicles in the intact mouse brain by disruption of the *ZnT3* gene. *Proc Natl Acad Sci USA* **96**, 1716–1721.

Conklin DS, Culbertson MR, Kung C. 1994 Interactions between gene products involved in divalent cation transport in *Saccharomyces cerevisiae*. *Mol Gen Genet* **244**, 303–311.

Conklin DS, McMaster JA, Culbertson MR, Kung C. 1992 *COT1*, a gene involved in cobalt accumulation in *Saccharomyces cerevisiae*. *Mol Cell Biol* **12**, 3678–3688.

Costello LC, Liu Y, Zou J, Franklin RB. 1999 Evidence for a zinc uptake transporter in human prostate cancer cells which is regulated by prolactin and testosterone. *J Biol Chem* **274**, 17499–17504.

Eide D, Bridgham JT, Zhong Z, Mattoon J. 1993 The vacuolar H^+-ATPase of *Saccharomyces cerevisiae* is required for efficient copper detoxification, mitochondrial function, and iron metabolism. *Mol Gen Genet* **241**, 447–456.

Eide D, Broderius M, Fett J, Guerinot ML. 1996 A novel iron-regulated metal transporter from plants identified by functional expression in yeast. *Proc Natl Acad Sci USA* **93**, 5624–5628.

Eng BH, Guerinot ML, Eide D, Saier MH. 1998 Sequence analyses and phylogenetic characterization of the ZIP family of metal ion transport proteins. *J Membr Biol* **166**, 1–7.

Fredrickson CJ, Suh SW, Silva D, Fredrickson CJ, Thompson RB. 2000 Importance of zinc in the central nervous system: the zinc-containing neuron. *J Nutr* **130**, 1471S–1483S.

Fuhrmann GF, Rothstein A. 1968 The transport of Zn^{2+}, Co^{2+} and Ni^{2+} into yeast cells. *Biochim Biophys Acta* **163**, 325–330.

Gaither LA, Eider DJ. 2001 The human ZIP1 transporter mediates zinc uptake in human K56Z erythroleukemia cells. *J Biol Chem* **276**, 22258–22264.

Gaither LA, Eide DJ. 2000 Functional characterization of the human hZIP2 zinc transporter. *J Biol Chem* **275**, 5560–5564.

Gitan RS, Eide DJ. 2000 Zinc-regulated ubiquitin conjugation signals endocytosis of the yeast ZRT1 zinc transporter. *Biochem J* **346**, 329–336.

Gitan RS, Lou H, Rodgers J, Broderius M, Eide D. 1998 Zinc-induced inactivation of the yeast ZRT1 zinc transporter occurs through endocytosis and vacuolar degradation. *J Biol Chem* **273**, 28617–28624.

Grotz N, Fox T, Connolly E, Park W, Guerinot ML, Eide D. 1998 Identification of a family of zinc transporter genes from *Arabidopsis* that respond to zinc deficiency. *Proc Natl Acad Sci USA* **95**, 7220–7224.

Guerinot ML. 2000 The ZIP family of metal transporters. *Biochim Biophys Acta* **1465**, 190–198.

Gunshin H, Mackenzie B, Berger UV, Gunshin Y, Romero MF, Boron WF, Nussberger S, Gollan JL, Hediger MA. 1997 Cloning and characterization of a mammalian proton-coupled metal-ion transporter. *Nature* **388**, 482–488.

Hamer DH. 1986 Metallothionein. *Annu Rev Biochem* **55**, 913–951.

Hicke L. 1997 Ubiquitin-dependent internalization and down-regulation of plasma membrane proteins. *FASEB J* **11**, 1215–1226.

Huang L, Gitschier J. 1997 A novel gene involved in zinc transport is deficient in the lethal milk mouse. *Nature Genet* **17**, 292–297.

Kamizono A, Nishizawa M, Teranishi Y, Murata K, Kimura A. 1989 Identification of a gene conferring resistance to zinc and cadmium ions in the yeast *Saccharomyces cerevisiae*. *Mol Gen Genet* **219**, 161–167.

Kobayashi S, Miyabe S, Izawa S, Inoue Y, Kimura A. 1996 Correlation of the OSR1/ZRC1 gene product and the intracellular glutathione levels in *Saccharomyces cerevisiae*. *Biotech Appl Biochem* **23**, 3–6.

Kobayashi T, Beuchat M, Lindsay M, Frias S, Palmiter RD, Sakuraba H, Parton RG, Gruenberg J. 1999 Late endosomal membranes rich in lysobisphosphatidic acid regulate cholesterol transport. *Nature Cell Biol* **1**, 113–118.

Koh J, Suh SW, Gwag BJ, He YY, Hsu CY, Choi DW. 1996 The role of zinc in selective neuronal death after transient global cerebral ischemia. *Science* **272**, 1013–1016.

Korshunova YO, Eide D, Clark WG, Guerinot ML, Pakrasi HB. 1999 The IRT1 protein from *Arabidopsis thaliana* is a metal transporter with a broad substrate range. *Plant Mol Biol* **40**, 37–44.

Langmade SJ, Ravindra R, Daniels PJ, Andrews GK. 2000 The transcription factor MTF-1 mediates metal regulation of the mouse ZnT1 gene. *J Biol Chem* **275**, 34803–34809.

Lasat MM, Pence NS, Garvin DF, Ebbs SD, Kochian LV. 2000 Molecular physiology of zinc transport in the Zn hyperaccumulator *Thlaspi caerulescens*. *J Exp Biol* **51**, 71–79.

Li L, Kaplan J. 1998 Defects in the yeast high affinity iron transport system result in increased metal sensitivity because of the increased expression of transporters with a broad transition metal specificity. *J Biol Chem* **273**, 22181–22187.

Li L, Kaplan J. 2000 The yeast gene MSC2, a member of the cation diffusion facilitator family, affects the cellular distribution of zinc. *J Biol Chem* **275**, 5036–5043.

Lioumi M, Ferguson CA, Sharpe PT, Freeman T, Marenholz I, Mischke D, Heizmann C, Ragoussis J. 1999 Isolation and characterization of human and mouse *ZIRTL*, a member of the IRT1 family of transporters, mapping within the epidermal differentiation complex. *Genomics* **62**, 272–280.

Liuzzi JP, Blanchard RK, Cousins RJ. 2001 Differential regulation of zinc transporter 1, 2, and 4 mRNA expression by dietary zinc in rats. *J Nutr* **131**, 46–52.

MacDiarmid CW, Gaither LA, Eide D. 2000 Zinc transporters that regulate vacuolar zinc storage in *Saccharomyces cerevisiae*. *EMBO J* **19**, 2845–2855.

Magneson GR, Puvathingal JM, Ray WJ. 1987 The concentrations of free Mg^{2+} and Zn^{2+} in equine blood plasma. *J Biol Chem* **262**, 11140–11148.

Manning DL, McClelland RA, Knowlden JM, Bryant S, Gee JM, Green CD, Robertson JF, Blamey RW, Sutherland RL, Ormandy CJ, Nicholson RI. 1995 Differential expression of oestrogen regulated genes in breast cancer. *Acta Oncol* **34**, 641–646.

McClung JP, Bobilya DJ. 1999 The influence of zinc status on the kinetics of zinc uptake into cultured endothelial cells. *J Nutr Biochem* **10**, 484–489.

McGowan SJ, Gorham HC, Hodgson DA. 1993 Light-induced carotenogenesis in *Myxococcus xanthus*: DNA sequence analysis of the *carR* region. *Mol Microbiol* **10**, 713–735.

McMahon RJ, Cousins RJ. 1998a Mammalian zinc transporters. *J Nutr* **128**, 667–670.

McMahon RJ, Cousins RJ. 1998b Regulation of the zinc transporter ZnT-1 by dietary zinc. *Proc Natl Acad Sci USA* **95**, 4841–4846.

Menard MP, Cousins RJ. 1983 Zinc transport by brush border membrane vesicles from rat intestine. *J Nutr* **113**, 1434–1442.

Mowll JL, Gadd GM. 1983 Zinc uptake and toxicity in the yeast *Sporobolomyces roseus* and *Saccharomyces cerevisiae*. *J Gen Microbiol* **129**, 3421–3425.

Murgia C, Vespignani I, Cerase J, Nobili F, Perozzi G. 1999 Cloning, expression, and vesicular localization of zinc transporter Dri 27/ZnT4 in intestinal tissue and cells. *Am J Physiol* **277**, G1231–G1239.

Nies DH, Silver S. 1995 Ion efflux systems involved in bacterial metal resistances. *J Ind Microbiol* **14**, 186–199.

Nishimura K, Igarashi K, Kakinuma Y. 1998 Proton gradient-driven nickel uptake by vacuolar membrane vesicles of *Saccharomyces cerevisiae*. *J Bacteriol* **180**, 1962–1964.

Palmiter RD, Cole TB, Findley SD. 1996a ZnT-2, a mammalian protein that confers resistance to zinc by facilitating vesicular sequestration. *EMBO J* **15**, 1784–1791.

Palmiter RD, Cole TB, Quaife CJ, Findley SD. 1996b ZnT-3, a putative transporter of zinc into synaptic vesicles. *Proc Natl Acad Sci USA* **93**, 14934–14939.

Palmiter RD, Findley SD. 1995 Cloning and functional characterization of a mammalian zinc transporter that confers resistance to zinc. *EMBO J* **14**, 639–649.

Patzer SI, Hantke K. 1998 The ZnuABC high affinity zinc uptake system and its regulator Zur in *Escherichia coli*. *Mol Microbiol* **28**, 1199–1210.

Paulsen IT, Saier MH. 1997 A novel family of ubiquitous heavy metal ion transport proteins. *J Membr Biol* **156**, 99–103.

Pence NS, Larsen PB, Ebbs SD, Letham DL, Lasat MM, Garvin DF, Eide D, Kochian LV. 2000 The molecular physiology of heavy metal transport in the Zn/Cd hyperaccumulator *Thlaspi caerulescens*. *Proc Natl Acad Sci USA* **97**, 4956–4960.

Piletz JE, Ganschow RE. 1978 Zinc deficiency in murine milk underlies expression of the lethal milk (*lm*) mutation. *Science* **199**, 181–183.

Ramsay LM, Gadd GM. 1997 Mutants of *Saccharomyces cerevisiae* defective in vacuolar function confirm a role for the vacuole in toxic metal ion detoxification. *FEMS Microbiol Lett* **152**, 293–298.

Raskin I. 1995 Plant genetic engineering may help with environmental cleanup. *Proc Natl Acad Sci USA* **93**, 3164–3166.

Rensing CBM, Rosen BP. 1997 The zntA gene of *Escherichia coli* encodes a Zn(II)-translocating P-type ATPase. *Proc Natl Acad Sci USA* **94**, 14326–14331.

Reyes JG. 1996 Zinc transport in mammalian cells. *Am J Physiol* **270**, C401–C410.

Rhodes D, Klug A. 1993 Zinc fingers. *Sci Am* **268**, 56–65.

Rogers EE, Eide DJ, Guerinot ML. 2000 Altered selectivity in an *Arabidopsis* metal transporter. *Proc Natl Acad Sci USA* **97**, 12356–12360.

Sacher A, Cohen A, Nelson N. 2001 Properties of the mammalian and yeast metal-ion transporters DCT1 and Smf1p expressed in *Xenopus laevis* oocytes. *J Exp Biol* **204**, 1053–1061.

Stein WD. 1990 *Channels, Carriers, and Pumps: An Introduction to Membrane Transport*. Academic Press, San Diego; pp. 35–38.

Suhy D, O'Halloran TV. 1995 Metal responsive gene regulation and the zinc metalloregulatory model. In: Sigel H, ed. *Metal Ions in Biological Systems*. Marcel Dekker, New York; Vol. 32, pp. 557–578.

Sytkowski AJ. 1977 Metal stoichiometry, coenzyme binding, and zinc and cobalt exchange in highly purified yeast alcohol dehydrogenase. *Arch Biochem Biophys* **184**, 505–517.

Taylor KM. 2000 LIV-1 breast cancer protein belongs to a new family of histidine-rich membrane proteins with potential to control intracellular zinc homeostasis. *Life* **49**, 249–253.

Thompson DA, Stahl FW. 1999 Genetic control of recombination partner preference in yeast meiosis: isolation and characterization of mutants elevated for meiotic unequal sister-chromatid recombination. *Genetics* **153**, 621–641.

Thompson JD, Gibson TJ, Plewniak F, Jeanmougin F, Higgins DG. 1997 The ClustalX windows interface: flexible strategies for multiple sequence alignment aided by quality analysis tools. *Nuclc Acids Res* **25**, 4876–4882.

Tsuda M, Imaizumi K, Katayama T, Kitagawa K, Wanaka A, Tohyama M, Takagi T. 1997 Expression of zinc transporter gene ZnT-1 is induced after transient ischemia in the gerbil. *J Neurosci* **17**, 6678–6684.

Vallee BL, Auld DS. 1990 Zinc coordination, function, and structure of zinc enzymes and other proteins. *Biochemistry* **29**, 5647–5659.

van der Zaal BJ, Neuteboom LW, Pinas JE, Chardonnens AN, Schat H, Verkleij JAC, Hooykaas PJJ. 1999 Overexpression of a novel *Arabidopsis* gene related to putative zinc-transporter genes from animals can lead to enhanced zinc resistance and accumulation. *Plant Physiol* **119**, 1047–1055.

Victery W, Smith JM, Vander AJ. 1981 Renal tubular handling of zinc in the dog. *Am J Physiol* **241**, F532–F539.

Welch RM, Norvell WA, Schaefer SC, Shaff JE, Kochian LV. 1993 Induction of iron(III) and copper(II) reduction in pea (*Pisum sativum* L.) roots by Fe and Cu status: does the root-cell plasmalemma Fe(III)-chelate reductase perform a general role in regulating cation uptake? *Planta* **190**, 555–561.

White C, Gadd GM. 1987 The uptake and cellular distribution of zinc in *Saccharomyces cerevisiae*. *J Gen Microbiol* **133**, 727–737.

Zhang P, Allen JC. 1995 A novel dialysis procedure measuring free Zn^{2+} in bovine milk and plasma. *J Nutr* **125**, 1904–1910.

Zhao H, Butler E, Rodgers J, Spizzo T, Duesterhoeft S. 1998 Regulation of zinc homeostasis in yeast by binding of the ZAP1 transcriptional activator to zinc-responsive promoter elements. *J Biol Chem* **273**, 28713–28720.

Zhao H, Eide D. 1996a The yeast ZRT1 gene encodes the zinc transporter of a high affinity uptake system induced by zinc limitation. *Proc Natl Acad Sci USA* **93**, 2454–2458.

Zhao H, Eide D. 1996b The ZRT2 gene encodes the low affinity zinc transporter in *Saccharomyces cerevisiae*. *J Biol Chem* **271**, 23203–23210.

Zhao H, Eide D. 1997 Zap1p, a metalloregulatory protein involved in zinc responsive transcriptional regulation in *Saccharomyces cerevisiae*. *Mol Cell Biol* **17**, 5044–5052.

BioMetals **14:** 271–313, 2001.

Review

Zinc coordination sphere in biochemical zinc sites

David S. Auld
Center for Biochemical and Biophysical Sciences and Medicine and Department of Pathology, Harvard Medical School, Boston, Massachusetts 02115, USA (Fax: 617-566-3137; E-mail: David_Auld@hms.harvard.edu)

Received 15 April 2001; accepted 28 May 2001

Key words: crystal structure, metalloenzyme, NMR, protein sequence, X-ray crystallography, XAFS or X-ray absorption fine structure

Abstract

Zinc is known to be indispensable to growth and development and transmission of the genetic message. It does this through a remarkable mosaic of zinc binding motifs that orchestrate all aspects of metabolism. There are now nearly 200 three dimensional structures for zinc proteins, representing all six classes of enzymes and covering a wide range of phyla and species. These structures provide standards of reference for the identity and nature of zinc ligands in other proteins for which only the primary structure is known. Three primary types of zinc sites are apparent from examination of these structures: *structural, catalytic* and *cocatalytic*. The most common amino acids that supply ligands to these sites are His, Glu, Asp and Cys. In catalytic sites zinc generally forms complexes with water and any three nitrogen, oxygen and sulfur donors with His being the predominant amino acid chosen. Water is always a ligand to such sites. Structural zinc sites have four protein ligands and no bound water molecule. Cys is the preferred ligand in such sites. Cocatalytic sites contain two or three metals in close proximity with two of the metals bridged by a side chain moiety of a single amino acid residue, such as Asp, Glu or His and sometimes a water molecule. Asp and His are the preferred amino acids for these sites. No Cys ligands are found in such sites. The scaffolding of the zinc sites is also important to the function and reactivity of the bound metal. The influence of zinc on quaternary protein structure has led to the identification of a fourth type of zinc binding site, *protein interface*. In this case zinc sites are formed from ligands supplied from amino acid residues residing in the binding surface of two proteins. The resulting zinc site usually has the coordination properties of a catalytic or structural zinc binding site.

Abbreviations: ABC – ATP-binding cassette; AAP – *Aeromonas proteolytica* aminopeptidase; ADA – adenosine deaminase; ADAM – A disintegrin and metalloprotease domain; ADH – alcohol dehydrogenase; ALA – 5-aminolevulinic acid; ALAD – 5-aminolevulinic acid dehydratase; Apo2L or TRAIL – apoptosis-inducing ligand 2; BIR – baculovirus inhibitor of apoptosis repeat; BLAP – bovine lens leucine aminopeptidase; CA – carbonic anhydrase; CAM – γ-carbonic anhydrase; CPD A – carboxypeptidase A; CDA – cytidine deaminase; EDTA – ethylenediaminetetraacetic acid; eNOS or NOS-3 – endothelial nitric oxide synthase; FPP – farnesyl diphosphate; FTase – farnesyl transferase; H_4B – tetrahydrobiopterin; HIV – human immunodeficiency virus; GGPP – geranylgeranyl diphosphate; GSNO – S-nitrosoglutathione; HLA-DR – class II major histocompatibility molecule; huIFN – human interferon; IAP – inhibitor of apoptosis; iNOS or NOS-2 – inducible nitric oxide synthase; Im3 – *E. coli* immunity protein; IUB – International Union of Biochemistry; MEROPS – system for classification of peptidase sequences; MetAP-1 – methionine aminopeptidase-1; MetAP-2 – methionine aminopeptidase-2; MHC – major histocompatibility complex; MMP – matrix metalloproteinase; MPD – 2-methyl-2,4-pentanediol; NAD – nicotinamide adenine dinucleotide; NADH – reduced nicotinamide adenine dinucleotide; NADP – nicotinamide adenine dinucleotide phosphate; NGF – nerve growth factor; nNOS, NOS-1 – neuronal nitric oxide synthase; PAC – perturbed angular correlation of γ-rays; PAP – purple acid phosphatase; PBG – porphobilinogen;

PBGS – porphobilinogen synthase; Peptidase – enzyme acting on peptides; PLBP – periplasmic ligand-binding protein; PKC – protein kinase C; PMI – phosphomannose isomerase; PTS – signal transducing protein; PsaA – pneumococcal surface antigen; Proteinase – enzyme acting on proteins; SEA, B etc – staphylococcal enterotoxins type A, B etc; SPEA, C etc – streptococcal pyrogenic exotoxins type A, C etc; SMEZ – streptococcal mitogenic exotoxin; SOD – superoxide dismutase; TCR – T cell receptor; TL – thermolysin; TRAP – tartrate-resistant acid phosphatases; TNF – tumor necrosis factor; TACE – tumor necrosis factor-α-converting enzyme; TSST – toxic shock syndrome toxin; VanX – dipeptidase of vancomycin-resistant pathogenic *Enterococci*; XAFS – X-ray absorption fine structure.

Introduction

Zinc deficiency studies of microorganisms followed by those in plants and animals established the importance of zinc to growth and development in all forms of life (Vallee & Falchuk 1993). Technical advances in analytical methods that could detect the presence of zinc in minute amounts such as atomic absorption, fluorescence and microwave emission spectroscopy coupled with advances in the methodology for protein isolation and purification led to the establishment of zinc involvement in a wide variety of metabolic processes including carbohydrate, lipid, protein and nucleic acid synthesis and degradation (Vallee & Auld 1992a). Zinc is the only metal to have representatives in all six of the International Union of Biochemistry, IUB, classes of enzymes. The fact that it was demonstrated to be involved in transcription and translation of the genetic message gave new meaning to its known essentiality to life processes.

The molecular details of the participation of zinc in enzyme systems came first through replacing the spectroscopically silent zinc with the chromophoric metal cobalt. These studies in conjunction with kinetic studies of function gave information on the importance of the metal site to protein structure and function. Structural studies obtained by X-ray diffraction, NMR and X-ray absorption fine structure, XAFS, techniques gave detailed information of the metal ligands and the coordination geometry of the metal site and allowed formulation of mechanisms that could be tested by a combined approach using mutagenesis and kinetics (Auld 1997). Finally, the ability to determine primary protein structure through translation of DNA sequences now permits prediction of zinc binding sites and thereby enzyme function without even having an expressed protein (Auld 2001). This in turn heightens the awareness of the participation of zinc in metabolic processes.

The zinc ligands and coordination geometries of about *one and a half* dozen zinc enzymes led to the recognition of three types of zinc binding sites: *catalytic, cocatalytic* and *structural* (Figure 1) (Vallee & Auld 1990b, 1993a). Today there are about *fifteen* dozen zinc sites that have been reported, the great majority of which still fit into the original classification (see below). A new type of zinc binding site, *protein interface*, has also become apparent during the last few years (Auld 2001). In this case zinc binding sites are formed through ligands supplied from amino acid residues residing in the binding surface of two protein molecules.

Catalytic zinc sites

There are today about 7 dozen 3-dimensional structural references of *catalytic* zinc sites encompassing five of the six classes of enzymes (Table 1). The class III hydrolases has by far the greatest number of representatives. A catalytic zinc generally forms complexes with any three nitrogen, oxygen and sulfur donors of His, Glu, Asp and Cys with His being the predominant amino acid chosen. Histidine (usually the Nε2 nitrogen) may be chosen because of its capacity to disperse charge through H-bonding of its non-liganding nitrogen. The overall length of such sites can be as small as 11 amino acids as is observed in the astacin superfamily of zinc proteases and 5-aminolevulinic acid dehydratase (Table 1). The ligands are generally separated by short and long amino acid spacers. Short spacers of one and three are commonly found although spacers of 2, 4, 6 and 7 have also been observed (Table 1). The length of the short spacer is often dictated by the ligand support structure; 3 for an α-helix and 1 for a β-sheet. In the case of the alcohol dehydrogenase family the involvement of H-bonding interactions between residues in the short spacer (for example His47 and/or His51) and the cofactor NADH may have led to the extension of the 'short' spacer to 20 (versus a long spacer of about 106).

Table 1. Catalytic zinc sites[a].

Enzyme	PDB#	L_1	X	L_2	Y	L_3	Z	L_4	L_5	Ref.
				Class I: Oxidoreductases						
Alcohol dehydrogenase family										
Horse EE	8ADH & 3BTO	Cys_L	20	$His_{b\beta}$	106	Cys_L(C)	-	H_2O		(Cho *et al.* 1997; Colonna-Cesari *et al.* 1986; Eklund *et al.* 1981)
Mouse Class II	1E3E	Cys_L	20	$His_{b\beta}$	110	Cys_L(C)	-	H_2O		(Svensson *et al.* 2000)
Cod	1CDO	Cys_L	21	His_L	106	Cys_L	-	H_2O		(Ramaswamy *et al.* 1996)
Human $\beta_1\beta_1$	1HDZ	Cys_L	20	$His_{b\beta}$	106	Cys_L	-	H_2O		(Hurley *et al.* 1991)
Human $\beta_2\beta_2$	1HDY	Cys_L	20	$His_{b\beta}$	106	Cys_L	-	H_2O		(Hurley *et al.* 1994)
Human $\beta_3\beta_3$	1DEH	Cys_L	20	$His_{b\beta}$	106	Cys_L	-	H_2O		(Davis *et al.* 1996)
Human $\chi\chi$	1TEH	Cys_L	20	$His_{b\beta}$	106	$Cys_{2b\alpha}$	-	H_2O	Glu_β	(Yang *et al.* 1997)
Human $\sigma\sigma$	1AGN	Cys_L	20	$His_{b\beta}$	106	Cys_L	-	H_2O		(Xie *et al.* 1997)
Clostridium beijerinckii	1KEV	Cys_L	21	His_L	90	Asp_L(C)	-	H_2O	Glu_β	(Korkhin *et al.* 1998)
Thermoanaerobacter brockii	1YKF	Cys_L	21	$His_{b\beta}$	90	Asp_L	-	H_2O		(Korkhin *et al.* 1998)
				Class II: Transferases						
Rat farnesyl transferase	1FT1	Asp	1	Cys	62	His_α(C)	-	H_2O		(Park *et al.* 1997)
Rat Rab geranylgeranyltransferase	1DCE	$Asp_{b\alpha}$	1	Cys_α	49	His_α(C)	-	H_2O		(Zhang *et al.* 2000)
				Class III: Hydrolases						
Carboxypeptidase (CPD) family										
Bovine A	3CPA, 1CPX	His	2	$Glu_{2b\alpha}$	123	His_β(C)	-	H_2O		(Bukrinsky *et al.* 1998; Quiocho & Lipscomb 1971)
Bovine B	1CPB	His	2	Glu	123	His	-	H_2O		(Schmid & Herriott 1976)
Rat A_2		His	2	Glu	123	His	-	H_2O		(Faming *et al.* 1991)
Human A_2	1DTD	His	2	$Glu_{2b\alpha}$	124	His_β	-	H_2O		(Reverter *et al.* 2000)
Avian D	1QMU	His	2	$Glu_{2b\alpha}$	103	His_β	-	H_2O		(Gomis-Ruth *et al.* 1999)
Thermoactinomyces vulgaris T	1OBR	His	2	$Glu_{2b\alpha}$	131	His_β	-	H_2O		(Teplyakov *et al.* 1992)
Porcine PCPD A	1PCA	His	2	$Glu_{2b\alpha}$	123	His_β	-	H_2O		(Guasch *et al.* 1992)
Bovine PCPD A	1PYT	His	2	$Glu_{2b\alpha}$	123	His_β	-	H_2O		(Gomis-Ruth *et al.* 1995)
Porcine PCPD B	1NSA	His	2	$Glu_{2b\alpha}$	123	His_β	-	H_2O		(Coll *et al.* 1991)
Human PCPD A_2	1AYE	His	2	Glu	123	His_β	-	H_2O		(Garcia-Saez *et al.* 1997)

Table 1. Continued.

Enzyme	PDB#	L_1	X	L_2	Y	L_3	Z	L_4	L_5	Ref.
Thermolysin family										
Bacillus thermoproteolyticus	1LND	His_α	3	His_α	19	Glu_α(C)	-	H_2O		(Matthews *et al.* 1974)
Bacillus cereus	1ESP	His_α	3	His_α	19	Glu_α	-	H_2O		(Pauptit *et al.* 1988)
Pseudomonas aeruginosa	1EZM	His_α	3	His_α	19	Glu_α	-	H_2O		(Thayer *et al.* 1991)
Staphylococcus aureus	1BQB	His_α	3	His_α	19	Glu_α	-	H_2O		(Banbula *et al.* 1998)
Human leukotriene A_4 hydrolase	1HS6	His_α	3	His_α	18	Glu_α(C)	-	H_2O		(Thunnissen *et al.* 2001)
Human neutral endoprotease (Neprilysin)	1DMT	His_α	3	His_α	58	Glu_α(C)	-	H_2O		(Oefner *et al.* 2000)
Leishmania major surface proteinase	1LML	His_α	3	His_α	65	Glu_β(C)	-	H_2O		(Schlagenhauf *et al.* 1998)
Clostridium botulinum neurotoxin A	3BTA	His_α	3	His_α	34	Glu_L (C)	-	H_2O		(Lacy & Stevens 1999; Lacy *et al.* 1998)
Clostridium botulinum neurotoxin B	1EPW	His_α	3	His_α	33	Glu_α(C)	-	H_2O		(Swaminathan & Eswaramoorthy 2000)
Streptomyces albus G DD-CPD	1LBU	His	6	Asp_β	35	His_β (C)	-	H_2O		(Ghuysen 1988)
VanX D-Ala-D-Ala carboxypeptidase		His_α	6	Asp_β	60	His_β (C)	-	H_2O		(Bussiere *et al.* 1998)
Mouse Sonic Hedgehog	1VHH	His_α	6	Asp_β	34	His_β(C)	-	H_2O		(Hall *et al.* 1995)
Astacin superfamily	1AST	His_α	3	His_α	5	$His_{a\beta}$(C)	-	H_2O	Tyr	(Bode *et al.* 1992)
Serratia family										
Pseudomonas aeruginosa alkaline protease	1KAP	His_α	3	His_α	5	His(C)	-	H_2O	Tyr	(Baumann *et al.* 1993; Miyatake *et al.* 1995)
Serratia marcescens metalloprotease	1SAT	His_α	3	His_α	5	His	-	H_2O	Tyr	(Baumann *et al.* 1995)
Serratia sp. E-15 metalloprotease	1SRP	His_α	3	His_α	5	His	-	H_2O	Tyr	(Hamada *et al.* 1996)
Snake venom protease family										
Crotalus adamanteus or Adamalysin II	1IAG	His_α	3	His_α	5	His(C)	-	H_2O		(Gomis-Ruth *et al.* 1993a; Gomis-Ruth *et al.* 1994)
Trimeresurus flavoviridis H_2-Proteinase		His_α	3	His_α	5	His	-	H_2O		(Kumasaka *et al.* 1996)
Agkistrodon acutus, Acutolysin A	1BSW	His_α	3	His_α	5	His	-	H_2O		(Gong *et al.* 1998)
Agkistrodon acutus, Acutolysin C	1QUA	His_α	3	His_α	5	His	-	H_2O		(Zhu *et al.* 1999)
Human TNF-α-converting enzyme (TACE)	1BKC	His_α	3	His_α	5	His	-	H_2O		(Maskos *et al.* 1998)
Matrix metalloproteinase family										
Human fibroblast collagenase (MMP-1)	1CGL	His_α	3	His_α	5	His(C)	-	H_2O		(Lovejoy *et al.* 1994a)
Human fibroblast collagenase (MMP-1)	1AYK	His_α	3	His_α	5	His	-	H_2O		(Moy *et al.* 1999; Moy *et al.* 1998) NMR

Table 1. Continued.

Enzyme	PDB#	L_1	X	L_2	Y	L_3	Z	L_4	L_5	Ref.
Human matrilysin (MMP-7)	1MMP	His_α	3	His_α	5	His	-	H_2O		(Browner *et al.* 1995)
Human neutrophil collagenase (MMP-8)	1KBC	His_α	3	His_α	5	$His_{a\beta}$	-	H_2O		(Betz *et al.* 1997; Bode *et al.* 1994)
Human stromelysin-1 (MMP-3)	2SRT, 1BM6	His_α	3	His_α	5	His	-	H_2O		(Gooley *et al.* 1994; Li *et al.* 1998; Van Doren *et al.* 1995) NMR
Human stromelysin-1 (MMP-3)	1B3D	His_α	3	His_α	5	His	-	H_2O		(Chen *et al.* 1999; Dhanaraj *et al.* 1996)
Human prostromelysin-1 (MMP-3)	1SLM	His_α	3	His_α	5	His(C)	125	Cys_β(N)		(Becker *et al.* 1995)
Human collagenase-3 (MMP-13)	830C	His_α	3	His_α	5	His	-	H_2O		(Lovejoy *et al.* 1999)
Mouse collagenase-3 (MMP-13)	1CXV	His_α	3	His_α	5	His	-	H_2O		(Botos *et al.* 1999)
Human progelatinase 72kDa (MMP-2)	1CK7	His_α	3	His_α	5	His(C)	200	Cys_β (N)		(Morgunova *et al.* 1999)
Streptomyces caespitosus endopeptidase	1KUH	His_α	3	His_α	5	Asp(C)	-	H_2O		(Kurisu *et al.* 1997)
Murine adenosine deaminase	2ADA, 1A4L	His	1	His	196	His_β(C)	-	H_2O	Asp_L	(Wang & Quiocho 1998; Wilson *et al.* 1991)
Escherichia coli cytidine deaminase	1CTT	Cys_α	2	$Cys_{b\alpha}$	26	$His_{b\alpha}$(N)	-	H_2O		(Betts *et al.* 1994; Xiang *et al.* 1995)
Escherichia coli peptide deformylase	1BS4	His_α	3	His_α	41	Cys(N)	-	H_2O		(Becker *et al.* 1998; Chan *et al.* 1997; Meinnel *et al.* 1996)
E. coli colicin E7 (ColE7) DNase	7CEI	$His_{a\alpha}$	3	His_α	24	His_β(N)	-	H_2O		(Ko *et al.* 1999)
Human GTP cyclohydrolase I	1FB1	Cys_β	2	$His_{2b\beta}$	67	$Cys_{a\beta}$(C)	-	H_2O		(Auerbach *et al.* 2000)
S. cerevisiae π-Scei endonuclease	1EF0	Cys_β	1	His_β	372	Glu_β (N)	-	H_2O		(Poland *et al.* 2000)
Bacteriophage T7 lysozyme	1LBA	$Cys_{2b\beta}$	7	His_β	104	His_β(N)	-	H_2O		(Cheng *et al.* 1994)
β-Lactamase family										
Bacillus cereus	1BMC	$His_{2a\beta}$	1	His	60	His(C)	-	H_2O		(Carfi *et al.* 1998a; Carfi *et al.* 1995; Fabiane *et al.* 1998)
Bacteroides fraglis	1ZNB	His	1	His	60	His(C)	-	H_2O		(Concha *et al.* 1996)
Stenotrophomonas maltophilia	1SML	His	1	His	73	His(C)	-	H_2O		(Ullah *et al.* 1998)
Pseudomonas aeruginosa	1DD6	$His_{2a\beta}$	1	His	59	His (C)	-	H_2O		(Concha *et al.* 2000)
Escherichia coli alkaline phosphatase	1ALK	Asp_α	3	His_α	80	His(C)	-	H_2O		(Kim & Wyckoff 1991; Stec *et al.* 1998)
Bacillus cereus phospholipase C	1AH7	Glu_α	3	His_α	13	His_α(N)	-	H_2O		(Hough *et al.* 1989)
Penicillium citrinum P1 nuclease	1AK0	$Asp_{a\alpha}$	3	His_α	12	His(N)	-	H_2O		(Volbeda *et al.* 1991)
Escherichia coli Endonuclease IV	1QTW	His_β	3	$Asp_{b\beta}$	46	His_α(N)	-	H_2O		(Hosfield *et al.* 1999)
Hepatitis C virus proteinase	1AIQ, 1JXP	Cys	1	Cys	45	Cys_β(C)	-	H_2O		(Love *et al.* 1996; Yan *et al.* 1998)

Table 1. Continued.

Enzyme	PDB#	L_1	X	L_2	Y	L_3	Z	L_4	L_5	Ref.
Hepatitis C virus NS3 proteinase	1BT7	Cys	1	Cys	45	$Cys_{2a\beta}$(C)	-	H_2O		(Barbato *et al.* 1999) NMR
				Class IV: Lyases						
Carbonic anhydrase family										
Homo Sapiens CA I	2CAB	His_β	1	His_β	22	His_β(C)	-	H_2O		(Kannan *et al.* 1975)
Homo Sapiens CA II	1CA2	His_β	1	His_β	22	His_β(C)	-	H_2O		(Liljas *et al.* 1972)
Bovine CA III		His_β	1	His_β	22	His_β(C)	-	H_2O		(Eriksson & Liljas 1993)
Rattus Norvegius CA III	1FLJ	His_β	1	His_β	22	His_β(C)	-	H_2O		(Mallis *et al.* 2000)
Homo Sapiens membrane CA IV	1ZNC	His_β	1	His_β	22	His_β(C)	-	H_2O		(Stams *et al.* 1996)
Murine CA IV	2ZNC	His_β	1	His_β	22	His_β(C)	-	H_2O		(Stams *et al.* 1998)
Murine mitochrondrial CA V	1DMX	His_β	1	His_β	22	His_β(C)	-	H_2O		(Boriack-Sjodin *et al.* 1995)
Neisseria gonorrhoeae CA	1KOP	His_β	1	His_β	16	His_β(C)	-	H_2O		(Huang *et al.* 1998)
Pisum sativum β-CA	1EKJ	Cys_L	2	His_β	59	Cys_β(N)	-	Acetate		(Kimber & Pai 2000)
Porphyridium purpureum β-CA	1DDZ	Cys_L	2	His_β	55	Cys_β(N)	-	$Asp_{a\beta}$		(Mitsuhashi *et al.* 2000)
Methanosarcina thermophila β-CA	1G5C	Cys_L	2	His_β	54	Cys_β(N)	-	H_2O		(Strop *et al.* 2001)
Methanosarcina thermophila γ-CA	1THJ	His_β	4	$His_{\alpha\beta}$	35	His_β(N)	-	H_2O		(Kisker *et al.* 1996) see Table 4
Rat 6-pyruvoyl-tetrahydropterin synthase	1B6Z	His_β	1	His_β	24	His_β(N)	-	H_2O		(Burgisser *et al.* 1995; Ploom *et al.* 1999)
S. cerevisiae 5-aminolaevulinate dehydratase	1AW5,1QVN	Cys_β	1	$Cys_{a\beta}$	7	$Cys_{2b\beta}$(C)	-	H_2O		(Erskine *et al.* 2000; Erskine *et al.* 1997)
E. coli 5-aminolaevulinate dehydratase	1B4E	$Cys_{a\beta}$	1	Cys	7	Cys(C)	-	H_2O		(Erskine *et al.* 1999b)
E. coli fuculose 1-phosphate aldolase	4FUA,1DZU	His_β	1	$His_{\alpha\beta}$	60	$His_{b\beta}$(C)	-	H_2O	Glu_L	(Dreyer & Schulz 1993; Joerger *et al.* 2000)
				Class V: Isomerases						
Candida albicans phosphomannose isomerase	1PMI	Gln_β	1	$His_{a\beta}$	24	Glu_β	146	His_β	H_2O	(Cleasby *et al.* 1996)

[a]The amino acid spacer between ligands L_1 and L_2 is X; that between L_3 and nearest ligand L_1 or L_2 is Y and that between L_3 and L_4 is Z. The symbols N and C indicate that L_3 is located on the amino (N) or the carboxyl (C) side of L_2, respectively. The subscripts α, β, refer to the α-or 3_{10} helix and β-sheet structure which supplies the ligand. The letter subscript L denotes an amino acid sequence of ≤ 5 residues between two structural elements. The subscripts a and b indicate the ligand is either one (or two, 2) residues after or before the secondary structural element.

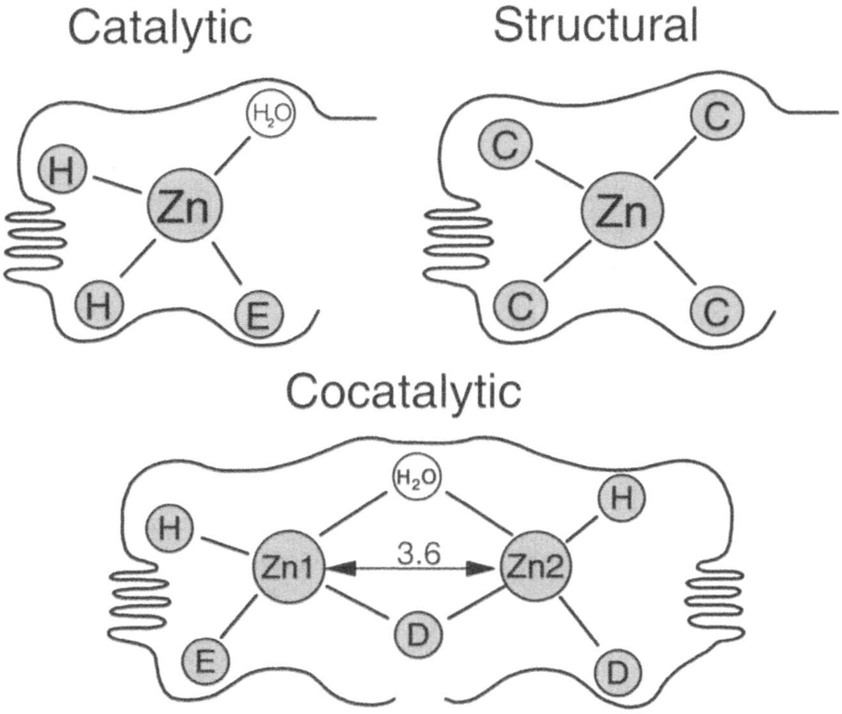

Fig. 1. Zinc binding sites in enzymes: catalytic (thermolysin (Matthews 1988)), structural (alcohol dehydrogenase (Eklund & Branden 1987)), cocatalytic (*Aeromonas proteolytica* aminopeptidase (Chevrier *et al.* 1994)). The letters D, E and H refer to the amino acids, aspartic acid, glutamic acid and histidine, respectively.

The coordination number for such sites is usually 4 or 5 and the geometry in the free state is frequently distorted-tetrahedral or trigonal-bipyramidal. Water is always a ligand to the catalytic zinc. The zinc-bound water is activated for ionization, polarization or displacement by the identity and arrangement of ligands coordinated to zinc (Vallee & Auld 1990a). Ionization of the activated water or its polarization brought about by a base form of an active-site amino acid provides hydroxide ions at neutral pH, and displacement of water or expansion of the coordination sphere results in Lewis acid catalysis by the catalytic zinc atom. The structure of the active site implies that the identity of the three protein ligands, their spacing and secondary interactions with neighboring amino acids in conjunction with the vicinal properties of the active center created by protein folding are all critical for the various mechanisms through which zinc can be involved in catalysis.

This group of zinc sites is too large for this perspective to comment on each individual zinc site. I will therefore generally restrict my comments to some of the larger and more well studied families.

Alcohol dehydrogenase

The dimeric alcohol dehydrogenases (ADH) (EC 1.1.1.1) contain both a catalytic and a structural zinc site (Tables 1 and 2). These zinc enzymes are NAD dependent and catalyze the reversible oxidation of alcohols to aldehydes. Seven human ADH genes have been identified and designated as ADH1 through ADH7 (Jornvall & Hoog 1995). The ADH1 to ADH5 and ADH7 encode 6 subunits of the ADH enzymes that are designated by the Greek letters, α, β, γ, π, χ and σ. The gene product of ADH6 has not been observed. Polymorphism occurs for the ADH2 (β) and ADH3(γ) loci resulting in nine distinct human subunits. These enzymes have also been classified, based on sequence identity and substrate specificity (Jornvall *et al.* 1987), as class I (α, β_1, β_2, β_3, γ_1, γ_2 containing isozymes), class II (π), class III (χ) and class IV (σ) ADHs. Class III ADH is also known as glutathione-dependent formaldehyde dehydrogenase (Yang *et al.* 1997). In addition, a protein that possesses the ability to metabolize S-nitrosoglutathione (GSNO) has been purified from *E. coli*, *S. cerevisiae* and mouse macrophages and identified as class III (χ) ADH (Liu *et al.* 2001).

The first crystal structure reported for this family of enzymes was for the horse enzyme, denoted EE, and considered to be part of class I ADH (Eklund *et al.* 1974, 1976). The three dimensional structure of several human isozymes as well as those from cod liver and mouse have been reported (Tables 1 and 3). A great deal of the structural and mechanistic studies have been performed on the horse liver enzyme. Each subunit of the dimeric enzyme is divided into a coenzyme binding domain and a catalytic domain that are separated by a cleft containing a deep pocket (Eklund & Branden 1987). Both zinc ions reside in the catalytic domain.

The catalytic zinc is ligated to the sulfurs of Cys46 and Cys174, the Nε2 nitrogen of His67 and a water molecule in a tetrahedral coordination geometry. When the coenzyme binds it triggers a major change in conformation of the enzyme (Eklund 1989). The two coenzyme-binding domains have similar orientations while the two catalytic domains are rotated relative to each other. When NADH is bound the cleft between the catalytic and coenzyme binding domains 'closes' around the coenzyme (Eklund *et al.* 1981). In the absence of the coenzyme the cleft is considered open (Eklund *et al.* 1976). Solvent is accessible to the catalytic zinc and the fourth ligand is water in the open conformation. The conformational changes places the zinc-bound substrate in the proper orientation to the C4 position of the cofactor nicotinamide ring for optimal hydride transfer (Klinman 1981).

The major changes in conformation upon cofactor binding have made it difficult to assign the pK_a values of about 7 and 9 in the pH profiles of coenzyme binding to functional groups in the enzyme (LeBrun & Plapp 1999). Candidates for these groups are the imidazolium of His51, the amino group of Lys228 and the catalytic zinc-bound water. The induction of the closed form by coenzyme permits the displacement of the water by the alcohol or aldehyde substrate and places the zinc substrate complex in a more hydrophobic environment (Eklund & Branden 1987).

The displacement of the zinc-bound water by substrate and role of the zinc as a Lewis acid catalyst has been a generally accepted mechanism although expansion of the coordination sphere to allow both Lewis acid catalysis by zinc and acid/base chemistry for zinc-bound water has been considered. A 1 Å resolution structure of the native zinc and cadmium substituted horse liver enzyme complexed with the cofactor NADH and 2-methyl-2,4-pentanediol (MPD) combined with quantum chemical calculations has given support to the involvement of a metal-bound water in catalysis (Meijers *et al.* 2001). The results suggest that a metal-bound hydroxide is part of the activation process for hydride transfer from the reduced NADH cofactor in the LADH•NADH complex. In the proposed mechanism the zinc-bound water is displaced towards the NADH in order to allow the aldehyde substrate to become the fifth ligand.

Metalloproteases

The carboxypeptidase family of exopeptidases and the thermolysin family of endopeptidases are likely examples of polarization assisted zinc water catalysis. The zinc containing pancreatic exopeptidases carboxypeptidase A & B (CPD A & B) were two of the earliest identified and studied zinc metalloenzymes (Auld, 1998a, 1998b; Aviles & Vendrel 1998). The carboxypeptidases catalyze the degradation of food proteins leading to the formation of amino acids. These enzymes complement the actions of chymotrypsin, pepsin and trypsin by allowing the production of essential amino acids such as Phe, Trp, Lys and Arg (Riordan 1974). The reader is directed to the Handbook of Proteolytic Enzymes for a presentation of evolutionary relationship of carboxypeptidases (Barrett *et al.* 1998) as well as several chapters on individual carboxypeptidases. The human mast cell, E, M, N carboxypeptidases are believed to be involved in immune/inflammatory and hormone processing (Auld 1998a; Vallee & Auld 1990b).

Three dimensional structures are available for several members of the CPD family (Table 1). The catalytic zinc site of CPD A is comprised of His69 (Nδ1), Glu72 (Oε1 and Oε2), His196 (Nδ1) and a water molecule. The first two ligands, separated by a short spacer of two, reside in a seven amino acid loop region between a β-sheet and an α-helix while His196 is the last residue in a β-pleated sheet extending from amino acids 191 to 196 (Rees *et al.* 1983). This site is highly conserved throughout the extended carboxypeptidase family (Vallee & Auld 1990b). There are also several crystalline derived structures of the thermolysin family (Table 1). In this case the amino acid ligands His142 (Nε2) and His146 (Nε2), separated by a short spacer of three, reside in an α-helix extending from amino acid 137 to 152 (Matthews *et al.* 1974). The amino acid residue Glu143, proposed to function as the general acid/general base in catalysis, resides within the short spacer (Holland *et al.* 1995; Matthews, 1988). The third ligand, Glu166 (Oε1), is supplied by a second α-helix extending from residue 160 to 180. The fourth ligand is water.

The immediate thermolysin family is composed of several bacterial endoproteases. However comparison of the properties of this zinc site to sequences of other proteins led to the prediction that the mono zinc aminopeptidases would also contain this same type of catalytic zinc site (Vallee & Auld 1990b). In addition these studies led to the prediction that human leukotriene A_4 hydrolase would be a zinc aminopeptidase with a thermolysin type zinc site. Both of these predictions have been confirmed (Haeggstrom *et al.* 1990; Medina *et al.* 1991; Thunnissen *et al.* 2001).

X-ray diffraction studies of the crystalline CPD A•inhibitor complexes (Christianson & Lipscomb 1989; Lipscomb & Strater 1996) in conjunction with spectroscopic studies on inhibitor and substrate binding on the cobalt substituted enzyme (Auld & Vallee 1987) and XAFS studies of the effect of pH and inhibitor bonding on the zinc coordination sphere (Auld 1997) have provided evidence for a mechanism involving the metal-bound water and a general acid/base role for Glu270 in catalysis assisted by Arg127 in the transition state (Auld 1987, 2001; Christianson & Lipscomb 1989; and references therein).

Astacin superfamily

This superfamily began with the identification of *Astacus* protease, or astacin, as a zinc protease in 1988 (Stocker *et al.* 1988). A potential zinc binding site signature existed within astacin, **H**E**xx**H**xx**G**xx**H, that was also present in a small number of proteases that would eventually make the four subclasses of this superfamily (Stocker *et al.* 1990) (Table 1). The recognition of the homology of both mouse kidney and human intestine meprin to *Astacus* protease resulted in the naming of the immediate astacin family of zinc proteases (Dumermuth *et al.* 1991). By 1992 the use of this putative zinc signature led to the identification of 33 proteases that defined four major groups of homologous proteins (Auld 1992). This discovery was rapidly followed by the X-ray crystallographic structure of astacin that confirmed the prediction of this new catalytic zinc site (Bode *et al.* 1992). Within two years the structures of a member of each of the astacin subfamilies was determined, i.e., matrix metalloproteinase-1 (Lovejoy *et al.* 1994b), the snake venom protease, adamalysin II (Gomis-Ruth *et al.* 1993a) and the alkaline protease of *Pseudomonas aeruginosa* (Baumann *et al.* 1993). As of Mar. of 2001 the MEROPS data base file lists 441 members of this superfamily (Barrett & Rawlings 2001) and there are 19 reported structures of individual members (Table 1). The human class of snake venom like proteases have been given the name ADAM, which stands for A disintegrin and metalloprotease domain. The tumor necrosis factor-α-converting enzyme or TACE is a member of this sub-group (Maskos *et al.* 1998). The involvement of these proteases in remodeling the extracellular matrix makes them targets of diseases that require these processes such as tumor progression, metastasis, arthritis and heart tissue instability.

The zinc binding site in this superfamily is the smallest site known since all zinc ligands and the presumed catalytic glutamate residue are supplied from an eleven amino acid segment (Table 1). The first two His ligands (Nε2) are part of a long (12 to 15 amino acids) α-helix in all members of this superfamily. This helix is broken by the highly conserved glycine residue residing three residues after the second His. The break in the helix allows the third His (Nε2) to complete the zinc binding site.

The role for the glutamate in catalysis is generally believed to be the same as in the carboxypeptidase A and thermolysin families. However comparison of the structures of matrilysin, thermolysin (TL) and carboxypeptidase A reveals both similarities and differences in their active sites (Auld 1997). A common feature is a catalytic zinc atom that is coordinated by three protein ligands and a nearby ionizable carboxylate group of a Glu residue that is considered to act as a nucleophile or general base. The fourth ligand is water in the active enzyme. However, the type of the ligand and the scaffolding of the zinc site is not the same. The catalytic zinc of matrilysin is made up of three His residues whereas the zinc atom of thermolysin and CPD-A contains 2 His and 1 Glu (Table 1). Furthermore, the secondary interactions of the zinc ligands with adjacent side chain carboxylate groups observed in TL and CPD-A is not observed in matrilysin. Mutagenesis studies of the proposed Glu residue in matrilysin suggest this residue may play a different role in catalysis in the MMPs (Cha & Auld 1997).

β-Lactamases

The majority of β-lactamases utilize an active site serine in the hydrolysis of the β-lactam ring. However there are now new pathogenic bacteria that have a metallo-β-lactamase that contains zinc. Several X-ray structures have recently appeared on this class of zinc enzymes from *Bacillus cereus* (Carfi *et al.* 1995, 1998a; Fabiane *et al.* 1998), *Bacteroides fragilis* (Carfi *et al.* 1998b; Concha *et al.* 1996), *Stenotrophomonas maltophilia* (Ullah *et al.* 1998) and *Pseudomonas aeruginosa* (Concha *et al.* 2000). These structures are similar in the presence of at least one zinc site that has the characteristics of a catalytic zinc site. The first structure of the *B. cereus* enzyme had one zinc coordinated by three histidines 86 (Nε2), 88 (Nδ1) and 149 (Nε2) and a water molecule in a tetrahedral arrangement (Carfi *et al.* 1995). These histidine residues are conserved in all members of this class of enzymes. These crystals were grown at pH 5.6 in 0.1 M $ZnSO_4$ in a citrate, cacodylate buffer. At this pH the binding constant for a second zinc is 29 mM (Baldwin *et al.* 1978). Growing the crystals at pH 7.0 in the presence of 0.5 mM $ZnSO_4$ in Tris buffer yields an enzyme with two zincs bound (Fabiane *et al.* 1998). The second zinc binds to Asp90, Cys168, and His210 and two water molecules in a trigonal bipyramidal coordination geometry. There is considerable variation in the ligand properties of the second zinc site raising questions about its role in catalysis (Auld 2001) (see below, cocatalytic sites).

Carbonic anhydrases

The class IV lyase carbonic anhydrase (CA) is likely the best example of an ionization activated zinc-bound water mechanism. This superfamily of enzymes, involved in the physiology of CO_2 transport, has been assigned to three independent gene families α, β and γ (Hewett-Emmett & Tashian 1996). The α-class contains all mammalian, as well as some CAs from algae and bacteria. Several distinct forms exist in mammals: three cytosolic forms, (CA I, II and III); two membrane-bound forms (CA IV and CA VII); a mitochondrial form (CA V) and a secreted salivary form (CA VI). Crystal structures of five different forms of the α-enzyme class (Table 1) has shown the catalytic zinc is located in a 15 Å deep active center cavity near the middle of the enzyme molecule. One part of the cavity is dominated by hydrophobic residues while another segment has a more hydrophilic nature (Silverman & Lindskog 1988). The catalytic zinc is tetrahedrally coordinated to the Nε2 of His94 and His96 supplied from a β-strand encompassing residues 88 to 108 and a Nδ1 of His119 from a second β-strand extending from 113 to 126 (Liljas *et al.* 1972) and a water molecule. The amino acid ligands and adjacent amino acids are highly conserved throughout this class of CAs.

The β-class has a strikingly different catalytic zinc coordination site (Table 1). This class includes CAs from plants, algae, bacteria and archaea (Hewett-Emmett & Tashian 1996; Smith & Ferry 2000). The higher plant and unicellular green algae use the β-CAs for photosynthetic CO_2 fixation. The crystal structures *Pisum sativum* β-CA (Kimber & Pai 2000), *Porphyridium purpureum* β-CA (Mitsuhashi *et al.* 2000) and *Methanosarcina thermophila* β-CA (Strop *et al.* 2001) have been reported recently. In all cases the catalytic zinc is coordinated by two Cys and a His as was anticipated by XAFS studies of the spinach CA (Bracey *et al.* 1994; Rowlett *et al.* 1994). Two of the ligands, His87 (Nε2) and Cys90, are supplied by a loop region and a β-sheet, respectively and are separated by a short spacer of two (Table 1). The third ligand, Cys32 (*M. thermophila* β-CA numbering), is supplied by a β-sheet from the N-terminal side. The reported structure for the *P. purpureum* β-CA (Mitsuhashi *et al.* 2000) also has an Asp coordinated to the zinc with no bound water molecules. It was therefore proposed that this β-CA and possible the γ-class does not use a zinc hydroxide mechanism in their function (Mitsuhashi *et al.* 2000). However in the other two reported structures the fourth ligand is a solvent molecule (acetate) for *P. sativum* β-CA (Kimber & Pai 2000) and a water molecule for *M. thermophila* β-CA (Strop *et al.* 2001). The potential Asp ligand exists in all three enzymes in a highly conserved loop region C-terminal to the third Cys ligand (CysXAs pSerArg). In the enzymes where the Asp is not a ligand the Arg guanidinium group is hydrogen-bonded to the Asp carboxyl group preventing it from coordinating the zinc (Strop *et al.* 2001). The zinc coordinated water molecule in the *M. thermophila* β-CA has therefore been displaced either by an acetate carboxylate in the *P. sativum* β-CA or the carboxylate of Asp151 in the *P. purpureum* β-CA. Further studies will be needed to determine what is the function of the Asp residue in question. However comparison of the α- and β-CA catalytic zinc sites is of interest in this regard. The positive charge on the zinc in the β-CAs should be reduced by the replacement of two His imidazole ligands by Cys sulfur ligands. The ionization of the zinc bound water molecule might therefore need the assistance of a neighboring carboxyl group as is postulated in metalloproteases (Auld 1987, 1997). A role for Asp151 in catalysis rather than zinc binding could be inferred from the results of a mutation of this residue to an Asn in spinach β-CA. The resulting enzyme still binds zinc but retains little CO_2 hydratase activity (Mitsuhashi *et al.* 2000).

The γ-class of CAs has one structural representative from *Methanosarcina thermophila* (Iverson *et al.* 2000; Kisker *et al.* 1996). While it retains three His ligands as in the α-class the spacing characteristics change (Table 1). The resulting trimeric enzyme forms a zinc site from the interface of its subunits (See protein interface and Table 4).

The α-class of enzymes is one of the best studied from the point of mechanism of action. The role of the zinc ligands, its bound water molecule, the 'gatekeeper' residue, Thr199 and orientating residues such as Glu106 and proton shuttle residues such as His64 have been carefully investigated (See reviews by Christianson & Cox 1999; Christianson & Fierke 1996; Coleman 1998; Lindskog & Liljas 1993; Silverman & Lindskog 1988).

Other catalytic zinc sites containing Cys ligands

While the predominant ligand found in these sites is His there are a number of sites that contain from one to three Cys residues. Several catalytic sites now exist where there is one Cys ligand in combination with two His or one His and an Asp or Glu residue (Table 1). In the likely best studied single Cys containing zinc site, the protein prenyltransferases catalyze the formation of a sulfur ether linkage between either the isoprenoid units of farnesyl diphosphate (FPP) or geranylgeranyl diphosphate (GGPP) and the cysteine residues of protein acceptor substrates such as Ras thereby attaching these proteins to the cell membrane (Zhang & Casey 1996). Since a number of human cancers are linked to Ras mutations, inhibition of these enzymes is a potential target for antitumor chemotherapy (Gibbs & Oliff 1997). In both the rat farnesyl transferase (FTase) and the Rab geranylgeranyl transferase the zinc site is found in the β-subunit coordinated to a Cys sulfur and an Asp carboxylate oxygen separated by one amino acid (Park *et al.* 1997; Zhang *et al.* 2000). The Nε2 nitrogen of a His residue and a water molecule complete the tetrahedral zinc site. Spectral kinetic studies on the cobalt substituted FTase (Huang *et al.* 1997) and an x-ray structure of a ternary complex containing a Cys peptide (Strickland *et al.* 1998) indicate that the Cys peptide thiol group displaces the metal-bound water. These results suggest the role of zinc in these enzymes is to activate the cysteine thiol of the protein substrate for nucleophilic attack on the C1 position of the FPP substrate.

The deamination mechanism of *E. coli* cytidine deaminase (CDA) is believed to be similar to that of *E. coli* adenosine deaminase (ADA) (Betts *et al.* 1994; Wang & Quiocho 1998). However the zinc binding site of ADA contains 3 His residues while that of CDA contains two Cys and one His residue (Table 1). Examination of the structures of transition state analogs show that the inhibitor complex is stabilized by a zinc hydroxide and an adjacent carboxyl group of an Asp or Glu residue in the active site. The differences in zinc ligands and tertiary structures of ADA and CDA has led to the proposal that the common features in the transition state stabilization has arisen from convergent evolution (Betts *et al.* 1994). The variation in the zinc binding site nature in this functional class of enzymes is also evident in the *Bacillus subtilis* cytidine deaminase. Sequence alignment has led to the proposal that the tetrameric *B. subtilis* enzyme has three Cys ligands, Cys89, Cys86 and Cys53 coordinated to the catalytic zinc (Carlow *et al.* 1999).

A three Cys ligated catalytic zinc site has been proposed for the 5-aminolevulinic acid dehydratase (ALAD) of *E. coli* (Erskine *et al.* 1999b) and *Saccharomyces cerevisiae* (Erskine *et al.* 1997). All ligands reside in a short amino acid sequence of eleven amino acids with the first two Cys ligands being separated by a spacer of one. ALAD or porphobilinogen synthase (PBGS) catalyzes the condensation of two 5-aminolevulinic acid (ALA) molecules to form the pyrrole porphobilinogen (PBG), one of the initial steps in the biosynthesis of tetrapyrroles. The zinc site lies next to the lysine residue in the active site that is believed to form a Schiff base with 5-aminolevulinic acid (Erskine *et al.* 1999a). The zinc ion is postulated to coordinate the C4 carbonyl of ALA. A water molecule is bound to the zinc in the *E. coli* enzyme (Erskine *et al.* 1999b) on the side of the two active site Lys residues. While the same tetrahedral geometry is found in the *S. cerevisiae* enzyme the fourth site is vacant on the Lys side (Erskine *et al.* 1997).

The biochemistry of this enzyme has received particular attention due an inactivating mutation in the human enzyme that is responsible for an inherited disease, porphyria (Doss *et al.* 1979) and the marked inhibition by lead ions, 70 fM (Simons 1995). Both lead and mercury have been shown to displace the active site zinc by the use of multiwavelength anomalous diffraction analysis (Erskine *et al.* 2000). The ligands involved in binding the zinc are not conserved in a number of plant species (Jaffe 2000). The three Cys ligands are replaced by either two Asp and an Ala or some combination of Asp and Cys residues. This finding has lead to the postulate that Mg may replace Zn in the plants. However the *Pseudomonas aeruginosa* ALAD (two Asp residues and an Ala) contains neither a Zn or Mg at this site (Frankenberg *et al.* 1999). Other binding sites for both zinc and magnesium have been observed in this class of enzymes (Jaffe 2000).

Another potential three Cys containing zinc site may exist in the NS3 protease from hepatitis C virus. This enzyme has a zinc site on the protein surface composed of Cys97, Cys99, Cys145 and a Zn-bound water molecule (Love *et al.* 1996). The zinc-bound water is further H-bonded to His149 in two of the three monomers in the asymmetrical unit cell. The His imidazole ring isolates the Zn site from solution. NMR studies of the enzyme also suggest a water or hydroxyl group is directly bound to the zinc and that the His imidazole nitrogen may H-bond to this water (Barbato *et al.* 1999; Urbani *et al.* 1998). This type of hydrogen bonding interaction is often seen in catalytic zinc sites but the zinc binding site in NS3 protease is assumed to play a structural role since it is 23 Å removed from the serine proteinase active site (Yan *et al.* 1998). The bound zinc can not be removed by chelators at pH values > 7 and its removal at lower pH values leads to protein unfolding and aggregation of the enzyme (De Francesco *et al.* 1996). However the k_{cat}/K_m value for the His149Ala mutant is decreased by 15,000-fold, suggesting there is some relationship between the coordination geometry of the metal site and catalytic activity (Urbani *et al.* 1998). The processing of the precursor protein NS2-NS3 is cleaved intramolecularly by an autocatalytic action that is chelator and zinc dependent (Pieroni *et al.* 1997). If this zinc site should be involved in the autoprocessing activity it may be more properly classified as a catalytic site.

Function of a fourth protein ligand

A fourth protein ligand has been observed in a few crystal structures of catalytic zinc sites. A highly conserved Cys is found in the zymogen form of the matrix metalloproteinases, MMPs, with no bound water molecule (Becker *et al.* 1995; Morgunova *et al.* 1999) (Table 1). The replacement of the water by the Cys residue was proposed to be cause of the lack of activity of the enzyme (Springman *et al.* 1990; Vallee & Auld 1990b). Upon activation, this metal site is converted into one that contains three protein ligands and a bound water, characteristic of a catalytic zinc site (Table 1). The only other example of a proposed catalytic zinc site lacking a water molecule is the *Porphyridium purpureum* β-carbonic anhydrase (Mitsuhashi *et al.* 2000). However other members of the β-carbonic anhydrases have 3 protein ligands and a solvent molecule bound to the zinc (see section on carbonic anhydrase).

Water is still bound to the site in a number of cases where a fourth protein ligand is found (Table 1). In these cases it appears that one ligand may dissociate from the zinc before or during catalysis. Thus the Oδ2 of the Asp295 carboxylate is 2.3 Å from the zinc in an adenosine deaminase inhibitor complex in which a zinc-activated water (hydroxide) binds to the metal (Wilson & Quiocho 1993). Conversion of this Asp residue to a Glu through mutagenesis results in the displacement of the metal-bound water by the γ-carboxyl group of Glu and inactivation of the enzyme (Sideraki *et al.* 1996). The role of the side chain carboxyl group Asp295 in catalysis is therefore not likely in binding to the zinc but rather in stabilizing the charge of the hydrate intermediate or orienting the zinc bound hydroxide oxygen for addition to the C6 position of adenosine (Wang & Quiocho 1998). The functionally similar zinc enzyme, cytidine deaminase, has a catalytic zinc bound to 3 protein ligands and a water or hydroxide ion (Betts *et al.* 1994) (Table 1).

In *Clostridium beijerinckii* alcohol dehydrogenase a Glu residue adjacent to the second catalytic zinc ligand is also bound to the zinc in the NADP free form of the enzyme (Korkhin *et al.* 1998). However, the distance between the carboxylate oxygen and the zinc increases from 2.35 to 3.94 Å upon binding the coenzyme. In the human $\chi\chi$ alcohol dehydrogenase, ADH, the equivalent residue, Glu68 is observed 2.9 Å and 2.0 Å from the catalytic zinc in the A and B subunits respectively (Yang *et al.* 1997). In other human and horse ADHs the carboxylate oxygens are at least 4.7 Å from the catalytic zinc atom. The binding of the Glu to the zinc in this case may be a reflection of the involvement of Glu68 in the displacement of water in a late step in catalysis as predicted by molecular dynamics calculations (Ryde 1995; Yang *et al.* 1997). This unusual binding of Glu to the metal may also be reflected in the presence of a pH independent band at 562 nm in Co(II) substituted $\chi\chi$ ADH (Maret 1989). Such an absorption band had been assigned to an anion binding in studies of horse EE ADH (Bertini *et al.* 1987; Maret & Zeppezauer 1986). The Glu68-carboxylate coordinated to zinc would be consistent with such a zinc-bound 'anion' species (Yang *et al.* 1997).

In *Escherichia coli* fuculose aldolase, both of the carboxylate oxygens of Glu73 bind to the zinc in the resting state (Dreyer & Schulz 1993). Upon binding of a transition state analog phosphoglycolohydroxamate, Glu73 rotates away providing room for the hydroxamate oxygen to bind to the zinc (Dreyer & Schulz 1996). In astacin, the gram-negative bacterium *Serratia marcescens* proteinase and the alkaline protease from *Pseudomonas aeruginosa* a conserved tyrosine residue following the Met turn in these proteins acts as a fourth protein zinc ligand with phenolic oxygen to Zn distances of 2.54 Å (Gomis-Ruth *et al.* 1993b), 2.75 Å (Baumann 1994; Hamada *et al.* 1996), and 3.01 Å (Miyatake *et al.* 1995) respectively, resulting in a trigonal-bipyramidal coordination sphere in the free enzymes. Binding of a phosphinic peptide inhibitor to astacin increases the distance between the phenolic oxygen and the zinc from 2.5 Å to 5.0 Å (Grams *et al.* 1996). Movement of the tyrosine away from the zinc has also been observed in inhibited forms of the *Serratia* protease (seralysin) (Baumann *et al.* 1995) and in the alkaline protease from *Pseudomonas aeruginosa* (Baumann *et al.* 1993; Miyatake *et al.* 1995).

The zinc binding site in *Candida albicans* phosphomannose isomerase, PMI, contains 4 protein ligands and a water molecule arranged in a distorted trigonal bipyramid (Cleasby *et al.* 1996). All the ligands are on three β-strands in the catalytic domain. Gln111 and His113 (Nε2) are on the same β-strand. The site is unusual for a catalytic zinc site in that one of the ligands is Gln, a residue found in cocatalytic sites but not previously in a catalytic zinc site. Gln111, Glu168 (Oε1), and His285 (Nε2) form the equatorial ligands and His113 and water the axial ligands. The pH dependence of PMI on activity was originally suggested to be due to a His residue on the basis of diethyl pyrocarbonate modification experiments (Cleasby *et al.* 1996). In view of some of the other 4 protein ligand sites the axial ligand His113 or one of the equatorial ligands, His285 or Glu168 may be displaced by the substrate in order to participate in the reversible isomerization of fructose-6-phosphate and mannose 6-phosphate.

In a number of the cases cited above the fourth protein ligand bonding distance is 2.3 to 3 Å which is long for inner sphere oxygen-zinc ligation. Such a distance maybe an indicator for an amino acid that dissociates from the zinc upon substrate binding and plays a role in catalysis.

Structural zinc sites

Structural zinc sites have 4 protein ligands and no bound water molecule. The first zinc enzymes recognized to have a structural zinc site were horse alcohol dehydrogenase (Eklund *et al.* 1974) and the regulatory subunit of aspartate carbamoyltransferase (Honzatko *et al.* 1982). In both cases the zinc is bound to four cysteines in a relatively short linear sequence of 15 to 33 amino acids with the ligands separated by 2, 2 and 7 or 4, 23 and 2 amino acids, respectively. There are today about 41/2 dozen zinc sites with representatives in five of the six IUB enzyme classes (Table 2). The overall length of these sites can vary from 15 to 209 but the majority are in the range of 20 to 40 amino acids. There is at least one short spacer, generally containing two amino acids, in the great majority of these sites. While Cys is still by far the preferred ligand for such sites other ligands are also found. The second most prevalent ligand is His which is usually found in combination with Cys. The presence of His and Cys in the same zinc binding site frequently gets some form of a 'zinc finger' nomenclature. Based on only 'loose' criteria some of the sites listed here are still referred as zinc finger domains even if they do not fit the criteria of having DNA binding properties or of the ligand nature (2His/2Cys), sequence length ($\sim$30 amino acids) and presence of both α-helix and β-sheets in the zinc binding site associated with a classical zinc finger (Vallee *et al.* 1991).

His and Asp/Glu sites

There are only a few examples where Asp and Glu in combination with His residues are found (Table 2). The matrix metalloproteinase class of enzymes contain such a non-Cys structural zinc site. Early metal analyses of the matrix metalloproteinases indicated that these enzymes contained a zinc site in addition to

Table 2. Structural zinc sites[a].

Enzyme	PDB#	L_1	X	L_2	Y	L_3	Z	L_4	Ref.
		Class I: Oxidoreductases							
Alcohol dehydrogenase									
Horse EE	8ADH & 3BTO	Cys	2	$Cys_{b\alpha}$	2	Cys_{α}	7	Cys	(Eklund *et al.* 1981)
Cod	1CDO	Cys	2	$Cys_{b\alpha}$	2	Cys_{α}	7	Cys	(Ramaswamy *et al.* 1996)
Mouse Class II	1E3E	Cys	2	$Cys_{b\alpha}$	2	Cys_{α}	7	Cys	(Svensson *et al.* 2000)
Human $\beta_1\beta_1$	1HDZ	Cys	2	$Cys_{b\alpha}$	2	Cys_{α}	7	Cys	(Hurley *et al.* 1991)
Human $\beta_2\beta_2$	1HDY	Cys	2	$Cys_{b\alpha}$	2	Cys_{α}	7	Cys	(Hurley *et al.* 1994)
Human $\beta_3\beta_3$	1DEH	Cys	2	$Cys_{b\alpha}$	2	Cys_{α}	7	Cys	(Davis *et al.* 1996)
Human $\chi\chi$	1TEH	Cys	2	$Cys_{b\alpha}$	2	Cys_{α}	7	Cys	(Yang *et al.* 1997)
Human $\sigma\sigma$	1AGN	Cys	2	$Cys_{b\alpha}$	2	Cys_{α}	7	Cys	(Xie *et al.* 1997)
Bovine heart cytochrome c oxidase	1OCC	Cys_{β}	1	$Cys_{2a\beta}$	19	Cys_{β}	2	Cys_{L}	(Karlin *et al.* 1998)
		Class II: Transferases							
Aspartate carbamoyltransferase	1AT1	Cys	4	$Cys_{2b\alpha}$	23	Cys_{β}	2	Cys_{L}	(Honzatko *et al.* 1982)
Human Bruton's tyrosine kinase	1BTK	$His_{a\beta}$	10	$Cys_{a\beta}$	0	$Cys_{2a\beta}$	9	$Cys_{b\beta}$	(Hyvonen & Saraste 1997)
Bacillus stearothermophilus adenylate kinase	1ZIP	Cys_{β}	2	Cys_{L}	16	Cys_{β}	2	Cys_{L}	(Berry & Phillips 1998)
Protein kinase family									
Human casein kinase, CK2β	1QF8	Cys_{β}	4	Cys_{L}	22	Cys_{β}	2	$Cys_{2b\beta}$	(Chantalat *et al.* 1999)
Rat C-α Cys-rich domain		Cys	2	Cys	21	His	2	Cys	(Hommel *et al.* 1994) NMR
Rat C-α		His	29	Cys	2	Cys	15	Cys	(Hommel *et al.* 1994)
Rabbit C-α Cys-rich domain		Cys_{β}	2	$Cys_{\alpha\beta}$	21	His_{β}	2	Cys_{α}	(Ichikawa *et al.* 1995) NMR
Rabbit C-α		$His_{b\beta}$	29	Cys_{β}	2	$Cys_{b\beta}$	15	Cys	(Ichikawa *et al.* 1995)
Mouse C-δ activator-binding domain	1PTQ	Cys_{β}	2	$Cys_{2b\beta}$	21	$His_{a\beta}$	2	Cys	(Zhang *et al.* 1995)
Mouse C-δ	1PTQ	$His_{b\beta}$	29	Cys_{β}	2	Cys_{L}	15	Cys	(Zhang *et al.* 1995)
Human Raf-1	1FAQ	Cys_{β}	2	Cys_{L}	17	$His_{a\beta}$	2	Cys	(Mott *et al.* 1996) NMR
Human Raf-1	1FAQ	His	25	Cys_{β}	2	Cys	15	Cys	(Mott *et al.* 1996)
Thermococcus celer RNA polymerase II RPB9	1QYP	Cys_{β}	2	$Cys_{b\beta}$	24	Cys_{β}	2	Cys_{L}	(Wang *et al.* 1998) NMR
Methanobacterium Thermautotrophicum									
RNA polymerase subunit RPB10	1EF4	Cys	2	Cys	33	Cys_{α}	0	Cys_{α}	(Mackereth *et al.* 2000) NMR
Thermus aquaticus RNA polymerase	1DDQ	$Cys_{a\alpha}$	81	Cys	6	$Cys_{b\alpha}$	2	Cys_{α}	(Zhang *et al.* 1999)
E. coli DNA polymerase III δ' subunit	1A5T	$Cys_{a\alpha}$	8	Cys	2	$Cys_{b\alpha}$	2	Cys_{α}	(Guenther *et al.* 1997)

Table 2. Continued.

Enzyme	PDB#	L_1	X	L_2	Y	L_3	Z	L_4	Ref.
Galactose-1-phosphate uridylyltransferase	1GUP	Cys	2	Cys	59	His	48	$His_{b\beta}$	(Geeganage & Frey 1999; Thoden *et al.* 1997)
Bacillus stearothermophilus DNA primase	1D0Q	$Cys_{2a\beta}$	2	His	17	Cys_{β}	2	Cys_{L}	(Pan & Wigley 2000)
Human O6-Alkylguanine-DNA alkyltransferase	1EH6	Cys	18	Cys_{β}	4	Cys_{β}	55	Cys	(Daniels *et al.* 2000)
				Class III: Hydrolases					
Matrix metalloproteinase family									
Human fibroblast collagenase (MMP-1)	1CGL	His	1	Asp	12	His_{β}	12	His_{β}	(Lovejoy *et al.* 1994a; Moy *et al.* 1998)
Human fibroblast collagenase (MMP-1)	3AYK	His	1	Asp	12	His_{β}	12	His_{β}	(Moy *et al.* 1999; Moy *et al.* 1998) NMR
Human matrilysin (MMP-7)	1MMP	His	1	Asp	12	His_{β}	12	His_{β}	(Browner *et al.* 1995)
Human neutrophil collagenase (MMP-8)	1KBC	His	1	Asp	12	His_{β}	12	His_{β}	(Betz *et al.* 1997; Bode *et al.* 1994)
Human stromelysin-1 (MMP-3)	2SRT, 1BM6	His	1	Asp	12	His_{β}	12	His_{β}	(Gooley *et al.* 1994; Li *et al.* 1998; Van Doren *et al.* 1995) NMR
Human stromelysin-1 (MMP-3)	1B3D	His	1	Asp	12	His_{β}	12	His_{β}	(Chen *et al.* 1999; Dhanaraj *et al.* 1996)
Human prostromelysin-1 (MMP-3)	1SLM	His	1	Asp	12	His_{β}	12	His_{β}	(Becker *et al.* 1995)
Human collagenase-3 (MMP-13)	830C	His	1	Asp	12	His_{β}	12	His_{β}	(Lovejoy *et al.* 1999)
Mouse collagenase-3 (MMP-13)	1CXV	His	1	Asp	12	His_{β}	12	His_{β}	(Botos *et al.* 1999)
Human progelatinase 72 kDa (MMP-2)	1CK7	His	1	Asp	12	His_{β}	12	His_{β}	(Morgunova *et al.* 1999)
Bacteriophage T4 endonuclease VII	1EN7	Cys_{β}	2	Cys_{L}	31	$Cys_{b\alpha}$	2	Cys_{α}	(Raaijmakers *et al.* 1999)
E. coli DNA mismatch endonuclease	1VSR	Cys_{α}	4	$Cys_{2a\alpha}$	1	Cys	42	Cys_{α}	(Tsutakawa *et al.* 1999)
Human picornavirus endoprotease 2A	2HRV	Cys	1	$Cys_{b\beta}$	57	Cys_{β}	1	His_{L}	(Petersen *et al.* 1999)
Thermus thermophilus HB8 MutM	1EE8	Cys_{β}	2	Cys_{L}	16	$Cys_{a\beta}$	2	Cys	(Sugahara *et al.* 2000)
				Class V: Isomerases					
Escherichia coli rhamnose isomerase	1DE5	Glu_{β}	32	Asp_{β}	26	His_{β}	39	Asp_{β}	(Korndorfer *et al.* 2000)
				Class VI: Ligases					
tRNA synthetase family									
Escherichia coli MetRS	1QQT	Cys_{β}	2	$Cys_{2b\beta}$	9	Cys_{β}	2	$Cys_{2b\beta}$	(Fourmy *et al.* 1993a; Fourmy *et al.* 1993b) NMR, (Mechulam *et al.* 1999)
Thermus thermophilus HB8 MetRS	1A8H	Cys_{β}	2	$Cys_{2b\beta}$	13	Cys	2	His	(Sugiura *et al.* 2000)

Table 2. Continued.

Enzyme	PDB#	L_1	X	L_2	Y	L_3	Z	L_4	Ref.
Thermus thermophilus HB8 ProRS		Cys	4	Cys_{α}	25	Cys	2	Cys	(Yaremchuk *et al.* 2000)
Thermus thermophilus IsoleucylRS	1ILE	Cys_{β}	2	$Cys_{2a\beta}$	37	Cys_{β}	1	$Cys_{2b\beta}$	(Nureki *et al.* 1998)
Thermus thermophilus IsoleucylRS		Cys_{β}	2	$Cys_{2b\beta}$	204	Cys_{β}	2	$Cys_{2b\beta}$	
Thermus thermophilus ValRS	1GAX	Cys_{β}	2	$Cys_{2a\beta}$	17	Cys	2	Cys	(Fukai *et al.* 2000)
		Cys_{β}	2	$Cys_{2b\beta}$	164	Cys_{β}	2	$Cys_{2b\beta}$	
Escherichia coli IsoleucylRS	1QU2	Cys_{β}	2	$Cys_{2b\beta}$	16	$Cys_{b\alpha}$	2	Cys_{α}	(Silvian *et al.* 1999)
HIV-1 Integrase	1WJA	His	3	His_{α}	33	$Cys_{2b\beta}$	2	Cys_{α}	(Cai *et al.* 1997) NMR
HIV-2 Integrase	1AUB	His_{α}	3	$His_{2\alpha}$	33	$Cys_{a\alpha}$	2	Cys	(Eijkelenboom *et al.* 1997; Eijkelenboom *et al.* 2000) NMR
Thermus filiformis DNA ligase	1DGS	Cys_{β}	2	Cys_L	12	Cys_{β}	4	Cys	(Lee *et al.* 2000)

[a]See footnote in Table 1 for other definitions.

the catalytic one proposed for the entire astacin family (Soler *et al.* 1994). This site is highly conserved in all the matrix metalloproteinases (Sang & Douglas 1996; Soler *et al.* 1995) and contains three His and an Asp residue in a linear sequence spanning 28 amino acids (Table 2). Thus in matrilysin the zinc is bound to His168 (Nε2), Asp170 (Oδ1), His183 (Nε2) and His196 (Nδ1). This zinc site has the characteristics of a structural zinc site since it is has four protein ligands, no bound water molecules and a relatively short sequence of the protein provides all four zinc ligands (Vallee & Auld 1990b). While this zinc site and the catalytic zinc of the MMPs (Table 1) are 12.5 Å apart several of the conserved amino acids adjacent to the third and fourth His ligands of the structural zinc site (e.g., Ala182,-184,-195 and Phe197) form the environment surrounding the predicted catalytic residue Glu219 (Auld 1997). The properties of the 'structural' zinc site in the MMPs and in other similar sites may therefore influence function indirectly through effects on local conformation or local environment.

The crystal structure of *Escherichia coli* rhamnose isomerase (Korndorfer *et al.* 2000) shows zinc bound to a 'structural' site formed from the amino acid sequence Glu (Oε1)-32X-Asp (Oδ1)-26X-His (Nδ1)-39X-Asp (Oδ1) (Table 2). In the presence of substrate a second 'catalytic' Mn binding site is found in close proximity to this zinc binding site. The catalytic site appears to be occupied only when the substrate is present suggesting the true substrate is the Mn bound form of it. Two very similar metal binding sites are observed in *Streptomyces rubiginsous* xylose isomerase (Whitlow *et al.* 1991) although the overall identity to rhamnose isomerase is only 13% (Korndorfer *et al.* 2000). Both metals are identified as Mn in the case of xylose isomerase (Whitlow *et al.* 1991). The structural site for xylose isomerase is made up of two Asp and two Glu ligands. It may be that a site composed of multiple Glu and Asp residues only is too flexible for zinc and leads to weak binding constants due to fast dissociation rates of the zinc from such sites. The usual presence of Ca^{2+} and Mg^{2+} in such acidic ligand sites does correlate with the weak binding constants of such ions for protein binding sites and fast off rate constants for such metals.

Thus far the amino acid residues Asp and Glu have not been found in combination with Cys residues in structural zinc binding sites. The zinc binding site in adenylate kinases is of interest in this regard. The *Bacillus stearothermophilus* adenylate kinase contains a zinc binding site composed of four Cys ligands a in linear 24 amino acid span (Berry & Phillips 1998). The adenylate kinase from *Bacillus subtilis* also contains a structural zinc site (Perrier *et al.* 1994). Sequence alignment and mutagenic studies suggest the fourth Cys has been replaced by an Asp residue in this structural zinc site. Based on the knowledge gained on these systems it was also possible to genetically engineer a structural zinc binding site equivalent to the *B. stearothermophilus* into the *E. coli* adenylate kinase that does not normally bind zinc (Perrier *et al.* 1998). The incorporation of the zinc site into the *E. coli* enzyme improved its thermal stability while not adversely affecting its catalytic properties.

Interwoven zinc binding site

An interwoven zinc site occurs in the protein kinase C, PKC, family. In this case one zinc site is formed from residues residing within the two long spacers of the other zinc site. These enzymes are Ser/Thr protein kinases that depend on phospholipids and diacyl glycerol and are known to play crucial roles in intracellular signal transduction events elicited by various extracellular signal transduction events summoned by extracellular stimuli such as growth factors, hormones and neurotransmitters (Quest *et al.* 1992) Rat brain PKC is a zinc metalloenzyme in which zinc is bound within the lipid binding domain (Quest *et al.* 1992). NMR studies indicate that the regulatory domain of rat PKC contains two Cys rich independent subdomains in which two zinc atoms each are bound in a C3H coordination (Hommel *et al.* 1994). This study is in agreement with an XAFS study (Hubbard *et al.* 1991) that favored 3S/1N ligand sites without bridging zinc sites. The metal ligands from the two zinc sites are interwoven. The ligands in site I are Cys26, Cys29, His51, Cys54 while those in site II are His13, Cys43, Cys46, Cys62. Thus the C2C ligation spacer of site I is contained within the first 29 amino acid spacer of site II while the H2C ligation spacer of site I resides within the last 15 amino acid spacer for site II. This type of site has been observed also in the solution structures of Raf-1 protein kinase (Mott *et al.* 1996) and rabbit PKC-α (Ichikawa *et al.* 1995) and the crystal structure of the cysteine rich activator domain of murine PKC-δ (Zhang *et al.* 1995). The zinc ligands are conserved in C1 domains of protein kinases encompassing a wide range of organisms (Hurley *et al.* 1997).

The role of the structural zinc sites in general is to maintain the structure of the protein in its immediate vicinity. These sites can have indirect effects on activity by affecting the chemical environment of the active center and/or influencing the alignment of active site residues for catalysis.

Cocatalytic zinc sites

While it was known for many years that a number of zinc enzymes required two or three metals for full activity the location of these metals relative to one another was often in debate. Prior to the X-ray structure analyses these metal atoms were referred to as 'modulating' or 'regulatory' (Vallee & Auld 1992a; 1993b). As the first three-dimensional structures of these enzymes became available it became apparent that the metals were in close proximity.

A novel feature of these sites is the bridging of two of the metal sites by a side chain moiety of a single amino acid residue, usually Asp and sometimes a water molecule (Vallee & Auld 1993a). In principle any sp^2 center containing two nucleophilic atoms should have a bridging potential. Thus the ring nitrogens of His and the carboxylate oxygens of Asp, Glu, $LysCO_2^-$ have been found to bridge such sites (Table 3). Such an interaction would of course require the metals to be in close proximity to each other. The distance between the metals in these sites depends on the bridging amino acid. In the case of an Asp or Glu it is generally between 3 and 4 Å apart (Table 3). In the case of His the distance increases to about 6 Å.

There are 3 dozen representatives of this type of zinc site with the great majority belonging to the Class III hydrolases (Table 3). Asp and His predominate as ligands in cocatalytic zinc sites where the frequency is Asp $\cong$ His > Glu. These sites also contain unusual zinc ligands such as amide carbonyls provided by Asn, Gln, and the peptide backbone; hydroxyl groups from Ser, Thr, and Tyr and the amine nitrogen of Lys or the N-terminal amino acid of the protein. With the possible exception of the β-lactamases (see below) there are no Cys ligands. The absence of Cys in this type of zinc site is perhaps surprising since there are bridging sulfur ligands in the zinc Cys clusters of metallothionein and fungal transcription factors such as GAL4 (Vallee *et al.* 1991). The ligands to these sites often come from nearly the entire length of the protein. The metals in these sites may therefore be important to the overall fold of the protein as well as catalytic function. The ligands are often part of a β-sheet or are provided by amino acids 1 or 2 residues before or after a β-sheet.

The bridging amino acids and H_2O could have critical roles in catalysis. Thus, their dissociation from either metal atom during catalysis could change the charge on the metal promoting its action as a Lewis acid or allowing interaction with an electronegative atom of the substrate. Alternatively, the bridging ligand might participate transiently in the reaction as a nucleophile or general acid/base catalyst. The flexibility of the arm supplying the bridging ligands (e.g., one C for Asp *vs* 5 C/N for $LysCO_2^-$) would be expected to influence the stability and reactivity of the two metal sites. In this manner the metal atoms and their associated ligands would play specific roles in each step of the reaction that works to bring about catalysis.

Only a few of these sites contain only zinc ions. Several contain zinc in combination with Cu, Fe or Mg. Zn/Mg are seen in alkaline phosphatase and lens aminopeptidase; Fe(III)/Zn in the purple acid phosphatase family and Cu(II)/Zn in the superoxide dismutase, SOD, family (Table 3).

Superoxide dismutases

This group of cocatalytic containing copper/zinc enzymes plays a critical role in the physiological control of oxygen radicals by catalyzing the dismutation of the superoxide anion into molecular oxygen and hydrogen peroxide (Bannister *et al.* 1987). There are x-ray structures available for several eukaryotic and bacterial sources of this enzyme (Table 3). This is the only cocatalytic site that has a bridging His residue thus far. The zinc site is composed of three His residues coordinated by their Nδ1 nitrogens and the Oδ1 oxygen of Asp81 while the Cu site has three His residues coordinated by their Nε2 nitrogens and only the first by Nδ1 (His44). Coordination by Nδ1 is not often observed in zinc sites and therefore maybe involved in fine tuning the function of the zinc. The role of zinc in the SOD family is generally considered supportive to that of the copper which undergoes oxidation/reduction during catalysis. However, zinc may be important to substrate specificity. Thus the zinc deficient SOD has been proposed to participate in both sporadic and familial amyotrophic lateral sclerosis by an oxidative mechanism involving nitric oxide (Estevez *et al.* 1999).

Phosphatases

Several of the zinc enzymes that catalyze phosphomonoester hydrolysis have cocatalytic zinc sites con-

Table 3. Cocatalytic zinc sites[a].

Enzyme	PDB#	metal	R,Å	L_1	X	L_2	Y	L_3	Z	L_4	L_5	Ref.
						Class I: Oxidoreductases						
Superoxide dismutase family												
Bovine	2SOD	Cu(II)	6.2	His$_{\beta}$	1	His$_{\beta}$	14	***His***$_{\beta}$	56	His$_{\beta}$(C)	Solv	(Tainer *et al.* 1982)
		Zn		***His***$_{\beta}$	7	His	8	His	2	Asp$_{2b\beta}$(C)	-	
Human	1SPD	Cu(II)	5.5	His$_{\beta}$	1	His$_{\beta}$	14	***His***	56	His$_{\beta}$(C)	-	(Parge *et al.* 1992)
		Zn		***His***	7	His	8	His	2	Asp$_{2b\beta}$(C)	-	
Frog (*Xenopus laevis*)	1XSO	Cu(II)	6.0	His$_{\beta}$	1	His$_{\beta}$	14	***His***$_{\beta}$	56	His$_{\beta}$(C)	Solv	(Carugo *et al.* 1994)
		Zn		***His***$_{\beta}$	7	His	8	His	2	Asp$_{2b\beta}$(C)	-	
Spinach (*Spinacia oleracea*)	1SRD	Cu(II)	6.0	His$_{\beta}$	1	His$_{\beta}$	14	***His***$_{\beta}$	56	His$_{\beta}$(C)	-	(Kitagawa *et al.* 1991)
		Zn		***His***$_{\beta}$	7	His	8	His$_{2b\beta}$	2	Asp$_{\beta}$(C)	-	
Yeast (*S. cerevisiae*)	1SDY	Cu(II)	6.1	His$_{\beta}$	1	His$_{\beta}$	14	***His***$_{\beta}$	56	His$_{\beta}$(C)	Solv	(Djinovic *et al.* 1992)
		Zn		***His***$_{\beta}$	7	His	8	His$_{2b\beta}$	2	Asp$_{\beta}$(C)	-	
Yeast (*Candida albicans*)	1YSO	Cu(I)	6.5	His$_{\beta}$	1	His$_{\beta}$	14	His$_{\beta}$	56	His$_{\beta}$(C)	Solv	(Ogihara *et al.* 1996)
		Zn		His$_{\beta}$	7	His	8	His$_{2b\beta}$	2	Asp$_{\beta}$(C)	-	
Escherichia coli	1ESO	Cu(II)	6.5	His$_{\beta}$	1	His$_{\beta}$	14	***His***$_{\beta}$	56	His$_{\beta}$(C)	-	(Pesce *et al.* 1997)
		Zn		***His***$_{\beta}$	7	His	8	His$_{2b\beta}$	2	Asp$_{\beta}$(C)	-	
Salmonella tryphimurium	1EQW	Cu(II)	6.5	His$_{\beta}$	1	His$_{\beta}$	14	***His***$_{\beta}$	56	His$_{\beta}$(C)	-	(Pesce *et al.* 2000)
		Zn		***His***$_{\beta}$	7	His	8	His$_{2b\beta}$	2	Asp$_{\beta}$(C)	-	
Photobacterium leiognathi	1YAI	Cu(II)	6.2	His$_{\beta}$	1	His$_{\beta}$	22	***His***$_{\beta}$	54	His$_{\beta}$(C)	-	(Bourne *et al.* 1996)
		Zn		***His***$_{\beta}$	8	His	8	His$_{2b\beta}$	2	Asp$_{\beta}$(C)	-	
Brucella abortus		Cu(II)		His	1	His	22	***His***	54	His(C)	-	(Chen *et al.* 1995) NMR
		Zn		***His***	8	His	7	His (C)	-	?		
Actinobacillus pleuropneumoniae	2APS	Cu(II)	6.4	His$_{\beta}$	1	His$_{\beta}$	22	***His***$_{\beta}$	55	His$_{\beta}$(C)	-	(Forest *et al.* 2000)
		Zn		***His***$_{\beta}$	8	His	8	His$_{2b\beta}$	2	Asp$_{\beta}$(C)	-	
						Class III: Hydrolases						
E. coli alkaline phosphatase	1ALK	Zn2	3.9	His	0	Asp$_{2a\beta}$	266	Ser$_{b\alpha}$	50	***Asp***$_{\beta}$(N)	-	
		Mg		***Asp***$_{\beta}$	103	Thr$_{\alpha}$	166	Glu$_{\beta}$ (C)	-	H_2O		(Kim & Wyckoff 1991; Stec *et al.* 1998)

Table 3. Continued.

Enzyme	PDB#	metal	R,Å	L_1	X	L_2	Y	L_3	Z	L_4	L_5	Ref.
B. cereus Phospholipase C1	AH7	Zn2	3.3	***Asp***$_{a\alpha}$	3	His$_{\alpha}$	48	His	13	Asp$_{a\alpha}$(N)	-	(Hough *et al.* 1989)
		Zn3		Trp$_{b\beta}$	12	His$_{\alpha}$	107	***Asp***$_{a\alpha}$(C)	-	H_2O		
P. citrinium P1 nuclease	1AK0	Zn2	3.2	***Asp***$_{\alpha}$	3	His$_{\alpha}$	55	His$_{a\alpha}$	14	Asp$_{\alpha}$(N)		(Volbeda *et al.* 1991)
		Zn3		Trp	4	His$_{\alpha}$	113	***Asp***$_{\alpha}$(C)	-	H_2O		
E. coli Endonuclease IV	1QTW	Zn2	3.4	***Glu***$_{\beta}$	33	Asp$_{\beta}$	36	His$_{\beta}$	44	Glu$_{\alpha\beta}$(C)	***H_2O***	(Hosfield *et al.* 1999)
		Zn3		***Glu***$_{\beta}$	35	His$_{\alpha\beta}$	39	His$_{\beta}$ (N)	-	***H_2O***		
Phosphotriesterase family												
Pseudomonas diminuta	1DPM	Zn1	3.3	His$_{2a\beta}$	1	His	111	***Lys***$_{\beta}$***CO_2^-***	131	Asp$_{2a\beta}$(C)	***OH***	
		Zn2		***Lys***$_{\beta}$***CO_2^-***	31	His$_{\beta}$	28	His$_{\beta}$ (C)	-	***OH***		(Benning *et al.* 1994;
												Vanhooke *et al.* 1996)
Escherichia coli	1BP6	Zn1	3.35	His$_{2a\beta}$	1	His	110	***Glu***$_{b}\beta$	117	Asp$_{2a\beta}$ (C)	***Unk***	(Buchbinder *et al.* 1998)
		Zn2		***Glu***$_{b}\beta$	32	His$_{\beta}$	27	His$_{\alpha\beta}$ (C)	-	***Unk***		
Clostridium perfringens α-toxin	1QM6	Zn1	3.2	***Asp***$_{\alpha}$	3	His$_{\alpha}$	57	His	11	Asp(N)		(Naylor *et al.* 1999)
		Zn2		Trp	10	His$_{\alpha}$	118	***Asp***$_{\alpha}$ (C)	-	H_2O		
Purple acid phosphatase family												
Kidney bean	1KBP	Fe(III)	3.3	Asp$_{\beta}$	28	***Asp***$_{\beta}$	2	Tyr$_{2a\beta}$	157	His$_{2a\beta}$(C)		
		Zn		***Asp***$_{\beta}$	36	Asn$_{b\alpha}$	84	His$_{a\beta}$	36	His$_{2a\beta}$(C)	H_2O	(Klabunde *et al.* 1996;
												Strater *et al.* 1995)
Rat	1QFC	Fe(III)	3.1	Asp$_{2a\beta}$	37	***Asp***$_{2a\beta}$	2	Tyr	167	His$_{b\beta}$(C)	H_2O	
		Fe(II)		***Asp***$_{2a\beta}$	38	Asn$_{b\alpha}$	94	His$_{a\beta}$	34	His$_{2a\beta}$(C)	H_2O	(Lindqvist *et al.* 1999;
												Uppenberg *et al.* 1999)
Porcine (uteroferrin)	1UTE	Fe(III)	3.3	Asp$_{2a\beta}$	37	***Asp***$_{2a\beta}$	2	Tyr	167	His$_{b\beta}$(C)	***OH***	(Guddat *et al.* 1999)
		Fe(II)		***Asp***$_{2a\beta}$	38	Asn$_{\alpha}$	94	His$_{a\beta}$	34	His$_{2a\beta}$(C)	***OH***	
Human protein phosphatase I		Fe(III)	3.5-4.0	Asp	1	His	25	***Asp***	179	Tyr(C)	***H_2O***	(Egloff *et al.* 1995)
		Mn		***Asp***	31	Asn	48	His	74	His(C)	***H_2O***	
Human brain calcineurin	1AUI	Fe(III)	3.1	Asp	1	His	25	***Asp***$_{2a\beta}$ (C)		H_2O	***H_2O***	
		Zn		***Asp***$_{2a\beta}$	31	Asn	48	His$_{\alpha\beta}$	81	His$_{2a\beta}$(C)	***H_2O***	(Griffith *et al.* 1995;
												Kissinger *et al.* 1995)
Escherichia coli UDP-sugar hydrolase	1USH	Zn1	3.3	Asp$_{2a\beta}$	1	His	40	***Asp***$_{2a\beta}$	169	Gln$_{b\beta}$(C)	***CO_2^-*** or ***H_2O***	(Knofel & Strater 1999)
		Zn2		***Asp***$_{2a\beta}$	31	Asn$_{L}$	100	His$_{\beta}$	34	His$_{a\beta}$(C)	***CO_2^-*** or ***H_2O***	

Table 3. Continued.

Enzyme	PDB#	metal	R,Å	L_1	X	L_2	Y	L_3	Z	L_4	L_5	Ref.
Pseudomonas sp. CPD G_2	1CG2	Zn1	3.3	***Asp***$_{2b\beta}$	34	Glu$_{\alpha\alpha}$	208	His (C)	-	***H₂O***		(Rowsell *et al.* 1997)
		Zn2		His$_\beta$	28	***Asp***$_{2b\beta}$	58	His$_\beta$ (C)	-	***H₂O***		
Aminopeptidase family												
bovine lens	1BLL	Zn1	2.9	***Glub***$_\alpha$	1	Asp	76	***Asp***$_{2a\beta}$(N)	-	***H₂O***		(Burley *et al.* 1992)
		Zn2		Lys$_\beta$	4	***Asp***$_{2a\beta}$	17	Asp$_{2b\beta}$	60	**Glu**$_{b\alpha}$(C)	***H₂O***	
Aeromonas proteolytica	1AMP	Zn1	3.5	***Asp***$_{2b\beta}$	34	Glu	104	His (C)	-	***H₂O***		(Chevrier *et al.* 1994)
		Zn2		His$_\beta$	19	***Asp***$_{2b\beta}$	61	Asp$_{a\beta}$(C)	-	***H₂O***		
Streptomyces griseus	1JXO, 1QQ9	Zn1	3.6	***Asp***$_{2b\beta}$	34	Glu	114	His (C)	-	***H₂O***		
		Zn2		His$_\beta$	11	***Asp***$_{2b\beta}$	62	Asp$_{a\beta}$ (C)	-	***H₂O***		(Gilboa *et al.* 2000; Greenblatt *et al.* 1997)
Escherichia coli methionine-1	1MAT	Co1	2.9	Asp$_\beta$	10	***Asp***$_\beta$	126	***Glu***$_{b\beta}$ (C)	-			
		Co2		***Asp***$_\beta$	62	His$_\beta$	32	Glu$_{\alpha\beta}$	30	***Glu***$_{b\beta}$(C)		(Lowther *et al.* 1999; Roderick & Matthews 1993)
Pyrococcus furiosus methionine-2	1XGM	Co1	2.8	Asp$_\beta$	10	***Asp***$_\beta$	186	***Glu***$_\beta$ (C)	-	***H₂O***		(Tahirov *et al.* 1998a)
		Co2		***Asp***$_\beta$	59	His$_\beta$	33	Glu$_{\alpha\beta}$	92	***Glu***$_\beta$ (C)	***H₂O***	
Human methionine-2	1B59	Co1	3.2	Asp$_\beta$	10	***Asp***$_{a\beta}$	196	***Glu***$_\beta$ (C)	-	H_2O		(Liu *et al.* 1998)
		Co2		***Asp***$_{a\beta}$	68	His$_\beta$	32	Glu$_\beta$	94	***Glu***$_\beta$ (C)	***H₂O***	
Human Glyoxalase II	1QH3,1QH5	Zn1	3.3-3.5	His$_{2a\beta}$	1	His	53	His	23	***Asp***$_{b\beta}$(C)	***H₂O***	(Cameron *et al.* 1999b)
		Zn2		Asp	0	His	74	***Asp***$_{b\beta}$	38	His$_\beta$(C)	***H₂O***	
***β*-lactamase family**												
Bacteroides fragilis	1ZNB	Zn1	3.5	His	1	His	60	His(C)	-	***H₂O***		
		Zn2		Asp	77	Cys$_{2a\beta}$	41	His(C)	-	***H₂O***	H_2O	(Carfi *et al.* 1998b; Concha *et al.* 1996)
Bacillus cereus	1BC2	Zn1	3.7-4.4	His$_{2a\beta}$	1	His$_{b\alpha}$	60	His (C)	-	H_2O		
		Zn2		Asp	77	Cys	41	His (C)	-	H_2O	H_2O	(Carfi *et al.* 1998a; Fabiane *et al.* 1998)
Stenotrophomonas maltophilia	1SML	Zn1	3.4	His	1	His	73	His (C)	-	***H₂O***		(Ullah *et al.* 1998)
		Zn2		Asp	0	His	135	His (C)	-	***H₂O***	H_2O	
Pseudomonas aeruginosa	1DD6	Zn1	3.5	His$_{2a\beta}$	1	His	59	His (C)	-	***H₂O***		(Concha *et al.* 2000)
		Zn2		Asp	76	Cys$_{2a\beta}$	38	His (C)	-	***H₂O***	H_2O	

[a]The amino acid residue which bridges the two metal sites is shown *in italic bold face*. R is the distance between the metal atoms. See footnote in Table 1 for other definitions. Unk is an unknown bound molecule.

taining two or three metal atoms in close proximity (Tables 1 and 3). *E. coli* alkaline phosphatase is the best studied representative of this group. It has a cocatalytic zinc site in both of its subunits composed of two zinc atoms and one magnesium that form a nonequilateral triangle with the metals as the apices (Kim & Wyckoff 1991). The ligands to these metals and the adjacent amino acids are highly conserved for a large family made up of representatives from bacteria, yeast and mammalian sources (Vallee & Auld 1993a). One metal site has the properties of a catalytic zinc site being formed from two ligands, Asp327 (Oδ1) and His331 (Nε2) supplied from a short α-helix, a third protein ligand His412 (Nε2) supplied by a β-strand and a water molecule (Table 1). The second zinc, Zn2, and the Mg are bridged by Asp51 (Table 3). This was the first zinc site where a reactant amino acid in catalysis, Ser102, was found to be a ligand to a metal (Zn2) in the resting state. The serine is reported to be bound as an alkoxide since the oxygen-Zn distance is 1.91 Å (Stec *et al.* 2000). A hydroxide bound to the Mg may be responsible for aiding in deprotonating the zinc-bound serine. Several other phosphate hydrolyzing enzymes also have cocatalytic sites resembling *E. coli* alkaline phosphatase (Tables 1 and 3).

A combination of X-ray crystallographic, NMR and kinetic studies on the Cd and Co substituted enzymes have aided in deciphering the mechanism of action of the *E. coli* enzyme (Coleman 1992; Kim & Wyckoff 1991; Vallee & Galdes 1984). The rate determining step is strongly pH dependent. In the alkaline pH region the release of the non covalently bound product phosphate (E•P $\rightarrow$ E + P) is rate limiting while in the acidic pH region the breakdown of the covalent phosphoryl intermediate (E-P $\rightarrow$E•P) is postulated to be rate limiting. Ser102, a ligand to Zn2, is the nucleophile in the first step of the reaction.

The breakdown of the Ser phosphoryl intermediate, E-P, is believed to be through a zinc-bound water/hydroxide on the catalytic zinc in the proposed mechanism. The enzyme•vanadate complex has been proposed to mimic the transition state complex (Holtz *et al.* 1999). The vanadate ion is bound in a trigonal bipyramidal geometry with the active site Ser102 and water molecule in opposite apical positions. The equatorial oxygens are stabilized by interaction with the guanidinium group of Arg166. Mutation of the catalytic zinc ligand, His331Gln yields an enzyme in which the covalent phosphate intermediate can be observed in the crystal structure (Murphy *et al.* 1997). The structure shows the zinc-bound water on the catalytic zinc is in a position for apical attack on the Ser102 phosphoryl bond.

In the E•P complex the phosphate ion is coordinated to both Zn and Zn2 and with two of its oxygens to the guanidinium group of Arg166. The phosphate is further hydrogen bonded to the amide of Ser102 and a water molecule that is coordinated to the Mg (Kim & Wyckoff 1991). Mutation of Ser102 to Gly, Ala or Cys decreases activity 10^4 to 10^5 fold with only the Cys mutant having an effect on the position of the phosphate (Stec *et al.* 1998).

The purple acid phosphatases are a group of nonspecific phosphomonoesterases that have been found in animal, plant and fungal sources (Klabunde & Krebs 1997). The characteristic purple color of this subclass of acid phosphatases comes from a Tyr phenolate-Fe(III) charge-transfer transition at 560 nm. The presence of Fe(III) is universally found in these enzymes. The 35 kDa mammalian purple acid phosphatases, PAP, or tartrate-resistant acid phosphatases, TRAP, contain a Fe(III)-Fe(II) iron center (Uppenberg *et al.* 1999) in contrast to the Fe(III)-Zn(II) center found in the 110 kDa kidney bean enzyme (Klabunde *et al.* 1996). The Ser/Thr human protein phosphatase 1 and calcineurin also contain a very similar cocatalytic Fe(III)-M(II) (where M is Zn or Mn) site to that of the PAPs (Egloff *et al.* 1995; Kissinger *et al.* 1995). In this case there is no Tyr-Fe(III) interaction but the general ligand nature, the distance between metals, and the presence of a bridging Asp residue, is common to all of these enzymes (Table 3). A mechanism has been proposed for the PAPs in which the phosphate ester binds to the Zn (II) site and the phosphate bond undergoes nucleophilic attack by an Fe (III)-coordinated hydroxide ion (Klabunde & Krebs 1997).

Aminopeptidases

Aminopeptidases containing two metals catalyze the hydrolysis of a wide variety of N-terminal peptides and amino acid derivatives. These enzymes are widely distributed in bacteria, yeast, plant and animal sources. The structures of several aminopeptidases containing cocatalytic zinc sites have been reported (Table 3). In addition, a cocatalytic zinc site has also been observed in a bacterial carboxypeptidase (Rowsell *et al.* 1997).

While several of the aminopeptidases have been characterized as the zinc or cobalt complex the native metal is sometimes still in question. The *E. coli* methionine aminopeptidase-1, MetAP-1, (Roderick

& Matthews 1993), the hyperthermophile *Pyrococcus furiosus* methionine aminopeptidase-2, MetAP-2 (Tahirov *et al.* 1998b) and human methionine aminopeptidase-2 (Liu *et al.* 1998) have been isolated as di-cobalt enzymes. The physiological metal for these enzymes is still not certain. Zinc works as well as cobalt in the yeast aminopeptidase-1 (Walker & Bradshaw 1998) and recent studies of the *E. coli* MetAP-1 indicate it functions as an Fe(II) enzyme (D'Souza & Holz 1999).

The cocatalytic zinc sites of two of the best studied aminopeptidases, bovine lens leucine aminopeptidase (BLAP) and *Aeromonas proteolytica* aminopeptidase (AAP) differ in several details (Chevrier *et al.* 1996; Strater & Lipscomb 1995). His ligands are bound to both sites in AAP while no His residues are involved in BLAP (Table 3). AAP uses both an Asp carboxylate and a water molecule as bridging ligands while BLAP uses the carboxylate oxygens of a Glu residue, one oxygen of an Asp residue and a bridging water molecule. BLAP uses a Lys residue to bind Zn at the tightly bound site. These combinations of ligands as well as the difference in interatomic distances of the two zinc atoms could lead to differences in the charge on the zinc that in turn could influence catalysis. The reader is directed to several reviews on this class of enzymes for more detailed comments on their mechanism of action (Auld 2001; Kim & Lipscomb 1994; Lipscomb & Strater 1996; Taylor 1993).

β-Lactamases

The first reported β-lactamase structure that contained two zinc sites was for the *B. fragilis* enzyme (Concha *et al.* 1996). This enzyme was crystallized at pH 7.0 in a 10 μM $ZnCl_2$, Hepes buffer. It is not yet clear if this is a true cocatalytic zinc site. If so it will be the first cocatalytic zinc site in which there is no bridging amino acid. The importance of the second zinc site to catalytic activity is still not clear. The second zinc site is not conserved in the few enzymes that have been examined. Thus the *Stenotrophomonas maltophilia* β-lactamase has no Cys in the second zinc site (Ullah *et al.* 1998). The mono zinc *B. cereus* enzyme is active and the *Aeromonas hydrophila* AE06 enzyme is inhibited by Zn with a K_i of 46 μM (Hernandez Valladares *et al.* 1997) while the catalytic zinc binds to the enzyme with a dissociation constant lower than 20 nM.

The kinetic parameters for the mono and di-zinc *B. fragilis* enzyme differ only slightly for 4 substrates (Paul-Soto *et al.* 1998). The mutation of Cys168 to Ser in the *B. fragilis* enzyme eliminates the second zinc site (Li *et al.* 1999c). The k_{cat} values for this enzyme are reduced 140 to 1,500 fold while the k_{cat}/K_m values are reduced 970 to 3,700 fold dependent on the substrate used (Yang *et al.* 1999). This reduction in activity could be due to the loss of a residue that played a role in the transition state in catalysis.

The mutant Cys168Ser *B. cereus* enzyme can bind a second zinc weakly (Paul-Soto *et al.* 1999). The results of kinetic, metal binding and XAFS studies of this enzyme indicates the mono-zinc enzyme is dependent on the Cys for optimal activity while the di-zinc enzyme is not. Thus, in summation, all the β-lactamases are dependent on the presence of one zinc that has the characteristics of a catalytic zinc site while the second zinc is not universally important to the activity of the β-lactamases.

Protein interface zinc sites

Zinc may also influence the quaternary structure of proteins. Thus zinc binding to ligands supplied from two protein molecules at their interface contact surface has been observed in increasing numbers in the last few years (Auld 2001). These interactions can lead to dimers or trimers of the same protein or link two different proteins through the intermolecular ligands. The amino acid residues His, Glu and Asp primarily supply the ligands to these sites but Cys containing sites are also found (Table 4). The ligands are generally contributed by some form of secondary structure with β-sheets predominating. The resulting sites can function like catalytic zinc sites as in γ-carbonic anhydrase (Iverson *et al.* 2000) or structural-like sites as in the proposed superantigen-MHC class II complexes (Papageorgiou & Acharya 1997) (Figure 2).

Protein interface catalytic zinc sites

The first member of the γ-class of carbonic anhydrases, CAM, was isolated and characterized from the methanogenic archaeon *Methanosarcina thermophila* (Alber & Ferry 1994). The trimeric enzyme catalyzes the interconversion of CO_2 and HCO_3^- with turnover numbers about 20-fold lower than the fastest of the α-class of carbonic anhydrases (Alber *et al.* 1999). Both classes of enzymes have three histidine ligands supplied by β-sheet secondary structures in the catalytic zinc site (Tables 1 and 4). However they differ in the length of their short spacers (one versus four). This

Table 4. Protein interface zinc sites.

Enzyme/Protein	PDB#	Class	L_1	X	L_2	Y	L_3	L'_1	Z	L'_2	Ref.
Nitric oxide synthase family		I									
Bovine endothelial	1NSE		Cys_β	4	Cys			Cys_β	4	Cys	(Raman *et al.* 1998)
Human endothelial	3NOS		Cys_β	4	Cys			Cys_β	4	Cys	(Fischmann *et al.* 1999)
Human inducible	1NSI		Cys	4	Cys			Cys	4	Cys	(Fischmann *et al.* 1999; Li *et al.* 1999b)
Mouse inducible	1DWV		Cys	4	Cys			Cys	4	Cys	(Ghosh *et al.* 1999)
Sulfolobus solfactaricus Cytochrome P450	1F4T	I	$Glu_{2b\alpha}$	39	$His_{a\alpha}$			$Glu_{2b\alpha}$	39	$His_{a\alpha}$	(Yano *et al.* 2000)
E. coli signal transducing protein, PTS IIA^{Glc}	1F3Z	II	His_β	14	His_β			Glu_β		H_2O	(Feese *et al.* 1997)
E. coli PTS IIA^{Glc}/glycerol kinase	1GLC	II	His_β	14	His_β			Glu_α		H_2O	(Feese *et al.* 1994)
Rat tonin	1TON	III	His	1	His	39	His (N)	Glu			(Fujinaga & James 1987)
Mouse 7S nerve growth factor	1SGF	III	Glu	6	His_β			$His_{\alpha\beta}$	4	Glu	(Bax *et al.* 1997)
Human glyoxalase I	1FRO	IV	Gln_β	65	Glu_β	-	H_2O	His_β	45	Glu_β	(Cameron *et al.* 1997)
E. coli glyoxalase I	1F9Z	IV	His_β	50	Glu_β	-	H_2O	His_β	47	Glu_β	(He *et al.* 2000)
Methanosarcina thermophila γ-CA	1THJ	IV	His_β	-	H_2O	-	H_2O	His_β	40	His_β	(Iverson *et al.* 2000)
Superantigen family											
Staphylococcus aureus enterotoxin (SEC2)	1STE		His	3	His_β	34	Asp_β(N)	Asp			(Papageorgiou *et al.* 1995)
Staphylococcus aureus enterotoxin (SEA)	1ESF		H_2O		His_β	27	Asp_β(N)	$Glu_{a\beta}$			(Schad *et al.* 1997)
Staphylococcus aureus enterotoxin (SEA)	1SXT		Asp_β	1	His_β	37	His_β (N)	His or H_2O			(Sundstrom *et al.* 1996b)
Staphylococcus aureus enterotoxin (SED)			Asp_β	1	His_β	37	Asp_β (N)	His_β			(Sundstrom *et al.* 1996a)
Staphylococcus aureus enterotoxin (SED)			His_β	3	Glu_β			His	3	Lys	(Sundstrom *et al.* 1996a)
Staphylococcus aureus enterotoxin (SEH)	1EWC		Asp_β	1	His_β			H_2O		H_2O	(Hakansson *et al.* 2000)
Streptococcus pyogenes exotoxin (SPE-C)	1AN8		Asp_β	1	His_β	33	His_β (N)	$His_{b\beta}$			(Roussel *et al.* 1997)
Staphylococcus aureus TSST-1	4TSS		His	60	His_β			His	60	His_β	(Prasad *et al.* 1997)
Human psoriasin (S100A7)	3PRS		His_α	3	$His_{a\alpha}$			Asp	6	His_α	(Brodersen *et al.* 1999)
Human interferon-$\alpha_{2\beta}$ dimer			Glu_α	0	Glu_α			Glu_α	0	Glu_α	(Radhakrishnan *et al.* 1996)
Human interferon β dimer	1AUI		His_α	3	His_α			$His_{b\alpha}$		H_2O	(Karpusas *et al.* 1997)
Human prolactin receptor/growth hormone	1BP3		Asp_β	0	$His_{a\beta}$			His_α	155	Glu_α	(Somers *et al.* 1994)
Mouse survivin	1F3H		Glu_β	3	$His_{2a\beta}$			Glu_β	3	$His_{2a\beta}$	(Muchmore *et al.* 2000)
Shaw T1 tetramer	3KVT		$Cys_{2b\beta}$	0	Cys	26	His (N)	$Cys_{b\beta}$			(Bixby *et al.* 1999)
Human Apo21L/TRAIL	1DG6		$(Cys_{a\beta})_3$		Solv						(Hymowitz *et al.* 2000)
E. coli colicin 3 immunity protein	3EIP		$Cys_{b\beta}$					$Cys_{b\beta}$			(Li *et al.* 1999a)

L'_1 and L'_2 are the second subunit or protein zinc ligands and Z is the spacer for these ligands. See footnote in Table 1 for other definitions.

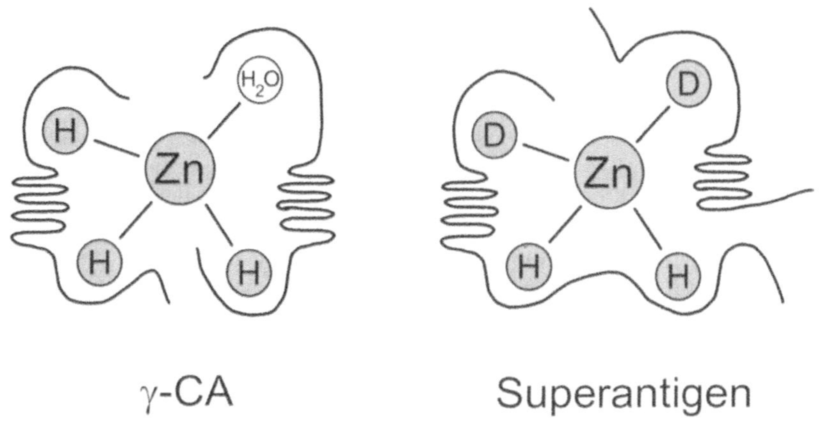

Fig. 2. Protein interface zinc binding sites: γ-CA, γ-carbonic anhydrase (Iverson *et al.* 2000), Superantigen, staphyloccus enterotoxin C2 (Papageorgiou *et al.* 1995)

in turn allows the second His of the CAM enzyme to become part of the zinc binding site in an adjacent monomer rather than bind to its own active site zinc. The crystal structures of the native zinc enzyme and the cobalt substituted enzyme show the metal binding site is formed from His81 (Nδ1) and His122 (Nε2) of one monomer and His117 (Nε2) of another monomer (Iverson *et al.* 2000; Kisker *et al.* 1996). In contrast to the tetrahedral zinc site observed in the α-class of enzymes that have only one zinc bound water molecule (Table 1), the zinc CAM contains two water molecules in a trigonal bipyramidal coordination geometry while the cobalt CAM contains three bound waters in an octahedral geometry. Both of these crystalline geometries are consistent with results obtained from XAFS studies performed in the solution state (Alber *et al.* 1999).

Another remarkable catalytic zinc site is found in human glyoxalase I where the C-terminal domain of one monomer interacts with the N-terminal domain of a second monomer to form two zinc binding sites at the interface of the subunits (Cameron *et al.* 1997). The active site zinc is coordinated by four protein residues and one water molecule in a square pyramidal coordination geometry. The base of the pyramid is formed from Gln33 and Glu 99 of one subunit, Glu172 from the second subunit and a water molecule. The apex of the pyramid uses His126 from the second subunit. Such a geometry could be viewed as octahedral with one axial ligand missing. Both EPR and XAFS studies suggest a distorted octahedral coordination in solution for the cobalt and zinc enzymes respectively (Garcia-Iniguez *et al.* 1984; Sellin *et al.* 1983). The sixth site may be a water molecule. Another remarkable feature of this metalloenzyme is that while zinc binds tighter than Mg (3×10^{-11} *vs* 1×10^{-6} M) the Mg enzyme is fully active (Sellin & Mannervik 1984). The octahedral metal coordination geometry and the functional replacement of Zn by Mg is unusual for catalytic zinc sites. These results raise the possibility that Mg or another metal is the functional metal *in vivo*. Nickel is believed to be the native metal in the *E. coli* glyoxalase I enzyme (Clugston *et al.* 1998). The nickel is bound octahedrally to His5 (Nε2) and Glu56 (Oε1) from one monomer and His74 (Nε2) and Glu122 (Oε1) from the other and two water molecules (He *et al.* 2000). Thus the human and *E. coli* enzymes do not have the identical protein ligands, even while they have homologous amino acid sequences and similar three-dimensional structures. Structure/function studies of *E. coli* glyoxalase I have shown that the nickel, cobalt and cadmium but not zinc enzymes are active (He *et al.* 2000). All of the former enzymes have an octahedral geometry as is observed in the human enzyme while the zinc *E. coli* glyoxalase I has a trigonal bipyramidal coordination.

Mutagenic studies of the zinc ligands of the human enzyme suggest that the metal ligand Glu172 may be directly involved in catalysis (Ridderstrom *et al.* 1998). Thus both the E172Q and Q33EE172Q double mutants still bind zinc but the catalytic activity is decreased by 10^5 and 10^8 fold, respectively. Crystallographic results of the enzyme complexed with a transition state analog S-(N-hydroxy-N-p-iodophenylcarbamoyl) glutathione that mimics the enediolate intermediate that should form along the reaction pathway are consistent with this hypothesis (Cameron *et al.* 1999a). In this structure the two oxygen atoms of the hydroxycarbamoyl moiety displace two zinc-bound water molecules that are observed in a non-transition state complex. In addition the Glu172 carboxylate oxygen-Zn distance has increased from 2.0 Å to 3.3 Å in the complex. The zinc ion is envisioned to play a Lewis acid role in catalysis by directly coordinating the enediol intermediate and the freed zinc ligand Glu172 is proposed to facilitate proton transfer between the adjacent carbon atoms of the substrate (Cameron *et al.* 1999a).

Zinc binding sites in superantigens

Staphylococcus aureus and *streptococcus pyogenes* secrete a number of enterotoxins, SE and pyrogenic exotoxins, SPE respectively. These toxins are known as superantigens, since they simultaneously form complexes with the major histocompatibility class II (MHC-II) molecules and T cell receptors (TCRs) enabling them to activate a number of T cell lymphocytes (Fraser *et al.* 2000; Papageorgiou & Acharya 1997; Proft *et al.* 1999). Thus superantigens stimulate up to 20% of the T cells while only 0.0001% to 0.001% T cells are stimulated upon normal antigen presentation. The massive T cell activation leads to cytokine release and systemic shock. The staphylococcal enterotoxins (SE) A, B, C1-C3, D and E, toxic shock syndrome toxin (TSST-1), the streptococcal pyrogenic exotoxins (SPE-) A, B, Cs, G, and H, the streptococcal mitogenic exotoxins, SMEZ and SMEZ-2 are the best structurally studied superantigens thus far (Tables 4 and 5).

Zinc is believed to be an important factor in the mechanism of T cell activation. Thus binding of staphylococcal enterotoxin A and enterotoxin E to HLA-DR1 is completely abolished by low concentrations of EDTA (Fraser *et al.* 1992). This binding is completely restored by the addition of a 2 μM excess of Zn^{2+} but not by Ca^{2+}, Mg^{2+} or other metal ions. Recent studies using EDTA $\pm$ $ZnCl_2$ show zinc dependent binding of SPE-C, SPE-G, SPE-H and the streptococcal mitogenic exotoxins, SMEZ and SMEZ-2, to the MHC class II molecule HLA-DR1 (affinities of 15 to 36 nM) (Proft *et al.* 1999). Based on this criteria the superantigens, SEB and the toxic shock syndrome toxin, TSST, do not show zinc dependent binding to the MHC class II molecule.

The first crystal structures of the superantigens indicated zinc could bind at the interface of two protein molecules in the crystalline state (Table 4). The first potential zinc binding site in superantigens was found in the cadmium substituted *Staphylococcus aureus* enterotoxin type A (SEA) that has a metal binding site composed of Ser1, His187, His225 and Asp227 (Schad *et al.* 1995). The amino acids involved in binding the Cd ion are arranged in an octahedral coordination geometry. An approximate square plane is formed from the α-amino group and γO of Ser1, and nitrogens of His187 and His225. Asp227 ligates from beneath the plane and a water from above. Mutations of the His and Asp ligands indicate this site is important to the binding of MHC class II molecules (Abrahmsen *et al.* 1995). Crystals of the zinc containing protein revealed Zn binding between two molecules in an asymmetric unit (Sundstrom *et al.* 1996b) (Table 4). In this case His187 (Nδ1), His225 (Nε2) and Asp227 (Oδ2) bind to the zinc in a tetrahedral geometry. The fourth ligand is His61 (Nε2) from a neighboring SEA molecule.

The finding of zinc binding at the interface of two enterotoxin molecules in several of the superantigens (Table 4) could mean that zinc influences the function by forming dimer forms of the superantigen. SED crystallizes as a Zn^{2+} dependent dimer with two high affinity zinc sites located between the C-terminal β-sheets of the two monomers (Sundstrom *et al.* 1996a). Each zinc is tetrahedrally coordinated by His118 from one SED molecule and Asp182, His220 and Asp222 from the other. Thus in the case of SED it has been suggested that zinc binding could induce dimer formation and each monomer could bind to the MHC class II molecule (Sundstrom *et al.* 1996a).

Two distinct zinc binding sites are apparent from the reported structures of the superantigens during the last five years (Tables 4 and 5). Both sites contain at least three protein ligands with each site having short and long spacers. The third ligand comes from the N-terminal side of the second ligand. In one case the short spacer is one (Asp (Oδ1)-X-His (Nε2)) with both ligands coming from a β-sheet, located in the C-terminal side of the protein. The third ligand is either an Asp or His residue located 33 to 39 amino acids away. SEA (Schad *et al.* 1995; Sundstrom *et al.* 1996b), SED (Sundstrom *et al.* 1996a) and SEH

(Hakansson *et al.* 2000) (partial) and SPE-C (Roussel *et al.* 1997), SPE-H and SMEZ-2 (Arcus *et al.* 2000) all contain this type of zinc binding site (Tables 4 and 5). The second type of zinc binding site has a short spacer of 3 and contains two His ligands in the N-terminal side of the superantigen with the third ligand located 27 to 34 amino acids away. SEC2 (Papageorgiou *et al.* 1995), SEA (Schad *et al.* 1997) (2nd Zn site) and SPEA and SPEA1 (Earhart *et al.* 2000; Papageorgiou *et al.* 1999) contain this type of site. The fourth ligand in these sites is quite variable. It may come from a neighboring protein molecule, from within the same molecule or be a bound water molecule. The variability in the fourth ligand could be due to the fact that this ligand should come from the MHC class II molecule. His81 of the β chain of the DR1 molecule has been proposed to be this ligand based on the fact that mutation of this ligand disrupts HLA-DR1 binding to SEA but not SEB or TSST-1 that do not bind zinc strongly (Karp & Long 1992).

A zinc binding site at a protein interface, somewhat similar to the superantigen sites, was observed in crystals of rat submaxillary gland tonin that had been grown in a zinc containing mother liquor (Fujinaga & James 1987). The zinc binds to the Nε2 nitrogens of His97, His99 and the catalytic His57 of one molecule and Glu148 of an adjacent one. The presence of zinc was reported to inhibit the enzyme at pH 6.5. The structure of the zinc binding sites in the dimeric form of psoriasin (S100A7) also contains three His and one Asp ligand (Brodersen *et al.* 1999). This 22.7 kDa homodimeric protein belongs to the S100 class of calcium binding EF-hand proteins. Two zinc binding sites occur per dimer through the formation of a tetrahedral site consisting of the N-terminal His17 (Nε2) and Asp24 (Oδ1) of one monomer and His86 (Nε2) and His90 (Nε2) of the other. This binding site is apparently weak since it is reported to have a dissociation constant of 100 μM.

Zinc-His/Glu sites at protein interfaces

There are several His/Glu combinations that lead to zinc binding at the interface of proteins (Table 4). In crystals of the high molecular weight form of nerve growth factor, 7S NGF a zinc is tetrahedrally bound between His82 (Nε2) and Glu75 (Oε1) of the α-subunit and His217 (Nε2) and Glu222 (Oε1) of the γ-subunit (Bax *et al.* 1997). The binding of zinc to this interface is consistent with the observation that zinc enhances the stability of the 7S complex.

A number of these zinc sites are symmetry related. Thus zinc binds between Glu139 (Oε1) and His178 (Nε2) in one molecule of the asymmetric crystal unit of the 4 phenylimidazole-bound *Sulfolobus solfataricus* cytochrome P450 and their symmetry residues in another molecule (Yano *et al.* 2000). The functional importance of a dimer and zinc's contribution to its stability is not known. However no zinc was added in the crystallization procedure. A His/Glu site with a short spacer of 3 is observed in the dimer interface of mouse survivin (Table 4) (Muchmore *et al.* 2000). The human survivin does not form a similar intermolecular zinc site (Verdecia *et al.* 2000).

An asymmetric dimer observed for human interferon β, huIFN-β, is obtained under conditions where no zinc is added to the crystallization buffer (Karpusas *et al.* 1997). His ligands 93(Nε2) and 97(Nε2) reside on an α-helix in one monomer. The second monomer provides His121 (Nε2) and a water molecule to complete the 4-coordinate zinc site. However gel permeation experiments in the presence of zinc do not show any evidence of a dimer in solution. The dimers form in the crystal on contact surfaces opposite to those found in the IFN-α_{2B} crystal structure. The residue Glu43, which H-bonds to zinc ligand His121 has been identified as part of the binding site of monoclonal antibodies (Redlich & Grossberg 1990). The presence of zinc might therefore modulate this interaction.

High zinc concentrations are used in the crystallization procedure in some of these cases making it more difficult to ascertain physiologically significance of the site. Thus crystals of the human interferon-α_{2B} dimers, huIFN-α_{2B}, obtained in the presence of 40 mM zinc acetate (Radhakrishnan *et al.* 1996), show that the most extensive interactions occur in the vicinity of the zinc binding site. This site is formed from adjacent Glu residues 41 and 42 located on a 3_{10} helix. A distorted tetrahedral zinc coordination sphere is completed by the identical glutamates from a 2-fold symmetry related molecule. However, gel permeation studies indicate huIFN-α_{2B} is a monomer up to 50 μM and the presence of 1 mM Zn does not shift the equilibrium toward a dimer (Radhakrishnan *et al.* 1996).

A tetrahedral zinc binding site is observed in the crystal form of the *E. coli* signal transducing protein PTS IIA^{Glc} (Feese *et al.* 1997) and in its complex with glycerol kinase (Feese *et al.* 1994). In both cases the signal transducing protein supplies two His residues with a spacer arm of 14 and the third Glu ligand comes from either a neighboring molecule in the crystal or the glycerol kinase. The fourth ligand in both cases is a water molecule (Table 4). The site is said to be geometrically equivalent to the zinc binding site in thermolysin (Table 1). If this site contained a suitable acid/base catalyst it might display hydrolytic activity. Although the biochemical effect of 0.01 mM zinc on the inhibition of glycerol kinase by PTS IIA^{Glc} has been demonstrated, the physiological role for zinc ions in PTS and protein interactions remains to be established (Feese *et al.* 1997).

Zinc-Cys protein interface sites

While His, Glu and Asp appear to be the predominate residues in forming these protein interface sites, Cys containing sites are also observed. A novel Cys site is found in the nitric oxide synthase enzymes (Table 4). In endothelial nitric oxide synthase, eNOS or NOS-3, a zinc ion is tetrahedrally coordinated to pairs of symmetry-related Cys residues near the bottom of the dimer interface (Raman *et al.* 1998) (Table 4). The Cys ligands, Cys96 and Cys101, are part of a small three-stranded antiparallel β-sheet that orientates the Cys ligands in the same direction directly across the antiparallel strands. The zinc site is further stabilized by H-bonds between the Cys ligands and the backbone amide NH of Leu102 and Gly103 as well as a H-bond of the amide NH of Cys101 to the carbonyl of Asn468. The zinc is positioned equidistant from each heme (21.6 Å) and each tetrahydrobiopterin, H_4B (12 Å). Ser104, 2 amino acids removed from one of the Cys ligands, H-bonds directly to the pterin side chain hydroxyl group. The metal center is believed to act in a structural capacity to maintain the integrity of the pterin-binding site. It is also centered in the most electropositive region of eNOS. It could therefore provide a binding site for the electronegative reductase domain. In addition, if one of the Cys ligands has nucleophilic capacity it could undergo S-nitrosylation (Stamler 1994). A precedent for this is the DNA repair protein *Escherichia coli* Ada in which a zinc bound Cys can be methylated irreversibly in the DNA complex (Myers *et al.* 1994). The nitrosylated Cys might then release zinc which may be controlled by the redox status in situ (Maret & Vallee 1998).

The crystal structure of the *E. coli* expressed human endothelial, eNOS, and the inducible form iNOS or NOS-2 also have this same zinc binding site (Fischmann *et al.* 1999). An independent study of human iNOS expressed in *E. coli* found the zinc site was not present after refolding (Li *et al.* 1999b), similar to an earlier study on the murine inducible, iNOS or NOS-2, where a disulfide was found (Crane *et al.* 1998). However in the presence of zinc and reducing agents the $Zn(Cys)_4$ site readily formed (Li *et al.* 1999b). These Cys residues are conserved in 20 mammalian e, i and n NOS enzymes suggesting the site may occur in all forms of NOS (Raman *et al.* 1998).

A highly conserved zinc protein interface Cys/His site occurs in the N-terminal, cytoplasmic tetramerization domain (T1) of voltage-gated K^+ channels (Bixby *et al.* 1999). The crystal structure of the *Shaw* T1 tetramers reveals a four-layered protein scaffolding. Within layer 4 on the proposed membrane side of the tetramer, there are 4 zinc ions coordinated by a His and two Cys from one monomer and one Cys from another. The zinc is tetrahedrally coordinated by Cys102 and Cys103 from layer 4 and His75 (Nδ1) from layer 3 of one monomer and Cys81 from layer 3 of another monomer. This site ($CysCysX_{20}CysX_5His$ of one monomer) is conserved for 62 members of the *Shaw, Shab* and *Shal* T1 domain sequences but not for the *Shaker* T1 domain (Bixby *et al.* 1999). The physiological function of this zinc site is unknown.

The apoptosis-inducing ligand 2 protein (Apo2L or TRAIL), a member of the trimeric tumor necrosis factor (TNF) superfamily, is a type-II transmembrane protein that can be cleaved at the cell surface to form a soluble protein (Mariani *et al.* 1997). The cell killing properties of these cytokines have made them particularly attractive for the design of drugs that might selectively kill tumor cells. Recent structural studies of Apo2L revealed a homotrimeric protein with a zinc ion binding at the interface of the three monomers to three symmetry-related Cys ligands, one each from the three monomers (Hymowitz *et al.* 2000). The fourth ligand appears to be a solvent molecule, possibly chloride. Removal of the bound metal by dialysis against chelating agents or replacement of the Cys by Ala results in about a 100-fold decrease in apoptotic activity. Further studies show that the zinc is important to the stability of the native protein. Other members of the TNF superfamily do not have this zinc binding site. The Apo2L appears to be unique among TNF-related cytokines in that it selectively induces apoptosis in tumor cells while not affecting normal tissue (Hymowitz *et al.* 2000). The relationship of this specific activity and the unique zinc binding site is not known.

The antibiotic-like protein colicin E3 of *E. coli* acts as an RNase that specifically cleaves 16S rRNA (Li *et al.* 1999a). *E. coli* is protected from the action of this enzyme by forming a tight complex with an immunity protein, Im3. The crystal structure of Im3 shows the residue that is considered the most important determinant to the formation of the colicin complex, Cys47 is bound covalently to zinc at the interface of two monomers. It is unclear at present whether this zinc mediated dimer of Im3 has biological significance. However if the zinc concentration in *E. coli* should allow this zinc mediated dimer of Im3 to form it would negate the protective action of Im3 inactivating the host colicin E3.

Zinc binding sites in other proteins

Structural studies in recent years have revealed a number of zinc binding sites in proteins of diverse function. In compiling Table 5 the large number of identified DNA binding transcription factors or zinc fingers is purposely left out since they have been reviewed in detail elsewhere (Berg & Shi 1996; Klug 1999; Laity *et al.* 2001; Mackay & Crossley 1998).

The great majority of these zinc sites are formed from four protein side chain ligands to yield a tetrahedral binding site (Table 5). Their function may therefore be related to the local and/or overall structure since the sites span from 22 to 227 amino acids. The vast majority of the sites contain at least one short spacer. The residues generally are supplied from β-sheets or reside within one or two residues of such a secondary structure (Table 5). The amino acids, His and Cys, are the predominate suppliers of ligands but several sites are made up of combinations of His, Asp and Glu residues.

A zinc binding site containing His and Asp ligands was recently discovered in the crystalline state dimers of the 20-kDa human endostatin, the angiogenesis inhibitor, that is produced by proteolytic processing of the C-terminal globular domain of the collagen-like protein XVIII (Ding *et al.* 1998). The tetrahedral zinc site is formed by His132 (Nδ1), His134 (Nε2), His142 (Nε2) and Asp207 (Oδ1) (precursor protein numbering) (Table 5). Atomic absorption spectrometry indicates this site also exists in the solution state of human and mouse endostatin. No dimer contacts are observed in the crystalline form of mouse endostatin (Hohenester *et al.* 2000). However two mouse crystal forms are obtained which bind zinc in the N-terminal region of the protein but with different ligand compositions. One crystal form uses the same zinc ligands as the human endostatin while the other form substitutes Asp136 for His132, and retains His134, His142 and Asp207 as the other ligands. The effects of mutating either His132 or Asp136 or both on zinc binding to the protein are consistent with the different zinc binding modes. Thus mutants His132Ala and Asp136Ala still bind zinc while the double mutation does not. In addition mutation of Asp207Ala leads to loss of zinc binding. A number of conditions are different (pH, PEG types, presence or absence of oligosaccharide and zinc) in the production of the two crystal forms (Hohenester *et al.* 2000). It is not known whether any one of these conditions alone could account for the different zinc binding modes. Since the site is immediately adjacent to the precursor cleavage site the zinc may be involved in the antiangiogenic activity of endostatin or regulation of it (Ding *et al.* 1998). However, the structural diversity observed in the zinc binding site has led to the conclusion that zinc is not likely involved in the anti-tumor activity of the protein (Hohenester *et al.* 2000).

A somewhat similar His/Asp zinc binding site is conserved in the thermoacidophilic archaea ferredoxin family where the zinc is tetrahedrally bound to His16, His19, His34 and Asp76 (Iwasaki *et al.* 1997). This zinc site is found in the unusually long N-terminal extension region of these proteins that at present has no known function.

Both the periplasmic zinc binding protein TroA from the human parasite *Treponema pallidum* (Lee *et al.* 1999) and pneumococcal surface antigen PsaA (Lawrence *et al.* 1998) contain zinc sites that span a great distance in the protein (Table 5). These sites are very similar. The bound zinc ion is coordinated by the Nε2 nitrogens of His68, His133, His199 and both oxygens of Asp279 in TroA and His67 (Nε2), His139 (Nε2), Glu205 (Oε1) and one oxygen of Asp280 (Oδ1) in PsaA. The resulting coordination geometries are described as tetrahedral for PsaA and distorted square pyramidal in TroA due to the interaction of both oxygens of Asp279. The apex of the pyramid is occupied by the nitrogen of His199 while the other His ligands and the two oxygens of Asp 279 make up the square plane. These metal-binding proteins are believed to be the ligand binding component of an ATP-binding cassette (ABC) transport system (Lee *et al.* 1999). The function of the proteins is presumed to be similar to that of members of the

Table 5. Miscellaneous zinc binding proteins.

Protein	Classification	PDB#	L_1	X	L_2	Y	L_3	Z	L_4	Ref.
Human endostatin	Angiogenesis Inhib.	1BNL	His	1	His	7	His_{β}	64	$Asp_{b\beta}$	(Ding *et al.* 1998)
Mouse endostatin	Angiogenesis Inhib.	1DY0	His	1	His	7	His_{β}	64	$Asp_{b\beta}$	(Hohenester *et al.* 2000)
			His	3	Asp	5	His_{β}	64	$Asp_{b\beta}$	
Staphylococcus aureus A (SEA)	Enterotoxin	1ESF	Asp_{β}	1	His_{β}	37	His_{β}	185	Ser (N)	(Schad *et al.* 1995)
Streptococcus pyogenes (SPE-H)	Exotoxin	1EU3	Asp_{β}	1	His_{β}	39	His_{β} (N)		H_2O	(Arcus *et al.* 2000)
Streptococcus pyogenes (SMEZ-2)	Exotoxin	1EU4	Asp_{β}	1	His_{β}	37	Asp_{β} (N)		H_2O	(Arcus *et al.* 2000)
Streptococcus pyogenes A(SPE-A1)	Exotoxin	1B1Z	His_{β}	3	His_{β}	28	Asp_{β} (N)		H_2O	(Papageorgiou *et al.* 1999)
Streptococcus pyogenes A(SPE-A)	Exotoxin	1FNU	His_{β}	3	His_{β}	28	Asp_{β}	43	Glu_{β} (N)	(Earhart *et al.* 2000)
Sulfolobus Sp. ferredoxin	Electron transport	1XER	His_{β}	2	His_{β}	14	His_{β}	41	Asp_{β}	(Fujii *et al.* 1996)
Treponema pallidum Troa A	Zn binding protein	1TOA	His	64	$His_{2a\beta}$	65	His_L	79	Asp	(Lee *et al.* 1999)
S. Pneumoniae surface antigen Psaa	Immune system	1PSZ	His	71	$His_{2a\beta}$	65	Glu_L	74	Asp	(Lawrence *et al.* 1998)
Human survivin	Apoptosis	1F3H	Cys_{β}	2	Cys_L	16	His_{α}	6	Cys_L	(Verdecia *et al.* 2000)
Mouse survivin	Apoptosis		Cys	2	Cys	16	His	6	Cys	(Muchmore *et al.* 2000)
Xenopus laevis Xnf7 Bbox	Development	1FRE	Cys_{β}	2	His	18	Cys	5	His	(Borden *et al.* 1995) NMR
Thermus thermophilus chain D	Ribosomal protein	1FJF	$Cys_{b\alpha}$	2	Cys_{α}	13	Cys	4	Cys	(Wimberly *et al.* 2000)
ARF-GAP domain PYK2-associated protein β		1DCQ	Cys_{β}	2	Cys_L	16	$His_{b\alpha}$	2	Cys_{α}	(Mandiyan *et al.* 1999)
Thermus thermophilus L36	Ribosomal protein	1DGZ	Cys	2	$Cys_{b\beta}$	12	$Cys_{a\beta}$	4	$His_{2b\beta}$	(Hard *et al.* 2000)
E. coli DnaJ protein	Chaperone	1EXK	Cys_{β}	2	Cys	49	Cys_{β}	2	Cys	
			Cys_{β}	2	Cys	18	Cys_{β}	2	Cys	(Martinez-Yamout *et al.* 2000)
E. coli ADA	Repair		Cys	3	Cys	26	Cys	2	Cys	(Myers *et al.* 1994)
E. coli threonyl-tRNA synthetase	Descriminating	1EVK	Cys_{α}	50	His_{β}	125	His_{β}		H_2O	(Sankaranarayanan *et al.* 2000; Sankaranarayanan *et al.* 1999)
Human Sex hormone-binding globulin	Signaling protein	1F5F	Asp_{β}	17	$His_{a\beta}$	52	$His_{b\beta}$		?	(Avvakumov *et al.* 2000)

periplasmic ligand-binding protein (PLBP) family that function as ligand binding receptors for active transport and chemotaxis. However these proteins do not have the flexible interdomain β-strands of the PLBPs. An argument has been proposed for a more rigid hinge (in this case a backbone α-helix) if the purpose of these proteins is 'small' metal ion transport (Lee *et al.* 1999). Thus since the free energy of zinc in solution and bound to a protein is considered to be small (Lipscomb & Strater 1996), the binding of zinc to a protein can not incur the large entropy changes that would be associated with the ordering of a very mobile hinge region.

The steroid-binding specificity of the human sex hormone-binding globulin may also be influenced through a zinc binding site (Avvakumov *et al.* 2000). In this case zinc binding to both oxygens of Asp65, His83 (Nε2) and His136 (Nε2) prevents Asp65 from interaction with steroid 17β-hydroxyl group and further disorders the binding site. Since this site is observed when soaking the crystals with 2.5 mM $ZnCl_2$ further experiments are needed to evaluate the physiological significance of this finding.

A number of multi Cys zinc binding sites, several of which contain one or two His ligands, have also been reported. Survivin is a relatively small, 16 kDa, protein that is an inhibitor of apoptosis, IAP. Both the human and mouse proteins contain a zinc binding site in the N-terminal, BIR-like (baculovirus IAP repeat) domain (Muchmore *et al.* 2000; Verdecia *et al.* 2000). Survivin as well as caspase-3 localize to the microtubule organization centers during mitosis (Verdecia *et al.* 2000). The zinc binds to Cys57, Cys60, His77 (Nε2) and Cys84 in both human and mouse proteins. No direct interaction of this region with caspase-3 is observed (Verdecia *et al.* 2000). However since free zinc inhibits caspase-3 with a K_i less than 10 nM (Maret *et al.* 1999) the effect could be indirect if some agent should cause release of the zinc from the survivin protein. Survivin's high expression in a number of malignant tissues makes it a novel target for cancer therapy (Verdecia *et al.* 2000). A somewhat similar zinc binding site (in terms of spacing, Table 5) has been proposed by a NMR structural analysis of a *Xenopus* nuclear factor XNF7 regulating early *Xenopus* development (Borden *et al.* 1995). In this case the zinc is bound to Cys6, His9 (Nε2), Cys28 and His34 (Nε2).

Zinc binding sites of varying length (Table 5) that are likely involved in protein structure are found in the ARF-GAP domain of the PYK2-associated protein (Mandiyan *et al.* 1999), the *Thermus thermophilus* ribsome proteins (Hard *et al.* 2000; Wimberly *et al.* 2000), and the cysteine-rich domain of the *E. coli* chaperone protein DnaJ (Martinez-Yamout *et al.* 2000).

Some zinc sites have evolved to perform some unusual functions. In response to the threat posed by aberrant methylation of DNA, organisms express proteins that recognize and repair the resulting lesions (Lindahl 1993). Studies of how the DNA repair proteins might function have revealed a unique function of zinc. The solution structure of the DNA methyl phosphotriester repair domain of *E. coli* Ada contains a zinc bound to Cys38, Cys42, Cys69 and Cys72 (Myers *et al.* 1994). The acceptor residue of the repair protein appears to be the zinc-ligated Cys69. Since this is an irreversible transfer the protein performs a suicide role. Upon methyl transfer, Ada acquires the ability to bind to a DNA sequence-specifically and thereby induce genes that confer resistance to methylating agents (Myers *et al.* 1994).

A zinc binding site in the *E. coli* threonyl tRNA synthetase is also a unique zinc site not seen in any of the other synthetase family members. In this case the zinc binds in the active site to Cys334, His385 (Nε2), His511 (Nδ1) and a water molecule (Sankaranarayanan *et al.* 1999). The zinc ion is directly involved in the threonine recognition/discrimination. Upon binding threonine the zinc coordination changes from tetrahedral to a square pyramidal pentacoordinate intermediate by replacing the zinc bound water with both the amino and side chain hydroxyl groups of threonine (Sankaranarayanan *et al.* 2000). This binding mode permits other H-bonding interactions with amino acid residues in the active site. Amino acid activation experiments show no activation with the isosteric valine and a 1000-fold decreased activation with serine.

Support structure of zinc binding sites

The secondary support and scaffolding in combination with the direct ligands of the zinc allows fine tuning of the role of the zinc and its neighboring amino acids in the function of the enzyme. The following of observations are made about the structural properties of zinc binding sites observed in enzymes and other biological zinc sites (Tables 1–5): (1) Nearly all the sites contain at least one secondary structural element; (2) β-sheets supply ligands to a wide variety of zinc binding sites; (3) Ligands frequently come from the first, last or 1 or 2 amino acids before or after a β-sheet or α-helix. This may allow more flexiblity in forming the zinc site. (4) Small loop regions (≤ 5 amino acids) between two types of secondary structure can also supply the ligands; (5) Short spacers of one generally use a β-sheet to supply the ligands while a spacer of three uses an α-helix; (6) An α-helix is used to supply the short spacer ligands most frequently in hydrolytic enzymes (in particular the metalloproteases). Overall this suggests that the stability of biological zinc sites requires some form of secondary structural support. However, the fact that short spacers are supplied by a variety of support types indicate this maybe one level of fine tuning the functional properties of such sites.

The stability and the function of the metal site is also likely influenced by the second shell of residues in the vicinity of the metal binding site. The secondary interactions of the ligands with hydrogen-bonding groups of the side chain groups of the amino acid residues or the carbonyl oxygen of the back-bone peptide chain may be critical to the formation and stabilization of the zinc sites containing oxygen, sulfur and nitrogen ligands. Comparative structural studies of four of the first known zinc enzymes, carbonic anhydrase, carboxypeptidase A, alcohol dehydrogenase and thermolysin led to the identification of carbonyl and carboxyl 'orienters' (Argos *et al.* 1978).

The interaction between a carboxylate anion orienter and a zinc-ligated histidine could have multiple effects on the reactivity and stability of the zinc site. Thus the negatively charged carboxylate could reduce the charge on the zinc which in turn could fine-tune the ability of the zinc to act as a Lewis acid or make it more difficult for zinc bound water to ionize. On the other hand the interaction would place constraints on the rotation of the His residue and might thus make zinc bind tighter and/or distort the geometry of the zinc site. Recent studies have shown some agents such as the drug D-pencillamine can facilitate the release of zinc, likely by disrupting stabilizing interactions between orienters and ligands (Chong & Auld 2000). Examination of the effect of the orienters or indirect ligands on catalysis has been most thoroughly examined in the carbonic anhydrase α-class of enzymes (Christianson & Cox 1999; Christianson & Fierke 1996). In addition the local environment provided by highly conserved residues surrounding the zinc and its ligands will likely be important to the function and stability of the zinc site. In the case of carbonic anhydrase, conserved hydrophobic residues are believed to be important to positioning the zinc ligands in a distorted tetrahedron (Hunt *et al.* 1999). The surrounding amino acids may also influence the off rate constants for the metal, thus influencing the stability of the metal site.

Concluding remarks

Information on zinc binding sites in metalloenzymes and related proteins has become increasingly available as the interest in how zinc affects biological function has accelerated. This is particularly apparent over the last decade. Zinc binding site motifs can be highly conserved, not only in the identity of the ligands and their spacing but in the neighboring amino acids in the linear sequence. This leads to the formation of a family of zinc enzymes that may have a similarity in their overall primary structure (sometimes low) and some biological function in common.

The availability of structural standards of reference for the types of zinc sites coupled with prediction of protein sequences from DNA sequencing has resulted in an explosion of information on potential biological zinc sites. In 1992 the number of zinc enzymes was estimated to be 300, based on the original criteria of determining the metal content and its relationship to the function of an enzyme (Vallee & Auld 1992a). If one now broadens the definition of zinc enzymes to include those proteins who likely will bind zinc due to the presence of zinc binding motifs found in zinc reference sites the number of zinc enzymes will be in the thousands. Thus the number of zinc proteases based on this criteria has increased from about 100 in 1989 to 2,169 (Barrett & Rawlings 2001) in March 2001. As the number of sequenced genomes increases this number will also increase. Of course the number of unique functional zinc sites that is used by the majority of species and phyla should be much smaller.

The challenge still remains to understand how these zinc sites function in detail. Future mechanistic studies of zinc metalloenzymes will continue to benefit from a combined use of structural, mutagenic, and transient state kinetics approaches to examine the system. Analyses of zinc proteins using methods that give both global structural information (e.g., X-ray diffraction and NMR) and dynamic local structural information (e.g., electronic absorption, XAFS and NMR) performed on specifically modified enzymes is beginning to reveal the manner in which the protein modulates the zinc to achieve both catalytic efficiency and specificity. These studies have established zinc as an integral component of numerous functional proteins involved in a multiplicity of tasks, thus accounting for its fundamental role in metabolism, growth and development.

There are likely a number of features that zinc possesses that make it suitable for such a wide variety of functions. Chief among this is the fact that it is a stable metal ion species in a biological medium whose redox potential is in constant flux. The filled d-shell of zinc prevents it from undergoing oxidation or reduction in contrast to some of its neighboring transition metal ions such as Cu and Fe where their oxido-reductive properties are essential to their function (Vallee & Auld 1992b). Redox changes in neighboring transition metals are major sources of change in coordination geometry and rate of ligand exchange. Zinc is amphoteric, existing in both hydrate and hydroxide forms even at neutrality. It has Lewis acid properties. It ligates nitrogen and oxygen compounds as readily as sulfur. While zinc can have coordination numbers from 2 to 8 in zinc complex ions, 4, 5 and 6 are most frequently found in biological systems. Its stereochemical flexibility likely contributes to catalysis since it can transiently accept different coordination geometries without impeding catalysis. This property allows expansion of the coordination sphere of the zinc at one step of catalysis and its contraction at another step.

Beyond the challenge of how zinc functions in detail in any given protein is the goal to understand how it is delivered and removed from proteins *in vivo*. The next frontier will likely begin to bring answers to how zinc is stored, transported and distributed and how does it influence the earliest stages of development.

References

Abrahmsen L, Dohlsten M, Segren S, Bjork P, Jonsson E, Kalland T. 1995 Characterization of two distinct MHC class II binding sites in the superantigen staphylococcal enterotoxin A. *EMBO J* **14**, 2978–2986.

Alber BE, Colangelo CM, Dong J, Stalhandske CM, Baird TT, Tu C, Fierke CA, Silverman DN, Scott RA, Ferry JG. 1999 Kinetic and Spectroscopic Characterization of the Gamma-Carbonic Anhydrase from the Methanoarchaeon *Methanosarcina thermophila*. *Biochemistry* **38**, 13119—13128.

Alber BE, Ferry JG. 1994 A carbonic anhydrase from the archaeon *Methanosarcina thermophila*. *Proc Natl Acad Sci USA* **91**, 6909–6913.

Arcus VL, Proft T, Sigrell JA, Baker HM, Fraser JD, Baker EN. 2000 Conservation and variation in superantigen structure and activity highlighted by the three-dimensional structures of two new superantigens from *Streptococcus pyogenes*. *J Mol Biol* **299**, 157–168.

Argos P, Garavito RM, Eventoff W, Rossmann MG, Branden CI. 1978 Similarities in active center geometries of zinc-containing enzymes, proteases and dehydrogenases. *J Mol Biol* **126**, 141–158.

Auerbach G, Herrmann A, Bracher A, Bader G, Gutlich M, Fischer M, Neukamm M, Garrido-Franco M, Richardson J, Nar H, Huber R, Bacher A. 2000 Zinc plays a key role in human and bacterial GTP cyclohydrolase I. *Proc Natl Acad Sci USA* **97**, 13567–13572.

Auld DS. 1987 Acyl group transfer- metalloproteinases. In: Page MI, Williams A, ed. *Enzyme Mechanisms*. London: Royal Society of Chemistry Burlington House; 241–258.

Auld DS. 1992 Astacin family of zinc proteases. *Faraday Discuss* **93**, 117–120.

Auld DS. 1997 Zinc catalysis in metalloproteases. *Structure and Bonding* **89**, 29–50.

Auld DS. 1998a Carboxypeptidase A. In: Barrett AJ, Rawlings ND, Woessner JF, eds. *Handbook of Proteolytic Enzymes*. London: Academic Press; 1321–1326.

Auld DS. 1998b Carboxypeptidase A2. In: Barrett AJ, Rawlings ND, Woessner JF, eds. *Handbook of Proteolytic Enzymes*. London: Academic Press; 1326–1328.

Auld DS. 2001 Zinc Sites in Metalloenzymes and Related Proteins. In: Bertini I, Sigel A, Sigel H, eds. *Handbook on Metalloproteins*. New York: M. Dekker; 881–959.

Auld DS, Vallee BL. 1987 Carboxypeptidase-A. In: Neuberger H, Brocklehurst K, ed. *Hydrolytic Enzymes*. Amsterdam: Elsevier; 201–255.

Aviles FX, Vendrel J. 1998 Carboxypeptidase B. In: Barrett AJ, Rawlings ND, Woessner JF, eds. *Handbook of Proteolytic Enzymes*. London: Academic Press; 1333–1335.

Avvakumov GV, Muller YA, Hammond GL. 2000 Steroid-binding specificity of human sex hormone-binding globulin is influenced by occupancy of a zinc-binding site. *J Biol Chem* **275**, 25920–25925.

Baldwin GS, Galdes A, Hill HA, Smith BE, Waley SG, Abraham EP. 1978 Histidine residues of zinc ligands in beta-lactamase II. *Biochem J* **175**, 441–447.

Banbula A, Potempa J, Travis J, Fernandez-Catalan C, Mann K, Huber R, Bode W, Medrano F. 1998 Amino-acid sequence and three-dimensional structure of the *Staphylococcus aureus* metalloproteinase at 1.72 Å resolution. *Structure* **6**, 1185–1193.

Bannister JV, Bannister WH, Rotilio G. 1987 Aspects of the structure, function, and applications of superoxide dismutase. *CRC Crit Rev Biochem* **22**, 111–180.

Barbato G, Cicero DO, Nardi MC, Steinkuhler C, Cortese R, De Francesco R, Bazzo R. 1999 The solution structure of the N-terminal proteinase domain of the hepatitis C virus (HCV) NS3 protein provides new insights into its activation and catalytic mechanism. *J Mol Biol* **289**, 371–384.

Barrett AJ, Rawlings ND. 2001 MEROPS the Peptidase Database. http://www.merops.co.uk.

Barrett AJ, Rawlings ND, Woessner JF. 1998 Introduction: Clan MC containing metallocarboxypeptidases. In: Barrett AJ, Rawlings ND, Woessner JF, eds. *Handbook of Proteolytic Enzymes*. London: Academic Press; 1318–1320.

Baumann U. 1994 Crystal structure of the 50 kDa metallo protease from *Serratia marcescens*. *J Mol Biol* **242**, 244–251.

Baumann U, Bauer M, Letoffe S, Delepelaire P, Wandersman C. 1995 Crystal structure of a complex between *Serratia marcescens* metallo-protease and an inhibitor from *Erwinia chrysanthemi*. *J Mol Biol* **248**, 653–661.

Baumann U, Wu S, Flaherty KM, McKay DB. 1993 Three-dimensional structure of the alkaline protease of *Pseudomonas aeruginosa*: a two-domain protein with a calcium binding parallel beta roll motif. *EMBO J* **12**, 3357–3364.

Bax B, Blundell TL, Murray-Rust J, McDonald NQ. 1997 Structure of mouse 7S NGF: a complex of nerve growth factor with four binding proteins. *Structure* **5**, 1275–1285.

Becker A, Schlichting I, Kabsch W, Groche D, Schultz S, Wagner AF. 1998 Iron center, substrate recognition and mechanism of peptide deformylase. *Nat Struct Biol* **5**, 1053–1058.

Becker JW, Marcy AI, Rokosz LL, Axel MG, Burbaum JJ, Fitzgerald PM, Cameron PM, Esser CK, Hagmann WK, Hermes JD, Springer JP. 1995 Stromelysin-1: three-dimensional structure of the inhibited catalytic domain and of the C-truncated proenzyme. *Protein Sci* **4**, 1966–1976.

Benning MM, Kuo JM, Raushel FM, Holden HM. 1994 Three-dimensional structure of phosphotriesterase: an enzyme capable of detoxifying organophosphate nerve agents. *Biochemistry* **33**, 15001–15007.

Berg JM, Shi Y. 1996 The galvanization of biology: a growing appreciation for the roles of zinc. *Science* **271**, 1081–1085.

Berry MB, Phillips GN, Jr. 1998 Crystal structures of *Bacillus stearothermophilus* adenylate kinase with bound Ap5A, Mg^{2+} Ap5A, and Mn^{2+} Ap5A reveal an intermediate lid position and six coordinate octahedral geometry for bound Mg^{2+} and Mn^{2+}. *Proteins* **32**, 276–288.

Bertini I, Lanini G, Luchinat C, Haas C, Maret W, Zeppezauer M. 1987 The influence of anions and inhibitors on the catalytic metal ion in Co(II)-substituted horse liver alcohol dehydrogenase. *Eur Biophys J* **14**, 431–439.

Betts L, Xiang S, Short SA, Wolfenden R, Carter CW, Jr. 1994 Cytidine deaminase. The 2.3 Å crystal structure of an enzyme: transition–state analog complex. *J Mol Biol* **235**, 635–656.

Betz M, Huxley P, Davies SJ, Mushtaq Y, Pieper M, Tschesche H, Bode W, Gomis-Ruth FX. 1997 1.8-Å crystal structure of the catalytic domain of human neutrophil collagenase (matrix metalloproteinase-8) complexed with a peptidomimetic hydroxamate primed-side inhibitor with a distinct selectivity profile. *Eur J Biochem* **247**, 356–363.

Bixby KA, Nanao MH, Shen NV, Kreusch A, Bellamy H, Pfaffinger PJ, Choe S. 1999 Zn^{2+}-binding and molecular determinants of tetramerization in voltage-gated K^{+} channels. *Nat Struct Biol* **6**, 38–43.

Bode W, Gomis-Ruth FX, Huber R, Zwilling R, Stocker W. 1992 Structure of astacin and implications for activation of astacins and zinc-ligation of collagenases. *Nature* **358**, 164–167.

Bode W, Reinemer P, Huber R, Kleine T, Schnierer S, Tschesche H. 1994 The X-ray crystal structure of the catalytic domain of human neutrophil collagenase inhibited by a substrate analogue reveals the essentials for catalysis and specificity. *EMBO J* **13**, 1263–1269.

Borden KL, Lally JM, Martin SR, O'Reilly NJ, Etkin LD, Freemont PS. 1995 Novel topology of a zinc-binding domain from a protein involved in regulating early *Xenopus* development. *EMBO J* **14**, 5947–5956.

Boriack-Sjodin PA, Heck RW, Laipis PJ, Silverman DN, Christianson DW. 1995 Structure determination of murine mitochondrial carbonic anhydrase V at 2.45-Å resolution: implications for catalytic proton transfer and inhibitor design. *Proc Natl Acad Sci USA* **92**, 10949–10953.

Botos I, Meyer E, Swanson SM, Lemaitre V, Eeckhout Y, Meyer EF. 1999 Structure of recombinant mouse collagenase-3 (MMP-13). *J Mol Biol* **292**, 837–844.

Bourne Y, Redford SM, Steinman HM, Lepock JR, Tainer JA, Getzoff ED. 1996 Novel dimeric interface and electrostatic recognition in bacterial Cu,Zn superoxide dismutase. *Proc Natl Acad Sci USA* **93**, 12774–12779.

Bracey MH, Christiansen J, Tovar P, Cramer SP, Bartlett SG. 1994 Spinach carbonic anhydrase: investigation of the zinc-binding ligands by site-directed mutagenesis, elemental analysis, and EXAFS. *Biochemistry* **33**, 13126–13131.

Brodersen DE, Nyborg J, Kjeldgaard M. 1999 Zinc-binding site of an S100 protein revealed. Two crystal structures of Ca^{2+}-bound human psoriasin (S100A7) in the Zn^{2+}-loaded and Zn^{2+}-free states. *Biochemistry* **38**, 1695–1704.

Browner MF, Smith WW, Castelhano AL. 1995 Matrilysin-inhibitor complexes: common themes among metalloproteases. *Biochemistry* **34**, 6602–6610.

Buchbinder JL, Stephenson RC, Dresser MJ, Pitera JW, Scanlan TS, Fletterick RJ. 1998 Biochemical characterization and crystallographic structure of an *Escherichia coli* protein from the phosphotriesterase gene family [published erratum appears in Biochemistry 1998 Jul 28;**37**(30):10860]. *Biochemistry* **37**, 5096–5106.

Bukrinsky JT, Bjerrum MJ, Kadziola A. 1998 Native carboxypeptidase A in a new crystal environment reveals a different conformation of the important tyrosine 248. *Biochemistry* **37**, 16555–16564.

Burgisser DM, Thony B, Redweik U, Hess D, Heizmann CW, Huber R, Nar H. 1995 6-Pyruvoyl tetrahydropterin synthase, an enzyme with a novel type of active site involving both zinc binding and an intersubunit catalytic triad motif; site-directed mutagenesis of the proposed active center, characterization of the metal binding site and modelling of substrate binding. *J Mol Biol* **253**, 358–369.

Burley SK, David PR, Sweet RM, Taylor A, Lipscomb WN. 1992 Structure determination and refinement of bovine lens leucine aminopeptidase and its complex with bestatin. *J Mol Biol* **224**, 113–140.

Bussiere DE, Pratt SD, Katz L, Severin JM, Holzman T, Park CH. 1998 The structure of VanX reveals a novel amino-dipeptidase involved in mediating transposon-based vancomycin resistance. *Mol Cell* **2**, 75–84.

Cai M, Zheng R, Caffrey M, Craigie R, Clore GM, Gronenborn AM. 1997 Solution structure of the N-terminal zinc binding domain of HIV-1 integrase [published erratum appears in Nat Struct Biol 1997 Oct; **4**(10): 839–840]. *Nat Struct Biol* **4**, 567–577.

Cameron AD, Olin B, Ridderstrom M, Mannervik B, Jones TA. 1997 Crystal structure of human glyoxalase I–evidence for gene duplication and 3D domain swapping. *EMBO J* **16**, 3386–3395.

Cameron AD, Ridderstrom M, Olin B, Kavarana MJ, Creighton DJ, Mannervik B. 1999a Reaction mechanism of glyoxalase I explored by an X-ray crystallographic analysis of the human enzyme in complex with a transition state analogue. *Biochemistry* **38**, 13480–13490.

Cameron AD, Ridderstrom M, Olin B, Mannervik B. 1999b Crystal structure of human glyoxalase II and its complex with a glutathione thiolester substrate analogue. *Structure Fold Des* **7**, 1067–1078.

Carfi A, Duee E, Galleni M, Frere JM, Dideberg O. 1998a 1.85 Å resolution structure of the zinc (II) beta-lactamase from *Bacillus cereus*. *Acta Cryst D* **54**, 313–323.

Carfi A, Duee E, Paul-Soto R, Galleni M, Frere JM, Dideberg O. 1998b X-ray structure of the ZnII beta-lactamase from *Bac-*

teroides fragilis in an orthorhombic crystal form. *Acta Cryst D Biol Cryst* **54**, 45–57.

Carfi A, Pares S, Duee E, Galleni M, Duez C, Frere JM, Dideberg O. 1995 The 3-D structure of a zinc metallo-beta-lactamase from *Bacillus cereus* reveals a new type of protein fold. *EMBO J* **14**, 4914–4921.

Carlow DC, Carter CW, Jr., Mejlhede N, Neuhard J, Wolfenden R. 1999 Cytidine Deaminases from *B. subtilis* and *E. coli*: Compensating effects of changing zinc coordination and quaternary structure. *Biochemistry* **38**, 12258–12265.

Carugo KD, Battistoni A, Carri MT, Polticelli F, Desideri A, Rotilio G, Coda A, Bolognesi M. 1994 Crystal structure of the cyanide-inhibited *Xenopus laevis* Cu,Zn superoxide dismutase at 98 K. *FEBS Lett* **349**, 93–98.

Cha J, Auld DS. 1997 Site-directed mutagenesis of the active site glutamate in human matrilysin: investigation of its role in catalysis. *Biochemistry* **36**, 16019–16024.

Chan MK, Gong W, Rajagopalan PT, Hao B, Tsai CM, Pei D. 1997 Crystal structure of the *Escherichia coli* peptide deformylase [published erratum appears in Biochemistry 1998 Sep 15; **37**(37): 13042]. *Biochemistry* **36**, 13904–13909.

Chantalat L, Leroy D, Filhol O, Nueda A, Benitez MJ, Chambaz EM, Cochet C, Dideberg O. 1999 Crystal structure of the human protein kinase CK2 regulatory subunit reveals its zinc finger-mediated dimerization. *EMBO J* **18**, 2930–2940.

Chen L, Rydel TJ, Gu F, Dunaway CM, Pikul S, Dunham KM, Barnett BL. 1999 Crystal structure of the stromelysin catalytic domain at 2.0 Å resolution: inhibitor-induced conformational changes. *J Mol Biol* **293**, 545–557.

Chen YL, Park S, Thornburg RW, Tabatabai LB, Kintanar A. 1995 Structural characterization of the active site of *Brucella abortus* Cu-Zn superoxide dismutase: a 15N and 1H NMR investigation. *Biochemistry* **34**, 12265–12275.

Cheng X, Zhang X, Pflugrath JW, Studier FW. 1994 The structure of bacteriophage T7 lysozyme, a zinc amidase and an inhibitor of T7 RNA polymerase. *Proc Natl Acad Sci USA* **91**, 4034–4038.

Chevrier B, D'Orchymont H, Schalk C, Tarnus C, Moras D. 1996 The structure of the *Aeromonas proteolytica* aminopeptidase complexed with a hydroxamate inhibitor. Involvement in catalysis of Glu151 and two zinc ions of the co–catalytic unit. *Eur J Biochem* **237**, 393–398.

Chevrier B, Schalk C, D'Orchymont H, Rondeau JM, Moras D, Tarnus C. 1994 Crystal structure of *Aeromonas proteolytica* aminopeptidase: a prototypical member of the co-catalytic zinc enzyme family. *Structure* **2**, 283–291.

Cho H, Ramaswamy S, Plapp BV. 1997 Flexibility of liver alcohol dehydrogenase in stereoselective binding of 3-butylthiolane 1-oxides. *Biochemistry* **36**, 382–389.

Chong CR, Auld DS. 2000 Inhibition of carboxypeptidase A by D-pencillamine: Mechanism and implications for drug design. *Biochemistry* **39**, 7580–7588.

Christianson DW, Cox JD. 1999 Catalysis by metal-activated hydroxide in zinc and manganese metalloenzymes. *Annu Rev Biochem* **68**, 33–57.

Christianson DW, Fierke CA. 1996 Carbonic anhydrase: Evolution of the design of the zinc binding site by nature and by design. *Acc Chem Res* **29**, 331–339.

Christianson DW, Lipscomb WN. 1989 Carboxypeptidase A. *Acc Chem Res* **22**, 62–69.

Cleasby A, Wonacott A, Skarzynski T, Hubbard RE, Davies GJ, Proudfoot AE, Bernard AR, Payton MA, Wells TN. 1996 The X-ray crystal structure of phosphomannose isomerase from *Candida albicans* at 1.7 angstrom resolution. *Nat Struct Biol* **3**, 470–479.

Clugston SL, Barnard JF, Kinach R, Miedema D, Ruman R, Daub E, Honek JF. 1998 Overproduction and characterization of a dimeric non-zinc glyoxalase I from *Escherichia coli*: evidence for optimal activation by nickel ions. *Biochemistry* **37**, 8754–8763.

Coleman JE. 1992 Structure and mechanism of alkaline phosphatase. *Annu Rev Biophys Biomol Struct* **21**, 441–483.

Coleman JE. 1998 Zinc enzymes. *Curr Opin Chem Biol* **2**, 222–234.

Coll M, Guasch A, Aviles FX, Huber R. 1991 Three-dimensional structure of porcine procarboxypeptidase B: a structural basis of its inactivity. *EMBO J* **10**, 1–9.

Colonna-Cesari F, Perahia D, Karplus M, Eklund H, Branden CI, Tapia O. 1986 Interdomain motion in liver alcohol dehydrogenase. Structural and energetic analysis of the hinge bending mode. *J Biol Chem* **261**, 15273–15280.

Concha NO, Janson CA, Rowling P, Pearson S, Cheever CA, Clarke BP, Lewis C, Galleni M, Frere JM, Payne DJ, Bateson JH, Abdel-Meguid SS. 2000 Crystal structure of the IMP-1 metallo beta-lactamase from *Pseudomonas aeruginosa* and its complex with a mercaptocarboxylate inhibitor: binding determinants of a potent, broad-spectrum inhibitor. *Biochemistry* **39**, 4288–4298.

Concha NO, Rasmussen BA, Bush K, Herzberg O. 1996 Crystal structure of the wide-spectrum binuclear zinc beta-lactamase from *Bacteroides fragilis*. *Structure* **4**, 823–836.

Crane BR, Arvai AS, Ghosh DK, Wu C, Getzoff ED, Stuehr DJ, Tainer JA. 1998 Structure of nitric oxide synthase oxygenase dimer with pterin and substrate. *Science* **279**, 2121–2126.

Daniels DS, Mol CD, Arvai AS, Kanugula S, Pegg AE, Tainer JA. 2000 Active and alkylated human AGT structures: a novel zinc site, inhibitor and extrahelical base binding. *EMBO J* **19**, 1719–1730.

Davis GJ, Bosron WF, Stone CL, Owusu-Dekyi K, Hurley TD. 1996 X-ray structure of human beta3beta3 alcohol dehydrogenase. The contribution of ionic interactions to coenzyme binding. *J Biol Chem* **271**, 17057–17061.

De Francesco R, Urbani A, Nardi MC, Tomei L, Steinkuhler C, Tramontano A. 1996 A zinc binding site in viral serine proteinases. *Biochemistry* **35**, 13282–13287.

Dhanaraj V, Ye QZ, Johnson LL, Hupe DJ, Ortwine DF, Dunbar JB, Jr., Rubin JR, Pavlovsky A, Humblet C, Blundell TL. 1996 X-ray structure of a hydroxamate inhibitor complex of stromelysin catalytic domain and its comparison with members of the zinc metalloproteinase superfamily. *Structure* **4**, 375–386.

Ding YH, Javaherian K, Lo KM, Chopra R, Boehm T, Lanciotti J, Harris BA, Li Y, Shapiro R, Hohenester E, Timpl R, Folkman J, Wiley DC. 1998 Zinc-dependent dimers observed in crystals of human endostatin. *Proc Natl Acad Sci USA* **95**, 10443–10448.

Djinovic K, Gatti G, Coda A, Antolini L, Pelosi G, Desideri A, Falconi M, Marmocchi F, Rotilio G, Bolognesi M. 1992 Crystal structure of yeast Cu,Zn superoxide dismutase. Crystallographic refinement at 2.5 Å resolution. *J Mol Biol* **225**, 791–809.

Doss M, von Tiepermann R, Schneider J, Schmid H. 1979 New type of hepatic porphyria with porphobilinogen synthase defect and intermittent acute clinical manifestation. *Klin Wochenschr* **57**, 1123–1127.

Dreyer MK, Schulz GE. 1993 The spatial structure of the class II L-fuculose-1-phosphate aldolase from *Escherichia coli*. *J Mol Biol* **231**, 549–553.

Dreyer MK, Schulz GE. 1996 Catalytic mechanism of the metal-dependent fuculose aldolase from *Escherichia coli* as derived from the structure. *J Mol Biol* **259**, 458–466.

D'Souza VM, Holz RC. 1999 The methionyl aminopeptidase from *Escherichia coli* can function as an iron(II) enzyme. *Biochemistry* **38**, 11079–11085.

Dumermuth E, Sterchi EE, Jiang WP, Wolz RL, Bond JS, Flannery AV, Beynon RJ. 1991 The astacin family of metalloendopeptidases. *J Biol Chem* **266**, 21381–21385.

Earhart CA, Vath GM, Roggiani M, Schlievert PM, Ohlendorf DH. 2000 Structure of streptococcal pyrogenic exotoxin A reveals a novel metal cluster. *Protein Sci* **9**, 1847–1851.

Egloff MP, Cohen PT, Reinemer P, Barford D. 1995 Crystal structure of the catalytic subunit of human protein phosphatase 1 and its complex with tungstate. *J Mol Biol* **254**, 942–959.

Eijkelenboom AP, van den Ent FM, Vos A, Doreleijers JF, Hard K, Tullius TD, Plasterk RH, Kaptein R, Boelens R. 1997 The solution structure of the amino-terminal HHCC domain of HIV-2 integrase: a three-helix bundle stabilized by zinc. *Curr Biol* **7**, 739–746.

Eijkelenboom AP, van den Ent FM, Wechselberger R, Plasterk RH, Kaptein R, Boelens R. 2000 Refined solution structure of the dimeric N-terminal HHCC domain of HIV-2 integrase. *J Biomol NMR* **18**, 119–128.

Eklund H. 1989 Coenzyme binding in alcohol dehydrogenase. *Biochem Soc Trans* **17**, 293–296.

Eklund H, Branden CI. 1987 Alcohol Dehydrogenase. In: Jurnak FA, McPherson A. eds. *Biological Macromolecules and Assemblies: Vol. 3, Active Sites of Enzumes*. New York: John Wiley & Sons, Inc; 73–143.

Eklund H, Nordstrom B, Zeppezauer E, Soderlund G, Ohlsson I, Boiwe T, Branden CI. 1974 The structure of horse liver alcohol dehydrogenase. *FEBS Lett.* **44**, 200–204.

Eklund H, Nordstrom B, Zeppezauer E, Soderlund G, Ohlsson I, Boiwe T, Soderberg BO, Tapia O, Branden CI, Akeson A. 1976 Three-dimensional structure of horse liver alcohol dehydrogenase at 2–4 Å resolution. *J Mol Biol* **102**, 27–59.

Eklund H, Samama JP, Wallen L, Branden CI, Akeson A, Jones TA. 1981 Structure of a triclinic ternary complex of horse liver alcohol dehydrogenase at 2.9 Å resolution. *J Mol Biol* **146**, 561–587.

Eriksson AE, Liljas A. 1993 Refined structure of bovine carbonic anhydrase III at 2.0 Å resolution. *Proteins* **16**, 29–42.

Erskine PT, Duke EM, Tickle IJ, Senior NM, Warren MJ, Cooper JB. 2000 MAD analyses of yeast 5-aminolaevulinate dehydratase: their use in structure determination and in defining the metal-binding sites. *Acta Cryst D* **56**, 421–430.

Erskine PT, Newbold R, Roper J, Coker A, Warren MJ, Shoolingin-Jordan PM, Wood SP, Cooper JB. 1999a The Schiff base complex of yeast 5-aminolaevulinic acid dehydratase with laevulinic acid. *Protein Sci* **8**, 1250–1256.

Erskine PT, Norton E, Cooper JB, Lambert R, Coker A, Lewis G, Spencer P, Sarwar M, Wood SP, Warren MJ, Shoolingin-Jordan PM. 1999b X-ray structure of 5-aminolevulinic acid dehydratase from *Escherichia coli* complexed with the inhibitor levulinic acid at 2.0 Å resolution. *Biochemistry* **38**, 4266–4276.

Erskine PT, Senior N, Awan S, Lambert R, Lewis G, Tickle IJ, Sarwar M, Spencer P, Thomas P, Warren MJ, Shoolingin-Jordan PM, Wood SP, Cooper JB. 1997 X-ray structure of 5-aminolaevulinate dehydratase, a hybrid aldolase. *Nat Struc Biol* **4**, 1025–1031.

Estevez AG, Crow JP, Sampson JB, Reiter C, Zhuang Y, Richardson GJ, Tarpey MM, Barbeito L, Beckman JS. 1999 Induction of nitric oxide-dependent apoptosis in motor neurons by zinc-deficient superoxide dismutase. *Science* **286**, 2498–2500.

Fabiane SM, Sohi MK, Wan T, Payne DJ, Bateson JH, Mitchell T, Sutton BJ. 1998 Crystal structure of the zinc-dependent beta-lactamase from *Bacillus cereus* at 1.9 Å resolution: binuclear active site with features of a mononuclear enzyme. *Biochemistry* **37**, 12404–12411.

Faming Z, Kobe B, Stewart CB, Rutter WJ, Goldsmith EJ. 1991 Structural evolution of an enzyme specificity. The structure of rat carboxypeptidase A2 at 1.9 Å- resolution. *J Biol Chem* **266**, 24606–24612.

Feese M, Pettigrew DW, Meadow ND, Roseman S, Remington SJ. 1994 Cation-promoted association of a regulatory and target protein is controlled by protein phosphorylation. *Proc Natl Acad Sci USA* **91**, 3544–3548.

Feese MD, Comolli L, Meadow ND, Roseman S, Remington SJ. 1997 Structural studies of the *Escherichia coli* signal transducing protein IIAGlc: implications for target recognition. *Biochemistry* **36**, 16087–16096.

Fischmann TO, Hruza A, Niu XD, Fossetta JD, Lunn CA, Dolphin E, Prongay AJ, Reichert P, Lundell DJ, Narula SK, Weber PC. 1999 Structural characterization of nitric oxide synthase isoforms reveals striking active-site conservation. *Nat Struct Biol* **6**, 233–242.

Forest KT, Langford PR, Kroll JS, Getzoff ED. 2000 Cu,Zn superoxide dismutase structure from a microbial pathogen establishes a class with a conserved dimer interface. *J Mol Biol* **296**, 145–153.

Fourmy D, Dardel F, Blanquet S. 1993a Methionyl-tRNA synthetase zinc binding domain. Three–dimensional structure and homology with rubredoxin and gag retroviral proteins. *J Mol Biol* **231**, 1078–1089.

Fourmy D, Meinnel T, Mechulam Y, Blanquet S. 1993b Mapping of the zinc binding domain of *Escherichia coli* methionyl-tRNA synthetase. *J Mol Biol* **231**, 1068–1077.

Frankenberg N, Jahn D, Jaffe EK. 1999 *Pseudomonas aeruginosa* contains a novel type V porphobilinogen synthase with no required catalytic metal ions. *Biochemistry* **38**, 13976–13982.

Fraser J, Arcus V, Kong P, Baker E, Proft T. 2000 Superantigens - powerful modifiers of the immune system. *Mol Med Today* **6**, 125–132.

Fraser JD, Urban RG, Strominger JL, Robinson H. 1992 Zinc regulates the function of two superantigens. *Proc Natl Acad Sci USA* **89**, 5507–5511.

Fujii T, Hata Y, Wakagi T, Tanaka N, Oshima T. 1996 Novel zinc-binding centre in thermoacidophilic archaeal ferredoxins [letter]. *Nat Struct Biol* **3**, 834–837.

Fujinaga M, James MN. 1987 Rat submaxillary gland serine protease, tonin. Structure solution and refinement at 1.8 Å resolution. *J Mol Biol* **195**, 373–396.

Fukai S, Nureki O, Sekine S, Shimada A, Tao J, Vassylyev DG, Yokoyama S. 2000 Structural basis for double-sieve discrimination of L-valine from L-isoleucine and L-threonine by the complex of tRNA(Val) and valyl-tRNA synthetase. *Cell* **103**, 793–803.

Garcia–Iniguez L, Powers L, Chance B, Sellin S, Mannervik B, Mildvan AS. 1984 X-ray absorption studies of the Zn^{2+} site of glyoxalase I. *Biochemistry* **23**, 685–689.

Garcia–Saez I, Reverter D, Vendrell J, Aviles FX, Coll M. 1997 The three-dimensional structure of human procarboxypeptidase A2. Deciphering the basis of the inhibition, activation and intrinsic activity of the zymogen. *EMBO J* **16**, 6906–6913.

Geeganage S, Frey PA. 1999 Significance of metal ions in galactose-1-phosphate uridylyltransferase: An essential structural zinc and a nonessential structural iron. *Biochemistry* **38**, 13398–133406.

Ghosh DK, Crane BR, Ghosh S, Wolan D, Gachhui R, Crooks C, Presta A, Tainer JA, Getzoff ED, Stuehr DJ. 1999 Inducible nitric oxide synthase: role of the N-terminal beta-hairpin hook and pterin-binding segment in dimerization and tetrahydrobiopterin interaction. *EMBO J* **18**, 6260–6270.

Ghuysen JM. 1988 Evolution of DD-peptidases and beta-lactamases. In: Actor P, Daneo-Moore L, Higgens ML, Salton MRJ, Shockman GD, ed. *Antibiotic Inhibition of Bacterial Cell Surface Assembly and Function*. Washington DC: American Society for Microbiology; 268–284.

Gibbs JB, Oliff A. 1997 The potential of farnesyltransferase inhibitors as cancer chemotherapeutics. *Annu Rev Pharmacol Toxicol* **37**, 143–166.

Gilboa R, Greenblatt HM, Perach M, Spungin-Bialik A, Lessel U, Wohlfahrt G, Schomburg D, Blumberg S, Shoham G. 2000 Interactions of *Streptomyces griseus* aminopeptidase with a methionine product analogue: a structural study at 1.53 Å resolution. *Acta Cryst D* **56**, 551–558.

Gomis-Ruth FX, Companys V, Qian Y, Fricker LD, Vendrell J, Aviles FX, Coll M. 1999 Crystal structure of avian carboxypeptidase D domain II: a prototype for the regulatory metallocarboxypeptidase subfamily. *EMBO J* **18**, 5817–5826.

Gomis-Ruth FX, Gomez M, Bode W, Huber R, Aviles FX. 1995 The three-dimensional structure of the native ternary complex of bovine pancreatic procarboxypeptidase A with proproteinase E and chymotrypsinogen C. *EMBO J* **14**, 4387–4394.

Gomis-Ruth FX, Kress LF, Bode W. 1993a First structure of a snake venom metalloproteinase: a prototype for matrix metalloproteinases/collagenases. *EMBO J* **12**, 4151–4157.

Gomis-Ruth FX, Kress LF, Kellermann J, Mayr I, Lee X, Huber R, Bode W. 1994 Refined 2.0 Å X-ray crystal structure of the snake venom zinc-endopeptidase adamalysin II. Primary and tertiary structure determination, refinement, molecular structure and comparison with astacin, collagenase and thermolysin. *J Mol Biol* **239**, 513–544.

Gomis-Ruth FX, Stocker W, Huber R, Zwilling R, Bode W. 1993b Refined 1.8 Å X-ray crystal structure of astacin, a zinc-endopeptidase from the crayfish *Astacus astacus* L. Structure determination, refinement, molecular structure and comparison with thermolysin. *J Mol Biol* **229**, 945–968.

Gong W, Zhu X, Liu S, Teng M, Niu L. 1998 Crystal structures of acutolysin A, a three-disulfide hemorrhagic zinc metalloproteinase from the snake venom of *Agkistrodon acutus*. *J Mol Biol* **283**, 657–668.

Gooley PR, O'Connell JF, Marcy AI, Cuca GC, Salowe SP, Bush BL, Hermes JD, Esser CK, Hagmann WK, Springer JP, Johnson BA. 1994 The NMR structure of the inhibited catalytic domain of human stromelysin-1. *Nat Struct Biol* **1**, 111–118.

Grams F, Dive V, Yiotakis A, Yiallouros I, Vassiliou S, Zwilling R, Bode W, Stocker W. 1996 Structure of astacin with a transition-state analogue inhibitor [letter]. *Nat Struct Biol* **3**, 671–675.

Greenblatt HM, Almog O, Maras B, Spungin-Bialik A, Barra D, Blumberg S, Shoham G. 1997 *Streptomyces griseus* aminopeptidase: X-ray crystallographic structure at 1.75 Å resolution. *J Mol Biol* **265**, 620–636.

Griffith JP, Kim JL, Kim EE, Sintchak MD, Thomson JA, Fitzgibbon MJ, Fleming MA, Caron PR, Hsiao K, Navia MA. 1995 X-ray structure of calcineurin inhibited by the immunophilin-immunosuppressant FKBP12-FK506 complex. *Cell* **82**, 507–522.

Guasch A, Coll M, Aviles FX, Huber R. 1992 Three-dimensional structure of porcine pancreatic procarboxypeptidase A. A comparison of the A and B zymogens and their determinants for inhibition and activation. *J Mol Biol* **224**, 141–157.

Guddat LW, McAlpine AS, Hume D, Hamilton S, de Jersey J, Martin JL. 1999 Crystal structure of mammalian purple acid phosphatase. *Structure Fold Des* **7**, 757–767.

Guenther B, Onrust R, Sali A, O'Donnell M, Kuriyan J. 1997 Crystal structure of the delta' subunit of the clamp–loader complex of *E. coli* DNA polymerase III. *Cell* **91**, 335–345.

Haeggstrom JZ, Wetterholm A, Shapiro R, Vallee BL, Samuelsson B. 1990 Leukotriene A4 hydrolase: a zinc metalloenzyme. *Biochem Biophys Res Commun* **172**, 965–700.

Hakansson M, Petersson K, Nilsson H, Forsberg G, Bjork P, Antonsson P, Svensson LA. 2000 The crystal structure of staphylococcal enterotoxin H: implications for binding properties to MHC class II and TcR molecules. *J Mol Biol* **302**, 527–537.

Hall TM, Porter JA, Beachy PA, Leahy DJ. 1995 A potential catalytic site revealed by the 1.7-Å crystal structure of the amino-terminal signalling domain of Sonic hedgehog. *Nature* **378**, 212–216.

Hamada K, Hata Y, Katsuya Y, Hiramatsu H, Fujiwara T, Katsube Y. 1996 Crystal structure of Serratia protease, a zinc-dependent proteinase from *Serratia sp.* E-15, containing a beta-sheet coil motif at 2.0 Å resolution. *J Biochem (Tokyo)* **119**, 844–851.

Hard T, Rak A, Allard P, Kloo L, Garber M. 2000 The solution structure of ribosomal protein L36 from *Thermus thermophilus* reveals a zinc-ribbon-like fold. *J Mol Biol* **296**, 169–180.

He MM, Clugston SL, Honek JF, Matthews BW. 2000 Determination of the structure of *Escherichia coli* glyoxalase I suggests a structural basis for differential metal activation. *Biochemistry* **39**, 8719–8727.

Hernandez Valladares M, Felici A, Weber G, Adolph HW, Zeppezauer M, Rossolini GM, Amicosante G, Frere JM, Galleni M. 1997 Zn(II) dependence of the *Aeromonas hydrophila* AE036 metallo-beta-lactamase activity and stability. *Biochemistry* **36**, 11534–11541.

Hewett-Emmett D, Tashian RE. 1996 Functional diversity, conservation, and convergence in the evolution of the alpha-, beta-, and gamma-carbonic anhydrase gene families. *Mol Phylogenet Evol* **5**, 50–77.

Hohenester E, Sasaki T, Mann K, Timpl R. 2000 Variable zinc coordination in endostatin. *J Mol Biol* **297**, 1–6.

Holland DR, Hausrath AC, Juers D, Matthews BW. 1995 Structural analysis of zinc substitutions in the active site of thermolysin. *Protein Sci* **4**, 1955–1965.

Holtz KM, Stec B, Kantrowitz ER. 1999 A model of the transition state in the alkaline phosphatase reaction. *J Biol Chem* **274**, 8351–8354.

Hommel U, Zurini M, Luyten M. 1994 Solution structure of a cysteine rich domain of rat protein kinase C. *Nat Struct Biol* **1**, 383–387.

Honzatko RB, Crawford JL, Monaco HL, Ladner JE, Ewards BF, Evans DR, Warren SG, Wiley DC, Ladner RC, Lipscomb WN. 1982 Crystal and molecular structures of native and CTP-liganded aspartate carbamoyltransferase from *Escherichia coli*. *J Mol Biol* **160**, 219–263.

Hosfield DJ, Guan Y, Haas BJ, Cunningham RP, Tainer JA. 1999 Structure of the DNA repair enzyme endonuclease IV and its DNA complex: double-nucleotide flipping at abasic sites and three-metal-ion catalysis. *Cell* **98**, 397–408.

Hough E, Hansen LK, Birknes B, Jynge K, Hansen S, Hordvik A, Little C, Dodson E, Derewenda Z. 1989 High-resolution (1.5 Å) crystal structure of phospholipase C from *Bacillus cereus*. *Nature* **338**, 357–360.

Huang CC, Casey PJ, Fierke CA. 1997 Evidence for a catalytic role of zinc in protein farnesyltransferase. Spectroscopy of Co^{2+}-

farnesyltransferase indicates metal coordination of the substrate thiolate. *J Biol Chem* **272**, 20–23.

Huang S, Xue Y, Sauer-Eriksson E, Chirica L, Lindskog S, Jonsson BH. 1998 Crystal structure of carbonic anhydrase from *Neisseria gonorrhoeae* and its complex with the inhibitor acetazolamide. *J Mol Biol* **283**, 301–310.

Hubbard SR, Bishop WR, Kirschmeier P, George SJ, Cramer SP, Hendrickson WA. 1991 Identification and characterization of zinc binding sites in protein kinase C. *Science* **254**, 1776–1179.

Hunt JA, Ahmed M, Fierke CA. 1999 Metal binding specificity in carbonic anhydrase is influenced by conserved hydrophobic core residues. *Biochemistry* **38**, 9054–9062.

Hurley JH, Newton AC, Parker PJ, Blumberg PM, Nishizuka Y. 1997 Taxonomy and function of C1 protein kinase C homology domains. *Protein Sci* **6**, 477–480.

Hurley TD, Bosron WF, Hamilton JA, Amzel LM. 1991 Structure of human beta 1 beta 1 alcohol dehydrogenase: catalytic effects of non-active-site substitutions. *Proc Natl Acad Sci USA* **88**, 8149–8153.

Hurley TD, Bosron WF, Stone CL, Amzel LM. 1994 Structures of three human beta alcohol dehydrogenase variants. Correlations with their functional differences. *J Mol Biol* **239**, 415–429.

Hymowitz SG, O'Connell MP, Ultsch MH, Hurst A, Totpal K, Ashkenazi A, de Vos AM, Kelley RF. 2000 A unique zinc-binding site revealed by a high-resolution X-ray structure of homotrimeric Apo2L/TRAIL. *Biochemistry* **39**, 633–640.

Hyvonen M, Saraste M. 1997 Structure of the PH domain and Btk motif from Bruton's tyrosine kinase: molecular explanations for X-linked agammaglobulinaemia. *EMBO J* **16**, 3396–3404.

Ichikawa S, Hatanaka H, Takeuchi Y, Ohno S, Inagaki F. 1995 Solution structure of cysteine-rich domain of protein kinase C alpha. *J Biochem (Tokyo)* **117**, 566–574.

Iverson TM, Alber BE, Kisker C, Ferry JG, Rees DC. 2000 A closer look at the active site of gamma-class carbonic anhydrases: high-resolution crystallographic studies of the carbonic anhydrase from *Methanosarcina thermophila*. *Biochemistry* **39**, 9222–9231.

Iwasaki T, Suzuki T, Kon T, Imai T, Urushiyama A, Ohmori D, Oshima T. 1997 Novel zinc-containing ferredoxin family in *thermoacidophilic archaea*. *J Biol Chem* **272**, 3453–3458.

Jaffe EK. 2000 The porphobilinogen synthase family of metalloenzymes. *Acta Cryst D* **56**, 115–128.

Joerger AC, Mueller-Dieckmann C, Schulz GE. 2000 Structures of l-fuculose-1-phosphate aldolase mutants outlining motions during catalysis. *J Mol Biol* **303**, 531–543.

Jornvall H, Hoog JO. 1995 Nomenclature of alcohol dehydrogenases. *Alcohol Alcohol* **30**, 153–161.

Jornvall H, Hoog JO, von Bahr-Lindstrom H, Vallee BL. 1987 Mammalian alcohol dehydrogenases of separate classes: intermediates between different enzymes and intraclass isozymes. *Proc Natl Acad Sci USA* **84**, 2580–2584.

Kannan KK, Notstrand B, Fridborg K, Lovgren S, Ohlsson A, Petef M. 1975 Crystal structure of human erythrocyte carbonic anhydrase B. Three-dimensional structure at a nominal 2.2-Å resolution. *Proc Natl Acad Sci USA* **72**, 51–55.

Karlin S, Zhu ZY, Karlin KD. 1998 Extended metal environments of cytochrome c oxidase structures. *Biochemistry* **37**, 17726–17734.

Karp DR, Long EO. 1992 Identification of HLA-DR1 beta chain residues critical for binding staphylococcal enterotoxins A and E. *J Exp Med* **175**, 415–424.

Karpusas M, Nolte M, Benton CB, Meier W, Lipscomb WN, Goelz S. 1997 The crystal structure of human interferon beta at 2.2-Å resolution. *Proc Natl Acad Sci USA* **94**, 11813–11818.

Kim EE, Wyckoff HW. 1991 Reaction mechanism of alkaline phosphatase based on crystal structures. Two-metal ion catalysis. *J Mol Biol* **218**, 449–464.

Kim H, Lipscomb WN. 1994 Structure and mechanism of bovine lens leucine aminopeptidase. *Adv Enzymol Relat Areas Mol Biol* **68**, 153–213.

Kimber MS, Pai EF. 2000 The active site architecture of *Pisum sativum* beta-carbonic anhydrase is a mirror image of that of alpha-carbonic anhydrases. *EMBO J* **19**, 1407–1418.

Kisker C, Schindelin H, Alber BE, Ferry JG, Rees DC. 1996 A left-hand beta-helix revealed by the crystal structure of a carbonic anhydrase from the archaeon *Methanosarcina thermophila*. *EMBO J* **15**, 2323–2330.

Kissinger CR, Parge HE, Knighton DR, Lewis CT, Pelletier LA, Tempczyk A, Kalish VJ, Tucker KD, Showalter RE, Moomaw EW, Gastinel LN, Habuka N, Chen X, Maldonado F, Baker JE, Bacquet R, Villefranco JE. 1995 Crystal structures of human calcineurin and the human FKBP12-FK506-calcineurin complex. *Nature* **378**, 641–644.

Kitagawa Y, Tanaka N, Hata Y, Kusunoki M, Lee GP, Katsube Y, Asada K, Aibara S, Morita Y. 1991 Three-dimensional structure of Cu,Zn-superoxide dismutase from spinach at 2.0 Å resolution. *J Biochem (Tokyo)* **109**, 477–485.

Klabunde T, Krebs B. 1997 The dimetal centre in purple acid phosphatases. *Structure and Bonding* **89**, 177–198.

Klabunde T, Strater N, Frohlich R, Witzel H, Krebs B. 1996 Mechanism of Fe(III)-Zn(II) purple acid phosphatase based on crystal structures. *J Mol Biol* **259**, 737–748.

Klinman JP. 1981 Probes of mechanism and transition-state structure in the alcohol dehydrogenase reaction. *CRC Crit Rev Biochem* **10**, 39–78.

Klug A. 1999 Zinc finger peptides for the regulation of gene expression. *J Mol Biol* **293**, 215–218.

Knofel T, Strater N. 1999 X-ray structure of the *Escherichia coli* periplasmic 5'-nucleotidase containing a dimetal catalytic site. *Nat Struct Biol* **6**, 448–453.

Ko TP, Liao CC, Ku WY, Chak KF, Yuan HS. 1999 The crystal structure of the DNase domain of colicin E7 in complex with its inhibitor Im7 protein. *Structure* **7**, 91–102.

Korkhin Y, Kalb AJ, Peretz M, Bogin O, Burstein Y, Frolow F. 1998 NADP-dependent bacterial alcohol dehydrogenases: crystal structure, cofactor-binding and cofactor specificity of the ADHs of *Clostridium beijerinckii* and *Thermoanaerobacter brockii*. *J Mol Biol* **278**, 967–981.

Korndorfer IP, Fessner WD, Matthews BW. 2000 The structure of rhamnose isomerase from *Escherichia coli* and its relation with xylose isomerase illustrates a change between inter and intra-subunit complementation during evolution. *J Mol Biol* **300**, 917–933.

Kumasaka T, Yamamoto M, Moriyama H, Tanaka N, Sato M, Katsube Y, Yamakawa Y, Omori-Satoh T, Iwanaga S, Ueki T. 1996 Crystal structure of H2-proteinase from the venom of *Trimeresurus flavoviridis*. *J Biochem (Tokyo)* **119**, 49–57.

Kurisu G, Kinoshita T, Sugimoto A, Nagara A, Kai Y, Kasai N, Harada S. 1997 Structure of the zinc endoprotease from *Streptomyces caespitosus*. *J Biochem (Tokyo)* **121**, 304–308.

Lacy DB, Stevens RC. 1999 Sequence homology and structural analysis of the clostridial neurotoxins. *J Mol Biol* **291**, 1091–1104.

Lacy DB, Tepp W, Cohen AC, DasGupta BR, Stevens RC. 1998 Crystal structure of botulinum neurotoxin type A and implications for toxicity. *Nat Struct Biol* **5**, 898–902.

Laity JH, Lee BM, Wright PE. 2001 Zinc finger proteins: new insights into structural and functional diversity. *Curr Opin Struct Biol* **11**, 39–46.

Lawrence MC, Pilling PA, Epa VC, Berry AM, Ogunniyi AD, Paton JC. 1998 The crystal structure of pneumococcal surface antigen PsaA reveals a metal-binding site and a novel structure for a putative ABC-type binding protein. *Structure* **6**, 1553–1561.

LeBrun LA, Plapp BV. 1999 Control of coenzyme binding to horse liver alcohol dehydrogenase. *Biochemistry* **38**, 12387–12393.

Lee JY, Chang C, Song HK, Moon J, Yang JK, Kim HK, Kwon ST, Suh SW. 2000 Crystal structure of NAD(+)-dependent DNA ligase: modular architecture and functional implications. *EMBO J* **19**, 1119–1129.

Lee YH, Deka RK, Norgard MV, Radolf JD, Hasemann CA. 1999 *Treponema pallidum* TroA is a periplasmic zinc-binding protein with a helical backbone. *Nat Struct Biol* **6**, 628–633.

Li C, Zhao D, Djebli A, Shoham M. 1999a Crystal structure of colicin E3 immunity protein: an inhibitor of a ribosome-inactivating RNase. *Structure Fold Des* **7**, 1365–1372.

Li H, Raman CS, Glaser CB, Blasko E, Young TA, Parkinson JF, Whitlow M, Poulos TL. 1999b Crystal structures of zinc-free and -bound heme domain of human inducible nitric-oxide synthase. Implications for dimer stability and comparison with endothelial nitric-oxide synthase. *J Biol Chem* **274**, 21276–21284.

Li YC, Zhang X, Melton R, Ganu V, Gonnella NC. 1998 Solution structure of the catalytic domain of human stromelysin-1 complexed to a potent, nonpeptidic inhibitor. *Biochemistry* **37**, 14048–14056.

Li Z, Rasmussen BA, Herzberg O. 1999c Structural consequences of the active site substitution Cys181 $\Longrightarrow$ Ser in metallo-beta-lactamase from *Bacteroides fragilis*. *Protein Sci* **8**, 249–252.

Liljas A, Kannan KK, Bergsten PC, Waara I, Fridborg K, Strandberg B, Carlbom U, Jarup L, Lovgren S, Petef M. 1972 Crystal structure of human carbonic anhydrase C. *Nat New Biol* **235**, 131–137.

Lindahl T. 1993 Instability and decay of the primary structure of DNA. *Nature* **362**, 709–715.

Lindqvist Y, Johansson E, Kaija H, Vihko P, Schneider G. 1999 Three-dimensional structure of a mammalian purple acid phosphatase at 2.2 Å resolution with a mu-(hydr)oxo bridged di-iron center. *J Mol Biol* **291**, 135–147.

Lindskog S, Liljas A. 1993 Carbonic anhydrose and the role of orientation in catalysis. *Cur Opin Struct Biol* **3**, 915–920.

Lipscomb WN, Strater N. 1996 Recent advances in zinc enzymology. *Chem Rev* **96**, 2375–2433.

Liu L, Hausladen A, Zeng M, Que L, Heitman J, Stamler JS. 2001 A metabolic enzyme for S-nitrosothiol conserved from bacteria to humans. *Nature* **410**, 490–494.

Liu S, Widom J, Kemp CW, Crews CM, Clardy J. 1998 Structure of human methionine aminopeptidase-2 complexed with fumagillin. *Science* **282**, 1324–1327.

Love RA, Parge HE, Wickersham JA, Hostomsky Z, Habuka N, Moomaw EW, Adachi T, Hostomska Z. 1996 The crystal structure of hepatitis C virus NS3 proteinase reveals a trypsin-like fold and a structural zinc binding site. *Cell* **87**, 331–342.

Lovejoy B, Cleasby A, Hassell AM, Longley K, Luther MA, Weigl D, McGeehan G, McElroy AB, Drewry D, Lambert MH, Jordan SR. 1994a Structure of the catalytic domain of fibroblast collagenase complexed with an inhibitor. *Science* **263**, 375–377.

Lovejoy B, Hassell AM, Luther MA, Weigl D, Jordan SR. 1994b Crystal structures of recombinant 19-kDa human fibroblast collagenase complexed to itself. *Biochemistry* **33**, 8207–8217.

Lovejoy B, Welch AR, Carr S, Luong C, Broka C, Hendricks RT, Campbell JA, Walker KA, Martin R, Van Wart H, Browner MF. 1999 Crystal structures of MMP-1 and -13 reveal the structural basis for selectivity of collagenase inhibitors. *Nat Struct Biol* **6**, 217–221.

Lowther WT, Zhang Y, Sampson PB, Honek JF, Matthews BW. 1999 Insights into the mechanism of *Escherichia coli* methionine aminopeptidase from the structural analysis of reaction products and phosphorus-based transition-state analogues. *Biochemistry* **38**, 14810–14819.

Mackay JP, Crossley M. 1998 Zinc fingers are sticking together. *Trends Biochem Sci* **23**, 1–4.

Mackereth CD, Arrowsmith CH, Edwards AM, McIntosh LP. 2000 Zinc-bundle structure of the essential RNA polymerase subunit RPB10 from *Methanobacterium thermoautotrophicum*. *Proc Natl Acad Sci USA* **97**, 6316–6321.

Mallis RJ, Poland BW, Chatterjee TK, Fisher RA, Darmawan S, Honzatko RB, Thomas JA. 2000 Crystal structure of S-glutathiolated carbonic anhydrase III. *FEBS Lett* **482**, 237–241.

Mandiyan V, Andreev J, Schlessinger J, Hubbard SR. 1999 Crystal structure of the ARF-GAP domain and ankyrin repeats of PYK2-associated protein beta. *EMBO J* **18**, 6890–6898.

Maret W. 1989 Cobalt(II)-substituted class III alcohol and sorbitol dehydrogenases from human liver. *Biochemistry* **28**, 9944–9949.

Maret W, Jacob C, Vallee BL, Fischer EH. 1999 Inhibitory sites in enzymes: zinc removal and reactivation by thionein. *Proc Natl Acad Sci USA* **96**, 1936–1940.

Maret W, Vallee BL. 1998 Thiolate ligands in metallothionein confer redox activity on zinc clusters. *Proc Natl Acad Sci USA* **95**, 3478–3482.

Maret W, Zeppezauer M. 1986 Influence of anions and pH on the conformational change of horse liver alcohol dehydrogenase induced by binding of oxidized nicotinamide adenine dinucleotide: binding of chloride to the catalytic metal ion. *Biochemistry* **25**, 1584–1588.

Mariani SM, Matiba B, Armandola EA, Krammer PH. 1997 Interleukin 1 beta-converting enzyme related proteases/caspases are involved in TRAIL-induced apoptosis of myeloma and leukemia cells. *J Cell Biol* **137**, 221–229.

Martinez-Yamout M, Legge GB, Zhang O, Wright PE, Dyson HJ. 2000 Solution structure of the cysteine-rich domain of the *Escherichia coli* chaperone protein DnaJ. *J Mol Biol* **300**, 805–818.

Maskos K, Fernandez-Catalan C, Huber R, Bourenkov GP, Bartunik H, Ellestad GA, Reddy P, Wolfson MF, Rauch CT, Castner BJ, Davis R, Clarke HR, Petersen M, Fitzner JN, Cerretti DP, March CJ, Paxton RJ, Black RA, Bode W. 1998 Crystal structure of the catalytic domain of human tumor necrosis factor-alpha-converting enzyme. *Proc Natl Acad Sci USA* **95**, 3408–3412.

Matthews BW. 1988 Structural basis of the action of thermolysin and related zinc proteases. *Acc Chem Res* **21**, 333–340.

Matthews BW, Weaver LH, Kester WR. 1974 The conformation of thermolysin. *J Biol Chem* **249**, 8030–8044.

Mechulam Y, Schmitt E, Maveyraud L, Zelwer C, Nureki O, Yokoyama S, Konno M, Blanquet S. 1999 Crystal structure of *Escherichia coli* methionyl-tRNA synthetase highlights species-specific features. *J Mol Biol* **294**, 1287–1297.

Medina JF, Wetterholm A, Radmark O, Shapiro R, Haeggstrom JZ, Vallee BL, Samuelsson B. 1991 Leukotriene A4 hydrolase: determination of the three zinc-binding ligands by site-directed mutagenesis and zinc analysis. *Proc Natl Acad Sci USA* **88**, 7620–7624.

Meijers R, Morris RJ, Adolph HW, Merli A, Lamzin VS, Cedergren-Zeppezauer ES. 2001 On the enzymatic activation of NADH. *J Biol Chem* **276**, 9316–9321.

Meinnel T, Blanquet S, Dardel F. 1996 A new subclass of the zinc metalloproteases superfamily revealed by the solution structure of peptide deformylase. *J Mol Biol* **262**, 375–386.

Mitsuhashi S, Mizushima T, Yamashita E, Yamamoto M, Kumasaka T, Moriyama H, Ueki T, Miyachi S, Tsukihara T. 2000 X-ray structure of beta-carbonic anhydrase from the red alga, *Porphyridium purpureum*, reveals a novel catalytic site for CO(2) hydration. *J Biol Chem* **275**, 5521–5526.

Miyatake H, Hata Y, Fujii T, Hamada K, Morihara K, Katsube Y. 1995 Crystal structure of the unliganded alkaline protease from *Pseudomonas aeruginosa* IFO3080 and its conformational changes on ligand binding. *J Biochem (Tokyo)* **118**, 474–479.

Morgunova E, Tuuttila A, Bergmann U, Isupov M, Lindqvist Y, Schneider G, Tryggvason K. 1999 Structure of human pro-matrix metalloproteinase-2: activation mechanism revealed [see comments]. *Science* **284**, 1667–1670.

Mott HR, Carpenter JW, Zhong S, Ghosh S, Bell RM, Campbell SL. 1996 The solution structure of the Raf-1 cysteine-rich domain: a novel ras and phospholipid binding site. *Proc Natl Acad Sci USA* **93**, 8312–8317.

Moy FJ, Chanda PK, Chen JM, Cosmi S, Edris W, Skotnicki JS, Wilhelm J, Powers R. 1999 NMR solution structure of the catalytic fragment of human fibroblast collagenase complexed with a sulfonamide derivative of a hydroxamic acid compound. *Biochemistry* **38**, 7085–7096.

Moy FJ, Chanda PK, Cosmi S, Pisano MR, Urbano C, Wilhelm J, Powers R. 1998 High-resolution solution structure of the inhibitor-free catalytic fragment of human fibroblast collagenase determined by multidimensional NMR. *Biochemistry* **37**, 1495–1504.

Muchmore SW, Chen J, Jakob C, Zakula D, Matayoshi ED, Wu W, Zhang H, Li F, Ng SC, Altieri DC. 2000 Crystal structure and mutagenic analysis of the inhibitor-of-apoptosis protein survivin. *Mol Cell* **6**, 173–182.

Murphy JE, Stec B, Ma L, Kantrowitz ER. 1997 Trapping and visualization of a covalent enzyme-phosphate intermediate [letter]. *Nat Struct Biol* **4**, 618–622.

Myers LC, Cushing TD, Wagner G, Verdine GL. 1994 Metal-coordination sphere in the methylated Ada protein-DNA co-complex. *Chem Biol* **1**, 91–97.

Naylor CE, Jepson M, Crane DT, Titball RW, Miller J, Basak AK, Bolgiano B. 1999 Characterisation of the calcium-binding C-terminal domain of *Clostridium perfringens* alpha-toxin. *J Mol Biol* **294**, 757–770.

Nureki O, Vassylyev DG, Tateno M, Shimada A, Nakama T, Fukai S, Konno M, Hendrickson TL, Schimmel P, Yokoyama S. 1998 Enzyme structure with two catalytic sites for double-sieve selection of substrate [see comments]. *Science* **280**, 578–582.

Oefner C, D'Arcy A, Hennig M, Winkler FK, Dale GE. 2000 Structure of human neutral endopeptidase (Neprilysin) complexed with phosphoramidon. *J Mol Biol* **296**, 341–349.

Ogihara NL, Parge HE, Hart PJ, Weiss MS, Goto JJ, Crane BR, Tsang J, Slater K, Roe JA, Valentine JS, Eisenberg D, Tainer JA. 1996 Unusual trigonal-planar copper configuration revealed in the atomic structure of yeast copper-zinc superoxide dismutase. *Biochemistry* **35**, 2316–2321.

Pan H, Wigley DB. 2000 Structure of the zinc-binding domain of *Bacillus stearothermophilus* DNA primase. *Structure* **8**, 231–239.

Papageorgiou AC, Acharya KR. 1997 Superantigens as immunomodulators: recent structural insights. *Structure* **5**, 991–996.

Papageorgiou AC, Acharya KR, Shapiro R, Passalacqua EF, Brehm RD, Tranter HS. 1995 Crystal structure of the superantigen enterotoxin C2 from *Staphylococcus aureus* reveals a zinc-binding site. *Structure* **3**, 769–779.

Papageorgiou AC, Collins CM, Gutman DM, Kline JB, O'Brien SM, Tranter HS, Acharya KR. 1999 Structural basis for the recognition of superantigen streptococcal pyrogenic exotoxin A (SpeA1) by MHC class II molecules and T-cell receptors. *EMBO J* **18**, 9–21.

Parge HE, Hallewell RA, Tainer JA. 1992 Atomic structures of wild-type and thermostable mutant recombinant human Cu,Zn superoxide dismutase [published erratum appears in Proc Natl Acad Sci USA 1992 Nov 15; **89**(22):11106]. *Proc Nat Acad Sci USA* **89**, 6109–6113.

Park HW, Boduluri SR, Moomaw JF, Casey PJ, Beese LS. 1997 Crystal structure of protein farnesyltransferase at 2.25 angstrom resolution [see comments] [published erratum appears in Science 1997 Apr 4; **276**(5309):21]. *Science* **275**, 1800–1804.

Paul-Soto R, Bauer R, Frere JM, Galleni M, Meyer-Klaucke W, Nolting H, Rossolini GM, de Seny D, Hernandez-Valladares M, Zeppezauer M, Adolph HW. 1999 Mono- and binuclear Zn^{2+}-beta-lactamase. Role of the conserved cysteine in the catalytic mechanism. *J Biol Chem* **274**, 13242–13249.

Paul-Soto R, Hernandez-Valladares M, Galleni M, Bauer R, Zeppezauer M, Frere JM, Adolph HW. 1998 Mono- and binuclear Zn-beta-lactamase from *Bacteroides fragilis*: catalytic and structural roles of the zinc ions. *FEBS Lett* **438**, 137–140.

Pauptit RA, Karlsson R, Picot D, Jenkins JA, Niklaus-Reimer AS, Jansonius JN. 1988 Crystal structure of neutral protease from *Bacillus cereus* refined at 3.0 Å resolution and comparison with the homologous but more thermostable enzyme thermolysin. *J Mol Biol 199*, 525–537.

Perrier V, Burlacu-Miron S, Bourgeois S, Surewicz WK, Gilles AM. 1998 Genetically engineered zinc-chelating adenylate kinase from *Escherichia coli* with enhanced thermal stability. *J Biol Chem* **273**, 19097–19101.

Perrier V, Surewicz WK, Glaser P, Martineau L, Craescu CT, Fabian H, Mantsch HH, Barzu O, Gilles AM. 1994 Zinc chelation and structural stability of adenylate kinase from *Bacillus subtilis*. *Biochemistry* **33**, 9960–9967.

Pesce A, Battistoni A, Stroppolo ME, Polizio F, Nardini M, Kroll JS, Langford PR, O'Neill P, Sette M, Desideri A, Bolognesi M. 2000 Functional and Crystallographic Characterization of *Salmonella typhimurium* Cu,Zn Superoxide Dismutase Coded by the sodCI Virulence Gene. *J Mol Biol* **302**, 465–478.

Pesce A, Capasso C, Battistoni A, Folcarelli S, Rotilio G, Desideri A, Bolognesi M. 1997 Unique structural features of the monomeric Cu,Zn superoxide dismutase from *Escherichia coli*, revealed by X-ray crystallography. *J Mol Biol* **274**, 408–420.

Petersen JF, Cherney MM, Liebig HD, Skern T, Kuechler E, James MN. 1999 The structure of the 2A proteinase from a common cold virus: a proteinase responsible for the shut-off of host-cell protein synthesis. *EMBO J* **18**, 5463–5475.

Pieroni L, Santolini E, Fipaldini C, Pacini L, Migliaccio G, La Monica N. 1997 In vitro study of the NS2–3 protease of hepatitis C virus. *J Virol* **71**, 6373–6380.

Ploom T, Thony B, Yim J, Lee S, Nar H, Leimbacher W, Richardson J, Huber R, Auerbach G. 1999 Crystallographic and kinetic investigations on the mechanism of 6-pyruvoyl tetrahydropterin synthase. *J Mol Biol* **286**, 851–860.

Poland BW, Xu MQ, Quiocho FA. 2000 Structural insights into the protein splicing mechanism of PI-SceI. *J Biol Chem* **275**, 16408–16413.

Prasad GS, Radhakrishnan R, Mitchell DT, Earhart CA, Dinges MM, Cook WJ, Schlievert PM, Ohlendorf DH. 1997 Refined structures of three crystal forms of toxic shock syndrome toxin-1 and of a tetramutant with reduced activity. *Protein Sci* **6**, 1220–1227.

Proft T, Moffatt SL, Berkahn CJ, Fraser JD. 1999 Identification and characterization of novel superantigens from *Streptococcus pyogenes*. *J Exp Med* **189**, 89–102.

Quest AF, Bloomenthal J, Bardes ES, Bell RM. 1992 The regulatory domain of protein kinase C coordinates four atoms of zinc. *J Biol Chem* **267**, 10193–10197.

Quiocho FA, Lipscomb WN. 1971 Carboxypeptidase A: a protein and an enzyme. *Adv Protein Chem* **25**, 1–78.

Raaijmakers H, Vix O, Toro I, Golz S, Kemper B, Suck D. 1999 X-ray structure of T4 endonuclease VII: a DNA junction resolvase with a novel fold and unusual domain-swapped dimer architecture. *EMBO J* **18**, 1447–1458.

Radhakrishnan R, Walter LJ, Hruza A, Reichert P, Trotta PP, Nagabhushan TL, Walter MR. 1996 Zinc mediated dimer of human interferon-alpha 2b revealed by X-ray crystallography. *Structure* **4**, 1453–1463.

Raman CS, Li H, Martasek P, Kral V, Masters BS, Poulos TL. 1998 Crystal structure of constitutive endothelial nitric oxide synthase: a paradigm for pterin function involving a novel metal center. *Cell* **95**, 939–950.

Ramaswamy S, el Ahmad M, Danielsson O, Jornvall H, Eklund H. 1996 Crystal structure of cod liver class I alcohol dehydrogenase: substrate pocket and structurally variable segments. *Protein Sci* **5**, 663–671.

Redlich PN, Grossberg SE. 1990 Immunochemical characterization of antigenic domains on human interferon-beta: spatially distinct epitopes are associated with both antiviral and antiproliferative activities. *Eur J Immunol* **20**, 1933–1939.

Rees DC, Lewis M, Lipscomb WN. 1983 Refined crystal structure of carboxypeptidase A at 1.54 Å resolution. *J Mol Biol* **168**, 367–387.

Reverter D, Fernandez-Catalan C, Baumgartner R, Pfander R, Huber R, Bode W, Vendrell J, Holak TA, Aviles FX. 2000 Structure of a novel leech carboxypeptidase inhibitor determined free in solution and in complex with human carboxypeptidase A2. *Nat Struct Biol* **7**, 322–328.

Ridderstrom M, Cameron AD, Jones TA, Mannervik B. 1998 Involvement of an active-site Zn^{2+} ligand in the catalytic mechanism of human glyoxalase I. *J Biol Chem* **273**, 21623–21628.

Riordan JF. 1974 Metal-containing exopeptidases. In: Whitaker JR, ed. *Food Related Enzymes*. Vol. 136. Washington DC: American Chemical Society; 220–240.

Roderick SL, Matthews BW. 1993 Structure of the cobalt-dependent methionine aminopeptidase from *Escherichia coli*: a new type of proteolytic enzyme. *Biochemistry* **32**, 3907–3912.

Roussel A, Anderson BF, Baker HM, Fraser JD, Baker EN. 1997 Crystal structure of the streptococcal superantigen SPE-C: dimerization and zinc binding suggest a novel mode of interaction with MHC class II molecules. *Nat Struct Biol* **4**, 635–643.

Rowlett RS, Chance MR, Wirt MD, Sidelinger DE, Royal JR, Woodroffe M, Wang YF, Saha RP, Lam MG. 1994 Kinetic and structural characterization of spinach carbonic anhydrase. *Biochemistry* **33**, 13967–13976.

Rowsell S, Pauptit RA, Tucker AD, Melton RG, Blow DM, Brick P. 1997 Crystal structure of carboxypeptidase G2, a bacterial enzyme with applications in cancer therapy. *Structure* **5**, 337–347.

Ryde U. 1995 On the role of Glu-68 in alcohol dehydrogenase. *Protein Sci* **4**, 1124–1132.

Sang QA, Douglas DA. 1996 Computational sequence analysis of matrix metalloproteinases. *J Protein Chem* **15**, 137–160.

Sankaranarayanan R, Dock-Bregeon AC, Rees B, Bovee M, Caillet J, Romby P, Francklyn CS, Moras D. 2000 Zinc ion mediated amino acid discrimination by threonyl-tRNA synthetase [see comments]. *Nat Struct Biol* **7**, 461–465.

Sankaranarayanan R, Dock-Bregeon AC, Romby P, Caillet J, Springer M, Rees B, Ehresmann C, Ehresmann B, Moras D. 1999 The structure of threonyl-tRNA synthetase-tRNA(Thr) complex enlightens its repressor activity and reveals an essential zinc ion in the active site. *Cell* **97**, 371–381.

Schad EM, Papageorgiou AC, Svensson LA, Acharya KR. 1997 A structural and functional comparison of staphylococcal enterotoxins A and C2 reveals remarkable similarity and dissimilarity. *J Mol Biol* **269**, 270–280.

Schad EM, Zaitseva I, Zaitsev VN, Dohlsten M, Kalland T, Schlievert PM, Ohlendorf DH, Svensson LA. 1995 Crystal structure of the superantigen staphylococcal enterotoxin type A. *EMBO J* **14**, 3292–3301.

Schlagenhauf E, Etges R, Metcalf P. 1998 The crystal structure of the *Leishmania major* surface proteinase leishmanolysin (gp63). *Structure* **6**, 1035–1046.

Schmid MF, Herriott JR. 1976 Structure of carboxypeptidase B at 2–8 Å resolution. *J Mol Biol* **103**, 175–190.

Sellin S, Eriksson LE, Aronsson AC, Mannervik B. 1983 Octahedral metal coordination in the active site of glyoxalase I as evidenced by the properties of Co(II)-glyoxalase I. *J Biol Chem* **258**, 2091–2093.

Sellin S, Mannervik B. 1984 Metal dissociation constants for glyoxalase I reconstituted with Zn^{2+}, Co^{2+}, Mn^{2+}, and Mg^{2+}. *J Biol Chem* **259**, 11426–11429.

Sideraki V, Mohamedali KA, Wilson DK, Chang Z, Kellems RE, Quiocho FA, Rudolph FB. 1996 Probing the functional role of two conserved active site aspartates in mouse adenosine deaminase. *Biochemistry* **35**, 7862–7872.

Silverman DN, Lindskog S. 1988 The catalytic mechanism of carbonic anhydrase: Implications of a rate limiting protolysis of water. *Acc Chem Res* **21**, 30–36.

Silvian LF, Wang J, Steitz TA. 1999 Insights into editing from an ile-tRNA synthetase structure with tRNAile and mupirocin. *Science* **285**, 1074–1077.

Simons TJ. 1995 The affinity of human erythrocyte porphobilinogen synthase for Zn^{2+} and Pb^{2+}. *Eur J Biochem* **234**, 178–183.

Smith KS, Ferry JG. 2000 Prokaryotic carbonic anhydrases. *FEMS Microbiol Rev* **24**, 335–366.

Soler D, Nomizu T, Brown WE, Chen M, Ye QZ, Van Wart HE, Auld DS. 1994 Zinc content of promatrilysin, matrilysin and the stromelysin catalytic domain. *Biochem Biophys Res Commun* **201**, 917–923.

Soler D, Nomizu T, Brown WE, Shibata Y, Auld DS. 1995 Matrilysin: expression, purification, and characterization. *J Protein Chem* **14**, 511–520.

Somers W, Ultsch M, De Vos AM, Kossiakoff AA. 1994 The X-ray structure of a growth hormone-prolactin receptor complex [see comments]. *Nature* **372**, 478–481.

Springman EB, Angleton EL, Birkedal-Hansen H, Van Wart HE. 1990 Multiple modes of activation of latent human fibroblast collagenase: evidence for the role of a Cys73 active-site zinc complex in latency and a 'cysteine switch' mechanism for activation. *Proc Natl Acad Sci USA* **87**, 364–368.

Stamler JS. 1994 Redox signaling: nitrosylation and related target interactions of nitric oxide. *Cell* **78**, 931–936.

Stams T, Chen Y, Boriack-Sjodin PA, Hurt JD, Liao J, May JA, Dean T, Laipis P, Silverman DN, Christianson DW. 1998 Structures of murine carbonic anhydrase IV and human carbonic anhydrase II complexed with brinzolamide: molecular basis of isozyme-drug discrimination. *Protein Sci* **7**, 556–563.

Stams T, Nair SK, Okuyama T, Waheed A, Sly WS, Christianson DW. 1996 Crystal structure of the secretory form of membrane-associated human carbonic anhydrase IV at 2.8-Å resolution. *Proc Natl Acad Sci USA* **93**, 13589–13594.

Stec B, Hehir MJ, Brennan C, Nolte M, Kantrowitz ER. 1998 Kinetic and X-ray structural studies of three mutant *E. coli* alkaline phosphatases: insights into the catalytic mechanism without the nucleophile Ser102. *J Mol Biol* **277**, 647-662.

Stec B, Holtz KM, Kantrowitz ER. 2000 A revised mechanism for the alkaline phosphatase reaction involving three metal ions. *J Mol Biol* **299**, 1303–1311.

Stocker W, Ng M, Auld DS. 1990 Fluorescent oligopeptide substrates for kinetic characterization of the specificity of Astacus protease. *Biochemistry* **29**, 10418–10425.

Stocker W, Wolz RL, Zwilling R, Strydom DJ, Auld DS. 1988 Astacus protease, a zinc metalloenzyme. *Biochemistry* **27**, 5026–5032.

Strater N, Klabunde T, Tucker P, Witzel H, Krebs B. 1995 Crystal structure of a purple acid phosphatase containing a dinuclear Fe(III)-Zn(II) active site. *Science* **268**, 1489–1492.

Strater N, Lipscomb WN. 1995 Transition state analogue L-leucinephosphonic acid bound to bovine lens leucine aminopeptidase: X-ray structure at 1.65 Å resolution in a new crystal form. *Biochemistry* **34**, 9200–9210.

Strickland CL, Windsor WT, Syto R, Wang L, Bond R, Wu Z, Schwartz J, Le HV, Beese LS, Weber PC. 1998 Crystal structure of farnesyl protein transferase complexed with a CaaX peptide and farnesyl diphosphate analogue. *Biochemistry* **37**, 16601–16611.

Strop P, Smith KS, Iverson TM, Ferry JG, Rees DC. 2001 Crystal structure of the 'cab type' beta class carbonic anhydrase from the archaeon *Methanobacterium thermoautotrophicum*. *J Biol Chem* **276**, 10299–10305.

Sugahara M, Mikawa T, Kumasaka T, Yamamoto M, Kato R, Fukuyama K, Inoue Y, Kuramitsu S. 2000 Crystal structure of a repair enzyme of oxidatively damaged DNA, MutM (Fpg), from an extreme thermophile, *Thermus thermophilus* HB8. *EMBO J* **19**, 3857–3869.

Sugiura I, Nureki O, Ugaji-Yoshikawa Y, Kuwabara S, Shimada A, Tateno M, Lorber B, Giege R, Moras D, Yokoyama S, Konno M. 2000 The 2.0 Å crystal structure of *Thermus thermophilus* methionyl-tRNA synthetase reveals two RNA-binding modules. *Structure Fold Des* **8**, 197–208.

Sundstrom M, Abrahmsen L, Antonsson P, Mehindate K, Mourad W, Dohlsten M. 1996a The crystal structure of staphylococcal enterotoxin type D reveals Zn^{2+}-mediated homodimerization. *EMBO J* **15**, 6832–6840.

Sundstrom M, Hallen D, Svensson A, Schad E, Dohlsten M, Abrahmsen L. 1996b The Co-crystal structure of staphylococcal enterotoxin type A with Zn^{2+} at 2.7 Å resolution. Implications for major histocompatibility complex class II binding. *J Biol Chem* **271**, 32212–32216.

Svensson S, Hoog JO, Schneider G, Sandalova T. 2000 Crystal Structures of mouse class II alcohol dehydrogenase reveal determinants of substrate specificity and catalytic efficiency. *J Mol Biol* **302**, 441–453.

Swaminathan S, Eswaramoorthy S. 2000 Structural analysis of the catalytic and binding sites of *Clostridium botulinum* neurotoxin B [see comments]. *Nat Struct Biol* **7**, 693–699.

Tahirov TH, Oki H, Tsukihara T, Ogasahara K, Yutani K, Libeu CP, Izu Y, Tsunasawa S, Kato I. 1998a High-resolution crystals of methionine aminopeptidase from *Pyrococcus furiosus* obtained by water-mediated transformation. *J Struct Biol* **121**, 68–72.

Tahirov TH, Oki H, Tsukihara T, Ogasahara K, Yutani K, Ogata K, Izu Y, Tsunasawa S, Kato I. 1998b Crystal structure of methionine aminopeptidase from hyperthermophile, *Pyrococcus furiosus*. *J Mol Biol* **284**, 101–124.

Tainer JA, Getzoff ED, Beem KM, Richardson JS, Richardson DC. 1982 Determination and analysis of the 2 Å-structure of copper, zinc superoxide dismutase. *J Mol Biol* **160**, 181–217.

Taylor A. 1993 Aminopeptidases: structure and function. *FASEB J* **7**, 290–298.

Teplyakov A, Polyakov K, Obmolova G, Strokopytov B, Kuranova I, Osterman A, Grishin N, Smulevitch S, Zagnitko O, Galperina O, Matz M, Stepanov V. 1992 Crystal structure of carboxypeptidase T from *Thermoactinomyces vulgaris*. *Eur J Biochem* **208**, 281–288.

Thayer MM, Flaherty KM, McKay DB. 1991 Three–dimensional structure of the elastase of *Pseudomonas aeruginosa* at 1.5-Å resolution. *J Biol Chem* **266**, 2864–2871.

Thoden JB, Ruzicka FJ, Frey PA, Rayment I, Holden HM. 1997 Structural analysis of the H166G site-directed mutant of galactose-1-phosphate uridylyltransferase complexed with either UDP-glucose or UDP-galactose: detailed description of the nucleotide sugar binding site. *Biochemistry* **36**, 1212–1222.

Thunnissen MM, Nordlund P, Haeggstrom JZ. 2001 Crystal structure of human leukotriene A(4) hydrolase, a bifunctional enzyme in inflammation. *Nat Struct Biol* **8**, 131–135.

Tsutakawa SE, Muto T, Kawate T, Jingami H, Kunishima N, Ariyoshi M, Kohda D, Nakagawa M, Morikawa K. 1999 Crystallographic and functional studies of very short patch repair endonuclease. *Mol Cell* **3**, 621–628.

Ullah JH, Walsh TR, Taylor IA, Emery DC, Verma CS, Gamblin SJ, Spencer J. 1998 The crystal structure of the L1 metallo–beta–lactamase from *Stenotrophomonas maltophilia* at 1.7 Å resolution. *J Mol Biol* **284**, 125–136.

Uppenberg J, Lindqvist F, Svensson C, Ek–Rylander B, Andersson G. 1999 Crystal structure of a mammalian purple acid phosphatase. *J Mol Biol* **290**, 201–211.

Urbani A, Bazzo R, Nardi MC, Cicero DO, De Francesco R, Steinkuhler C, Barbato G. 1998 The metal binding site of the hepatitis C virus NS3 protease. A spectroscopic investigation. *J Biol Chem* **273**, 18760–18769.

Vallee BL, Auld DS. 1990a Active-site zinc ligands and activated H_2O of zinc enzymes. *Proc Natl Acad Sci USA* **87**, 220–224.

Vallee BL, Auld DS. 1990b Zinc coordination, function, and structure of zinc enzymes and other proteins. *Biochemistry* **29**, 5647–5659.

Vallee BL, Auld DS. 1992a Active zinc binding sites of zinc metalloenzymes. *Matrix Suppl* **1**, 5–19.

Vallee BL, Auld DS. 1992b Functional zinc-binding motifs in enzymes and DNA-binding proteins. *Faraday Discuss* **93**, 47–65.
Vallee BL, Auld DS. 1993a New perspective on zinc biochemistry: cocatalytic sites in multi-zinc enzymes. *Biochemistry* **32**, 6493–6500.
Vallee BL, Auld DS. 1993b Zinc: Biological Functions and Coordination Motifs. *Acc Chem Res* **26**, 543–551.
Vallee BL, Coleman JE, Auld DS. 1991 Zinc fingers, zinc clusters, and zinc twists in DNA-binding protein domains. *Proc Natl Acad Sci USA* **88**, 999–1003.
Vallee BL, Falchuk KH. 1993 The biochemical basis of zinc physiology. *Physiol Rev* **73**, 79–118.
Vallee BL, Galdes A. 1984 The metallobiochemistry of zinc enzymes. *Adv Enzymol Relat Areas Mol Biol* **56**, 283–430.
Van Doren SR, Kurochkin AV, Hu W, Ye QZ, Johnson LL, Hupe DJ, Zuiderweg ER. 1995 Solution structure of the catalytic domain of human stromelysin complexed with a hydrophobic inhibitor. *Protein Sci* **4**, 2487–2498.
Vanhooke JL, Benning MM, Raushel FM, Holden HM. 1996 Three-dimensional structure of the zinc-containing phosphotriesterase with the bound substrate analog diethyl 4-methylbenzylphosphonate. *Biochemistry* **35**, 6020–6025.
Verdecia MA, Huang H, Dutil E, Kaiser DA, Hunter T, Noel JP. 2000 Structure of the human anti-apoptotic protein survivin reveals a dimeric arrangement [see comments]. *Nat Struct Biol* **7**, 602–608.
Volbeda A, Lahm A, Sakiyama F, Suck D. 1991 Crystal structure of *Penicillium citrinum* P1 nuclease at 2.8 Å resolution. *EMBO J* **10**, 1607–1618.
Walker KW, Bradshaw RA. 1998 Yeast methionine aminopeptidase I can utilize either Zn^{2+} or Co^{2+} as a cofactor: a case of mistaken identity? *Protein Sci* **7**, 2684–2687.
Wang B, Jones DN, Kaine BP, Weiss MA. 1998 High-resolution structure of an archaeal zinc ribbon defines a general architectural motif in eukaryotic RNA polymerases. *Structure* **6**, 555–569.
Wang Z, Quiocho FA. 1998 Complexes of adenosine deaminase with two potent inhibitors: X-ray structures in four independent molecules at pH of maximum activity. *Biochemistry* **37**, 8314–8324.
Whitlow M, Howard AJ, Finzel BC, Poulos TL, Winborne E, Gilliland GL. 1991 A metal-mediated hydride shift mechanism for xylose isomerase based on the 1.6 Å *Streptomyces rubiginosus* structures with xylitol and D-Xylose. *Proteins Struc Funct Genet* **9**, 153–173.
Wilson DK, Quiocho FA. 1993 A pre-transition-state mimic of an enzyme: X-ray structure of adenosine deaminase with bound 1-deazaadenosine and zinc-activated water. *Biochemistry* **32**, 1689–1694.
Wilson DK, Rudolph FB, Quiocho FA. 1991 Atomic structure of adenosine deaminase complexed with a transition-state analog: understanding catalysis and immunodeficiency mutations. *Science* **252**, 1278–1284.
Wimberly BT, Brodersen DE, Clemons WM, Jr., Morgan-Warren RJ, Carter AP, Vonrhein C, Hartsch T, Ramakrishnan V. 2000 Structure of the 30S ribosomal subunit [see comments]. *Nature* **407**, 327–339.
Xiang S, Short SA, Wolfenden R, Carter CW, Jr. 1995 Transition-state selectivity for a single hydroxyl group during catalysis by cytidine deaminase. *Biochemistry* **34**, 4516–4523.
Xie P, Parsons SH, Speckhard DC, Bosron WF, Hurley TD. 1997 X-ray structure of human class IV sigmasigma alcohol dehydrogenase. Structural basis for substrate specificity. *J Biol Chem* **272**, 18558–18563.
Yan Y, Li Y, Munshi S, Sardana V, Cole JL, Sardana M, Steinkuehler C, Tomei L, De Francesco R, Kuo LC, Chen Z. 1998 Complex of NS3 protease and NS4A peptide of BK strain hepatitis C virus: a 2.2 Å resolution structure in a hexagonal crystal form. *Protein Sci* **7**, 837–847.
Yang Y, Keeney D, Tang X, Canfield N, Rasmussen BA. 1999 Kinetic properties and metal content of the metallo-beta-lactamase CcrA harboring selective amino acid substitutions. *J Biol Chem* **274**, 15706–15711.
Yang ZN, Bosron WF, Hurley TD. 1997 Structure of human chi chi alcohol dehydrogenase: a glutathione-dependent formaldehyde dehydrogenase. *J Mol Biol* **265**, 330–343.
Yano JK, Koo LS, Schuller DJ, Li H, Ortiz De Montellano PR, Poulos TL. 2000 Crystal structure of a thermophilic cytochrome P450 from the archaeon *sulfolobus solfataricus*. *J Biol Chem* **275**, 31086–31092.
Yaremchuk A, Cusack S, Tukalo M. 2000 Crystal structure of a eukaryote/archaeon-like prolyl-tRNA synthetase and its complex with tRNA(Pro)(CGG). *EMBO J* **19**, 4745–4758.
Zhang FL, Casey PJ. 1996 Protein prenylation: molecular mechanisms and functional consequences. *Ann Rev Biochem* **65**, 241–269.
Zhang G, Campbell EA, Minakhin L, Richter C, Severinov K, Darst SA. 1999 Crystal structure of *Thermus aquaticus* core RNA polymerase at 3.3 Å resolution [see comments]. *Cell* **98**, 811–824.
Zhang G, Kazanietz MG, Blumberg PM, Hurley JH. 1995 Crystal structure of the cys2 activator-binding domain of protein kinase C delta in complex with phorbol ester. *Cell* **81**, 917–924.
Zhang H, Seabra MC, Deisenhofer J. 2000 Crystal structure of Rab geranylgeranyltransferase at 2.0 Å resolution. *Structure* **8**, 241–251.
Zhu X, Teng M, Niu L. 1999 Structure of acutolysin-C, a haemorrhagic toxin from the venom of *Agkistrodon acutus*, providing further evidence for the mechanism of the pH-dependent proteolytic reaction of zinc metalloproteinases. *Acta Crystallogr D Biol Cryst* **55**, 1834–1841.

BioMetals 14: 315–330, 2001.

Review

The role of zinc in caspase activation and apoptotic cell death

Ai Q. Truong-Tran, Joanne Carter, Richard E. Ruffin & Peter D. Zalewski*
*Department of Medicine, University of Adelaide, The Queen Elizabeth Hospital, Woodville, South Australia 5011, Australia; *Author for correspondence (Tel: (+61) 8 8222 7344; Fax: (+61) 8 8222 6042; E-mail: peter.zalewski@adelaide.edu.au)*

Received 5 January 2001; accepted 15 April 2001

Key words: apoptosis, caspase, ROS, sulfhydryl, zinc

Abstract

In addition to its diverse role in many physiological systems, zinc (Zn) has now been shown to be an important regulator of apoptosis. The purpose of this review is to integrate previously published knowledge on Zn and apoptosis with current attempts to elucidate the mechanisms of action of this biometal. This paper begins with an introduction to apoptosis and then briefly reviews the evidence relating Zn to apoptosis. The major focus of this review is the mechanistic actions of Zn and its candidate intracellular targets. In particular, we examine the cytoprotective functions of Zn which suppress major pathways leading to apoptosis, as well as the more direct influence of Zn on the apoptotic regulators, especially the caspase family of enzymes. These two mechanisms are closely related since a decline in intracellular Zn below a critical threshold level may not only trigger pathways leading to caspase activation but may also facilitate the process by which the caspases are activated. Studies by our laboratory in airway epithelial cells show that Zn is co-localized with the precursor form of caspase-3, mitochondria and microtubules, suggesting this Zn is critically placed to control apoptosis. Further understanding the different pools of Zn and how they interact with apoptotic pathways should have importance in human disease.

Abbreviations: AEC – Airway epithelial cells; Cu/Zn SOD – Copper zinc superoxide dismutase; z-DEVD-AFC – boc-Asp-Glu-Val-Asp 7-amino-4-trifluoromethyl coumarin; H_2O_2 – Hydrogen peroxide; HNE – Hydroxynonenal; L-NAME – NG-nitro-L-arginine methyl ester; PARP – Poly (ADP-ribose) polymerase; ROS – Reactive oxygen species; TPEN – N,N,N′,N′-tetrakis-2-pyridylmethyl-ethylenediamine; Zn – Zinc.

Introduction

In the last two and a half decades, substantial evidence *in vitro* and *in vivo* has accumulated linking Zn deficiency with a markedly increased susceptibility of cells and tissues to die by apoptosis. This form of cellular demise is an active, tightly-regulated process that involves a series of cytoskeletal, membrane, nuclear and cytoplasmic changes that culminate in condensation and fragmentation of the cell into apoptotic bodies, which are eventually cleared by phagocytosis. Increase in apoptosis in Zn deficiency is directly related to a decrease in intracellular Zn within the cells fated to die. On the other hand, supplementing cells with exogenous Zn *in vitro*, and possibly also *in vivo*, decreases the susceptibility of cells and tissues to spontaneous or toxin-induced apoptosis, even when the cells apparently have a normal Zn status. The pools of Zn which influence cell susceptibility to apoptosis are the more exchangeable (labile) Zn, which are readily depleted in Zn deficiency and augmented following Zn supplementation. The subcellular distribution of these labile pools is amenable to studies using Zn-specific fluorophores or the Neo-Timm's silver sulphide stain for electron microscopy (Frederickson 1989).

This review will discuss aspects of Zn physiology and its possible beneficial actions in regulating apoptosis. There are a number of issues concerning the role of Zn in protection against apoptosis which need to be considered in the context of normal cellular physiology. Amongst these issues are (1) the distribution and homeostasis of the subcellular pools of labile Zn that mediate cellular protection, (2) the mechanisms for delivery of this Zn to critical target(s) of the induction and/or effector pathways of apoptosis, (3) the precise mechanism by which Zn regulates the processing and/or catalytic activity of the caspases, (4) the relationship of Zn to other regulators of apoptosis e.g. Bcl-2 and (5) the role of enhanced apoptosis in increased vulnerability to disease (e.g. diabetes mellitus, Alzheimer's dementia and asthma), where there is often an underlying subclinical or overt state of Zn deficiency. For detailed reviews on the earlier papers relating Zn and apoptosis, the reader is referred to the following reports (Zalewski & Forbes 1993; Sunderman 1995; Fraker & Telford 1997; Chai *et al.* 1999; Truong-Tran *et al.* 2000b).

This review highlights more recent studies and aims to integrate this knowledge with current concepts of the cellular biology of this biometal. It also takes a critical look at some of the interpretations that have emerged and attempts to identify some of the gaps in our understanding of Zn and apoptosis. We begin with a brief overview of apoptosis and then explore the mechanisms by which it is influenced by Zn. Also discussed are the homeostatic mechanisms which control the intracellular levels and subcellular distribution of Zn.

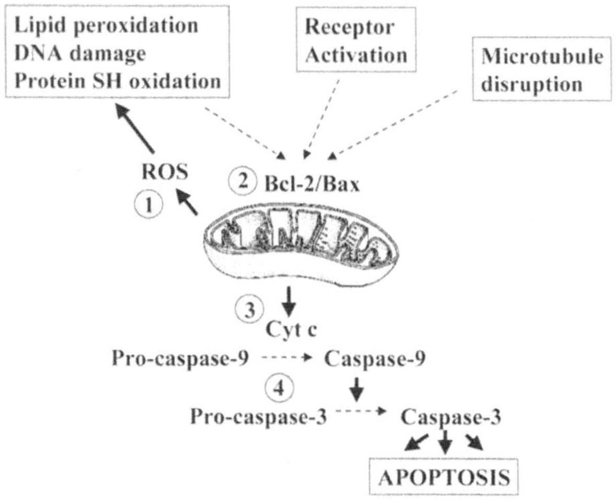

Fig. 1. The mitochondrion as a central controller of apoptosis. Figure shows the central role of the mitochondria in the regulation of apoptosis. (1) Mitochondria release ROS during aerobic respiration, potentially causing oxidation of lipids, DNA and protein sulfhydryls. Up to 5% of the oxygen consumed by mitochondria is thought to be converted to ROS, however under normal conditions these are rapidly removed by anti-oxidants (Halliwell 1991). (2) Mitochondria also influence the process of apoptosis by coordinating the various input signaling pathways and channeling them onto a central pathway which is governed by mitochondrial-associated anti-apoptotic (e.g., Bcl-2) and pro-apoptotic (e.g., Bax) families of regulators (Strasser *et al.* 2000). Whether the cell proceeds into apoptosis is determined by the ratio of Bcl-2/Bax proteins; this has also been referred to as the Bcl-2/Bax rheostat and serves as a major check-point for commitment of the cell to death (Korsmeyer *et al.* 1993). (3) Once a cell has passed through this checkpoint cytochrome c is released through mitochondrial pores, triggering (4) the activation of the caspase cascade leading to morphological changes of apoptosis.

Apoptosis (gene-directed cell death)

Although the finding that some cells in the body die as a part of normal development, instead of injury, emerged from studies several decades ago, it is only relatively recently that we have come to the realization that the mechanism of this programmed cell death shares many features with that of cells dying as a consequence of mild damage (e.g., by oxyradicals or microtubule poisons), starvation of growth factors and activation of certain membrane receptors (for review see Kerr *et al.* 1987). This process, referred to as apoptosis or gene-directed cell death, is the major mechanism of cell death in the body, enabling the removal of superfluous, mutant or moderately damaged cells. Distinct from necrosis, due to severe physical, chemical or osmotic cell damage, apoptosis deletes cells without release of their contents that would otherwise damage neighboring cells and provoke an inflammatory response.

Apoptosis is an active, energy-dependent process divided into two phases, biochemical and morphological. The biochemical phase involves diverse input signaling pathways originating from the plasma membrane (e.g., Fas receptor ligation, lipid peroxidation), the nucleus (e.g., newly-transcribed gene products such as *reaper*, DNA damage or mutations) or the cytoskeleton (e.g., disruption of microtubules) (Sun *et al.* 1999). As depicted in Figure 1, the mitochondrion is a major player in the induction, regulation and execution of apoptosis. Mitochondria coordinate apoptosis by channeling the input pathways onto a central pathway which is governed by mitochondrial-associated anti-apoptotic (e.g., Bcl-2) and pro-apoptotic (e.g., Bax) families of regulators and by providing a scaffolding for the proteolytic events that trigger processing and activation of various members of the caspase enzyme family (Strasser *et al.* 2000). Caspases are proteases which recognize and cleave their substrate proteins at tetra-peptide sites with characteristic sequences, e.g., the DXXD motif for caspase-3, where X is any amino acid (Thornberry & Lazebnik 1998). They comprise two families: (1) the initiator caspases (such as caspases-8 and -9) which act upstream and transduce signals from specific input pathways (e.g., Fas ligation results in caspase-8 activation while release of cytochrome c from mitochondria triggers caspase-9 activation) and (2) the executioner caspases (such as caspases-3 and -6) which are activated by initiator caspases and, in turn, cleave critical substrates leading to downstream events of apoptosis. For example, caspase-3 cleaves ICAD (inhibitor of calcium-activated DNAase) liberating the endonuclease and enabling it to cleave DNA at inter-nucleosomal sites (Janicke *et al.* 1998). Caspase-6, previously known as Mch2α, cleaves the nuclear lamin scaffold proteins as a prerequisite for fragmentation of the nucleus (Takahashi *et al.* 1996). Action of the caspases leads to the morphological changes such as cell shrinkage, condensation and fragmentation of the cytoplasm and nucleus and formation of membrane-enclosed apoptotic bodies, which contain the contents of the cell and are cleared *in vivo* by phagocytosis (Song & Steller 1999; Strasser *et al.* 2000).

Apoptosis is tightly regulated and its dysregulation is central to the pathogenesis of a number of diseases, being excessive in neurodegenerative disorders, AIDS and diabetes mellitus and inadequate in autoimmune disease and malignancies (Thompson 1995; Wyllie 1997). A number of therapeutic strategies are aimed at correcting these imbalances and, as a consequence, the factors regulating the induction and execution phases of apoptosis are being intensively studied. One such factor is zinc.

Zinc and apoptosis

This section considers the evidence relating labile pools of Zn to regulation of caspase-dependent apoptosis. Although high concentrations of Zn may, in some cells, trigger cell death either by apoptosis or necrosis (Sensi *et al.* 1999; Hamatake *et al.* 2000; Untergasser *et al.* 2000), the bulk of evidence indicates that Zn is a physiological suppressor of apoptosis.

Zn deficiency increases apoptosis

The most convincing evidence for a physiological role for Zn in suppression of apoptosis comes from Zn deprivation animal studies. Although systematic studies of apoptosis in Zn deficient animals are still lacking, it is clear that the frequency of apoptotic cells is markedly increased in certain tissues and organs, including the intestinal and retinal pigment epithelium, skin, thymic lymphocytes, testis and pancreatic acinar cells of adult animals (reviewed in Duvall & Wyllie 1986; Zalewski & Forbes 1993) as well as the neuroepithelium of foetal rats (Rogers *et al.* 1995). The latter, which occurred within 4 days of maternal Zn deficiency, was particularly evident in the neural crest cells and interfered with neural tube closure leading to severe congenital abnormalities of the nervous system. In nearly all examples of Zn deficiency-induced apoptosis *in vivo*, increased cell death appears to be a direct consequence of a lowering of intracellular levels of Zn in the affected tissues. One exception is the involution of the thymus in Zn deficient mice which is at least partly due to increased circulating glucocorticoids produced in response to the stress associated with Zn deficiency (Fraker & Telford 1997).

Numerous *in vitro* studies have now shown that depletion of intracellular Zn by culture of cells in Zn-depleted medium or by treatment of cells with N,N,N′,N′-tetrakis-2-pyridylmethyl-ethylenediamine (TPEN) results in apoptosis (Martin *et al.* 1991; Zalewski *et al.* 1991,1993; Treves *et al.* 1994). Zn deficiency-induced apoptosis, *in vitro* and *in vivo*, has all of the major morphological features of apoptosis, including DNA and nuclear fragmentation, chromatin condensation and apoptotic body formation (Truong-Tran *et al.* 2000b). In addition, Zn deficiency-induced apoptosis is dependent on caspase-3 activation, since cytosolic caspase-3 activity is increased in Zn deficient cells while the specific caspase-3 inhibitor z-DEVD-fmk can partially suppress apoptosis (Chai *et al.* 2000). There is also now at least one *in vivo* example of this where Zn deficiency-induced apoptosis in rat embryos was associated with increased caspase-3 activity (Jankowski-Hennig *et al.* 2000).

Zn depletion renders cells more susceptible to apoptosis by toxins

Zn depletion not only increases the rate of apoptosis but there is a potent synergy, in the induction of apoptosis, between Zn depletion and other apoptotic inducers such as colchicine, tumour necrosis factor and HIV-1 Tat protein (Zalewski *et al.* 1993; Seve *et al.* 1999; Meerarani *et al.* 2000). Furthermore, studies by our laboratory have shown that Zn deficiency increases the susceptibility of respiratory epithelial cells to H_2O_2-induced apoptosis (Truong-Tran *et al.* 2000a). Thus, a severe reduction of intracellular labile Zn can directly induce apoptosis while smaller decreases may simply render cells more vulnerable to apoptosis by other toxins. The latter has important clinical implications for the relationships between mild or sub-clinical states of Zn deficiency and disease severity and incidence (see later).

Exogenous Zn suppresses apoptosis

There are several examples of increased resistance of Zn-supplemented animals to toxic apoptotic inducers. These include the protective effects of Zn salts against whole body irradiation in mice (Floersheim *et al.* 1992), sporidesmin-induced immunosuppression in sheep (Waring *et al.* 1990), neuronal apoptosis following transient forebrain ischemia in the hippocampus of primates (Matsushita *et al.* 1996) and apoptosis of the anterior and stromal keratinocytes in the eye following superficial keratectomy in rabbits (Kuo *et al.* 1997). Recently, Kown and colleagues (2000) demonstrated that Zn suppresses caspase-3 activity and apoptosis *in vivo* using rats transplanted in the abdomen with allogeneic hearts. They found that i.p. injections of 1–5 mg/kg of $ZnCl_2$ suppressed caspase-3 activity by up to 3.7-fold and apoptosis by up to 2-fold. The latter was assessed by (99m)Tc-Annexin V labelling in the allografts. Survival rates of the allografts were also increased from 6.4 days to 11.5 days ($P \leq 0.005$). These findings are consistent with the cytoprotective and anti-apoptotic effects of exogenous Zn *in vitro* in diverse cellular models. In particular, it should be noted that Zn blocks apoptosis induced by all apoptosis-inducing treatments tested, indicating that it suppresses a common event (reviewed in Zalewski & Forbes 1993; Sunderman 1995).

However, interpretation of the Zn supplementation studies remains unclear. While the *in vivo* studies suggest that intracellular Zn levels can be manipulated sufficiently to influence vulnerability of cells and tissues to apoptosis-inducing agents, there has been no attempt as yet to correlate changes in intracellular Zn content with resistance to toxins *in vivo*. Another concern is that most studies of Zn suppression of apoptosis *in vitro* have used supraphysiological concentrations of Zn salts in the medium (e.g., 5 mM in the study of Barbieri *et al.* 1992). One of the reasons for this is the relatively poor uptake of ionic Zn across the plasma cell membrane. However, mM concentrations of Zn will cross-link proteins non-specifically and no useful information can be gained from such experiments. Unfortunately, this includes most published studies. Our laboratory has attempted to address this by showing suppression of apoptosis using low concentrations of Zn sulphate (typical of those found in extracellular fluids) and given in combination with a Zn ionophore such as pyrithione (Zalewski *et al.* 1991). However, this approach also has the limitation in that intracellular levels of Zn in these Zn-loaded cells may very likely be much higher than would occur *in vivo*. This is not a trivial issue since metabolically available Zn is distributed non-uniformly throughout the cell, existing in nM-pM concentrations in the cytosol and up to mM concentrations within vesicles (Frederickson 1989). It is not known whether the pools of Zn responsible for suppression of apoptosis are compartmentalized and, if so, what the local available concentrations of Zn are. Finally, it is not known whether Zn supplementation affects the same pools and molecular targets within the apoptosis pathway as does Zn depletion. The various supplementation and depletion studies reported in the literature may be measuring different aspects of the influence of Zn on apoptosis, especially if there are multiple targets of Zn such as the nucleases, caspases and the Bcl-2 family (see below).

Zn and necrosis

While Zn depletion triggers caspase activation leading to apoptosis, other cell death can also occur. In some cells, e.g., T cell leukaemic Molt-3 cells, Zn deprivation resulted in necrosis (Martin *et al.* 1991). The reason for this is not clear but may depend, in part, on the functional state of the caspases. For example, in a recent study of TPEN-induced Zn deficiency in human renal cell carcinoma cell lines, mutant cell lines lacking caspases-3, -7, -8 and -10 died by necrosis rather than apoptosis (Kolenko *et al.* 1999). Therefore, we should not view Zn as a specific regulator of apoptosis but rather as a cytoprotectant that, when lacking, renders the cell vulnerable to death both by apoptosis and necrosis. A similar argument has been advanced for cytoprotection by Bcl-2 against both necrosis and apoptosis (Okuno *et al.* 1998).

It has been proposed that commitment to cell death can be regulated at a point upstream from caspases, probably at the level of the mitochondria since pharmacological agents which cause increased permeability of mitochondrial membranes cause both necrosis and apoptosis, depending on the activity of the caspases (reviewed in Kroemer & Reed 2000). Caspase inhibitors often fail to prevent cell death, rather shifting the mechanism of death from apoptosis to necrosis. For agents to be cytoprotective, they must act at the pre-mitochondrial or mitochondrial stages of apoptosis, rather than at the level of the caspases (Jacotot *et al.* 1999). This issue needs addressing in relation to the mechanism of action of Zn. Even if increase in intracellular Zn does specifically suppress apoptosis-related biochemical events, the cells may still die in the longer term. The evidence that Zn fails to block cell death in many systems has recently been reviewed (Fraker & Telford 1997). There are two separate issues to consider here. First, by suppressing apoptosis, Zn may simply divert irreversibly-damaged cells into necrosis. Secondly, the apoptotic program may be regulated by multiple factors and complete cytoprotection may require all of these in addition to Zn. A critical issue that has scarcely received attention is how the different anti-apoptotic factors cooperate with each other. Rather than simply testing the effect of high, pharmacological or even toxic, concentrations of Zn in isolation, experiments are now required to test the effects of smaller, more physiological Zn fluxes in combination with different levels of other cellular survival factors.

Mechanisms of the anti-apoptotic actions of zinc

There are very likely to be two aspects to the anti-apoptotic mechanisms of action of Zn. Firstly, it limits the extent of damage induced by oxyradicals and other toxins and thereby suppresses some of the signaling pathways leading to caspase activation and apoptosis. Secondly, it directly affects some of the apoptotic regulators, principally the caspase enzymes. These two actions may, in fact, be related since both proteolytic processing of precursor caspases and caspase enzymatic activity are influenced by the redox state of the cell; and part of the cytoprotective role of Zn may be to protect (mask) essential sulfhydryls of the caspases.

The intention of this section is to bring together the evidence for the role of Zn as both a cytoprotectant and an anti-apoptotic agent, principally acting at the level of the mitochondrion. The final parts of this section concern the possible inter-relationships between Zn and other anti-apoptotic factors in cells.

Zn is a cytoprotectant against oxidative stress

Zn is well known as a cytoprotectant, protecting and stabilizing cellular molecules (e.g., proteins and DNA), macromolecular complexes (e.g., microtubules) and subcellular organelles (e.g., membranes) (Vallee & Falchuk 1993). Central to this role is its capacity to minimize oxidative damage. It has been proposed that Zn became especially important at the time of evolution of cellular respiration when the hazards of oxidative stress became manifest (Da Silva & Williams 1991).

Apoptosis is closely linked with oxidative stress and may well have evolved primarily to rid the body of oxidatively-damaged cells (Frade & Michaelidis 1997). Many agents which induce apoptosis are either oxidants or stimulators of cellular oxidative metabolism. These include reactive oxygen species (ROS), such as superoxide anion, hydroxyl radical and peroxynitrite which are highly unstable compounds with unpaired electrons, capable of oxidizing lipids, proteins and nucleic acids (Halliwell 1991). Likewise, some anti-apoptotic factors (e.g., Bcl-2) also protect cellular organelles from oxidative damage and typical anti-oxidants (e.g., vitamin E and vitamin C and N-acetylcysteine) are often found to be anti-apoptotic (Dimmeler *et al.* 1997; Hennig *et al.* 1999).

Many previous studies have linked Zn deficiency in animals with enhanced rates of oxidative damage such as in testes and erythrocytes; supplementation with other anti-oxidants (vitamin C, vitamin E or β-carotene) reversed these effects and protected against the development of skin lesions and some other manifestations of Zn deficiency (Bettger and O'Dell 1981; Taylor *et al.* 1990; Oteiza *et al.* 1995; Kraus *et al.* 1997). Similarly, Zn supplementation also protects against oxidative damage. Thus, Zn acexamate, an anti-ulcer agent, inhibited lipid peroxidation and lesions in rat gastric mucosa *in vitro* and *in vivo* (Tsutsui *et al.* 1999).

Two recent studies have further contributed to our understanding of the link between Zn deficiency, oxidative stress and apoptosis. Firstly, Oteiza and colleagues (2000) investigated the effects of Zn deficiency on oxidative stress in 3T3 cells using a novel fluorescent probe (carboxy-2′7′-dichlorodihydrofluorescein diacetate) to assess for oxidative damage. When cells were exposed to Zn manipulated media the fluorescence intensity was increased up to 15-fold in the Zn deficient cells over that of untreated cells. Hence, they concluded that oxidative stress is induced by Zn deficiency in 3T3 cells. Secondly, Cui and colleagues (1999, 2000) have provided a strong link between (1) increased oxidative stress, due to increased levels and/or activity of inducible nitric oxide synthase and consequent nitric oxide production, (2) increased tissue damage, manifested by enhanced microvascular permeability and appearance of inflammatory lesions, and (3) increased apoptosis (as assessed by terminal deoxynucleotidyl transferase-mediated dUTP-biotin nick end labeling), in the skin and intestinal villi of Zn deficient mice and rats. The nitric oxide synthase inhibitor, NG-nitro-L-arginine methyl ester (L-NAME), given in the drinking water suppressed all of these changes. In airway epithelial cells (AEC) and in the malignant human lung epithelial cell lines A549 and NCI-H292, we have shown that TPEN acts synergistically with low concentrations of ROS (peroxynitrite or H_2O_2) in the induction of caspase-3 activity (Truong-Tran *et al.* 2000a; unpublished observations). Furthermore, the effects of TPEN were decreased by 50%–60% when cells were pre-incubated with various anti-oxidants (vitamin C, vitamin E or N-acetylcysteine, unpublished observations). In preliminary studies, we have found, using immunocytochemistry with antibody to hydroxynonenal (a lipid peroxidation marker), increased lipid peroxidation in the membranes of TPEN-treated primary AEC (see later).

Mechanisms by which Zn blocks oxidative damage

We believe it is unlikely that Zn protects against oxidative damage via its role in Cu/Zn superoxide dismutase (Cu/Zn SOD), an enzyme which removes the superoxide anion radical. The role of Zn in Cu/Zn SOD may simply be structural since severe depletion of intracellular Zn in keratinocytes by treatment with TPEN *in vitro* resulted in apoptosis but did not affect the Cu/Zn SOD activity of cells (Parat *et al.* 1997); in addition, lung Cu/Zn SOD was not responsible for the cytoprotection of Zn in a mouse model of hyperoxia; in fact, activity of this enzyme was paradoxically increased in Zn deficiency (Oteiza *et al.* 1996).

It is more likely that the cytoprotective Zn is to be found in the more labile and exchangeable cellular pools of Zn. This Zn may both directly and indirectly protect cells from oxidative damage. Directly, Zn is a well-known stabilizer of lipids and proteins. Thus, Zn protects cellular membranes and macromolecules against oxidative damage (Bettger and O'Dell 1981; Hennig *et al.* 1992, 1993; Kraus *et al.* 1997; Taylor *et al.* 1997; Powell 2000). Zn protects sulfhydryl groups in proteins from oxidation by forming strong, yet readily reversible, thiolate complexes (Williams 1987). Zn, thereby, affords protection to thiol-dependent enzymes such as δ-aminolevulinic acid dehydratase and dihydroorotase. It will also protect other cellular proteins with essential thiols such as tubulin, where sulfhydryls are required for polymerization into microtubules (Roychowdhury *et al.* 2000). The finding of large amounts of Zn within the microtubular structures of basal bodies and cilia in airway epithelial cells (Truong-Tran *et al.* 2000a) raises the question of whether Zn is there to protect their sulfhydryls against ROS released from their abundant mitochondria. Zn is a stabilizer of microtubules (Hesketh 1982) and microtubular disruption occurs in Zn deficiency (Nickolson & Veldstra 1972), oxidative stress (Banan *et al.* 2000) and in the early stages of apoptosis (Martin & Cotter 1990).

Indirectly, Zn may act via effects on glutathione, the main intracellular anti-oxidant. Recent studies have implicated glutathione in protection against inappropriate apoptosis. In view of its general cellular role as a redox regulator, glutathione could influence a number of events in the apoptotic pathway. One of these is the release of cytochrome c from mitochondria, which serves to initiate caspase-9-dependent processing and activation of caspase-3 (Grutter 2000). Appearance of cytochrome c in the cytosol is a cellular response to the depletion of glutathione, induced by inhibition of glutathione biosynthesis (Ghibelli *et al.* 1999). In this context, the reports by Parat and colleagues (1997) and Nakatani and colleagues (2000), showing that TPEN depletes intracellular glutathione in cultured keratinocytes and hepatocytes, respectively, are interesting. In keratinocytes, TPEN reduced glutathione content from 65 to 8 μmoles/g protein. In the study by Nakatani *et al.*, the glutathione depletion, caspase-3 activation and apoptosis that occurred in TPEN-treated hepatocytes were not only suppressed by Zn supplementation but also by addition of N-acetylcysteine, a precursor of glutathione. This suggests that the increase in caspase-dependent apoptosis in response to Zn depletion, was due to the decline in glutathione levels. However, this hypothesis needs substantiation from time course studies that show glutathione depletion *follows* the decrease in cellular Zn, but *precedes* the rise in caspase-3 levels. Furthermore, the kinetic relationships between depletion of Zn and glutathione, and onset of oxidative damage, mitochondrial cytochrome c release and caspase-3 activation need to be determined.

Effects on the processing and catalytic activity of caspase-3

In addition to blocking oxidative stress, Zn may also influence apoptosis by directly interacting with other apoptotic regulators. Historically, Zn was first shown to be a potent inhibitor of the endonuclease responsible for apoptotic DNA fragmentation and it was assumed that this was the critical target affected by Zn deficiency (Duvall & Wyllie 1986). At the time this seemed logical, since our whole understanding of the biochemistry of apoptosis centered on this enzyme and the nuclear changes. Furthermore, data presented showed that treatment of isolated nuclei with Zn salts could suppress Ca^{2+}-induced apoptotic changes within nuclei (Cohen & Duke 1984). For a variety of reasons, we now know that the endonucleases are only a secondary target. Apoptosis can occur without DNA fragmentation and even in cytoplasts lacking a nucleus; yet Zn is still suppressive in these models (reviewed in Truong-Tran *et al.* 2000b). Moreover, the concentrations of Zn required to inhibit the endonuclease are relatively high.

The possibility that Zn may suppress a step prior to activation of the endonucleases was first shown by Lazebnik and colleagues (1993), who used a cell-free model of apoptosis in which cytosols prepared from cells primed to undergo apoptosis were added to intact nuclei to produce nuclear condensation and DNA fragmentation. Importantly, the apoptosis-associated morphological changes of the nuclei were suppressed by concentrations of Zn lower than those required to suppress the fragmentation of DNA by the endonucleases. In addition, experiments in which cytosols and nuclei were separately pre-treated with Zn, indicated that the primary target of Zn was cytoplasmic rather than nuclear. Subsequent studies identified the cytosolic target of Zn as an aspartate-specific protease CPP-32 (renamed caspase-3), although they did not establish whether Zn blocked caspase-3 enzymatic activity or the steps leading to its activation (Faleiro *et al.* 1997).

This remains a controversial issue. We, and others, have found caspase-3 to be relatively insensitive to exogenous Zn (Takahashi *et al.* 1996; Truong-Tran *et al.* 2000b), although the activation of pro-caspase-3 in cell-free models was very sensitive to Zn. Using a cell-free system (described in Liu *et al.* 1996) in which addition of cytochrome c to cytosol of healthy cells triggers proteolytic conversion of pro-caspase-3 to the active enzyme, Mesner and colleagues (1999) found that Zn (>500 μM total concentration) inhibited pro-caspase-3 activation. We observed that addition of 800 nM *free* Zn blocked activation of caspase-3 by 50% in this model; there was no effect of Zn when added 90 min after cytochrome c but prior to addition of fluorogenic caspase-3 substrate, confirming that Zn blocks the process of caspase-3 activation rather than the already activated enzyme (Truong-Tran *et al.* 2000b). A similar finding using Western blotting to track caspase-3 processing in HL60 cells has also been reported (Aiuchi *et al.* 1998).

Caspase-3 requires an essential sulfhydryl for enzymatic activity (Nicholson *et al.* 1995). Its failure to be directly inhibited by Zn in the above mentioned studies may be a consequence of the use of metal chelators and thiol-reducing agents (e.g., dithiothreitol) in the cell extraction buffers, since these would otherwise bind Zn and lower the available Zn concentration. Using a cell-free system consisting of purified bovine poly (ADP) ribose polymerase (PARP) as a substrate and an apoptotic extract or recombinant caspase-3, Perry and colleagues (1997) reported that Zn inhibits PARP-proteolysis by caspase-3 in the low μM range. Furthermore, they found that cleavage of the caspase-3 tetrapeptide substrate was even more sensitive to Zn, being inhibited in the nM range. It is relevant that the extraction buffer used in this study lacked metal chelators and thiol reducing agents. Stennicke and Salvesen (1997) derived an inhibitory value of 0.15 μM for caspase-3 and concluded that Zn is a good caspase inhibitor, albeit very dependent on the thiol content and therefore presumably the redox potential of the cell. Another important paper supporting the role of Zn as a physiological caspase inhibitor was that of Maret *et al.* (1999) who found a 50% inhibition of caspase-3 activity by 1.7 nM Zn, under conditions where they used highly purified metal free buffers and no thiol reducing agents and a 1:1 stoichiometric relationship in the inhibition of caspase-3 by exogenous Zn. The latter implies that Zn may interact with only one sulfhydryl group in caspase-3, presumably the sulfhydryl required for catalytic activity.

Effects on caspases-6 and -9

Since the proteolytic processing and activation of caspase-3 is dependent on other caspases, including caspase-6 and caspase-9 (see earlier), these enzymes, and or their mechanisms of activation, may be the critical targets of Zn. It is of interest, that both of these caspases are reported to be very sensitive to inhibition by Zn. At least two recent studies have shown that although Zn inhibits all of the different caspases that have been tested, at lower, more physiological concentrations, it is a selective inhibitor of caspase-6/Mch2-α; complete inhibition was observed at 10 μM Zn (Takahashi *et al.* 1996; Stennicke & Salvesen 1997). Thus, Zn was found to block cleavage of lamin A by pro-apoptotic cell extracts and by recombinant caspase-6 but did not block cleavage of PARP by caspase 3. A 50% inhibition occurred at about 400 μM total added Zn, however the cell extracts contained metal chelators and the actual free Zn concentration was not determined. These effects on caspase-6 are interesting in light of studies by Cohen and colleagues (1992) where, using electron microscopy, they showed that Zn supplementation of dexamethasone-treated thymocytes *in vitro* blocked the transition from peripheral nuclear chromatin condensation to nuclear collapse. Since caspase-6 is responsible for nuclear lamin cleavage leading to nuclear fragmentation (see earlier), Zn-mediated suppression of this caspase may be responsible for the morphological effects.

Zn may also suppress caspase-3 activation by initially blocking caspase-9. At least one study has reported Zn-dependent suppression of cleavage of the specific tetrapeptide fluorogenic substrate LEHD-AFC by recombinant human caspase-9 in the 300–1000 μM range. Zn was also found to inhibit the generation of active caspase-9 as detected by immunoblotting and enzymatic activity in intact cells (Mesner *et al.* 1999; Wolf & Eastman 1999). Interestingly, both caspase-6 and caspase-9 belong to the group III subset of caspases which preferentially cleave at (I/V/L)E*X*D tetrapeptide sequences (Earnshaw *et al.* 1999). Other members of this group include caspase-8 and granzyme B. There is no information yet on effects of Zn on caspase-8 but Zn does not block recombinant murine granzyme B, although neither do peptide caspase inhibitors (Pham *et al.* 1998).

TPEN might act by stripping Zn from either or both of caspases 6 and 9, enhancing their activity. However, we were unable to show any significant activation of caspase-6 prior to activation of caspase-3. Caspase-6 was activated in TPEN-treated cells, but only at time-points later than that of caspase-3 activation (Truong-Tran *et al.* 2000a). It is possible that only small amounts of caspase-6 are required for processing of caspase-3 and that these levels were too low to be detected by our fluorogenic substrate assay.

New insights into the compartmentalization of caspase processing and activation in cells coupled with colocalization studies of labile Zn and caspases (especially caspase-6) may provide further clues. How many of the other 14 caspases are as sensitive to Zn as caspase-6 and caspase-9 is not known. There is also a particular need for determining whether Zn deprivation *in vitro* and *in vivo* directly activates other suicide enzymes, as it does caspase-3.

Does Zn protect caspase sulfhydryls from oxidation?

In cells in which the caspases are dysfunctional, due to knock-out of the genes or after blockade by specific caspase peptide inhibitors, damage often results in necrosis rather than apoptosis (Kolenko *et al.* 1999). Necrosis is injurious to the body since the plasma membrane of the dying cell is disrupted, allowing release of phospholipases, proteases and other agents that can damage neighbouring cells and trigger an inflammatory response. Caspases, therefore, play a critical role in the commitment to apoptosis rather than necrosis. In cells damaged by ROS, there is an increased risk of cell death by necrosis since the caspase enzymes, themselves, contain an essential sulfhydryl (Nicholson *et al.* 1995) which is susceptible to oxidation (Sen *et al.* 1999) and S-nitrosation (Rossig *et al.* 1999). How then can caspases operate effectively in cells that are undergoing apoptosis triggered by an onslaught of ROS? We speculate that Zn may interact in a reversible manner with the essential sulfhydryl of Cys163 of the human pro-caspase-3 molecule (Figure 2). This would have two effects: the sulfhydryl would be protected from oxidation to a disulphide and the caspase activity would be temporarily suppressed.

In cells committed to apoptose, the masking Zn may be removed by changes in the redox state of the cell and/or by exchange of Zn with thionein. Maret *et al.* (1999) reported that Zn can be transferred from metallothionein to the apoforms of Zn metalloenzymes forming thionein. In addition, thionein was able to reactivate some Zn-inhibited enzymes, suggesting that it is an effective, intracellular chelator and may regulate a number of Zn-dependent cellular processes. Of particular relevance here, Maret and colleagues (1999) found that thionein was a potent activator of Zn-inhibited caspase-3; 100 nM thionein was as effective as 1 mM EDTA in reactivating this caspase (W. Maret, personal communication). It will be impor-

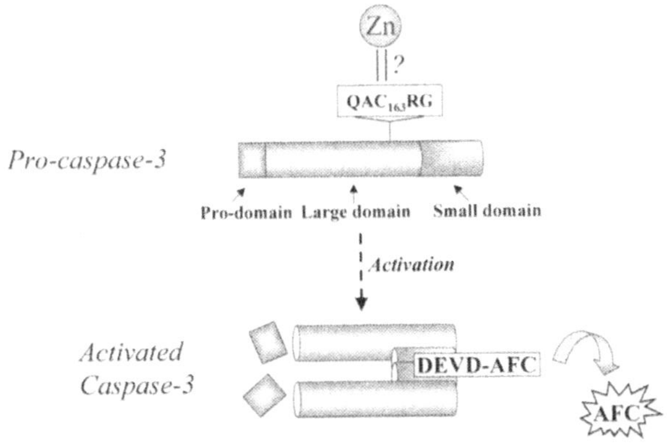

Fig. 2. Model for inactivation of caspases by Zn. The diagram shows the pro-caspase-3 polypeptide containing three domains (N-terminal prodomain, central large domain and C-terminal short domain) and the highly conserved pentapeptide $QAC_{163}RG$ within the long domain. Activation of caspase-3 involves the removal of the pro-domain and separation of large and short domains which reassemble to form the active enzyme. The activated enzyme is now able to cleave substrates such as the fluorogenic tetrapeptide DEVD-AFC releasing the coumarin derivative AFC. Cys_{163} is essential for catalytic activity (Nicholson *et al.* 1995; Earnshaw *et al.* 1999) and may be inactivated by oxidation to a disulfide. There are several other cysteines within 20 amino acids of this Cys_{163} which could potentially form a disulfide bond with this cysteine; however its not clear whether these do form in the tertiary structure of the molecule. It is proposed that Zn binds to this sulfhydryl in Cys_{163}, protecting it from oxidation and reversibly inhibiting enzymatic activity.

tant now to determine whether thionein is capable of triggering caspase activation and downstream events of apoptosis in cell-free models and to what extent thionein and the redox state of cells influences the activity of the caspases *in vivo*.

Other potential targets of Zn

Pharmacological concentrations of Zn and other divalent cations have been shown to block the mitochondrial pore through which cytochrome c is released (Zamzami *et al.* 1996). However, it is not clear whether this suppression occurs at a physiological level of Zn within the vicinity of the mitochondria. This mechanism can not explain the suppressive effects of Zn in cell-free models which lack mitochondria and contain exogenous cytochrome c. Furthermore, Wolf & Eastman (1999) recently confirmed that Zn prevents the activation of downstream caspases but does not prevent release of mitochondrial cytochrome c, as assessed by Western blotting studies to measure release of this cofactor.

Alternatively, Zn may influence the levels of auxillary factors in the activation of caspase-3. Candidates include members of the Bcl-2 family of proteins which suppress caspase-3 activation. Fukamachi and colleagues (1998) showed that Zn supplementation of cells *in vitro* increased the Bcl-2/Bax ratio in a monocyte cell line U937, thereby increasing the resistance of the cells to apoptosis. However, a criticism of this study is that a very high concentration (1 mM) of Zn salt was used to supplement the cells and therefore the increase in the Bcl-2/Bax ratio may be partially due to a response to the stress of excessive levels of Zn rather than a physiological effect of Zn on the regulation of apoptosis. Further experiments using a more physiological concentration of Zn are required to confirm these findings. However, there may be an interesting synergistic relationship between Zn and Bcl-2 (see below).

Zn may also influence events at the level of gene expression since Zn deprivation-induced apoptosis was suppressed by cycloheximide in neurons (Ahn *et al.* 1998). One possible target of Zn in gene-dependent apoptosis may be the Zn finger proteins. Previously it was thought that the Zn in these Zn finger domains is so tightly bound that it is not influenced by Zn deficiency. We now know that this Zn is a kinetically-labile pool especially in an oxidative microenvironment (Berendji *et al.* 1997).

Relationship of Zn to other anti-apoptotic regulators

A key issue is how the different physiological anti-apoptotic factors in cells cooperate to suppress apoptosis. Decreases in the concentrations of one factor may be compensated for by increases in another factor, if there is a level of redundancy.

Vitamin E (alpha-tocopherol) and Zn may at least partially substitute for each other. Vitamin E, which occurs in membranes and lipoproteins, blocks the chain reaction of lipid peroxidation by scavenging intermediate peroxyl radicals, particularly in membranes that are rich in polyunsaturated lipids (Liebler 1993). Like Zn deficiency, depletion of vitamin E disrupts membranes (Hennig *et al.* 1993) and causes neurodegeneration (Vatassery *et al.* 1992). As discussed earlier, vitamin E can reduce oxidative damage and apoptosis in Zn-depleted cells *in vivo* and *in vitro*. Hennig and colleagues (1993) have interpreted the decrease in plasma concentrations of vitamin E in dietary Zn deficiency to indicate greater utilization of this vitamin as a substitute for Zn. Vitamin E, however, cannot mimic Zn in the protection of protein sulfhydryls from oxidation and this may explain why vitamin E could only partially block apoptosis in Zn-depleted cells (unpublished observations).

The anti-apoptotic Bcl-2 family of proteins share a number of properties with Zn. Both are antagonists of a central mechanism in apoptotic cell death and therefore suppress apoptosis in response to a variety of inducers acting via diverse pathways. Like Zn, Bcl-2 is localized around mitochondria and protects against apoptosis and necrosis by multiple mechanisms including its protective effects against oxyradicals and interference with caspase processing (Korsmeyer *et al.* 1993; Reed *et al.* 1998; Thornberry & Lazebnik 1998; Esposti *et al.* 1999). Bcl-2 knock-out mice, like Zn-deficient rodents, exhibit massive apoptotic involution of thymus and spleen associated with depletion of $CD4^+$ T cells, are growth stunted and exhibit abnormal skin pigmentation (suggested, in Bcl-2 deficient mice, to be due to a defect in redox-regulated melanin synthesis (Veis *et al.* 1993)).

Zinc homeostasis and apoptosis

The relationship between Zn homeostasis and apoptosis needs further investigation and, in particular, the following questions need to be addressed. Firstly, why does Zn deficiency primarily increase apoptosis in those tissues undergoing rapid cell turnover (e.g., bone, thymus, epidermis, esophagus, testis, intestinal crypts and developing tissues of fetus)? We believe that cycling cells are highly prone to undergo apoptosis following Zn depletion. One mechanism may involve the cell cycle regulator $p21^{waf1/cip1}$ which is cleaved immediately following TPEN treatment of cells (Chai *et al.* 2000). This loss immediately followed activation of caspase-3 in Zn-depleted cells and proceeded in two steps: initial cleavage of p21 at D_{112} was followed by complete degradation of the protein. Since the turnover of p21 is regulated by the proteasome complex of proteases (Blagosklonny *et al.* 1996), the initial cleavage to p15 may expose sites that facilitate its ubiquitination and subsequent targeting to the proteasome. Loss of p21 leads to a dramatic induction of cdk2 activity which may result in premature entry of the cells into S-phase and apoptotic cell death, as occurs in some other forms of apoptosis (King & Cidlowski 1998). This may, in part, explain why *in vivo* Zn deficiency preferentially increases apoptotic cells in tissues where there is rapid cell turnover.

Secondly, do those cells in the body which are rich in exchangeable Zn (e.g., mast cells, pancreatic islet cells, ejaculated spermatozoa, certain hippocampal neurons and AEC) have increased resistance to apoptosis (Frederickson 1989; Zalewski *et al.* 1994a; Zalewski *et al.* 1996; Truong-Tran *et al.* 2000a)? By virtue of their specialized functions, these cells may be more exposed to ROS or other noxious agents and therefore need extra Zn for protection. The answer remains unknown but our laboratory is currently exploring this question using cell types with different concentrations of labile Zn. Only when the cellular biology of labile Zn is better understood will the full implications of Zn-related apoptosis become apparent. Thirdly, which of the subcellular pools of Zn participate in the suppression of apoptosis? Zn has diverse functions in cells which extend from structural and/or catalytic roles within metalloenzymes and Zn finger proteins to transient interactions with cellular signaling pathways (Vallee & Falchuk 1993). These functions can be sub-divided into those dependent on a largely fixed pool of cellular Zn which is poorly exchangeable (e.g., stoichiometric amounts of Zn that are tightly bound within the tertiary protein structure of metalloenzymes) and the more dynamic, labile Zn pools which are either loosely bound or tightly bound but kinetically labile (Frederickson 1989; Vallee & Falchuk 1993; Zalewski *et al.* 1993). These labile pools are sufficiently accessible and as a result are able to be visualized by specific Zn fluorophores (Zalewski *et al.* 1993) and by electron microscopy with the Neo-Timm's stain (Frederickson 1989). Since apoptosis is readily influenced by Zn deprivation or supplementation, it is likely that the more labile pools of Zn are involved in this regulation. Studies with mouse thymocytes and human chronic lymphocytic leukaemia cells revealed a strong inverse correlation between the level of intracellular labile Zn (increased or decreased by treatment with Zn ionophore or chelator, respectively) and the extent of apoptotic DNA fragmentation. Relatively small changes in labile Zn were able to cause large changes in susceptibility of cells to apoptosis and a threshold concentration in intracellular Zn may exist, below which apoptosis is induced (Zalewski *et al.* 1993).

Further clues to the identity of the key targets of Zn may come from a better understanding of the local interactions between Zn and apoptotic regulators in discrete subcellular pools. In addition to interactions between Zn and caspases, there may be distinct apoptosis-regulating functions of other cellular pools of Zn including intranuclear pools (Longin *et al.* 1997), microtubular and other cytoskeletal pools (Hesketh 1982; Zalewski *et al.* 1990), mitochondrial Zn (Untergasser *et al.* 2000), intravesicular Zn (Zalewski *et al.* 1993; Palmiter *et al.* 1996) and Zn within the cell membranes (Bettger & O'Dell 1981).

It is not clear whether intravesicular pools of Zn mediate protection against oxyradicals and interact with caspases and other molecular targets within the apoptosis pathway or whether this Zn is simply in the process of being transported to such sites. Vesicular traffic of molecules originating from the post-Golgi and endosomal sorting compartments is thought to be directed to their final destinations in the apical or basolateral domains and retained in place by microtubules and the intermediate filaments (Salas 1999). In this context, investigations performed by our laboratory using AEC may be relevant to studies concerning the importance of labile intracellular pools of Zn and its role in apoptosis.

Co-localization of mitochondria, lipid peroxides, Zn and procaspase-3 in the apical region of AEC

In our laboratory, we have utilized AEC as models to study Zn-regulated, oxidant-triggered, apoptosis (Truong-Tran *et al.* 2000a). AEC have a well-differentiated structure that is especially advantageous for study of the subcellular localization of the regulatory molecules and factors that govern apoptosis. They are polarized, with an apical plasma membrane that contains numerous beating cilia anchored to basal bodies (Figure 3A); the apical membrane faces the tracheobronchial airway lumen, *in vivo*, and the cilia function by propelling foreign particles, trapped in the extracellular mucin, back up the airways. To provide energy for cilial beating, AEC contain abundant mitochondria in the apical cytoplasm immediately below the basal bodies (Mills *et al.* 1999). Of clinical importance, AEC become fragile and easily damaged in asthma (Laitinen & Laitinen 1994), a disease with an underlying hypozincaemia (Di Toro *et al.* 1987; el-Kholy *et al.* 1990; Kadrabova *et al.* 1996). Although it is still controversial as to whether this hypozincaemia is a reflection of true Zn deficiency or rather a secondary response to the airway inflammation, an understanding of the mechanisms by which Zn protects these cells from apoptosis may have implications for the pathogenesis of a disease which affects 10% of adults and 25% of children in certain regions of the world (Adams *et al.* 1997).

The sub-cellular distribution of Zn was studied in AEC, using Zinquin, which detects the more labile pools of cellular Zn (Zalewski *et al.* 1993; 1994a, b; 1996). Labile Zn was found to be highly concentrated in the apical cytoplasm, both around the mitochondria and within the microtubular-derived basal bodies and cilia of sheep, pig and human AEC (Figure 3B) (Truong-Tran *et al.* 2000a). A similar pattern of Zn distribution was observed when cryostat sections of sheep or pig trachea and bronchi were labelled with Zinquin (Truong-Tran *et al.* 2000a). The cytoplasmic Zn was especially concentrated in perinuclear and apically-distributed vesicles as well as in basal bodies and cilia. Zn may be acquired from capillaries at the basal membranes of these cells, then packaged into vesicles and transported to the apical cytoplasm where at least some of it is incorporated into the microtubules of basal bodies and cilia (Truong-Tran *et al.* 2000a). Other Zn may be disposed around the mitochondria.

This same apical region was also rich in pro-caspase-3 as shown by immunocytochemistry (Figure 3C,D), using antibody to the precursor form of human caspase-3 (Krajewska *et al.* 1997). This suggests that pro-caspase-3 is strategically placed to trigger apoptosis should mitochondrial- or lumenal-derived ROS or other oxidants overwhelm the local defences. By immunocytochemistry with antibody to hydroxynonenal, there was some lipid peroxidation in the cell membranes, even in freshly isolated AEC from sheep trachea (Figure 3E). This was particularly evident in the apical membranes. Addition of exogenous H_2O_2 greatly increased lipid peroxidation throughout the cell.

In support of the hypothesis that the apical Zn is protecting the cells against ROS and caspase-3 activation, we found that depletion of Zn by treatment of AEC with TPEN resulted in increased lipid peroxidation (Figure 3F) and rapid activation of caspase-3 and morphological changes of apoptosis (Truong-Tran *et al.* 2000a). Interestingly, onset of caspase-3 activation followed a lag period of about 60 min after decline in intracellular Zn. It is essential now to understand the factors that govern the subcellular distribution of this family of enzymes. The precursor forms of caspases appear to exist in high molecular weight, multi-protein complexes (Cain *et al.* 1999), and may be anchored to components of the mitochondrial membranes and microtubules. Two recent studies (Zhivotovsky *et al.* 1999; Krebs *et al.* 2000) have also reported the localization of this caspase in mitochondria and/or associated membranes. Dual-labelling electron microscopy of AEC and other types of cells using Neo-Timm's stain for labile Zn and gold-

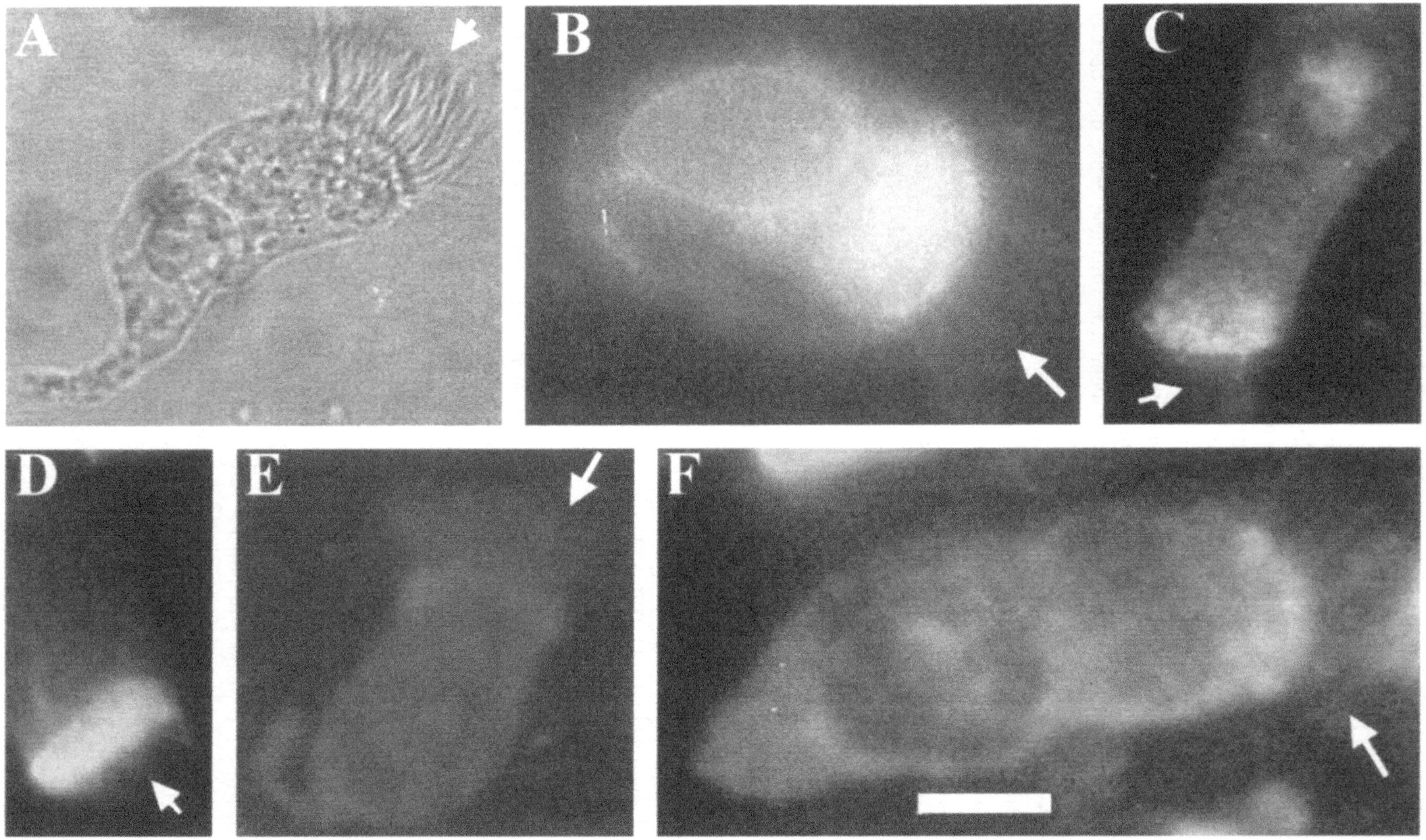

Fig. 3. Subcellular distribution of Zn, pro-caspase-3 and lipid peroxide in airway epithelial cells. Figure shows the co-localization of Zn, pro-caspase-3 and lipid peroxide HNE in the apical cytoplasm of AEC. A: Morphology of sheep AEC showing the columnar shape with distinct cilia at the apical membrane and the basally-located nucleus. B: Labile Zn concentrated in the apical cytoplasm and within the microtubular basal bodies of the cilia in a typical sheep AEC. Note also the perinuclear distribution of Zn. Zn was detected using 25 μM Zinquin and UV fluorescence microscopy. C: Pro-caspase-3 distribution in TPEN-treated sheep AEC showing predominate apical localization with some staining also in the nucleus. Cells were treated with 25 μM TPEN for 90 min and pro-caspase-3 was detected using anti-rabbit directed towards human pro-caspase-3 and visualized with goat anti-rabbit Ig FITC conjugated secondary antibody. D: Pro-caspase-3 distribution in untreated human AEC showing intense fluorescence at the apical region below the cilia. Same method of detection was used as in C. E: Low levels of lipid peroxide HNE in the apical membranes of untreated sheep AEC. HNE was detected using rabbit-anti human HNE and visualized with goat anti-rabbit Ig FITC conjugated secondary antibody. F: Increased levels of HNE especially in the apical region of TPEN-treated sheep AEC. Same method of detection was used as in E. Scale bar indicates 10 μm in panels A, C, D and E and 6 μm in panels B and F.

conjugated antibodies to detect caspases may provide important new information on the regulation of these enzymes by Zn.

Changes in Zn homeostasis during apoptosis

When considering the functions of anti-apoptotic regulators it is pertinent to ask what happens to their levels or distribution when cells do undergo apoptosis. In the case of Bcl-2, it appears to be down-regulated, probably by caspase-dependent cleavage (Fujita *et al.* 1998). Similarly, there may be a decline in intracellular Zn since several apoptosis-inducing agents cause a decrease in total intracellular Zn prior to induction of apoptosis (Treves *et al.* 1994). Somewhat paradoxically, the levels of intracellular labile Zn may rise dramatically during apoptosis as shown by an intense reaction with Zinquin (Zalewski *et al.* 1994b). We have proposed that the new pools of Zinquin-reactive Zn arise as a result of a change in the redox state of the cell which releases Zn bound to protein sulfhydryls when cells are in the later stages of apoptosis (Zalewski *et al.* 1994b). Since Zn is a structural building block in cells (Vallee & Falchuk 1993), the dismantling of cellular microtubules, chromatin and other components during apoptosis is likely to liberate large amounts of Zn which may be recycled or excreted from the body.

Conclusions

Life has evolved around metals like Zn and so apparently has the regulation of cellular death processes. We now need to understand the relationship between the physiological and pathological changes in altered Zn homeostasis and their control of apoptosis. There is relatively little knowledge on the differences in content and distribution of pools of Zn in different cells, tissues and organs at different stages of development, in different metabolic states and in local or systemic disease. It is essential that investigations are now focused on the cellular homeostasis of Zn in the normal and diseased states and how this directly or indirectly impacts on the pathways governing apoptosis.

This review has attempted to integrate the available information concerning Zn physiology and its involvement in regulating apoptosis. The advent of new technologies, such as the visualization of labile pools of Zn by Zn fluorophores, will enable new insights into the mechanisms by which Zn exerts these anti-apoptotic effects.

References

Adams R, Ruffin R, Wakefield M, Campbell D, Smith B. 1997 Asthma prevalence, morbidity and management practices in South Australia, 1992–1995. *Aust N Z J Med* **27**, 672–679.

Ahn YH, Kim YH, Hong SH, Koh JY. 1998 Depletion of intracellular zinc induces protein synthesis-dependent neuronal apoptosis in mouse cortical culture. *Exp Neurol* **154**, 47–56.

Aiuchi T, Mihara S, Nakaya M, Masuda Y, Nakajo S, Nakaya K. 1998 Zinc ions prevent processing of caspase-3 during apoptosis induced by geranylgeraniol in HL-60 cells. *J Biochem (Tokyo)* **124**, 300–303.

Banan A, Fields JZ, Decker H, Zhang Y, Keshavarzian A. 2000 Nitric oxide and its metabolites mediate ethanol-induced microtubule disruption and intestinal barrier dysfunction. *J Pharmacol Exp Ther* **294**, 997–1008.

Barbieri D, Troiano L, Grassilli E, Agnesini C, Cristofalo EA, Monti D, Capri M, Cossarizza A, Franceschi C. 1992 Inhibition of apoptosis by zinc: a reappraisal. *Biochem Biophys Res Commun* **187**, 1256–1261.

Berendji D, Kolb-Bachofen V, Meyer KL, Grapenthin O, Weber H, Wahn V, Kröncke KD. 1997 Nitric oxide mediates intracytoplasmic and intranuclear zinc release. *FEBS Lett* **405**, 37–41.

Bettger WJ, O'Dell BL. 1981 A critical physiological role of zinc in the structure and function of biomembranes. *Life Sci* **28**, 1425–1438.

Blagosklonny MV, Wu GS, Omura S, el-Deiry WS. 1996 Proteasome-dependent regulation of p21wafi1/cip1 expression. *Biochem Biophys Res Commun* **227**, 564–569.

Cain K, Brown DG, Langlais C, Cohen GM. 1999 Caspase activation involves the formation of the aposome, a large (approximately 700 kDa) caspase-activating complex. *J Biol Chem* **274**, 22686–22692.

Chai F, Truong-Tran AQ, Ho LH, Zalewski PD. 1999 Regulation of caspase activation and apoptosis by cellular zinc fluxes and zinc deprivation: A review. *Immunol Cell Biol* **77**, 272–278.

Chai F, Truong-Tran AQ, Evdokiou, A Young GP, Zalewski PD. 2000 Intracellular Zinc Depletion Induces Caspase Activation and $p21^{Waf1/Cip1}$ Cleavage in Human Epithelial Cell Lines. *J Infect Diseases* **182**, S85–S92.

Chautan M, Chazal G, Cecconi F, Gruss P, Golstein P. 1999 Interdigital cell death can occur through a necrotic and caspase-independent pathway. *Curr Biol* **9**, 967–970.

Cohen JJ, Duke RC. 1984 Glucocorticoid activation of a calcium-dependent endonuclease in thymocyte nuclei leads to cell death. *J Immunol* **132**, 38–42.

Cohen G, Sun X-M, Snowden RT, Dinsdale D, Skilleter DN. 1992 Key morphological features of apoptosis may occur in the absence of internucleosomal DNA fragmentation. *Biochem J* **286**, 331–334.

Cui L, Takagi Y, Wasa M, Sando K, Khan J, Okada A. 1999 Nitric oxide synthase inhibitor attenuates intestinal damage induced by zinc deficiency in rats. *J Nutr* **129**, 792–798.

Cui L, Takagi Y, Sando K, Wasa M, Okada A. 2000 Nitric oxide synthase inhibitor attenuates inflammatory lesions in the skin of zinc-deficient rats. *Nutrition* **16**, 34–41.

Da Silva JJRF, Williams. RJP: 1991 Zinc: Lewis acid catalysis and regulation. In: Williams, RJP ed. *The Biological Chemistry of the Elements.* Oxford, UK: Clarendon Press: 299–318.

Dimmeler S, Haendeler J, Galle J, Zeiher AM. 1997 Oxidized low-density lipoprotein induces apoptosis of human endothelial cells by activation of CPP32-like proteases. A mechanistic clue to the 'response to injury' hypothesis. *Circulation* **95**, 1760–1763.

Di Toro R, Galdo CG, Gialanella G, Miraglia DE, Giudice M, Moro R, Perrone L. 1987 Zinc and copper status of allergic children. *Acta Paediatr Scand* **76**, 612–617.

Duvall E, Wyllie AH. 1986 Death and the cell. *Immunol Today* **7**, 115–119.

Earnshaw WC, Martins LM, Kaufmann SH. 1999 Mammalian caspases: structure, activation, substrates, and functions during apoptosis. *Annu Rev Biochem* **68**, 383–424.

el-Kholy MS, Gas Allah MA, el-Shimi S, el-Baz F, el-Tayeb H, Abdel-Hamid MS. 1990 Zinc and copper status in children with bronchial asthma and atopic dermatitis. *J Egypt Public Health Assoc* **65**, 657–668.

Esposti MD, Hatzinisiriou I, McLennan H, Ralph S. 1999 Bcl-2 and mitochondrial oxygen radicals. New approaches with reactive oxygen species-sensitive probes. *J Biol Chem* **274**, 29831–29837.

Faleiro L, Kobayashi R, Fearnhead H, Lazebnik Y. 1997 Multiple species of CPP32 and Mch2 are the major active caspases present in apoptotic cells. *EMBO J* **16**, 2271–2281.

Floersheim GL, Christ A, Koenig R, Racine C, Gudat F. 1992 Radiation-induced lymphoid tumors and radiation lethality are inhibited by combined treatment with small doses of zinc aspartate and WR 2721. *Int J Cancer* **52**, 604–608.

Frade JM, Michaelidis TM. 1997 Origin of eukaryotic programmed cell death: a consequence of aerobic metabolism? *Bioessays* **19**, 827–832.

Fraker PJ, Telford WG. 1997 A reappraisal of the role of zinc in life and death decisions of cells. *Proc Soc Exp Biol Med* **215**, 229–236.

Frederickson CJ. 1989 Neurobiology of zinc and zinc-containing neurons. *Int Rev Neurobiol* **31**, 145–238.

Fujita N, Nagahashi A, Nagashima K, Rokudai S, Tsuruo T. 1998 Acceleration of apoptotic cell death after the cleavage of Bcl-XL protein by caspase-3-like proteases. *Oncogene* **17**, 1295–1304.

Fukamachi Y, Karasaki Y, Sugiura T, Itoh H, Abe T, Yamamura K, Higashi K. 1998 Zinc suppresses apoptosis of U937 cells induced by hydrogen peroxide through an increase of the Bcl-2/Bax ratio. *Biochem Biophys Res Commun* **246**, 364–369.

Ghibelli L, Coppola S, Fanelli C, Rotilio G, Civitareale P, Scovassi AI, Ciriolo MR. 1999 Glutathione depletion causes cytochrome c release even in the absence of cell commitment to apoptosis. *FASEB J* **13**, 2031–2036.

Grutter MG. 2000 Caspases: key players in programmed cell death. *Curr Opin Struct Biol* **10**, 649–55.

Halliwell B. 1991 Reactive oxygen species in living systems: source, biochemistry, and role in human disease. *Am J Med* **91**, 14S–22S.

Hamatake M, Iguchi K, Hirano K, Ishida R. 2000 Zinc Induces Mixed Types of Cell Death, Necrosis, and Apoptosis, in Molt-4 Cells. *J Biochem (Tokyo)* **128**, 933–939.

Hennig B, Wang Y, Ramasamy S, McClain CJ. 1992 Zinc deficiency alters barrier function of cultured porcine endothelial cells. *J Nutr* **122**, 1242–1247.

Hennig B, McClain CJ, Diana JN. 1993 Function of vitamin E and zinc in maintaining endothelial integrity. Implications in atherosclerosis. *Ann NY Acad Sci* **686**, 99–109.

Hennig B, Meerarani P, Ramadass P, Toborek M, Malecki A, Slim R, McClain CJ. 1999. Zinc nutrition and apoptosis of vascular endothelial cells: implications in atherosclerosis. *Nutrition* **15**, 744–748.

Hesketh JE. 1982. Zinc-stimulated microtubule assembly and evidence for zinc binding to tubulin. *Int J Biochem* **14**, 983–990.

Jacotot E, Costantini P, Laboureau E, Zamzami N, Susin SA, Kroemer G. 1999 Mitochondrial membrane permeabilization during the apoptotic process. *Ann NY Acad Sci* **887**, 18–30.

Janicke RU, Sprengart ML, Wati MR, Porter AG. 1998 Caspase-3 is required for DNA fragmentation and morphological changes associated with apoptosis. *J Biol Chem* **273**, 9357–9360.

Jankowski-Hennig MA, Clegg MS, Daston GP, Rogers JM, Keen CL. 2000 Zinc-deficient rat embryos have increased caspase 3-like activity and apoptosis. *Biochem Biophys Res Commun* **271**, 250–256.

Kadrabova J, Mad'aric A, Podivinsky F, Gazdik F, Ginter F. 1996 Plasma zinc, copper and copper/zinc ratio in intrinsic asthma. *J Trace Elem Med Biol* **10**, 50–53.

Kerr JFR, Searle J, Harmon BV, Bishop CJ. 1987 Apoptosis. In: Potten CS. ed. *Perspectives of Mammalian Cell Death*. Oxford, UK: Oxford Uni. Press: 93–128.

King KL, Cidlowski JA. 1998 Cell cycle regulation and apoptosis. *Annu Rev Physiol* **60**, 601–617.

Kolenko V, Uzzo RG, Bukowski R, Bander NH, Novick AC, His ED, Finke JH. 1999 Dead or dying: necrosis versus apoptosis in caspase-deficient human renal cell carcinoma. *Cancer Res* **59**, 2838–2842.

Korsmeyer SJ, Shutter JR, Veis DJ, Merry DE, Oltvai ZN. 1993 Bcl-2/Bax: a rheostat that regulates an antioxidant pathway and cell death. *Semin Cancer Biol* **4**, 327–332.

Kown MH, Van Der Steenhoven T, Blankenberg FG, Hoyt G, Berry GJ, Tait JF, Strauss HW, Robbins RC. 2000 Zinc-mediated reduction of apoptosis in cardiac allografts. *Circulation* **102 (Suppl 3)**, 228–232.

Krajewska M, Wang HG, Krajewski S, Zapata JM, Shabaik A, Gascoyne R, Reed JC. 1997 Immunohistochemical analysis of *in vivo* patterns of expression of CPP32 (Caspase-3), a cell death protease. *Cancer Res* **57**, 1605–1613.

Kraus A, Roth HP, Kirchgessner M. 1997 Supplementation with vitamin C, vitamin E or beta-carotene influences osmotic fragility and oxidative damage of erythrocytes of zinc-deficient rats. *J Nutr* **127**, 1290–1296.

Krebs JF, Srinivasan A, Wong AM, Tomaselli KJ, Fritz LC, Wu JC. 2000 Heavy membrane-associated caspase 3: Identification, isolation, and characterization. *Biochemistry* **39**, 16056–16063.

Kroemer G. 1998 The mitochondrion as an integrator/coordinator of cell death pathways. *Cell Death Differ* **5**, 547.

Kroemer G, Reed JC. 2000 Mitochondrial control of cell death. *Nat Med* **6**, 513–519.

Kuo IC, Seitz B, LaBree L, McDonnell PJ. 1997 Can zinc prevent apoptosis of anterior keratocytes after superficial keratectomy? *Cornea* **16**, 550–555.

Laitinen A, Laitinen LA. 1994 Airway morphology: epithelium/basement membrane. *Am J Respir Crit Care Med* **150**, S14–S17.

Lazebnik YA, Cole S, Cooke CA, Nelson WG, Earnshaw WC. 1993 Nuclear events of apoptosis *in vitro* in cell-free mitotic extracts: A model system for analysis of the active phase of apoptosis. *J Cell Biol* **123**, 7–22.

Liebler DC. 1993 The role of metabolism in the anti-oxidant function of vitamin E. *Crit Rev Toxicol* **23**, 147–169.

Liu X, Kim CN, Yang J, Jemmerson R, Wang X. 1996 Induction of apoptotic program in cell-free extracts: requirement for dATP and cytochrome c. *Cell* **86**, 147–157.

Longin AS, Mezin P, Favier A, Verdetti J. 1997 Presence of zinc and calcium permeant channels in the inner membrane of the nuclear envelope. *Biochem Biophys Res Commun* **235**, 236–241.

Maret W, Jacob C, Vallee BL, Fischer EH. 1999 Inhibitory sites in enzymes: zinc removal and reactivation by thionein. *Proc Natl Acad Sci USA* **96**, 1936–1940.

Martin SJ, Cotter TG. 1990 Specific loss of microtubules in HL- 60 cells leads to programmed cell death (apoptosis). *Biochem Soc Trans* **18**, 299–301.

Martin SJ, Mazdai G, Strain JJ, Cotter TG, Hannigan BM. 1991 Programmed cell death (apoptosis) in lymphoid and myeloid cell lines during zinc deficiency. *Clin Exp Immunol* **83**, 338–343.

Matsushita K, Kitagawa K, Matsuyama T, Ohtsuki T, Taguchi A, Mandai K, Mabuchi T, Yagita Y, Yanagihara T, Matsumoto M. 1996 Effect of systemic zinc administration on delayed neuronal death in the gerbil hippocampus. *Brain Res* **743**, 362–365.

Meerarani P, Ramadass P, Toborek M, Bauer HC, Bauer H, Hennig B. 2000 Zinc protects against apoptosis of endothelial cells induced by linoleic acid and tumor necrosis factor alpha. *Am J Clin Nutr* **71**, 81–87.

Mesner PW, Bible KC, Martins LM, Kottke TJ, Srinivasula SM, Svingen PA, Chilcote TJ, Basi GS, Tung JS, Krajewski S, Reed JC, Alnemri ES, Earnshaw WC, Kaufmann SH. 1999 Characterization of caspase processing and activation in HL-60 cell cytosol under cell-free conditions. Nucleotide requirement and inhibitor profile. *J Biol Chem* **274**, 22635–22645.

Mills PR, Davies RJ, Devalia JL. 1999 Airway epithelial cells, cytokines, and pollutants. *Am J Respir Crit Care Med* **160**, S38–S43.

Nakatani T, Tawaramoto M, Opare Kennedy D, Kojima A, Matsui-Yuasa I. 2000 Apoptosis induced by chelation of intracellular zinc is associated with depletion of cellular reduced glutathione level in rat hepatocytes. *Chem Biol Interact* **125**, 151–163.

Nicholson DW, Ali A, Thornberry NA, Vaillancourt JP, Ding CK, Gallant M, Gareau Y, Griffin PR, Labelle M, Lazebnik YA, Munday A, Raju SM, Smulson E, Yamin T-T, Yu VL, Miller DK. 1995 Identification and inhibition of the ICE/CED-3 protease necessary for mammalian apoptosis. *Nature (Lond)* **376**, 37–43.

Nickolson VJ, Veldstra H. 1972 The influence of various cations on the binding of colchicine by rat brain homogenates. Stabilization of intact neurotubules by zinc and cadmium ions. *FEBS Lett* **23**, 309–314.

Okuno S, Shimizu S, Ito T, Nomura M, Hamada E, Tsujimoto Y, Matsuda H. 1998 Bcl-2 prevents caspase-independent cell death. *J Biol Chem* **273**, 34272–34277.

Oteiza PI, Olin KL, Fraga CG, Keen CL. 1995 Zinc deficiency causes oxidative damage to proteins, lipids and DNA in rat testes. *J Nutr* **125**, 823–829.

Oteiza PI, Olin KL, Fraga CG, Keen CL. 1996 Oxidant defense systems in testes from zinc-deficient rats. *Proc Soc Exp Biol Med* **213**, 85–91.

Oteiza PI, Clegg MS, Zaga MP, Keen CL. 2000 Zinc deficiency induces oxidative stress and AP-1 activation in 3T3 cells. *Free Radic Biol Med* **28**, 1091–1099.

Palmiter RD, Cole TB, Findley SD. 1996 ZnT-2, a mammalian protein that confers resistance to zinc by facilitating vesicular sequestration. *EMBO J* **15**, 1784–1791.

Parat MO, Richard MJ, Beani JC, Favier A. 1997 Involvement of zinc in intracellular oxidant/anti-oxidant balance. *Biol Trace Elem Res* **60**, 187–204.

Perry DK, Smyth MJ, Stennicke HR, Salvesen GS, Duriez P, Porier GG, Hannun YA. 1997 Zinc is a potent inhibitor of the apoptotic protease caspase-3. A novel target for zinc in the inhibition of apoptosis. *J Biol Chem* **272**, 18530–18533.

Pham CT, Thomas DA, Mercer JD, Ley TJ. 1998 Production of fully active recombinant murine granzyme B in yeast. *J Biol Chem* **273**, 1629–1633.

Powell SR. 2000 The antioxidant properties of zinc. *J Nutrition* **130**, 1447S–1454S.

Reed JC, Jurgensmeier JM, Matsuyama S. 1998 Bcl-2 family proteins and mitochondria. *Biochim Biophys Acta* **1366**, 127–137.

Rogers JM, Taubeneck MW, Daston GP, Sulik KK, Zucker RM, Elstein KH, Jankowski MA, Keen CL. 1995 Zinc deficiency causes apoptosis but not cell cycle alterations in organogenesis-stage rat embryos: effect of varying duration of deficiency. *Teratology* **52**, 149–159.

Rossig L, Fichtlscherer B, Breitschopf K, Haendeler J, Zeiher AM, Mulsch A, Dimmeler S. 1999 Nitric oxide inhibits caspase-3 by S-nitrosation *in vivo*. *J Biol Chem* **274**, 6823–6826.

Roychowdhury M, Sarkar N, Manna T, Bhattacharyya S, Sarkar T, Basusarkar P, Roy S, Bhattacharyya B. 2000 Sulfhydryls of tubulin. A probe to detect conformational changes of tubulin. *Eur J Biochem* **267**, 3469–3476.

Salas PJ. 1999 Insoluble gamma-tubulin-containing structures are anchored to the apical network of intermediate filaments in polarized CACO-2 epithelial cells. *J Cell Biol* **146**, 645–658.

Sen CK, Sashwati R, Packer L. 1999 Fas mediated apoptosis of human Jurkat T-cells: intracellular events and potentiation by redox-active alpha-lipoic acid. *Cell Death Differ* **6**, 481–491.

Sensi SL, Yin HZ, Carriedo SG, Rao SS, Weiss JH. 1999 Preferential Zn^{2+} influx through Ca^{2+}-permeable AMPA/kainate channels triggers prolonged mitochondrial superoxide production. *Proc Natl Acad Sci USA* **96**, 2414–2419.

Seve M, Favier A, Osman M, Hernandez D, Vaitaitis G, Flores NC, McCord JM, Flores SC. 1999 The human immunodeficiency virus-1 tat protein increases cell proliferation, alters sensitivity to zinc chelator-induced apoptosis, and changes Sp1 DNA binding in HeLa cells. *Arch Biochem Biophys* **361**, 165–172.

Song Z, Steller H. 1999 Death by design: mechanism and control of apoptosis. *Trends Cell Biol* **9**, 49–52.

Stennicke HR, Salvesen GS. 1997 Biochemical characteristics of caspases-3, -6, -7, and -8. *J Biol Chem* **272**, 25719–25723.

Strasser A, O'Connor L, Dixit VM. 2000 Apoptosis signaling. *Annu Rev Biochem* **69**, 217–245.

Sun XM, MacFarlane M, Zhuang J, Wolf BB, Green DR, Cohen GM. 1999 Distinct caspase cascades are initiated in receptor-mediated and chemical-induced apoptosis. *J Biol Chem* **274**, 5053–5060.

Sunderman FW. 1995 The influence of zinc on apoptosis. *Ann Clin Lab Sci* **25**, 134–142.

Takahashi A, Alnemri ES, Lazebnik YA, Fernandes-Alnemri T, Litwack G, Moir RD, Goldman RD, Poirier GG, Kaufmann SH, Earnshaw WC. 1996 Cleavage of lamin A by Mch2 alpha but not CPP32: multiple interleukin 1 beta-converting enzyme-related proteases with distinct substrate recognition properties are active in apoptosis. *Proc Natl Acad Sci USA* **93**, 8395–8400.

Taylor CG, Towner RA, Janzen EG, Bray TM. 1990 MRI detection of hyperoxia-induced lung edema in Zn deficient rats. *Free Radic Biol Med* **9**, 229–233.

Taylor CG, McCutchon TL, Boermans HJ, DiSilvestro RA, Bray TM. 1997 Comparison of Zn and vitamin E for protection against hyperoxia-induced lung damage. *Free Radic Biol Med* **22**, 543–550.

Thompson CB. 1995 Apoptosis in the pathogenesis and treatment of disease. *Science* **267**, 1456–1462.

Thornberry NA, Lazebnik Y. 1998 Caspases: enemies within. *Science* **281**, 1312–1316.

Treves S, Trentini PL, Ascanelli M, Bucci G, Di Virgilio F. 1994 Apoptosis is dependent on intracellular zinc and independent of intracellular calcium in lymphocytes. *Exp Cell Res* **211**, 339–343.

Truong-Tran AQ, Ruffin RE, Zalewski PD. 2000a Visualization of labile zinc and its role in apoptosis of primary airway epithelial cells and cell lines. *Am J Physiol* **279**, L1172–L1183.

Truong-Tran AQ, Ho LH, Chai F, Zalewski PD. 2000b Cellular zinc fluxes and the regulation of apoptosis/gene-directed cell death. *J Nutrition* **130 (Suppl)**, 1459S–1466S.

Truong-Tran AQ, Carter J, Ruffin RE, Zalewski PD. 2001 New Insights Into The Role of Zinc in The Respiratory Epithelium. *Immunol Cell Biol* **79**, 170–177.

Tsutsui Y, Nakamura Y, Yamaguchi S, Kawanaka N, Sato M. 1999 Effects of zinc acexamate (NAS-501) on superoxide radicals and lipid peroxidation of rat gastric mucosa. *Pharmacology* **58**, 209–219.

Untergasser G, Rumpold H, Plas E, Witkowski M, Pfister G, Berger P. 2000 High Levels of Zinc Ions Induce Loss of Mitochondrial Potential and Degradation of Anti-apoptotic Bcl-2 Protein in *in vitro* Cultivated Human Prostate Epithelial Cells. *Biochem Biophys Res Commun* **279**, 607–614.

Vallee BL, Falchuk KH. 1993 The biochemical basis of zinc physiology. *Physiol Rev* **73**, 79–118.

Vatassery GT. 1992 Vitamin E. Neurochemistry and implications for neurodegeneration in Parkinson's disease. *Ann NY Acad Sci* **669**, 97–109.

Veis DJ, Sorenson CM, Shutter JR, Korsmeyer SJ. 1993 Bcl-2-deficient mice demonstrate fulminant lymphoid apoptosis, polycystic kidneys, and hypopigmented hair. *Cell* **75**, 229–240.

Waring P, Egan M, Braithwaite A, Mullbacher A, Sjaarda A. 1990 Apoptosis induced in macrophages and T blasts by the mycotoxin sporidesmin and protection by Zn^{2+} salts. *Int J Immunopharmac* **12**, 445–457.

Williams RJP. 1987 The biochemistry of zinc. *Polyhedron* **6**, 61–69.

Wolf CM, Eastman A. 1999 The temporal relationship between protein phosphatase, mitochondrial cytochrome c release, and caspase activation in apoptosis. *Exp Cell Res* **247**, 505–513.

Wyllie AH. 1997 Apoptosis: an overview. *Br Med Bull* **53**, 451–465.

Zalewski PD, Forbes IJ, Giannakis C, Cowled PA, Betts WH. 1990 Synergy between zinc and phorbol ester in translocation of protein kinase C to cytoskeleton. *FEBS Lett* **273**, 131–134.

Zalewski PD, Forbes IJ, Giannakis C. 1991 Physiological role for zinc in prevention of apoptosis (gene-directed death). *Biochem Inter* **24**, 1093–1101.

Zalewski PD, Forbes IJ. 1993 Intracellular zinc and the regulation of apoptosis. In: Lavin M, Watters D. eds. *Programmed Cell Death: The Cellular and Molecular Biology of Apoptosis*. Melbourne: Harwood Academic Publishers: 73–86.

Zalewski PD, Forbes IJ, Betts WH. 1993 Correlation of apoptosis with change in intracellular labile Zn, using Zinquin, a new specific fluorescent probe for zinc. *Biochem J* **296**, 403–408.

Zalewski PD, Millard SH, Forbes IJ, Kapaniris O, Slavotinek A, Betts WH. 1994a Video image analysis of labile zinc in viable pancreatic islet cells using a specific fluorescent probe for zinc. *J Histochem Cytochem* **42**, 877–884.

Zalewski PD, Forbes IJ, Seamark RF, Borlinghaus R, Betts WH, Lincoln SF, Ward AD. 1994b Flux of intracellular labile zinc during apoptosis (gene-directed cell death) revealed by a specific chemical probe, Zinquin. *Chem Biol* **1**, 153–161.

Zalewski PD, Jian X, Soon LLL, Breed WG, Seamark RF, Lincoln SF, Ward AD, Sun F-Z. 1996 Changes in distribution of labile zinc in mouse spermatozoa during maturation in the epididymis assessed by the fluorophore Zinquin. *Reprod Fertil Develop* **8**, 1097–1105.

Zamzami N, Susin SA, Marchetti P, Hirsch T, Gomez-Monterrey I, Castedo M, Kroemer G. 1996 Mitochondrial control of nuclear apoptosis. *J Exp Med* **183**, 1533–1544.

Zhivotovsky B, Samali A, Gahm A, Orrenius S. 1999 Caspases: their intracellular localization and translocation during apoptosis. *Cell Death Differ* **6**, 644–651.

BioMetals **14**: 331–341, 2001.

Review

Functions of zinc in signaling, proliferation and differentiation of mammalian cells

Detmar Beyersmann* & Hajo Haase
*Department of Biology and Chemistry, University of Bremen, D-28334 Bremen, Germany; *Author for correspondence (Tel: +49-421-2182377; Fax: +49-421-2187433; E-mail: beyers@chemie.uni-bremen.de)*

Received 18 January 2001; accepted 17 May 2001

Key words: differentiation, metallothionein, proliferation, signal transduction, zinc

Abstract

Zinc is essential for cell proliferation and differentiation, especially for the regulation of DNA synthesis and mitosis. On the molecular level, it is a structural constituent of a great number of proteins, including enzymes of cellular signaling pathways and transcription factors. Zinc homeostasis in eukaryotic cells is controlled on the levels of uptake, intracellular sequestration in zinc storing vesicles ('zincosomes'), nucleocytoplasmic distribution and elimination. These processes involve the major zinc binding protein metallothionein as a tool for the regulation of the cellular zinc level and the nuclear translocation of zinc in the course of the cell cycle and differentiation. In addition, there is also increasing evidence for a direct signaling function for zinc on all levels of signal transduction. Zinc can modulate cellular signal recognition, second messenger metabolism, protein kinase and protein phosphatase activities, and it may stimulate or inhibit activities of transcription factors, depending on the experimental systems studied. Zinc has been shown to modify specifically the metabolism of cGMP, the activities of protein kinase C and mitogen activated protein kinases, and the activity of transcription factor MTF-1 which controls the transcription of the genes for metallothionein and the zinc transporter ZnT-1. As a conclusion of these observations new hypotheses regarding regulatory functions of zinc ions in cellular signaling pathways are proposed.

Abbreviations: ATF-2 – Activating transcription factor-2; CAF – CREB associated factor; cAMP – Cyclic adenosine monophosphate; CaMPK-2 – Calcium/calmodulin dependent protein kinase-2; cGMP – Cyclic guanosine monophosphate; CREB– Cyclic AMP response element binding factor; DTPA – Diethylenetriaminepentaacetic acid; EGF – Epidermal growth factor; ERK – Extracellular signal-regulated kinase; GABA – γ-Aminobutyric acid; IGF-1 – Insulin-like growth factor-1; JNK – Jun-kinase; MAPK – Mitogen activated protein kinase; MT – Metallothionein; MTF-1 – Metal response element-binding transcription factor-1; NMDA – N-Methyl-D-aspartic acid; NO – Nitric monoxide; PDE – Cyclic nucleotide phosphodiesterase; PI3K – Phosphatidylinositol 3-kinase; PKC – Protein kinase C; TPEN, N,N,N′,N′-Tetrakis(2-pyridylmethyl) ethylenediamine; ZnT-1 – Zinc transporter-1.

Introduction

The essentiality of zinc for growth and proliferation was recognized by the observation that zinc deficiency caused growth retardation in all organisms investigated (Vallee & Falchuk 1993). Growth and differentiation of eukaryotic cells generally are induced by growth hormones/growth factors that trigger cascades of intracellular signaling elements. These include hormone receptors, intracellular second messengers, cascades of protein kinases, protein phosphatases and transcription factors binding to promoters of the genes to be addressed. On all levels of cellular signal transduction zinc is involved, either as a structural element or a regulatory factor or both. Thus zinc is an essential prerequisite for the progress of many signaling processes in eukaryotes. But there is also evidence for a direct signaling function of zinc: It modulates the GABA and NMDA receptors in mammalian brain cells (Cuajungco & Lees 1997; Canzoniero *et al.* 1997) and it binds to zinc sensing domains as in the case of the metal-regulated transcription factor MTF-1 (Andrews, this issue). Further, transcription factors have been detected that require zinc in the medium for binding to enhancers or their associated factors and probably are subject to modulation by zinc (Berg *et al.* 1997; Inada *et al.* 1997). All crucial decisions in the life of mammalian cells are involving zinc in its ionic or protein-bound form: be it cell growth and proliferation, differentiation or programmed cell death.

Intracellular distribution of zinc

A regulatory function of zinc requires a strict regulation of the cellular zinc content and its distribution. 30 to 40% of the cellular zinc is localized in the nucleus, 50% in the cytosol and cytosolic organelles and the remainder is associated with membranes (Vallee & Falchuk 1993). In the last few years, some of the zinc transport proteins that maintain this distribution have been identified in mammalian cells. These include plasma membrane zinc importer and zinc exporter proteins as well as transporters that mediate the sequestration of Zn^{2+} into intracellular vesicles. For a detailed discussion of zinc transporters, see Gaither & Eide (this issue).

Most of the cellular zinc is bound to or at least associated with proteins or complexed by anions, hence the level of free intracellular zinc is very low. The major Zn^{2+} binding protein in mammalian systems is metallothionein which donates Zn^{2+} to enzymes (Udom & Brady 1980, Jacob *et al.* 1998) and transcription factors with zinc finger domains (Zeng *et al.* 1991, Cano-Gauci & Sarkar 1996, Maret *et al.* 1997) wheras the apoprotein thionein accepts Zn^{2+} from binding sites in proteins with moderate affinity for Zn^{2+} (Maret *et al.* 1999). The latter authors have estimated from enzyme inhibition constants that inhibition of crucial enzymes by zinc may become significant with free Zn^{2+} above 10^{-8} M. Hence, the free cellular Zn^{2+} concentration may be estimated to be of this order of magnitude.

Several techniques have been used to determine the free zinc concentration experimentally. Measurements with radioactive ^{65}Zn yielded a concentration of 24 pM (Simons 1991), and ^{19}F-NMR spectroscopy with 5-F-BAPTA gave a concentration of 0.5 nM free intracellular zinc (Benters *et al.* 1997). With fluorescent probes traditionally employed for Ca^{2+} analysis, intracellular free Zn^{2+} concentrations were estimated to be 1 nM with FURA-2 (Atar *et al.* 1995) and 2 nM with Mag-Fura-2 (Sensi *et al.* 1997). Whereas these probes are not specific for Zn^{2+} and thus their signals are difficult to separate from those for Ca^{2+}, the zinc-specific fluorescent probe Zinquin allows a specific detection of intracellular zinc, but a quantitative analysis is hampered by the lack of a ratio-technique with the Zinquin spectrum. An estimation of cellular Zn^{2+} levels from the Zinquin fluorescence results in micromolar concentrations. These high values are at least in part due to a complexation of protein-bound zinc as has been shown for metallothionein (Coyle *et al.* 1994). Therefore Zinquin cannot be used for a quantification of free cytosolic zinc ions but it has been shown to be an excellent tool to investigate the intracellular distribution of loosely bound zinc, i.e. the amount of the ion that can be easily exchanged and therefore should be considered as the mediator of a signaling function for zinc. Recently, a new fluorescent Zn^{2+} chelator, Zinpyr-1, has been developed with an apparent dissociation constant of 2 nM (Walkup *et al.* 2000). However, this new probe like Zinquin does not allow the ratio-technique, which would allow precise quantification of intracellular Zn^{2+}.

In a number of cell lines it has been demonstrated with Zinquin that there is a typical pattern of zinc-specific fluorescence (Zalewski *et al.* 1993, Coyle *et al.* 1994). First it consists of a fluorescent cytosol, second a nearly nonfluorescent nucleus and third of zincosomes, vesicular structures containing high amounts of loosely bound zinc (Figure 1). Furthermore, the total cellular zinc level is tightly regulated. For example, in C6 rat glioma cells the total cellular zinc content is independent of the extracellular zinc concentration up to a threshold of 100 μM (Haase & Beyersmann 1999) mediated by an export mechanism for zinc that can be inhibited by lanthanum(III) ions in this cell line (Haase 2001). These observations demonstrate the existence of a complex regulation of the total zinc and its intracellular distribution, which both are a necessary basis for a function of zinc in signal transduction.

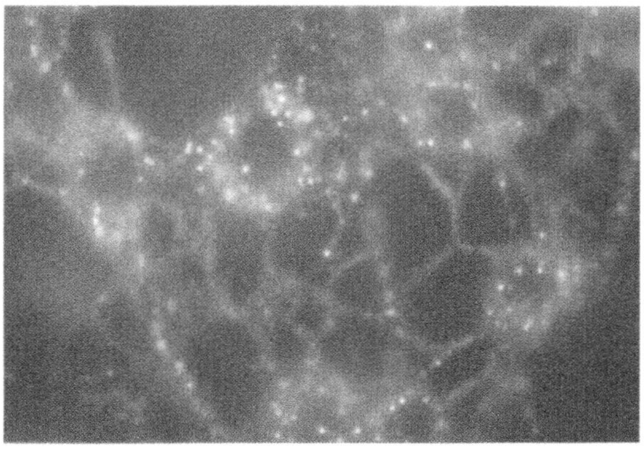

Figure 1. Intracellular distribution of loosely bound zinc ions. C6 rat glioma cells were loaded with the zinc-specific fluorescent probe Zinquin. Typical observations are a fluorescent cytosol, a nearly nonfluorescent nucleus and vesicular structures of high fluorescence intensity, so-called zincosomes. Magnification of original photograph is 400×.

Effect of zinc on major cellular signaling mechanisms

Zinc is involved in extracellular signal recognition, second messenger metabolism, protein phosphorylation and dephosphorylation and activity of transcription factors. In several instances, zinc ions have been shown to directly modulate cellular signaling. Figure 2 depicts the main interactions of zinc with receptors, protein kinases and transcription factors, which are described in detail below.

Calcium signaling

Zinc interferes with different aspects of calcium regulation. Electrical stimulation of heart cells evoked Zn^{2+} entry through voltage-dependent Ca^{2+} channels, and the addition of extracellular Zn^{2+} to spontaneously depolarizing pituitary tumor cells induced the expression of a reporter gene driven by the metallothionein promoter (Atar *et al.* 1995). In some cell types, elevation of extracellular zinc evoked intracellular Ca^{2+} mobilization. E.g., in primary hepatocyte cultures, 100 μM Zn^{2+} caused an increase in the intracellular free calcium concentration by stimulation of hormone sensitive intracellular calcium stores (McNulty & Taylor 1999). The authors of this report postulate the existence of a hepatic heavy metal stimulated receptor which is only expressed in hepatocytes and hypothesized that zinc might function as a local hormone that is secreted with insulin from pancreatic β-cells and participates in the regulation of liver metabolism. Employing the fluorescent probe Zinquin, a release of zinc from secretory vesicles of pancreatic islet cells was indeed shown to be induced by insulin secretagogues (Qian *et al.* 2000).

For the calcium/calmodulin-dependent protein kinase-2 (CaMPK-2), opposite effects of low and elevated zinc concentrations were observed. Whereas low zinc concentrations resulted in an increase of calmodulin-independent activity, high levels of zinc inhibited the binding of Ca^{2+}-calmodulin and the activity of the kinase (Lengyel *et al.* 2000).

Cyclic nucleotide metabolism

The cellular contents of the second messengers cyclic adenosine monophosphate (cAMP) and cyclic guanosine monophosphate (cGMP) are regulated via their synthesis by cyclases and their degradation by cyclic nucleotide phosphodiesterases (PDE). Both groups of enzymes consist of large families of different subtypes. Zinc has been shown to affect these signal transduction pathways by modulating PDE activities. After the first observation of an inhibition of cAMP and cGMP hydrolysis by zinc in bovine heart (Donelly 1978), there have been several reports of both activating and inhibiting effects of zinc on different PDE subfamilies. Cyclic nucleotide phosphodiesterases are assumed to be zinc hydrolases, because two typical zinc-binding sequences ($HX_3HX_{24\text{-}26}E$) have been identified in the catalytic domain of PDE V (Francis *et al.* 1994). Zn^{2+} seems to regulate PDE activities, because the same authors found that binding of cGMP to cGMP-specific PDE (type V) is activated by zinc concentrations up to 1 μM whereas zinc concentrations above 1 μM had inhibitory effects. Also for PDE IV A (Percival *et al.* 1997) and for PDE VI (He *et al.* 2000), a dependence of catalytic activity on zinc and an inhibition at higher zinc concentrations was shown, but these results cannot be generalized for all types of PDE. The cGMP-inhibited PDE (Omburo *et al.* 1995) and the cAMP-specific PDE (Kovala *et al.* 1997) are inhibited by zinc, but no activation at low zinc concentrations could be detected. In the latter case the authors discussed the possibility that zinc remained bound to the enzyme during the isolation process, so that zinc binding would not have been necessary for its activation. In conclusion, low concentrations of zinc seem to be essential for PDE activity while high concentrations have inhibitory effects, indicating a regulatory function of zinc for cyclic nucleotide phosphodiesterases.

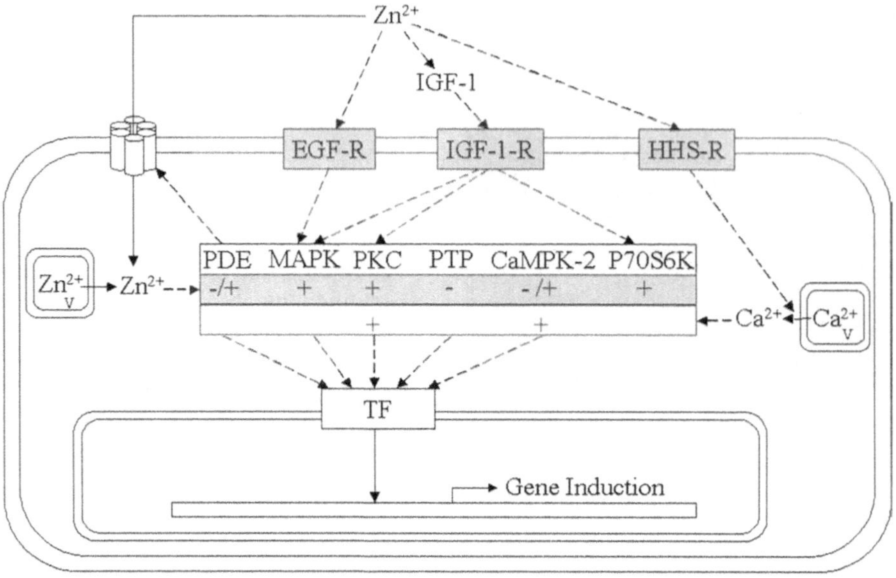

Figure 2. Effects of zinc on signal transduction pathways. Extracellular zinc can increase the formation of insulin-like growth factor (IGF), and stimulate the epidermal growth factor-receptor (EGR-R). The activation of a hepatic heavy metal ion stimulated-receptor (HHS-R) causes the intracellular release of Ca^{2+} in hepatocytes. At the level of protein phosphorylation, Zn^{2+} taken up and/or Zn^{2+} released from zincosomes can modulate the activity of cyclic nucleotide phosphodiesterase (PDE), mitogen-activated protein kinase (MAPK), protein kinase C (PKC), protein tyrosine phosphatases (PTP), Ca^{2+}-calmodulin activated protein kinase-2 (CaMPK-2), and P70S6 kinase (P70S6K). Activation of protein kinases or phosphatases leads to changes in the phosphorylation state of transcription factors (TF) and gene activities. Activating and inhibitory interactions are represented by + and −, respectively. Index V: vesicular localization.

In investigations with intact cells, these effects of zinc on the cGMP content were confirmed. Incubation of PC12 rat pheochromocytoma cells with zinc led to an increase of the cellular cGMP concentration, and in the homogenate of these cells a zinc-mediated inhibition of cGMP hydrolysis was identified as the cause for this effect (Waetjen *et al.* 2001). Recently, a new link between cGMP and zinc homeostasis was established. We observed, that not only zinc modulates cGMP signaling, but that cGMP also modulates the uptake of zinc (Haase 2001). In C6 rat glioma cells, zinc uptake was saturated after 1 h of incubation with 150 μM $ZnCl_2$. This type of saturation was not observed in the presence of LY 83583 or methylene blue, specific inhibitors of guanylate cyclase, and its appearance was accelerated by nitric oxide, an activator of soluble guanylate cyclase (Haase 2001). These observations have led us to the following hypothesis (Figure 3): Elevation of cellular Zn^{2+} results in a rise in cGMP, which inhibits further zinc import. This effect is only effective under conditions of ongoing cGMP synthesis and is abolished if guanylate cyclase is inhibited. At present we investigate whether the impairment of zinc import is caused by direct interaction of cGMP with an ion transporter or mediated by a cGMP-dependent protein kinase.

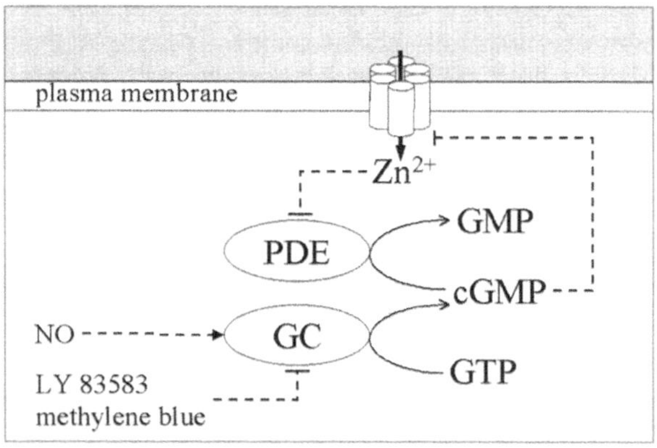

Figure 3. Mutual effects of zinc and cGMP-signaling. Increased intracellular zinc concentrations inhibit cyclic nucleotide phosphodiesterase (PDE), the cGMP-degrading enzyme. The resulting increase in cellular cGMP leads to a downregulation of zinc uptake, but only if the guanylate cyclase (GC) is not inhibited (e.g., by LY 83583 or methylene blue) and cGMP formation still can take place. Nitric monoxide (NO), an activator of GC, augments the downregulation of zinc uptake.

Receptor tyrosine kinases and mitogen-activated protein kinases

Several laboratories have investigated effects of zinc deficiency, chelation and/or zinc supplementation on mitogenic signaling pathways evoked by growth factors. Zinc deficiency specifically caused diminished levels of insulin-like growth factor-1 (IGF-1) both in humans (Cossack 1991) and rats (Dorup *et al.* 1991). In rat liver, zinc deficiency led to a decreased expression of growth hormone and IGF-1 genes (McNall *et al.* 1995). There is some controversy whether zinc stimulates the IGF-1 receptor activity directly. With NIH3T3 murine fibroblasts transfected with the IGF-1 receptor promoter, the zinc chelator DTPA did not affect phosphorylation of the IGF-1 receptor and of insulin receptor substrates (MacDonald 2000). However, zinc increased the affinity of both IGF-1 and IGF-2 to the type 1 IGF receptor on murine myoblasts with a half-maximal concentration of about 50 μM (McCusker *et al.* 1998). In accordance with this effect, zinc chelation by DTPA partially abolished the stimulation of mitogen-activated protein kinases (MAPK) by IGF-1 in rat fibroblasts, and this inhibition was reversed by addition of equimolar concentrations of zinc sulfate (Lefebvre *et al.* 1999). Conversely, elevation of extracellular zinc stimulated protein tyrosine phosphorylation and mitogen-activated protein kinase activity in murine fibroblasts (Hansson 1996). In human bronchial epithelial cells, various toxic metals and zinc induced EGF receptor phosphorylation and MAP kinase activation (Wu *et al.* 1999). The stimulating effect of zinc on tyrosine phosphorylation may be caused at least partially by interference with tyrosine dephosphorylation, because zinc inhibited various protein tyrosine phosphatases in human airway epithelial cells (Samet *et al.* 1999). This interpretation is in accordance with the observation that isolated recombinant protein tyrosine phosphatase was inhibited by zinc with an IC_{50} of 200 nM (Maret *et al.* 1999).

Incubation of mouse cortical cells with toxic zinc concentrations caused an activation of the MAP kinase ERK, leading to increased expression of the immediate early gene egr-1 (Park & Koh 1999). Treatment of human bronchial epithelial cells with subtoxic concentrations of Zn^{2+} activated the MAP kinases ERK, JNK and p38, and evoked increased phosphorylation of the transcription factors Jun and ATF-2, which are substrates of MAP kinases (Samet *et al.* 1998). A synergistic effect of zinc and calcium on DNA synthesis and mitogenic signaling was detected with NIH3T3 murine fibroblasts. When these cells were cultured in a medium containing 1.8 mM Ca^{2+} and starved from serum, supplementation with additional 1-2 mM Ca^{2+} and 40 μM Zn^{2+} markedly stimulated DNA synthesis and prolonged activation of MAPKs and P70S6 kinase (Huang *et al.* 1999). In Swiss 3T3 cells, phosphorylation and activation of P70S6 kinase was observed after treatment with zinc, and experiments with kinase inhibitors indicated activation through the phosphatidylinositol 3-kinase (PI3K) signaling pathway (Kim *et al.* 2000). Since PI3K is regulated by zinc proteins containing Zn^{2+} bound to a so-called FYVE domain (Wurmser *et al.* 1999), it may be speculated that this enzyme might be activated through Zn^{2+} binding to these PI3K regulating proteins. However, it is very difficult to predict effects of zinc on this system *in vivo*, because in the absence of Mg^{2+} micromolar concentrations of several transition metal ions including Zn^{2+} inhibited P70S6 kinase (Ferrari *et al.* 1991).

Protein kinase C structure and activity

The amino acid sequences of most isoforms of PKC contain two identical zinc-binding motifs at the N-terminus, the regulatory region of this enzyme (Parker *et al.* 1986). By site-directed mutagenesis it was demonstrated that at least one intact sequence is necessary for the binding of phorbol esters (Ono *et al.* 1989). Also other isoforms of PKC (α, βII, γ) were shown to contain four zinc atoms (Quest *et al.* 1992). EXAFS spectra of PKCβI indicated a complexation of each of the four zinc atoms by three cysteines and one histidine residue (Hubbard *et al.* 1991), and NMR spectroscopy of PKCα demonstrated that two non-consecutive sets of zinc-binding residues form two separate metal-binding sites (Hommel *et al.* 1994). A regulatory function of zinc for PKC is inferred from the observation that nanomolar concentrations of zinc can activate PKC and cause a translocation to the plasma membrane, a central event in the activation of PKC (Csermely *et al.* 1988a, b). Zinc also seems to regulate the translocation of PKC to the cytoskeleton (Zalewski *et al.* 1990). Furthermore, it was shown that a chelatable pool of intracellular zinc increases the binding of the PKC activator phorbol dibutyrate (Forbes *et al.* 1990) and that the zinc-binding cysteines are essential for phorbol ester binding (Ono *et al.* 1989).

It has been suggested that the above-mentioned four zinc-binding sites mediate the regulatory effects of zinc on PKC. But on the other hand, the zinc finger-bound zinc could not even be removed by high affinity heavy metal ion chelators, making regulation by free zinc at these sites unlikely (Hubbard *et al.* 1991). Nevertheless, the chelators TPEN (Csermely *et al.* 1988b) and 1,10-phenanatroline (Forbes *et al.* 1990) were able to inhibit PKC activation. So the activation of PKC by zinc is mediated by a chelatable pool of zinc that is not identical with the metal ions bound to the zinc finger structures.

Furthermore, zinc was found to modulate the autonomous activity of PKC, i.e., the activity in the absence of activating cofactors. The oxidation of the zinc-binding cysteine residues led to a release of zinc and to an increase of the autonomous PKC activity, but to a loss of sensitivity to regulating cofactors (Knapp & Klann 2000). This indicates a possible involvement of the cellular redox state in PKC signaling mediated by the zinc finger structures.

Transcription factors

From gene sequencing data it is estimated that zinc is a structural element of more than a thousand transcription factors containing zinc finger domains (Berg & Shi 1996). With respect to regulatory functions of zinc, it is of special interest whether Zn^{2+} ions actually activate transcription factors to adopt specific promoter binding conformations. The best-characterized zinc-activated transcription factor is the metal response element-binding transcription factor-1 (MTF-1), which induces the metallothionein promoter in response to cellular zinc. MTF-1 contains six finger structures of which the first binds zinc with low affinity (Bittel *et al.* 1998). After complexing zinc, MTF-1 is translocated from the cytoplasm to the nucleus (Smirnova *et al.* 2000) where it binds to metal-response elements of MT promoters and the promoter from the zinc transporter ZnT-1 (Langmade *et al.* 2000). The activation of transcription factors by zinc is not restricted to MTF-1. In thyroid nuclear extracts, the human thyroglobulin enhancer is induced by the transcription factor CREB in synergism with a further factor, the CREB associated factor (CAF). CAF binding to DNA is abolished by 0.5 mM 1,10-phenanthroline and thus seems to depend on zinc (Berg *et al.* 1997). Furthermore, the binding of the negative regulator QM to the transcription factor Jun requires 1 μM Zn^{2+} and thus appears to be a zinc-regulated process (Inada *et al.* 1997). Zinc may also inhibit transcription factor activities directly. The activation of nuclear factor-κB in bovine cerebral cells is suppressed if cellular zinc is elevated by application of the zinc ionophore pyrithione (Kim *et al.* 1999). Also, the binding of steroids to the murine glucocorticoid receptor is reversibly inhibited by Zn^{2+} (Telford & Fraker 1997).

Zinc in cell proliferation, differentiation and apoptosis

Proven and putative regulatory functions of zinc in cell proliferation

Biochemical mechanisms for the function of zinc in cell proliferation were detected when zinc was shown to be a structural element in enzymes involved in DNA synthesis (Springgate *et al.* 1973; Chesters *et al.* 1989), transcription (Wu *et al.* 1992), aminoacyl-tRNA synthesis (Hicks & Wallwork 1987) and ribosomal function (Hard *et al.* 2000). Furthermore, zinc is present in the zinc finger structures of transcription factors that control the activity of genes responding to growth factors (Berg & Shi 1996). Zinc is not only a structural element but is also involved in regulatory mechanisms of cell proliferation. Based on observations that serum addition to mammalian cell cultures enhanced the cellular uptake of zinc, and that zinc deprivation by metal chelators caused decreased growth and DNA synthesis, zinc was proposed to be a second messenger of mitogenic signaling (Grummt *et al.* 1986). Taking the above aspects together, zinc was even suggested to act as a master hormone of growth and proliferation (Frausto da Silva & Williams 1991). Although this hypothesis cannot be substantiated in this general manner, there are several lines of evidence showing the central role of this element in the control of growth and proliferation. Zinc and calcium synergistically stimulated DNA synthesis and mitogenic signaling in murine fibroblasts (Huang *et al.* 1999). And conversely, treatment of Swiss 3T3 cells with the zinc chelator DTPA inhibited thymidine incorporation into DNA (Chesters *et al.* 1989) and impaired the transcription of the thymidine kinase gene (Chesters *et al.* 1990). The latter effect could be ascribed to the increased binding of an inhibitory transcription factor to the promoter of the thymidine kinase gene (Chesters *et al.* 1995).

Both zinc deficiency and zinc chelation caused impaired availability of growth hormones. In experiments with rats, zinc deficiency led to decreased secretion of growth hormone from the pituitary gland (Roth

& Kirchgessner 1994). As discussed above in the section on receptor tyrosine kinases, zinc deficiency specifically caused diminished levels of insulin-like growth factor-1 (IGF-1) both in humans and rats, and zinc deficiency led to a decreased expression of growth hormone and IGF-1 genes in the liver. The effects of zinc deficiency on the metabolism of IGF-1 are of special relevance, because this factor mediates the transition from G1- to S-phase of the cell cycle in cultured cells.

Several laboratories have investigated effects of zinc chelation and/or zinc supplementation on mitogenic signaling pathways evoked by growth factors (see section on signaling). Zinc stimulated protein tyrosine phosphorylation in murine fibroblasts and phosphorylation of the epidermal growth factor in human lung epithelial cells; it also stimulated various MAPK activities in murine fibroblasts and human lung epithelial cells, whereas zinc chelation interfered with the stimulation of MAP kinases by IGF-1 in rat fibroblasts. Treatment of human lung epithelial cells with Zn^{2+} activated the phosphorylation of the transcription factors Jun and ATF-2 (Wu *et al.* 1999). The enhancement of mitogen-activated kinase activities may be at least partially caused by the inhibition of protein phosphatases by Zn^{2+} (Maret *et al.* 1999, Samet *et al.* 1999).

A role for metallothionein in zinc-mediated cell proliferation

The cellular homeostasis of zinc is at least partially controlled by metallothionein, which has been shown to play a role in the regulation of cell proliferation. MT is overexpressed in proliferating tissues, e.g. in regenerating rat liver (Tsujikawa *et al.* 1994), in developing rat liver (Andrews *et al.* 1987) and in various tumors (Panemangalore *et al.* 1983; Cai *et al.* 1998). In cultured human colonic cancer cells (HT-29) cellular MT levels oscillated with the cell cycle and reached a maximum in successive G1-phases and at the G1- to S-transitions (Nagel & Vallee 1995). Hence, the cellular MT level could be an instrument of proliferation control. Not only the total cellular level but also the subcellular location of this protein is remarkably variable in the course of the cell cycle. Whereas MT is normally found in the cytoplasm and not in the nucleus, it accumulated transiently in the nuclear fraction of fetal and newborn rat livers, followed by redistribution to the cytoplasma 2–3 weeks post partum (Templeton *et al.* 1985; Nartey *et al.* 1987a). A translocation

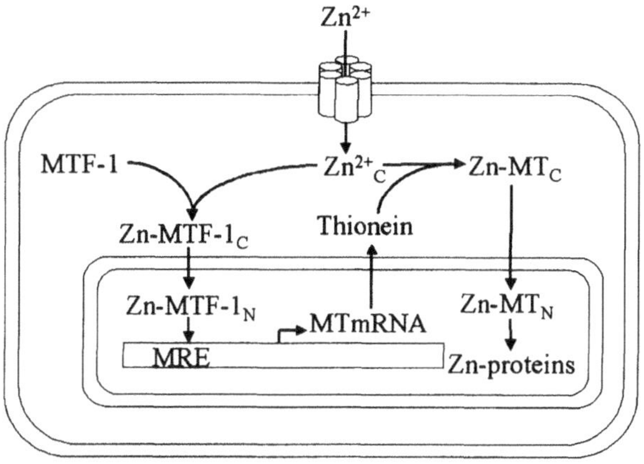

Figure 4. Model for interrelations between zinc and metallothionein. Binding of Zn^{2+} to the metal response element-binding transcription factor-1 (MTF-1) causes translocation of $Zn\text{-}MTF\text{-}1_C$ into the nucleus. There, $Zn\text{-}MTF\text{-}1_N$ binds a metal response element (MRE) in the promoter of the metallothionein (MT) gene. This interaction results in increased transcription and synthesis of thionein, which binds intracellular available Zn^{2+}. Index C: cytoplasmic localization; Index N: nuclear localization.

of MT into the nucleus was also observed during the early S-phase of growth factor-stimulated primary rat hepatocytes (Tsujikawa *et al.* 1991), in regenerating rat liver (Tsujikawa *et al.* 1994), certain tumors (Nartey *et al.* 1987b; Kuo *et al.* 1994; Woo & Lazo 1997) and in rat keratinocytes after irradiation with UV-B (Hanada *et al.* 1998). The nuclear translocation of MT probably is a vehicle for the achievement of a high nuclear zinc level in the S-phase of the cell cycle. Synchronized 3T3L1 fibroblasts show a nuclear accumulation of MT during the transition from G1- to S-phase of the cell cycle. When DNA synthesis was inhibited by aphidicolin, the cells were blocked at the G1- to S-transition, and zinc as well as MT were retained in the nucleus (Apostolova & Cherian 2000). After removal of aphidicolin, the cells reentered the S-phase, and both zinc and MT were relocated to the cytoplasm. The role of MT in these transitions seems to be to provide the increased amount of nuclear zinc required for DNA synthesis and mitogenic gene induction. The interrelations between zinc and MT during the onset of the S-phase of the cell cycle are summarized in Figure 4.

Cell differentiation

The overexpression of MT in developing tissues and at the transition from fetal to newborn rat development suggests a role for MT in differentiation, too. The differentiation of myoblasts to myotubes was inhibited by the lack of zinc (Petrie *et al.* 1991). A novel role for zinc mediated by MT was found in the process of differentiation of 3T3L1 preadipocytes (Schmidt & Beyersmann 1999). After stimulation of differentiation by insulin and dexamethasone, these cells enter into a phase of rapid proliferation with a concomitant rise in cellular zinc and MT contents. Simultaneously MT is translocated from the cytoplasm into the nucleus. Upon entry of the cells into the subsequent actual differentiation, the elevated levels of zinc and MT return to the initial amounts, and a redistribution of MT to the cytoplasm occurs. Similar changes in subcellular localization of zinc and MT were also observed in the course of differentiation of two myoblast cell lines to myotubes (Apostolova *et al.* 1999). Induction of differentiation by lowering serum or addition of IGF-1 caused nuclear translocation of zinc and MT during early differentiation, whereas in fully differentiated myoblasts, MT was relocated to the cytoplasm and the total cellular MT content declined. At least three different protein kinases are involved in the nuclear translocation of MT in this system. Inhibitors of MAP-kinase kinase (PD98059), PI3-kinase (LY294092) and P70S6-kinase (rapamycin) all retained MT in the cytoplasm after induction of differentiation (Apostolova *et al.* 2000). Because MT itself is not phosphorylated by protein kinases, it has to be assumed that phosphorylation of a mediator protein not yet identified is required for nuclear translocation of MT. From these experiments it may be concluded that a high level of zinc is required for nuclear functions during the early stage of differentiation of some cell systems. It is tempting but premature to generalize that these roles of zinc and MT are not limited to the control of differentiation of preadipocytes and myoblasts, but occur in other systems, too.

Apoptosis

Generally, zinc protects from apoptosis induced by various agents. On the other hand, concentrations of extracellular zinc which exceed the capacity of homeostatic control may also induce programmed cell death in several mammalian cell lines. The mechanisms of induction of apoptosis by zinc are not well understood yet. For a discussion of the role of zinc in apoptosis, see Truong-Tran *et al.* (this issue).

Perspectives

Zn^{2+} ions should be regarded as possible cellular signaling factors. In some aspects, they exhibit features similar to those of regulatory free Ca^{2+} ions. As with Ca^{2+} there is a relatively strict control of the intracellular concentration of labile zinc available for binding to signaling proteins. Also similar to Ca^{2+}, there is a mechanism for sequestration of excess zinc in cytoplasmic vesicles and a control of nuclear translocation of the metal ion. In other aspects, zinc is different from calcium as a second messenger. Whereas the hormone-induced increases in free Ca^{2+} are short-lived transients of a few seconds, the changes in labile cellular zinc concentrations are much slower and longer lasting than those of Ca^{2+}. Furthermore, whereas for Ca^{2+} there exists the specific sensing protein calmodulin, which transmits the message from Ca^{2+} to the corresponding protein kinase, there exists no such general zinc sensor, but a host of zinc-dependent enzymes and transcription factors linked to DNA synthesis and gene expression. Zinc binding to and release from metallothionein is a tool to control the availability of zinc and its nucleocytoplasmic localization, but MT is not a signal transducing molecule in the strict sense as calmodulin is for Ca^{2+}.

There are many aspects of zinc functions that need further elucidation. E.g., we know too little about the physiology and biochemistry of zinc storage in vesicles and zinc translocation to and from the nucleus. Which signals regulate the total cellular zinc level and its intracellular distribution and the changes that can be observed during proliferation or differentiation? The level of intranuclear loosely bound zinc seems to be very low compared to the cytosolic level. The maintenance of this gradient requires an active and energy consuming but yet unidentified mechanism for the control of nuclear zinc, but we still lack knowledge about the nuclear functions of zinc. Possibly the nuclear availability of zinc is a mechanism for a control of gene expression, and MT may serve as suppliers for a controlled transfer of zinc to nuclear proteins.

If zinc serves as a signaling factor in the regulation of cell proliferation, differentiation and death, it is crucial to understand how specificity is achieved. Again, a comparison with the far better established regulatory mechanisms of calcium might bear the answer. Above all, the tissue-specific equipment of cells with receptors and signal processing intracellular proteins provides specificity. Within cells, zinc-mediated signals may be discriminated by compartmentation.

The cellular zinc storing vesicles, the so-called zincosomes, and the nucleocytoplasmic distribution of zinc are potential tools to secure target-specificity of zinc signals. At present, we know much less about the mechanisms of signaling by zinc than by calcium, and there is a need for future research in the field of specific zinc distribution and binding to functional targets.

References

Andrews GK, Gallant KR, Cherian MG. 1987 Regulation of the ontogeny of rat liver metallothionein mRNA by zinc. *Eur J Biochem* **166**, 527–531.

Apostolova MD, Ivanova IA, Cherian MG. 1999 Metallothionein and apoptosis during differentiation of myoblasts to myotubes: protection against free radical toxicity. *Toxicol Appl Pharmacol* **159**, 175–184.

Apostolova MD, Cherian MG. 2000 Delay of M-phase onset by aphidicolin can retain the nuclear localization of zinc and metallothionein in 3T3L1 fibroblasts. *J Cell Physiol* **183**, 247–253.

Apostolova MD, Ivanova IA, Cherian MG. 2000 Signal transduction pathways, and nuclear translocation of zinc and metallothionein during differentiation of myoblasts. *Biochem Cell Biol* **78**, 27–37.

Atar D, Backx PH, Appel MM, Gao WD, Marban E. 1995 Excitation-transcription coupling mediated by zinc influx through voltage-dependent calcium channels. *J Biol Chem* **270**, 2473–2477.

Benters J, Flögel U, Schäfer T, Leibfritz D, Hechtenberg S, Beyersmann D. 1997 Study of the interactions of cadmium and zinc ions with cellular calcium homeostasis using ^{19}F-NMR spectroscopy. *Biochem J* **322**, 793–799.

Berg JM, Shi Y. 1996 The galvanization of biology: A growing appreciation for the roles of zinc. *Science* **271**, 1081–1085.

Berg V, Vassart G, Christophe D. 1997 A zinc-dependent DNA-binding activity co-operates with cAMP-responsive-element-binding protein to activate the human thyroglobulin enhancer. *Biochem J* **323**, 349–357.

Bittel D, Dalton T, Samson SL, Gedamu L, Andrews GK. 1998 The DNA binding activity of metal response element-binding transcription factor-1 is activated in vivo and in vitro by zinc, but not by other transition metals. *J Biol Chem* **273**, 7127–7133.

Cai L, Wang GJ, Xu ZL, Deng DX, Chakrabarti S, Cherian MG. 1998 Metallothionein and apoptosis in primary human hepatocellular carcinoma (HCC) from northern China. *Anticancer Res* **18**, 4667–4672.

Cano-Gauci DF, Sarkar B. 1996 Reversible zinc exchange between metallothionein and the estrogen receptor zinc finger. *FEBS Lett* **386**, 1–4.

Canzoniero LMT, Sensi SL, Choi DW. 1997 Measurement of intracellular free zinc in living neurons. *Neurobiol Dis* **4**, 275–279.

Chesters JK, Petrie L, Vint H. 1989 Specificity and timing of the Zn^{2+} requirement for DNA synthesis by 3T3 cells. *Exp Cell Res* **184**, 499–508.

Chesters JK, Petrie L, Travis AJ. 1990 A requirement for Zn^{2+} for the induction of thymidine kinase but not ornithine decarboxylase in 3T3 cells stimulated from quiescence. *Biochem J* **272**, 525–527.

Chesters JK, Boyne R, Petrie L, Lipson KE. 1995 Role of the promoter in the sensitivity of human tymidine kinase to lack of Zn^{2+}. *Biochem J* **308**, 659–664.

Cossack ZT. 1991 Decline in somatomedin-C, insulin-like growth factor-1 with experimentally induced zinc deficiency in human subjects. *Clin Nutr* (Edinb.) **10**, 284–291.

Coyle P, Zalewski PD, Philcox JC, Forbes IJ, Ward AD, Lincoln SF, Mahadevan I, Rofe AM. 1994 Measurement of zinc in hepatocytes by using a fluorescent probe, Zinquin: Relationship to metallothionein and intracellular zinc. *Biochem J* **303**, 781–786.

Csermely P, Szamel M, Resch K, Somogyi J. 1988a Zinc increases the affinity of phorbol ester receptor in T lymphocytes. *Biochem Biophys Res Comm* **154**, 578–583.

Csermely P, Szamel M, Resch K, Somogyi J. 1988b Zinc can increase the activity of protein kinase C and contributes to its binding to the plasma membrane in T lymphocytes. *J Biol Chem* **263**, 6487–6490.

Cuajungco MP, Lees GJ. 1997. Zinc metabolism in the brain: relevance to human neurodegenerative disorders. *Neurobiol Dis* **4**, 137–169.

Donnelly TE. 1978 Effects of zinc chloride on the hydrolysis of cyclic GMP and cyclic AMP by the activator-dependent cyclic nucleotide phosphodiesterase from bovine heart. *Biochim Biophys Acta* **522**, 151–160.

Dorup I, Flyvbjerg A, Everts ME, Clausen T. 1991 Role of insulin-like growth factor-1 and growth hormone in growth inhibition induced by magnesium and zinc deficiencies. *Br J Nutr* **66**, 505–521.

Ferrari S, Bandi HR, Hofsteenge J, Bussian BM, Thomas G. 1991 Mitogen-activated 70K S6 kinase. Identification of in vitro 40 S ribosomal S6 phosphorylation sites. *J Biol Chem* **266**, 22770–22775.

Forbes IJ, Zalewski PD, Giannakis C, Petkoff HS, Cowled PA. 1990 Interaction between protein kinase C and regulatory ligand is enhanced by a chelatable pool of cellular zinc. *Biochim Biophys Acta* **1053**, 113–117.

Francis SH, Colbran JL, McAllister-Lucas LM, Corbin JD. 1994 Zinc interactions and conserved motifs of the cGMP-binding cGMP specific phosphodiesterase suggest that it is a zinc hydrolase. *J Biol Chem* **269**, 22477–22480.

Frausto da Silva JJR, Williams RJP. 1991 The biological chemistry of the elements. *Clarendon Press, Oxford,* 299–318.

Grummt F, Weinmann-Dorsch C, Schneider-Schaulies J, Lux A. 1986 Zinc as a messenger of mitogenic induction. Effects on diadenosine tetraphosphate (Ap_4A) and DNA synthesis. *Exp Cell Res* **163**, 191–200.

Haase H. 2001 Zinkhomöostase in Säugerzellen: Untersuchungen zur Aufnahme, intrazellulären Verteilung und Toxizität. *GCA-Verlag, Herdecke*, ISBN 3-89863-024-2.

Haase H, Beyersmann D. 1999 Uptake and intracellular distribution of labile and total Zn(II) in C6 rat glioma cells investigated with fluorescent probes and atomic absorption. *Biometals* **12**, 247–254.

Hanada K, Tamai K, Sawamura D, Hashimoto I, Muramatsu T. 1998 Dynamic changes in intracellular location of metallothionein in rat keratinocytes after ultraviolet-B irradiation. *J Invest Dermatol* **110**, 98–100.

Hansson A. 1996 Extracellular zinc ions induce mitogen-activated protein kinase activity and protein tyrosine phosphorylation in bombesin-sensitive Swiss 3T3 fibroblasts. *Arch Biochem Biophys* **328**, 233–238.

Hard T, Rak A, Allard P, Kloo L, Garber M. 2000 The solution structure of ribosomal protein L36 from Thermus thermophilus reveals a zinc-ribbon-like fold. *J Mol Biol* **296**, 169–180.

He F, Seryshev AB, Cowan CW, Wensel TG. 2000 Multiple zinc binding sites in retinal rod cGMP phosphodiesterase, PDE6$\alpha\beta$. *J Biol Chem* **275**, 20572–20577.

Hicks SE, Wallwork JC. 1987 Effect of dietary zinc deficiency on protein synthesis in cell-free systems isolated from rat liver. *J Nutr* **117,** 1234–1240.

Hommel U, Zurini M, Luyten M. 1994 Solution structure of a cysteine rich domain of rat protein kinase C. *Nat Struct Biol* 1, 383–387.

Huang JS, Mukherjee JJ, Chung T, Crilly KS, Kiss Z. 1999 Extracellular calcium stimulates DNA synthesis in synergism with zinc, insulin and insulin-like growth factor I in fibroblasts. *Eur J Biochem* **266**, 943–951.

Hubbard SR, Bishop WR, Kirschmeier P, George SJ, Cramer SP, Hendrickson WA. 1991 Identification and characterization of zinc binding sites in protein kinase C. *Science* **254**, 1776–1779.

Inada H, Mukai J, Matsushima S, Tanaka T. 1997 QM is a novel zinc-binding transcriptional regulatory protein: its binding to c-jun is regulated by zinc ions and phosphorylation by protein kinase C. *Biochem Biophys Res Commun* **230**, 331–334.

Jacob C, Maret W, Vallee BL. 1998 Control of zinc transfer between thionein, metallothionein, and zinc proteins. *Proc Natl Acad Sci USA* **95**, 3489–3494.

Kim CH, Kim JH, Moon SJ, Chung KC, Hsu CY, Seo JT, Ahn YS. 1999 Pyrithione, a zinc ionophore, inhibits NF-κB activation. *Biochem Biophys Res Commun* **259**, 505–509.

Kim S, Jung Y, Kim D, Koh H, Chung J. 2000 Extracellular zinc activates p70S6 kinase through the phosphatidylinositol 3-kinase signaling pathway. *J Biol Chem* **275**, 25979–25984.

Knapp LT, Klann E. 2000 Superoxide-induced stimulation of protein kinase C via thiol modification and modulation of zinc content. *J Biol Chem* **275**, 24136–24145.

Kovala T, Sanwal BD, Ball EH. 1997 Recombinant expression of a Type IV, cAMP-specific phosphodiesterase: characterization and structure-function studies of deletion mutants. *Biochemistry* **36**, 2968–2976.

Kuo SM, Kondo Y, DeFillipo JM, Einstorff MS, Bahuson RR, Lazo JS. 1994 Subcellular localization of metallothionein IIA in human bladder tumor cells using a novel epitope-specific antiserum. *Toxic Appl Pharmacol* **125**, 104–110.

Langmade SJ, Ravindra E, Daniels PJ, Andrews GK. 2000 The transcription factor MTF-1 mediates metal regulation of the mouse ZnT1 gene. *J Biol Chem* **275**, 34803–34809.

Lefebvre D, Boney CM, Ketelslegers JM, Thissen JP. 1999 Inhibition of insulin-like growth factor-I mitogenic action by zinc chelation is associated with a decreased mitogen-activated protein kinase activation in RAT 1 fibroblasts. *FEBS Lett* **449**, 284–288.

Lengyel I, Fieuw-Makaroff S, Hall AL, Sim ATR, Rostas JAP, Dunkley PR. 2000 Modulation of the phosphorylation and activity of calcium/calmodulin-dependent protein kinase II by zinc. *J Neurochem* **75**, 594–605.

MacDonald RS. 2000 The role of zinc in growth and cell proliferation. *J Nutr* **130**, 1500S–1508S.

Maret W, Larsen KS, Vallee BL. 1997 Coordination dynamics of biological zinc 'clusters' in metallothioneins and in the DNA binding domain of the transcription factor Gal4. *Proc Natl Acad Sci USA* **94**, 2233–2237.

Maret W, Jacob C, Vallee BL, Fischer EH. 1999 Inhibitory sites in enzymes: Zinc removal and reactivation by thionein. *Proc Natl Acad Sci USA* **96**, 1936–1940.

McCusker RH, Kaleko M, Sackett RL 1998 Multivalent cations and ligand affinity of the type 1 insulin-like growth factor receptor on P2A2-LISN muscle cells. *J Cell Physiol* **176**, 392–401.

McNall AD, Etherton TD, Fosmire GJ. 1995 The impaired growth induced by zinc deficiency in rats is associated with decreased expression of the hepatic insulin-like growth factor I and growth hormone receptor genes. *J Nutr* **125**, 874–879.

McNulty TJ, Taylor CW. 1999 Extracellular heavy-metal ions stimulate Ca^{2+} mobilization in hepatocytes. *Biochem J* **339**, 555–561.

Nagel WW, Vallee BL. 1995 Cell cycle regulation of metallothionein in human colon cancer cells. *Proc Natl Acad Sci USA* **92**, 579–583.

Nartey NO, Banerjee D, Cherian MG. 1987a Immunohistochemical localization of metallothionein in cell nucleus and cytoplasm of fetal human liver and kidney and its changes during development. *Pathology* **119**, 233–238.

Nartey NO, Cherian MG, Banerjee D. 1987b Immunohistochemical localization of metallothionein in human thyroid tumors. *Am J Pathol* **129**, 177–182.

Omburo GA, Brickus T, Gazaleh FA, Colman RW. 1995 Divalent metal cation requirement and possible classification of cGMP-inhibited phosphodiesterase as a metallohydrolase. *Arch Biochem Biophys* **323**, 1–5.

Ono Y, Fujii T, Igarashi K, Kuno T, Tanaka C, Kikkawa U, Nishizuka Y. 1989 Phorbol ester binding to protein kinase C requires a cysteine-rich zinc-finger-like sequence. *Proc Natl Acad Sci USA* **86**, 4868–4871.

Panemangalore M, Banerjee D, Okosaka S, Cherian MG. 1983 Changes in the intracellular accumulation and distribution of metallothionein in rat liver and kidney during postnatal development. *Dev Biol* **97**, 95–102.

Park JA, Koh JY. 1999 Induction of an immediate early gene egr-1 by zinc through extracellular signal-related kinase activation in cortical culture: its role in zinc-induced neuronal death. *J Neurochem* **73**, 450–456.

Parker P, Coussens L, Totty N, Rhee L, Young S, Chen E, Stabel S, Waterfield MD, Ullrich A. 1986 The complete primary structure of protein kinase C – the major phorbol ester receptor. *Science* **233**, 853–859.

Percival MD, Yeh B, Falgueyret JP. 1997 Zinc dependent activation of cAMP-specific phosphodiesterase (PDE4A). *Biochem Biophys Res Commun* **241**, 175–180.

Petrie L, Chesters JK, Franklin M. 1991 Inhibition of myoblast differentiation by lack of zinc. *Biochem J* **276**, 109–111.

Qian WJ, Aspinwall CA, Battiste MA, Kennedy RT. 2000 Detection of secretion from single pancreatic β-cells using extracellular fluorogenic reactions and confocal fluorescence microscopy. *Anal Chem* **72**, 711–717.

Quest AFG, Bloomenthal J, Bardes ESG, Bell RM. 1992 The regulatory domain of protein kinase C coordinates four atoms of zinc. *J Biol Chem* **267**, 10193–1197.

Roth HP, Kirchgessner M. 1994 Influence of alimentary zinc deficiency on the concentration of growth hormone (GH), insulin-like growth factor I (IGF-I) and insulin in the serum of force-fed rats. *Horm Metab Res* **26**, 404–408.

Samet JM, Silbajoris R, Wu W, Graves LM. 1999 Tyrosine phosphatases as targets in metal-induced signaling in human airway epithelial cells. *Am J Resp Cell Mol Biol* **21**, 357–364.

Samet JM, Graves LM, Quay J, Dailey LA, Devlin RB, Ghio AJ, Wu W, Bromberg PA, Reed W. 1998 Activation of MAPKs in human bronchial epithelial cells exposed to metals. *Am J Physiol* **275**, L551–L558.

Schmidt C, Beyersmann D. 1999 Transient peaks in zinc and metallothionein levels during differentiation of 3T3L1 cells. *Arch Biochem Biophys* **364**, 91–98.

Sensi SL, Canzoniero LMT, Yu SP, Ying HS, Koh JY, Kerchner GA, Choi DW. 1997 Measurement of intracellular free zinc in living cortical neurons: Routes of entry. *J Neurosci* **17**, 9554–9564.

Simons TJ. 1991 Intracellular free zinc and zinc buffering in human red blood cells. *J Membr Biol* **123**, 63–71.

Smirnova IV, Bittel DC, Ravindra R, Jiang H, Andrews GK. 2000 Zinc and cadmium can promote rapid nuclear translocation of metal response element-binding transcription factor-1. *J Biol Chem* **275**, 9377–9384.

Springgate CF, Mildvan AS, Abramson R, Engle JL, Loeb LA. 1973 Escherichia coli deoxyribonucleic acid polymerase I, a zinc metalloenzyme. Nuclear quadrupole relaxation studies of the role of bound zinc. *J Biol Chem* **248**, 5987–5993.

Telford WG, Fraker PJ. 1997 Zinc reversibly inhibits steroid binding to murine glucocorticoid receptor. *Biochem Biophys Res Commun* **238**, 86–89.

Templeton DM, Banerjee MD, Cherian MG. 1985 Metallothionein synthesis and localization in relation to metal storage in rat liver during gestation. *Can J Biochem Cell Biol* **63**, 16–22.

Tsujikawa K, Imai T, Kakutani M, Kayamori Y, Mimura T, Otak N, Kimura M, Fukuyama R, Shimizu N. 1991 Localization of metallothionein in nuclei of growing primary cultured adult rat hepatocytes. *FEBS Lett* **283**, 239–242.

Tsujikawa K, Suzuki N, Sagawa K, Itoh M, Sugiyama T, Kohama Y, Otaki N, Kimura M, Mimura T. 1994 Induction and subcellular localization of metallothionein in regenerating rat liver. *Eur J Cell Biol* **63**, 240–246.

Udom AO, Brady FO. 1980 Reactivation in vitro of zinc-requiring apo-enzymes by rat liver zinc-thionein. *Biochem J* **187**, 329–335.

Vallee BL, Falchuk KH. 1993 The biochemical basis of zinc physiology. *Physiol Rev* **73**, 79–118.

Waetjen W, Benters J, Haase H, Schwede F, Jastorff B, Beyersmann D. 2001 Zn^{2+} and Cd^{2+} increase the cyclic GMP level in PC 12 cells by inhibition of the cyclic nucleotide phosphodiesterase. *Toxicology* **157**, 167–175.

Walkup GK, Burdette SC, Lippard SJ, Tsien RY. 2000 A new cell-permeable fluorescent probe for Zn^{2+}. *J Am Chem Soc* **122**, 5644–5645.

Woo ES, Lazo JS. 1997 Nucleocytoplasmic functionality of metallothionein. *Cancer Res* **57**, 4236–4241.

Wu FY, Huang WJ, Sinclair RB, Powers L. 1992 The structure of the zinc sites of Escherichia coli DNA-dependent RNA polymerase. *J Biol Chem* **267**, 25560–25567.

Wu W, Graves LM, Jaspers I, Devlin RB, Reed W, Samet JM. 1999 Activation of the EGF receptor signaling pathway in human airway epithelial cells exposed to metals. *Am J Physiol* **277**, L924–L931.

Wurmser AE, Gary JD, Emr SD. 1999. Phosphoinositide 3-kinases and their FYVE domain-containing effectors as regulators of vacuolar/lysosomal membrane trafficking pathways. *J Biol Chem* **274**, 9129–9132.

Zalewski PD, Forbes IJ, Betts WH. 1993 Correlation of apoptosis with change in intracellular labile Zn(II) using Zinquin [(2-methyl-8-*p*-toluenesulphonamido-6-quinolyloxy) acetic acid], a new specific fluorescent probe for Zn(II). *Biochem J* **296**, 403–408.

Zalewski PD, Forbes IJ, Giannakis C, Cowled PA, Betts WH. 1990 Synergy between zinc and phorbol ester in translocation of protein kinase C to cytoskeleton. *FEBS Lett* **273**, 131–134.

Zeng J, Heuchel R, Schaffner W, Kägi JHR. 1991 Thionein (apometallothionein) can modulate DNA binding and transcription activation by zinc finger containing factor Sp1. *FEBS Lett* **279**, 310–312.

BioMetals 14: 343–351, 2001.

Review

Zinc homeostasis and functions of zinc in the brain

Atsushi Takeda
Department of Radiobiochemistry, School of Pharmaceutical Sciences, University of Shizuoka, 52-1 Yada, Shizuoka 422-8526, Japan (Tel: 81-54-264-5700; Fax: 81-54-264-5705; E-mail: takedaa@u-shizuoka-ken.ac.jp)

Received 18 December 2000; accepted 1 March 2001

Key words: brain function, limbic system, vesicular zinc, zinc deprivation, zinc homeostasis

Abstract

The brain barrier system, i.e., the blood-brain and blood-cerebrospinal fluid barriers, is important for zinc homeostasis in the brain. Zinc is supplied to the brain via both barriers. A large portion of zinc serves as zinc metalloproteins in neurons and glial cells. Approximately 10% of the total zinc in the brain, probably ionic zinc, exists in the synaptic vesicles, and may serve as an endogenous neuromodulator in synaptic neurotransmission. The turnover of zinc in the brain is much slower than in peripheral tissues such as the liver. However, dietary zinc deprivation affects zinc homeostasis in the brain. Vesicular zinc-enriched regions, e.g., the hippocampus, are responsive to dietary zinc deprivation, which causes brain dysfunctions such as learning impairment and olfactory dysfunction. Olfactory recognition is reversibly disturbed by the chelation of zinc released from amygdalar neuron terminals. On the other hand, the susceptibility to epileptic seizures, which may decrease vesicular zinc, is also enhanced by zinc deficiency. Therefore, zinc homeostasis in the brain is closely related to neuronal activity. Even in adult animals and probably adult humans, adequate zinc supply is important for brain functions and prevention of neurological diseases.

Introduction

Zinc, an essential nutrient, is the second most abundant trace element in the body and powerfully influences cell division and differentiation (Vallee & Falchuk 1981; Coleman 1992). In microorganisms, plants and animals, over 300 enzymes require zinc for their functions. Zinc has three functions in zinc enzymes: catalytic, coactive (or cocatalytic) and structural (Vallee & Auld 1992; Vallee & Falchuk 1993).

In the brain, zinc is necessary for the maturation and function. Approximately 90% of the total brain zinc is bound in zincproteins (Frederickson 1989). The rest is in the presynaptic vesicles and histochemically reactive (as revealed by Timm's sulfide-silver staining method) (Haug 1973; Frederickson 1989; Howell & Frederickson 1989). Vesicular zinc, probably ionic zinc, may play a role in synaptic neurotransmission in the mammalian brain and serve as an endogenous neuromodulator of several important receptors including the α-amino-3-hydroxy-5-methyl-4-isoxazolepropionic acid (AMPA)/kainate receptor, N-methyl-D-aspartate (NMDA) and γ-aminobutyric acid (GABA) receptors (Harrison *et al.* 1993; Smart *et al.* 1994; Huang 1997).

The presence of zinc-containing neurons that sequester zinc in the presynaptic vesicles and release it in a calcium- and impulse-dependent manner has been demonstrated in the brain, especially the telencephalon (Assaf & Chung 1984; Howell *et al.* 1984; Perez-Clausell & Danscher 1985; Frederickson & Danscher 1990). The hippocampal and amygdalar regions may possess zinc-containing neuron terminals in high densities (Christensen & Frederickson 1998; Frederickson *et al.* 2000). All zinc-containing neurons reported so far are considered to be glutamatergic (Crawford & Connor 1973). However, not all glutamatergic neurons are zinc-containing (Frederickson & Moncrieff 1994). Neural circuits of the zinc-containing glutamatergic neurons are considered to be associated with the episodic memory function and are important for behavior, emotional expression and cognitive-mnemonic operations (Frederickson & Danscher 1990; Takeda *et al.* 1995). Thus, zinc serves not only intraneuronal and intraglial functions but also in synaptic neurotransmission.

Zinc concentration in the brain increases with growth after birth (Sawashita *et al.* 1997) and is maintained constant in the adult brain (Markesbery *et al.* 1984). Zinc is also supplied to the adult brain, probably as a required component for neural functions (Pullen *et al.* 1991; Takeda *et al.* 1994a). The turnover of zinc in the brain is slower than in peripheral tissues such as the liver (Kasarskis 1984; Takeda *et al.* 1995). The slow turnover of zinc is due to the presence of the brain barrier system, i.e., the blood-brain and the blood-cerebrospinal fluid (CSF) barriers.

The brain barrier system is important for zinc homeostasis in the brain, and its alteration may be associated with brain dysfunctions and neurological diseases (Takeda 2000). Dietary zinc deprivation causes alteration of zinc homeostasis in the brain (Takeda *et al.* 2000d; Takeda *et al.* 2001) and brain dysfunctions, e.g., mental disorders (Dreosti 1983; Golub *et al.* 1995).

This review summarizes that brain zinc homeostasis is closely related to neural functions.

Zinc transport across the brain barrier system

^{65}Zn is concentrated in the choroid plexus of young adult mice and rats 1 h after intravenous injection of $^{65}ZnCl_2$ and then concentrated in the brain parenchyma with decrease in choroidal ^{65}Zn (Takeda *et al.* 1994a, 2000c). The maximum uptake of ^{65}Zn is probably 6–10 days after parenteral injection to rats (Kasarskis 1984; Takeda *et al.* 1994a). In the brain of adult rats, ^{65}Zn is concentrated and retained in the hippocampal CA3 and dentate gyrus. ^{65}Zn is also concentrated and retained in the amygdala, especially the amygdalopiriform transition and the amygdalo-hippocampal transition areas (Takeda *et al.* 1995). In the case of intracerebroventricular injection of $^{65}ZnCl_2$, ^{65}Zn is concentrated in the brain parenchyma, e.g., the hippocampus and hypothalamus, in young adult rats (Takeda *et al.* 1994b). Zinc is transported into the brain across the blood-CSF barrier, in addition to the blood-brain barrier (Pullen *et al.* 1990; Franklin *et al.* 1992) (Figure 1). The main supply path of zinc to the brain is the blood-brain barrier.

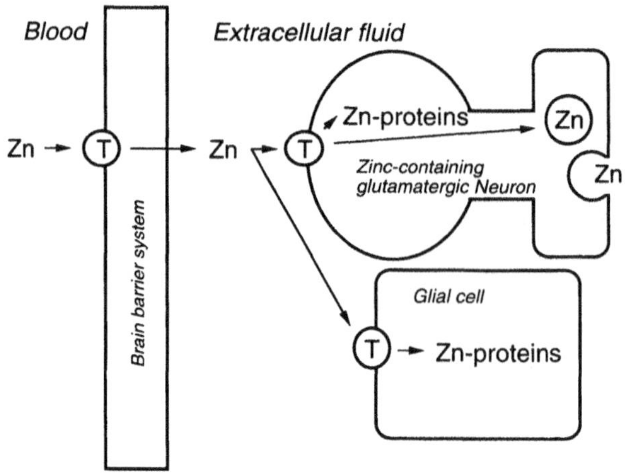

Fig. 1. Zinc transport into the brain. Zinc bound to histidine (and to albumin), which serve as the exchangeable zinc pool in the plasma, may be involved in zinc transport into the brain via transporters on the blood-brain and the blood-CSF barriers. Some transporters such as DMT1, ZIP2 and PHT1 are candidates for zinc transport. They might be also involved in zinc uptake in neurons and glial cells. A large portion of zinc functions as zinc metalloproteins. A portion of zinc is sequestered in the synaptic vesicles and functions as a neuromodulator. T; transporter.

The choroid plexus may participate in slow supply of some trace metals such as zinc and manganese to the brain (Takeda *et al.* 1994a, b, 2000a).

Zinc-binding affinity for ligands in the plasma is important for understanding the mechanism of zinc transport into the brain across the brain barrier system. Plasma zinc (approximately 15 μM) is partitioned between high molecular weight and low molecular weight fractions (Prasad & Oberleas 1970; Henkin 1979). The former is a protein-bound form (98%) and the latter is a low molecular weight ligand-bound form (1–2%) and ionic zinc, which is estimated to be as low as 10^{-9}–10^{-10} M (Magneson *et al.* 1987).

The largest component of exchangeable zinc in the plasma is albumin. A brain autoradiogram with ^{65}Zn in the Nagase analbuminemic rat, which has a genetic mutation affecting albumin mRNA processing and lacks plasma albumin, demonstrated that albumin is not essential for zinc transport into the brain (Takeda *et al.* 1997a). However, plasma albumin may participate in zinc transport as a large pool of exchangeable zinc in normal animals. Zinc is also known to bind to other plasma proteins such as transferrin and α_2–macroglobulin. Although zinc firmly binds to α_2-macroglobulin, its functional significance is unknown (Giroux & Henkin 1972).

The next largest components of exchangeable zinc in the plasma are complexes of amino acids, i.e., histidine and cysteine (Hallman *et al.* 1971; Harris & Keen 1989). Aiken *et al.* (1992) report that ^{65}Zn uptake in the brain as well as in other tissues, expressed relative to plasma ^{65}Zn level is enhanced by L-histidine infusion. Buxani-Rice *et al.* (1994) report that ^{65}Zn transport into the brain during a short cerebrovascular perfusion is enhanced by addition of 100 μM L-histidine. Brain distribution of ^{65}Zn-His is consistent with the data of the L-histidine infusion experiment (Takeda *et al.* 2000c). On the other hand, supplementation with L-histidine during dietary zinc repletion improves short-term memory in zinc-restricted young adult male mice (Keller *et al.* 2000). L-histidine is probably involved in zinc transport into the brain across the brain barrier system. A rat brain peptide/histidine transporter (PHT1) has been cloned (Yamashita *et al.* 1997). PHT1 mRNA is intensely expressed in the choroid plexus. However, it is obscure whether histidine-bound forms actually pass across the plasma membranes of the choroidal epithelial cells (and brain capillary endothelial cells). On the other hand, DMT1, a divalent metal transporter, is expressed in brain capillary endothelial cells and choroidal epithelial cells (Gunshin *et al.* 1997). Histidine might serve to transfer zinc to DMT1. There is also the possibility that other zinc transporters, e.g., hZIP (ZRT1, IRT1-like protein), are involved in zinc transport across the brain barrier system (Grotz *et al.* 1998; Gaither & Eide, 2000).

The mechanism of zinc secretion from brain capillary endothelial cells and choroidal epithelial cells to the brain extracellular fluid and the CSF, respectively, is unknown.

The half-time for elimination of ^{65}Zn from the rat brain is in the range of 16–43 days (Takeda *et al.* 1995). There are considerable differences in elimination of ^{65}Zn between the brain regions. Zinc might be eliminated from the arachnoid villi and/or arachnoid granulations via the CSF.

Uptake and release of zinc in neurons and glial cells

Four putative zinc transporters, known as ZnT-1 through ZnT-4, have been cloned (McMahon and Cousins 1998). ZnT-1 is ubiquitously expressed and associated with zinc efflux (Tsuda *et al.* 1997; McMahon & Cousins 1998). The mechanism by which zinc is taken up in neurons and glial cells is poorly understood (Figure 1). DMT1 is present in the hippocampal pyramidal and granule cells, cerebellar granule cells, the preoptic nucleus and pyramidal cells of the piriform cortex in high densities (Gunshin *et al.* 1997). This transporter appears to be involved in zinc uptake into neurons (Colvin *et al.* 2000).

CSF zinc is approximately 0.15 μM and zinc concentrations into the brain parenchyma cells are estimated to be approximately 150 μM, judging from the average total brain zinc concentration (Frederickson 1989). To study zinc uptake into the brain parenchyma cells via CSF, ^{65}Zn-His and $^{65}ZnCl_2$ are injected intracerebroventricularly to rats. The radioactivity from ^{65}Zn-His is distributed extensively in the brain compared to that from $^{65}ZnCl_2$ (Takeda *et al.* 2000b). PHT1 mRNA is widely expressed in the brain (Yamashita *et al.* 1997). Especially, the intense hybridization signals are found in the hippocampus, cerebellum and pontine nucleus. There is the possibility that PHT1 are involved in zinc uptake in neurons and glial cells. On the other hand, the finding that histidine decreases ^{65}Zn uptake in the synaptosomal preparation suggests that histidine does not participate in a carrier-mediated uptake by neuron terminals (Wensink *et al.* 1988). There might be differences in the mechanism of zinc uptake between neuron terminal and the cell body. Moreover, a unique mechanism of zinc uptake might exist in zinc-containing glutamatergic neuron terminals, where zinc is concentrated in the synaptic vesicles via ZnT-3, which transports cytosolic zinc into the synaptic vesicles (Palmiter *et al.* 1996; Wenzel *et al.* 1997; Cole *et al.* 1999).

Zinc taken up by neurons is transported anterogradely and retrogradely via the axonal transport system (Takeda *et al.* 1997b, 1998). In zinc-containing glutamatergic neurons, zinc is transported to synaptic vesicles (Figure 1). Zinc concentration in the vesicles in the giant boutons of hippocampal mossy fibers is estimated to be 300–350 μM (Frederickson *et al.* 1983), and is higher than in the cell body. Zinc sequestered in the synaptic vesicles is released with glutamate, and may modulate excitatory neurotransmission via glutamate (Spiridon *et al.* 1998; Traynelis *et al.* 1998; Vogt *et al.* 2000). The glutamate released into the synaptic clefts is primarily removed by transport into glial cells via a glutamate transporter (Zerangue & Kavanaugh 1996). In the case of zinc released into the synaptic cleft, a portion of the zinc may be taken up by postsynaptic neurons, in addition to presynaptic neurons (Takeda *et al.* 1997b). Especially, zinc uptake in postsynaptic neurons is observed after excessive excitation of zinc-containing glutamatergic neurons, followed by degeneration of postsynaptic neurons (Sloviter 1985; Choi *et al.* 1988; Tonder *et al.* 1990; Koh *et al.* 1996; Choi & Koh 1998). Zinc may be taken up in neurons by mechanisms via the voltage-gated calcium channel (Wang & Quastel 1990; Weiss *et al.* 1993), NMDA receptor (Koh & Choi, 1994; Koh *et al.* 1996) and calcium-permeable AMPA/kainate receptor (Yin & Weiss 1995; Sensi *et al.* 1997; Yin *et al.* 1998).

Brain zinc homeostasis

Zinc homeostasis in the brain is very complex. Metallothioneins, known as MT-I through MT-IV, are metal-binding proteins and may be involved in intracellular zinc homeostasis (Ebadi *et al.* 1995; Vallee 1995). However, the function of MTs in zinc homeostasis in the brain is poorly understood. MT-I and MT-II are ubiquitously expressed. In the brain, MT-I and MT-II in astrocytes are induced by heavy metals such as zinc and copper, glucocorticoid and interleukin-1 (Ebadi *et al.* 1995; Hidalgo & Carrasco 1998). MT-III is abundantly expressed in the brain and suppresses neurite formation (Uchida *et al.* 1991). Although the induction of MT-III is not noticeable in comparison with MT-I and MT-II, MT-III is highly expressed in zinc-containing glutamatergic neurons (Palmiter *et al.* 1992; Masters *et al.* 1994). This metalloprotein may be an important regulator of neuromodulatory zinc in the brain (Erickson *et al.* 1997). Uchida *et al.* (1991) and Tsuji *et al.* (1992) reported that MT-III is deficient in Alzheimer's disease brain. On the other hand, there are some reports that the regulation of MT-III is not impaired in Alzheimer's disease brain (Erickson *et al.* 1994; Amoureux *et al.* 1997). The implication of MT-III in the development of Alzheimer's disease is controversial.

Alteration of zinc homeostasis in the brain may be involved in neurological diseases such as Alzheimer's disease, Parkinson's disease, amyotrophic lateral sclerosis, in which oxidative stress has been implicated as a cause (Cuajungco & Lees, 1997a, b). Zinc can reduce oxidative stress by binding to thiol groups, decreasing their oxidation. MT is a good scavenger of free radicals, resulting in zinc release from MT by disulfides formed under oxidative stress (Sato & Bremner 1993; Maret 1994, 2000; Maret & Vallee 1998). Zinc is also a structural component of superoxide dismutase (SOD), and copper, zinc-SOD is a cytosolic antioxidant. Mutations in SOD cause one form of familiar amyotrophic lateral sclerosis (Rosen *et al.* 1993).

Alteration of brain zinc homeostasis by zinc deficiency

Zinc concentration in the extracellular fluid in the brain is estimated to be almost the same as that in the CSF (Hershey *et al.* 1983; Palm *et al.* 1986). There is a 100-fold difference in zinc concentration between the extracellular fluid and plasma. Dietary zinc deprivation causes a decrease in plasma zinc concentration (Prohaska *et al.* 1974), probably in the exchangeable zinc pool, resulting in a decrease of zinc in peripheral tissues such as the liver. On the other hand, several researchers failed to find any change in brain zinc concentration during dietary zinc deprivation. They demonstrated that zinc concentration is tightly regulated in the brain (Prohaska *et al.* 1974; O'Dell *et al.* 1976; Szerdahelyi *et al.* 1982; Wallwork *et al.* 1983; Prohaska 1987; Golub *et al.* 1986, 1995). However, inadequate dietary zinc supplies cause changes in behavior such as reduced activity and responsiveness (Shagal 1980; Dreosti 1983; Golub *et al.* 1995). In the case of dietary zinc deprivation in infancy, the improvement of learning behavior is not observed in zinc-deficient diet-treated rats (Takeda *et al.* 2000d). Especially, periods of rapid growth such as pregnancy and infancy are susceptible to dietary zinc deprivation (Favier 1992; Sandstead *et al.* 2000). Even in adult animals, learning behavior is impaired by dietary zinc deprivation (Takeda *et al.* 2000d). Thus, it is likely that dietary zinc deprivation affects zinc homeostasis in the brain.

Exogeneous zinc uptake in the brain under zinc deficiency is an index of endogeneous zinc status in the brain. Kasarskis (1984) reported a insufficient gain in brain weight and a increase of ^{65}Zn uptake in the brain of rats, which were fed a zinc-deficient diet for 45 days after wean, suggesting that brain development is suppressed by dietary zinc deprivation. Zinc concentration and ^{65}Zn uptake in the brain were also examined using adult rats fed a halfzinc-deficient diet for 12 weeks (Takeda *et al.* 2000d). Zinc concentrations in the brain, except for the hippocampal formation, did not decrease significantly in zinc-deficient rats, although zinc concentration in the liver of the zinc-deficient rats was approximately one of that of control rats (Figure 2). In an experiment of brain autoradiography with ^{65}Zn, ^{65}Zn concentration in any brain

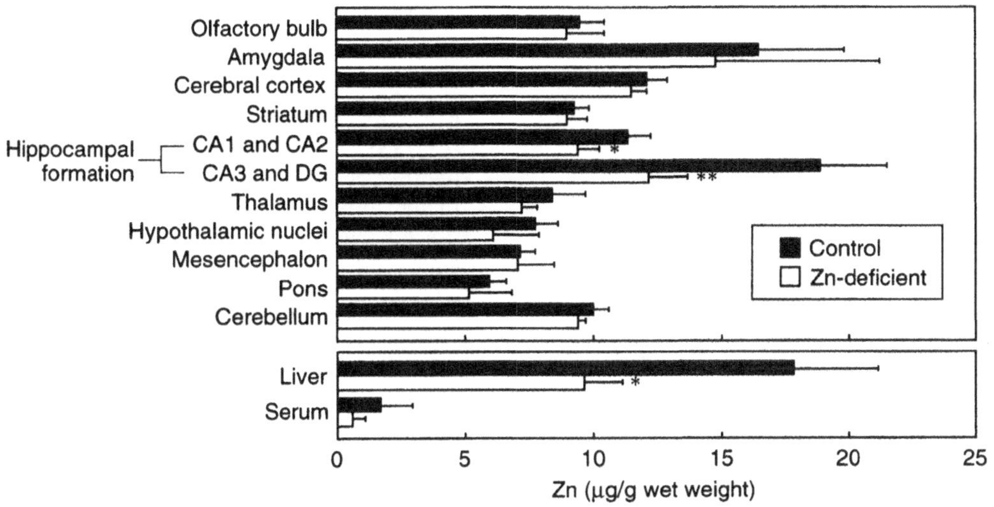

Fig. 2. Zinc concentrations in rats fed zinc-deficient diet. Rats were fed control or zinc-deficient diet for 12 weeks. Each value represents the mean ± SD ($n = 5$). Asterisks indicate significant difference ($^{*}P < 0.05$; $^{**}P < 0.01$; t-test) from control. In the brain, zinc concentration in the hippocampal formation was significantly decreased by dietary zinc deprivation, implying that dietary zinc deprivation affects vesicular zinc levels. [Reprinted with permission from Takeda *et al.* 2001.]

region of the zinc-deficient rats is significantly higher than in control rats 6 days after intravenous injection of $^{65}ZnCl_2$ (Takeda *et al.* 2001). The increase rate of ^{65}Zn concentration in the brain by the zinc deprivation is approximately 150%. Interestingly, the increase rate is similar to that in the liver. The increase rate of the serum is also approximately 150%. Zinc concentration in the brain, unlike in the liver, is not increased after parenteral injection of zinc (Ebadi 1986), indicating that zinc concentration in the brain is tightly regulated by the brain barrier system in the case of an increase of zinc in the serum. The ^{65}Zn supply to the brain of zinc-deficient rats reflects a requirement for brain functions. On the other hand, the half-time for elimination of ^{65}Zn from the rat brain is in the range of 16–43 days (Takeda *et al.* 1995). Thus, dietary zinc deprivation can cause a scarcity of zinc in the brain. Even in adult rats and probably in adult humans, adequate zinc supply is important for brain functions and also for prevention of neurological diseases.

Hippocampal functions under zinc deficiency

In the brain of zinc-deficient rats, the increase rate of ^{65}Zn concentration in the hippocampal CA3 and the dentate gyrus seems to be the lowest in the brain in spite of the significant decrease of zinc concentration in the hippocampal CA3 and the dentate gyrus (Takeda *et al.* 2001). Dietary zinc deprivation for 12 weeks in rats also demonstrated a 30% maximum reduction of zinc concentration in hippocampal mossy fibers (Wensink *et al.* 1987). The hippocampal formation, especially hippocampal mossy fibers, may be the most responsive to dietary zinc deprivation in the brain (Figure 2). The hippocampal mossy neuropil is a region with the highest density of zinc-containing neuron terminals (Frederickson & Danscher 1990). It is considered that vesicular zinc is responsive to dietary zinc deprivation. If zinc replenishment to the synaptic vesicles of zinc-containing glutamatergic neurons depends on a supply from the plasma zinc pool in addition to reutilization of zinc released into synaptic clefts, vesicular zinc could be decreased by dietary zinc deprivation (Dreosti *et al.* 1981; Lu *et al.* 2000).

Long-term potentiation (LTP), which is involved in learning and memory function, has been observed in zinc-containing glutamatergic neuron-rich areas such as the hippocampus and amygdala. Hesse (1979) demonstrated abnormal hippocampal mossy fiber synaptic responses during low-frequency stimulation in zinc-deficient rats. Vesicular zinc might be involved in the induction of LTP (Weiss *et al.* 1989) and dietary zinc deprivation might affect the development of LTP. Learning behavior is reversibly impaired by chelation of zinc in the synapses; Frederickson *et al.* (1990) demonstrated disturbance of hippocampal-dependent spatial-working memory function by diethyldithiocarbamate infusion into the hippocampus.

Amygdalar functions under zinc deficiency

Dietary zinc deprivation can decrease zinc concentration in the amygdala, in addition to the hippocampus (unpublished data), although zinc concentration in the amygdala is not decreased appreciably in Figure 2. Amygdalar functions seem to be affected by dietary zinc deprivation, which disturbs olfactory appreciation in humans and animals (Henkin *et al.* 1975; Mackay-Sim & Dreosti 1989). The amygdala is a region with high densities of zinc-containing neuron terminals and is a part of the olfactory cortices. Vesicular zinc-enriched neuron terminals also exist in the olfactory bulb (Jo *et al.* 2000). There is the possibility that dietary zinc deprivation decreases vesicular zinc in zinc-containing neurons in the olfactory areas and that its decrease is involved in olfactory dysfunction. When the amygdalae are perfused with diethyldithiocarbamate during behavioral test for odor recognition, the recognition of aversive odor is reversibly disturbed by the perfusion (Takeda *et al.* 1999b). Therefore, vesicular zinc is probably involved in amygdalar functions e.g., olfactory appreciation. The relationship between vesicular zinc and functions in the limbic system indicates that zinc homeostasis in zinc-containing glutamatergic synapses is important for the excitatory neurotransmission via glutamate (Takeda 2000). On the other hand, because olfactory epithelium (receptor cell) is differentiated from the basal cell, the regrowth of olfactory epithelium might be inhibited by zinc deficiency (Shigihara *et al.* 1987).

Zinc and epilepsy

Zinc has been reported to act either as an anticonvulsant (Williamson and Spencer, 1995) or a proconvulsant (Pei *et al.* 1983). Alteration of zinc homeostasis in the brain may be associated with the etiology and manifestation of epileptic seizures (Sterman *et al.* 1988). Elimination of zinc from the brain of El (epilepsy) mice is facilitated during induction of seizures (Takeda *et al.* 1999b). Seizure susceptibility of El mice is decreased by dietary zinc loading, while it is increased by dietary zinc deficiency (Fukahori & Itoh 1990). Zinc concentration in the hippocampal dentate area of seized El mice is significantly lower than that of control mice (Fukahori *et al.* 1988), suggesting that the decrease of the hippocampal zinc is involved in the pathophysiology of convulsive seizures in the El mice.

A selective loss of Timm's stain in the hippocampal mossy fiber was observed after electrical stimulation of the perforant path, which evoked hippocampal granule spikes and epileptiform discharges (Sloviter 1985). The loss of zinc stain with N-(6-methoxy-8-quinolyl)-para-toluenesulfonamide (TSQ) in the mossy fibers was also caused by administration of kainate, a seizure-inducing agent (Frederickson *et al.* 1988, 1989). Zinc homeostasis in zinc-containing glutamatergic synapses is altered by the excess excitation (Frederickson *et al.* 2000; Takeda 2000). This alteration may influence the degree and balance of inhibition-excitation (Xie & Smart 1991) and cause an increase in susceptibility to epileptic seizures (Takeda *et al.* 1999b). Actually, ZnT-3-null mice, which lack histochemically reactive zinc in synaptic vesicles, are more sensitive than control mice to seizures induced by kainate (Cole *et al.* 2000). Vesicular zinc may be important for the regulation of excitatory neurotransmission via glutamate.

Perspective on the future

Zinc homeostasis in the brain is closely related to brain functions. The alteration of zinc homeostasis in the brain seems to be a cause of brain dysfunctions and some neurological diseases. However, the mechanism of the alteration is poorly understood. The decrease of zinc metalloproteins affects gene expression and enzyme activity, followed by the alteration of various physiological responses. The decrease of vesicular zinc may affect the balance of inhibition-excitation in zinc-containing glutamatergic synapses. The clarification of neuronal and glial zinc homeostasis is important for understanding zinc functions in the brain.

References

Aiken SP, Horn NR, Saunders NR. 1992 Effect of histidine on tissue zinc distribution in rats. *BioMetals* **5**, 235–243.

Amoureux MC, Van Gool D, Herrero MT, Dom R, Colpaert FC, Pauwels PJ. 1997 Regulation of metallothionein-III (GIF) mRNA in the brain of patients with Alzheimer disease is not impaired. *Mol Chem Neuropathol* **32**, 101–121.

Assaf SY, Chung S-H. 1984 Release of endogeneous Zn^{2+} from brain tissue during activity. *Nature* **308**, 734–735.

Buxani-Rice S, Ueda F, Bradbury MWB. 1994 Transport of zinc-65 at the blood-brain barrier during short cerebrovascular perfusion in the rat: its enhancement by histidine. *J Neurochem* **62**, 665–672.

Choi DW, Koh JY. 1998 Zinc and brain injury. *Annu Rev Neurosci* **21**, 347–375.
Choi DW, Yokoyama M, Koh JY. 1988 Zinc neurotoxicity in cortical cell culture. *Neuroscience* **24**, 67–79.
Cole TB, Robbins CA, Wenzel HJ, Schwartzkroin PA, Palmiter RD. 2000 Seizures and neuronal damage in mice lacking vesicular zinc. *Epilepsy Res* **39**, 153–169.
Cole TB, Wenzel HJ, Kafer KE, Schwartzkroin PA, Palmiter RD. 1999 Elimination of zinc from synaptic vesicles in the intact mouse brain by disruption of the ZnT-3 gene. *Proc Natl Acad Sci USA* **96**, 1716–1721.
Coleman JE. 1992 Zinc proteins: enzymes, storage proteins, transcription factors, and replication proteins. *Annu Rev Biochem* **61**, 897–946.
Colvin RA, Davis N, Nipper RW, Carter PA. 2000 Zinc transport in the brain: routes of zinc influx and efflux in neurons. *J Nutr* **130**, 1484S–1487S.
Christensen MK, Frederickson CJ. 1998 Zinc-containing afferent projections to the rat corticomedial amygdaloid complex: a retrograde tracing study. *J Comp Neurol* **400**, 375–390.
Crawford IL, Connor JD. 1973 Localization and release of glutamic acid in relation to the hippocampal mossy fiber pathway. *Nature* **244**, 422–423.
Cuajungco MP, Lees GJ. 1997a Zinc metabolism in the brain: Relevance to human neurodegenerative disorders. *Neurobiol Dis* **4**, 137–169.
Cuajungco MP, Lees GJ. 1997b Zinc and Alzheimer's disease: is there a direct link? *Brain Res Rev* **23**, 219–236.
Dreosti IE. 1983 Zinc and the central nervous system. In: Dreosti IE, Smith RM, eds. *Neurobiology of the Trace Elements*. Vol. 1, Clifton: Humana: 135–162.
Dreosti IE, Manuel SJ, Buckley RA, Fraser FJ, Record IR. 1981 The effect of late prenatal and/or early postnatal zinc deficiency on the development and some biochemical aspects of the cerebellum and hippocampus in rats. *Life Sci* **28**, 2133–2141.
Ebadi M. 1986 Biochemical characterization of a metallothionein-like protein in rat brain. *Biol Trace Elem Res* **11**, 101–116.
Ebadi M, Iversen PL, Hao R, Cerutis DR, Rojas P, Happe HK, Murrin LC, Pfeiffer RF. 1995 Expression and regulation of brain metallothionein. *Neurochem Int* **27**, 1–22.
Erickson JC, Hollopeter G, Thomas SA, Froelick GJ, Palmiter RD. 1997 Disruption of the metallothionein-III gene in mice: Analysis of brain zinc, behavior, and neuron vulnerability to metals, aging, and seizures. *J Neurosci* **17**, 1271–1281.
Erickson JC, Sewell AK, Jensen LT, Winge DR, Ralmiter RD. 1994 Enhanced neurotrophic activity in Alzheimer's disease cortex is not associated with down-regulation of metallothionein-III (GIF). *Brain Res* **649**, 297–304.
Favier AE. 1992 Hormonal effects of zinc on growth in children. *Biol Trace Elem Res* **32**, 383–398.
Franklin PA, Pullen RGL, Hall GH. 1992 Blood-brain exchange routes and distribution of ^{65}Zn in rat brain. *Neurochem Res* **17**, 767–771.
Frederickson CJ. 1989 Neurobiology of zinc and zinc-containing neurons. *Int Rev Neurobiol* **31**, 145–238.
Frederickson CJ, Hernandez MD, McGinty JF. 1989 Translocation of zinc may contribute to seizure-induced death of neurons. *Brain Res* **480**, 317–321.
Frederickson CJ, Danscher G. 1990 Zinc-containing neurons in hippocampus and related CNS structures. *Prog Brain Res* **83**, 71–84.
Frederickson RE, Frederickson CJ, Danscher G. 1990 In situ binding of bouton zinc reversibly disrupts performance on a spatial memory task. *Behav Brain Res* **38**, 25–33.
Frederickson CJ, Hernandez MD, Goik SA, Morton JD, McGinty JF. 1988 Loss of zinc staining from hippocampal mossy fibers during kainic acid induced seizures: a histofluorescence study. *Brain Res* **446**, 383–386.
Frederickson CJ, Klitenick MA, Manton WI, Kirkpatrick JB. 1983 Cytoarchitectonic distribution of zinc in the hippocampus of man and rat. *Brain Res* **273**, 335–339.
Frederickson CJ, Moncrieff DW. 1994 Zinc-containing neurons. *Biol Signals* **3**, 127–139.
Frederickson CJ, Suh SW, Silva D, Thompson RB. 2000 Importance of zinc in the central nervous system: the zinc-containing neuron. *J Nutr* **130**, 1471S–1483S.
Fukahori M, Itoh M. 1990 Effects of dietary zinc status on seizure susceptibility and hippocampal zinc content in the El (epilepsy) mouse. *Brain Res* **529**, 16–22.
Fukahori M, Itoh M, Oomagari K, Kawasaki H. 1988 Zinc content in discrete hippocampal and amygdaloid areas of the epilepsy (El) mouse and normal mice. *Brain Res* **455**, 381–384.
Gaither LA, Eide DJ. 2000 Functional expression of the human hZIP2 zinc transporter. *J Biol Chem* **275**, 5560–5564.
Giroux EL, Henkin RI. 1972 Competition for zinc among serum albumin and amino acids. *Biochim Biophys Acta* **273**, 64–72.
Golub MS, Keen CL, Gershwin ME, Vijayan VK. 1986 Growth, development and brain zinc levels in mice marginally or severely deprived of zinc during postembryonic brain development. *Nutr Behav* **3**, 169–180.
Golub MS, Keen CL, Gershwin ME, Hendrickx AG. 1995 Developmental zinc deficiency and behavior. *J Nutr* **125**, 2263–2271.
Grotz N, Fox T, Connolly E, Park W, Guerinot ML, Eide D. 1998 Identification of a family of zinc transporter genes from Arabidopsis that respond to zinc deficiency. *Proc Natl Acad Sci USA* **95**, 7220–7224.
Gunshin H, Mackenzie B, Berger UV, Gunshin Y, Romero MF, Boron WF, Nussberger S, Gollan JL, Hediger MA. 1997 Cloning and characterization of a mammalian protein-coupled metal-ion transporter. *Nature* **388**, 482–488.
Hallman PS, Perrin DD, Watt AE. 1971 The computed distribution of copper (II) and zinc (II) ions among seventeen amino acids in human blood plasma. *Biochem J* **121**, 549–555.
Harris WR, Keen CL. 1989 Calculations of the distribution of zinc in a computer model of human serum. *J Nutr* **119**, 1677–1682.
Harrison NL, Gibbons SJ. 1994 Zn^{2+}: an endogenous modulator of ligand- and voltage-gated ion channels. *Neuropharmacology* **33**, 935–952.
Haug F-MS. 1973 Heavy metals in the brain. A light microscope study of the rat with Timms' sulphide silver method. Methodological considerations and cytological and regional staining patterns. *Adv Anat Embryol Cell Biol* **47**, 1–71.
Henkin RI. 1979 Zinc. Baltimore: Univ. Park Press.
Henkin RI, Patten BM, Re PK, Bronzert DA. 1975 A syndrome of acute zinc loss. *Arch Neurol* **32**, 745–751.
Hershey CO, Hershey LA, Varnes A, Vibhakar SD, Lavin P, Strain WH. 1983 Cerebrospinal fluid trace element content in dementia: clinical, radiologic, and pathologic correlations. *Neurology* **33**, 1350–1353.
Hesse GW. 1979 Chronic zinc deficiency alters neuronal function of hippocampal mossy fibers. *Science* **205**, 1005–1007.
Hidalgo J, Carrasco J. 1998 Regulation of the synthesis of brain metallothionein. *J Neurotoxicol* **19**, 661–666.
Howell GA, Frederickson CJ. 1989 A retrograde transport method for mapping zinc-containing fiber systems in the brain. *Brain Res* **515**, 277–286.

Howell GA, Welch MG, Frederickson CJ. 1984 Stimulation-induced uptake and release of zinc in hippocampal slices. *Nature* **308**, 736–738.

Huang EP. 1997 Metal ions and synaptic transmission: think zinc. *Proc Natl Acad Sci USA* **94**, 13386–13387.

Jo SM, Won MH, Cole TB, Jensen MS, Palmiter RD, Danscher G. 2000 Zinc-enriched (ZEN) terminals in mouse olfactory bulb. *Brain Res* **865**, 227–236.

Kasarskis EJ. 1984 Zinc metabolism in normal and zinc-deficient rat brain. *Exp Neurol* **85**, 114–127.

Keller KA, Chu Y, Grider A, Coffield JA. 2000 Supplementation with L-histidine during dietary zinc repletion improves short-term memory in zinc-restricted young adult male rats. *J Nutr* **130**, 1633–1640.

Klaassen CD. 1999 *Metallothionein IV*. Basel: Birkhauser.

Koh JY, Choi DW. 1994 Zinc toxicity on cultured cortical neurons: Involvement of N-methyl-D-aspartate receptors. *Neuroscience* **60**, 1049–1057.

Koh JY, Suh SW, Gwag BJ, He YY, Hsu CY, Choi DW. 1996 The role of zinc in selective neuronal death after transient global cerebral ischemia. *Science* **272**, 1013–1016.

Lee JY, Cole TB, Palmiter RD, Koh JY. 2000 Accumulation of zinc in degenerating hippocampal neurons of ZnT3-null mice after seizures: evidence against synaptic vesicle origin. *J Neurosci* **20**, RC79.

Lu YM, Taverna FA, Tu R, Ackerley CA, Wang YT, Roder J. 2000 Endogenous Zn(2+) is required for the induction of long-term potentiation at rat hippocampal mossy fiber-CA3 synapses. *Synapse* **38**, 187–197.

Mackay-Sim A, Dreosti IE. 1989 Olfactory function in zinc-deficient adult mice. *Exp Brain Res* **76**, 207–212.

Magneson GR, Puvathingal JM, Ray Jr, WJ. 1987 The concentrations of free Mg^{2+} and free Zn^{2+} in equine blood plasma. *J Biol Chem* **262**, 11140–11148.

Maret W. 1994 Oxidative metal release from metallothionein via zinc-thiol/disulfide interchange. *Proc Natl Acad Sci USA* **91**, 237–241.

Maret W. 2000 The function of zinc metallothionein: a link between cellular zinc and redox state. *J Nutr* **130**, 1455S–1458S.

Maret W, Vallee BL. 1998 Thiolate ligands in metallothionein confer redox activity on zinc clusters. *Proc Natl Acad Sci USA* **95**, 3478–3482.

Markesbery WR, Ehmann WD, Alauddin M, Hossain TIM. 1984 Brain trace element concentrations in aging. *Neurobiol Aging* **5**, 19–28.

Masters BA, Quaife CJ, Erickson JC, Kelly EJ, Froelick GJ, Zambrowicz BP, Brinster RL, Palmiter RD. 1994 Metallothionein III is expressed in neurons that sequester zinc in synaptic vesicles. *J Neurosci* **14**, 5844–5857.

McMahon RJ, Cousins RJ. 1998 Mammalian zinc transporters. *J Nutr* **128**, 667–670.

McMahon RJ, Cousins RJ. 1998 Regulation of the zinc transporter ZnT-1 by dietary zinc. *Proc Natl Acad Sci USA* **95**, 4841–4846.

O'Dell BL, Reeves PG, Morgan RF. 1976 Interrelationships of tissue copper and zinc concentrations in rats nutritionally deficient in one or the other of these elements. In: Hemphill, DD, ed. *Trace substances in environmental health*. Columbia MO: University of Missouri: 411–521.

Palm R, Strand T, Hallmans G. 1986 Zinc, total protein, and albumin in CSF of patients with cerebrovascular diseases. *Acta Neurol Scand* **74**, 308–313.

Palmiter RD, Cole TB, Quaife CJ, Findley SD. 1996 ZnT-3, a putative transporter of zinc into synaptic vesicles. *Proc Natl Acad Sci USA* **93**, 14934–14939.

Palmiter RD, Findley SD, Whitmore TE, Durnam DM. 1992 MT-III, a brain-specific member of the metallothionein gene family. *Proc Natl Acad Sci USA* **89**, 6333–6337.

Pei Y, Zhao D, Huang J, Cao L. 1983 Zinc-induced seizures: a new experimental model of epilepsy. *Epilepsia* **24**, 169–176.

Perez-Clausell J, Danscher G. 1985 Intravesicular localization of zinc in rat telencephalic boutons. A histochemical study. *Brain Res* **337**, 91–98.

Peters DP. 1979 Effects of prenatal nutritional deficiency on affiliation and aggression in rats. *Physiol Behav* **20**, 359–362.

Prasad AS, Oberleas, D. 1970 Binding of zinc to amino acids and serum proteins *in vitro*. *J Lab Clin Med* **76**, 416–425.

Prohaska JR. 1987 Functions of trace elements in brain metabolism. *Physiol Rev* **67**, 858–901.

Prohaska JR, Luecke RW, Jasinski R. 1974 Effect of zinc deficiency from day 18 of gestation and/or during lactation on the development of some rat brain enzymes. *J Nutr* **104**, 1525–1531.

Pullen RGL, Franklin PA, Hall GH. 1990 65Zinc uptake from blood into brain and other tissues in the rat. *Neurochem Res* **15**, 1003–1008.

Pullen RGL, Franklin PA, Hall GH. 1991 ^{65}Zn uptake from blood into brain in the rat. *J Neurochem* **56**, 485–489.

Rosen DR, Siddique T, Patterson D, Figlewicz DA, Sapp P, Hentati A, Donaldson D, Goto J, O'Regan JP, Deng HX, Rahmani Z, Krizus A, Mckenna-Yasek D, Cayabyab A, Gaston SM, Berger R, Tanzi RE, Halperin JJ, Herzfeldt B, Van den Bergh R, Hung WY, Bird T, Deng G, Mulder DW, Smyth C, Laing NG, Soriano E, Pericak-Vance MA, Haines J, Rouleau GA, Gusella JS, Horvitz HR, Brown RHJ. 1993 Mutations in Cu/Zn superoxide dismutase gene are associated with familial amyotrophic lateral sclerosis. *Nature* **362**, 59–62.

Sandstead HH, Frederickson CJ, Penland JG. 2000 History of zinc as related to brain function. *J Nutr* **130**, 496S–502S.

Sato M, Bremner I. 1993 Oxygen free radicals and metallothionein. *Free Radicals Biol Med* **14**, 325–337.

Sawashita J, Takeda A, Okada S. 1997 Change of zinc distribution in rat brain with increasing age. *Dev Brain Res* **102**, 295–298.

Sensi SL, Canzoniero LMT, Yu SP, Ying HS, Koh JY, Kerchner GA, Choi DW. 1997 Measurement of intracellular free zinc in living cortical neurons: Routes of entry. *J Neurosci* **15**, 9554–9564.

Shagal A. 1980 Functions of the hippocampal system. *Trends Neurosci* **3**, 116–119.

Shigihara S, Ikui A, Shigihara J, Tomita H, Okano M. 1987 Electron microscopy of the olfactory epithelium in zinc-deficient rats. *Ann NY Acad Sci* **510**, 606–609.

Sloviter R. 1985 A selective loss of hippocampal mossy fiber Timm stain accompanies granule cell seizure activity induced by perforant path stimulation. *Brain Res* **330**, 150–153.

Smart TG, Xie X, Krishek BJ. 1994 Modulation of inhibitory and excitatory amino acid receptor ion channels by zinc. *Prog Neurobiol* **42**, 393–441.

Spiridon M, Kamm D, Billups B, Mobbs P, Attwell D. 1998 Modulation by zinc of the glutamate transporters in glial cells and cones isolated from the tiger salamander retina. *J Physiol* **506**, 363–376.

Sterman MB, Shouse MN, Fairchild MD. 1988 Zinc and seizure mechanisms. In: Morley JE, Sterman MB, Walsh JH, eds. *Nutritional modulation of neural function*. San Diego: Academic Press: 307–319.

Szerdahelyi P, Kozma M, Ferke A. 1982 Zinc-deficiency-induced trace element concentration and localization changes in the central nervous system of albino rat during postnatal development. *Acta Histochem* **70**, 173–182.

Takeda A. 2000 Movement of zinc and its functional significance in the brain. *Brain Res Rev* **34**, 137–148.

Takeda A, Akiyama T, Sawashita J, Okada S. 1994a Brain uptake of trace metals, zinc and manganese, in rats. *Brain Res* **640**, 341–344.

Takeda A, Ishiwatari S, Okada S. 2000a Influence of transferrin on manganese uptake in rat brain. *J Neurosci Res* **59**, 542–552.

Takeda A, Hanajima T, Ijiro H, Ishige A, Iizuka S, Okada S, Oku N. 1999a Release of zinc from the brain of El (epilepsy) mice during seizure induction. *Brain Res* **828**, 174–178.

Takeda A, Kawai M, Okada S. 1997a Zinc distribution in the brain of Nagase analbuminemic rat and enlargement of the ventricular system. *Brain Res* **769**, 193–195.

Takeda A, Kodama Y, Ohnuma M, Okada S. 1998 Zinc transport from the striatum and substantia nigra. *Brain Res Bull* **47**, 103–106.

Takeda A, Minami A, Takefuta S, Tochigi M, Oku N. 2001 Zinc homeostasis in the brain of adult rats fed zinc-deficient diet. *J Neurosci Res* **63**, 447–452.

Takeda A, Ohnuma M, Sawashita J, Okada S. 1997b Zinc transport in the rat olfactory system. *Neurosci Lett* **225**, 69–71.

Takeda A, Sawashita J, Okada S. 1994b Localization in rat brain of the trace metals, zinc and manganese, after intracerebroventricular injection. *Brain Res* **658**, 252–254.

Takeda A, Sawashita J, Okada S. 1995 Biological half-lives of zinc and manganese in rat brain. *Brain Res* **695**, 53–58.

Takeda A, Sawashita J, Takefuta S, Ohnuma M, Okada S. 1999b Role of zinc released by stimulation in rat amygdala. *J Neurosci Res* **57**, 405–410.

Takeda A, Suzuki M, Okada S, Oku N. 2000b ^{65}Zn localization in rat brain after intracerebroventricular injection of ^{65}Zn-histidine. *Brain Res* **863**, 241–244.

Takeda A, Suzuki M, Okada S, Oku N. 2000c Influence of histidine on zinc transport into rat brain. *J Health Sci* **46**, 209–213.

Takeda A, Takefuta S, Okada S, Oku N. 2000d Relationship between brain zinc and transient learning impairment of adult rats fed zinc-deficient diet. *Brain Res* **859**, 352–357.

Tonder N, Johansen FF, Frederickson CJ, Zimmer J, Diemer NH. 1990 Possible role of zinc in the selective degeneration of dentate hilar neurons after cerebral ischemia in the adult rat. *Neurosci Lett* **109**, 247–252.

Traynelis SF, Burgess MF, Zheng F, Lyuboslavsky P, Powers JL. 1998 Control of voltage-independent zinc inhibition of NMDA receptors by the NR1 subunit. *J Neurosci* **18**, 6163–6175.

Tsuda M, Imaizumi K, Katayama T, Kitagawa K, Wanaka A, Tohyama M, Takagi T. 1997 Expression of zinc transporter gene, ZnT-1, is induced after transient forebrain ischemia in the gerbil. *J Neurosci* **17**, 6678–6684.

Tsuji S, Kobayashi H, Uchida Y, Ihara Y, Miyatake T. 1992 Molecular cloning of human growth inhibitory factor cDNA and its down-regulation in Alzheimer's disease. *EMBO J* **11**, 4843–4850.

Uchida Y, Takiko K, Titani K, Ihara Y, Tomonaga M. 1991 The growth inhibitory Factor that is deficient in the Alzheimer's disease brain is a 68 amino acid metallothionein-like protein. *Neuron* **7**, 337–347.

Vallee BL. 1995 The function of metallothionein. *Neurochem Int* **27**, 23–33.

Vallee BL, Auld DS. 1992 Active zinc binding sites of zinc metalloenzymes. In: Birkedal-Hansen H, Werb Z, Welgus H, Wart HV, eds. *Matrix Metalloproteinases and Inhibitors*. Matrix Supplement 1. Stuttgart: Fischer: 5–19.

Vallee BL, Falchuk KH. 1981 Zinc and gene expression. *Philos. Trans R Soc Lond B Biol Sci* **294**, 185–197.

Vallee BL, Falchuk KH. 1993 The biological basis of zinc physiology. *Physiol Rev* **73**, 79–118.

Vogt K, Mellor J, Tong G, Nicoll R. 2000 The actions of synaptically released zinc at hippocampal mossy fiber synapses. *Neuron* **26**, 187–196.

Wallwork JC, Milne DB, Sims RL, Sandstead HH. 1983 Severe zinc deficiency: Effects on the distribution of nine elements (potassium, phosphorus, sodium, magnesium, calcium, iron, zinc, copper and manganese) in regions of the rat brain. *J Nutr* **113**, 1895–1905.

Wang Y-X, Quastel DMJ. 1990 Multiple actions of zinc on transmitter release at mouse end-plates. *Eur J Physiol* **415**, 582–587.

Weiss JH, Hartley DM, Koh JY, Choi DW. 1993 AMPA receptor activation potentiates zinc neurotoxicity. *Neuron* **10**, 43–49.

Weiss JH, Koh J, Christine CW, Choi DW. 1989 Zinc and LTP. *Nature* **338**, 212.

Wensink J, Lenglet WJM, Vis RD, Van den Hamer CJA. 1987 The effect of dietary zinc deficiency on the mossy fiber zinc content of the rat hippocampus. *Histochemistry* **87**, 65–69.

Wensink J, Molenaar AJ, Woroniecka UD, Van den Hamer CJA. 1988 Zinc uptake into synaptosomes. *J Neurochem* **50**, 782–789.

Wenzel HJ, Cole TB, Born DD, Schwartzkroin PA, Palmiter RD. 1997 Ultrastructural localization of zinc transporter-3 (ZnT-3) to synaptic vesicle membranes within mossy fiber boutons in the hippocampus of mouse and monkey. *Proc Natl Acad Sci USA* **94**, 12676–12681.

Williamson A, Spencer D. 1995 Zinc reduces dentate granule cell hyperexcitability in epileptic humans. *NeuroReport* **6**, 1562–1564.

Xie X, Smart TG. 1991 A physiological role for endogeneous zinc in rat hippocampal synaptic neurotransmission. *Nature* **349**, 521–524.

Yamashita T, Shimada S, Guo W, Sato K, Kohmura E, Hayakawa T, Takagi T, Tohyama M. 1997 Cloning and functional expression of a brain peptide/histidine transporter. *J Biol Chem* **272**, 10205–10211.

Yin HZ, Weiss JH. 1995 Zn^{2+} permeates Ca^{2+} permeable AMPA/kainate channels and triggers selective neural injury. *NeuroReport* **6**, 2553–2556.

Yin HZ, Ha DH, Carriedo SG, Weiss JH. 1998 Kainate-stimulated Zn^{2+} uptake labels cortical neurons with Ca^{2+}-permeable AMPA/kainate channels. *Brain Res* **781**, 45–56.

Zerangue N, Kavanaugh MP. 1996 Flux coupling in a neuronal glutamate transporter. *Nature* **383**, 634–637.

BioMetals 14: 353–366, 2001.

Review

Synaptically released zinc: Physiological functions and pathological effects

Christopher J. Frederickson[1,*] & Ashley I. Bush[2]
[1]*NeuroBioTex, Inc., Biomedical Engineering and Anatomy and Neuroscience, The University of Texas Medical Branch, Galveston, Texas, USA;* [2]*Laboratory for Oxidation Biology, Harvard Medical School, Boston, Massachusetts, USA; *Author for correspondence (Tel: 409-762-0678; Fax: 866-422-4403; E-mail: cjfrederickson@hotmail.com)*

Received 17 April 2001; accepted 8 May 2001

Abstract

In addition to its familiar role as a component of metalloproteins, zinc is also sequestered in the presynaptic vesicles of a specialized type of neurons called 'zinc-containing' neurons. Here we review the physiological and pathological effects of the release of zinc from these zinc-containing synaptic terminals. The best-established physiological role of synaptically released zinc is the tonic modulation of brain excitability through modulation of amino acid receptors; prominent pathological effects include acceleration of plaque deposition in Alzheimer's disease and exacerbation of excitotoxic neuron injury. Synaptically released zinc functions as a conventional synaptic neurotransmitter or neuromodulator, being released into the cleft, then recycled into the presynaptic terminal. Beyond this, zinc also has the highly unconventional property that it passes into postsynaptic neurons during synaptic events, functioning analogously to calcium in this regard, as a transmembrane neural signal. To stimulate comparisons of zinc signals with calcium signals, we have compiled a list of the important parameters of calcium signals and zinc signals. More speculatively, we hypothesize that zinc signals may loosely mimic phosphate 'signals' in the sense that signal zinc ions may commonly bind to proteins in a lasting manner (i.e., 'zincylating' the proteins) with consequential changes in protein structure and function.

Overview

Although it was Maske (1955) who first discovered 'stainable' (i.e., weakly bound) zinc in the brain (Figure 1), it was actually Turner McLardy who introduced the notion of synaptic zinc. For it was he who realized in the late 1950's that the swath of bright red zinc-dithizonate staining in the hippocampal formation of the brain was exactly coextensive with the peculiar 'mossy' axons that connect the dentate gyrus to ammon's horn (see McLardy 1970). The electronmicrographs of Finn-Mogens Haug (1967) later showed that this peculiar, 'stainable', pool of zinc was located exactly and exclusively in the presynaptic vesicle regions of synaptic terminals (Figure 2).

Two developments since have carried the concept of 'synaptic zinc signals' from neuroscience heresy to dogma (Baranano *et al.* 2001; Weiss & Sensi 2000; Weiss *et al.* 2000). First, Danscher and colleagues have adduced incontrovertible evidence that the stainable zinc in the brain is virtually without exception located in the synaptic vesicles of presynaptic boutons, a finding which has allowed the histochemical methods to be used to study the synaptic pool (Danscher 1996; Franco-Pons *et al.* 2000). Second, Palmiter and his co-workers have identified a specific zinc-transporter protein (ZnT-3) that pumps zinc into vesicles and is co-localized with vesicular zinc, on the membranes of synaptic vesicles in the Golgi apparatus, axons, and presynaptic boutons of zinc-containing CNS neurons (Palmiter *et al.* 1996). Because the ZnT-3 knockouts have no stainable zinc in their presynaptic vesicles (Cole *et al.* 1999), there is essentially no remaining doubt that the stainable zinc in synaptic terminals and the ZnT-3-dependent, presynaptic vesicular zinc pool are one and the same.

In the brain, the only neurons that have vesicular zinc (i.e., stainable zinc) are glutamatergic (Beaulieu

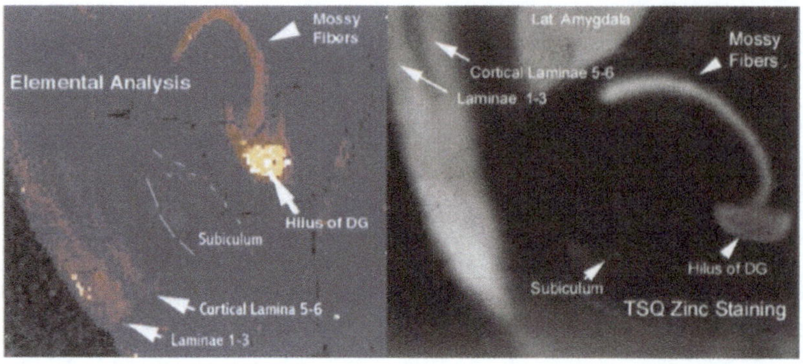

Fig. 1. The left image shows synchrotron-radiation-induced x-ray fluorescence of a rat temporal lobe (horizontal section) (Courtesy of Dr Jane Flinn. Morvan et al. 2000), and the right shows a similar section (somewhat more ventral) stained for free zinc by TSQ. The pseudo color on the left shows total elemental zinc, and the peak intensity (yellow-white) corresponds to approximately 400 ppm (dry weight), with the red (CA3 mossy fibers) corresponding to about 200 ppm. These values for total zinc are in reasonable agreement with prior averages obtained by micro dissection (Frederickson et al. 1983), with the present ($\sim 2 \times$ higher) values presumably more accurate reflections of true peak levels. Background total zinc is about 60 ppm. DG, dentate gyrus; TSQ, toluene-sulfonamide quinoline zinc staining method (Frederickson et al. 1987a).

et al. 1992) (Figure 3). Not all glutamatergic neurons are of this 'zinc-containing' flavor, but all zinc-containing neurons are glutamatergic (Frederickson 1989; Frederickson *et al.* 2000 for review). In the spinal cord of the lamprey (Birinyi *et al.* 2001) and the mouse (Danscher, personal communication), zinc is co-localized with glycine and/or GABA, and spinal vesicular zinc may have different roles than cerebral Zn^{2+} (Kovacks & Larson 1997).

The parcellation of cerebral glutamatergic neurons into the zinc-containing and the non-zinc-containing variety is definitely non-random. As a general rule, the large-neuron, long-axon systems of the brain are non-zinc-containing. Specific examples include all of the first-order sensory fibers of the cranial nerves, most of the thalamocortical fibers of those same, ascending sensory pathways, and essentially all of the long-fiber pathways descending, for example, from the cerebral cortex to the brain stem or spinal cord, or descending pathways from the diencephalon or mesencephalon (reviews in Frederickson *et al.* 2000; Franco-Pons *et al.* 2000; Frederickson & Moncrieff 1994).

Another view of the segregation of zinc-containing versus non-zinc-containing glutamatergic neurons is that almost all of the zinc-containing neuronal somata that have been identified in the brain have been found in either cortical (including allo and isocortex) or amygdalar brain regions. But for a few exceptions (granule neurons in the dorsal cochlear nucleus (Rubio & Juiz 1998), scattered medial thalamic neurons (Long & Frederickson 1994)) the overwhelming preponderance of the zinc-containing neurons of origin are cerebrocortical or amygdalar (Casanovas-Aguila *et al.* 1998; Christensen & Frederickson 1998).

There are some thought provoking changes in the zinc-containing innervation during early brain development. For example, in the hippocampal formation of the newborn rat pup, transient columnar 'patches' of zinc-containing innervation are conspicuous in the CA1-CA3 fields. These patches disappear quickly thereafter, as the entire blank spaces in the neuropil fill in with a quite uniform staining (Frederickson *et al.* 1991). Another example occurs in the striatum, where patches that are destined to become neurochemically distinct from the surrounding 'matrix' of the striatum are first identifiable in the zinc staining, which distinctly labels the patches even on post natal day 1, before the other neurochemical and anatomical markers of the patches have arrived (Vincent & Semba 1989). A third example is found in the lateral geniculate nucleus of the albino rat, where there are transient patches of zinc-containing axonal boutons that correspond to the ingrowing uncrossed retino-geniculate pathway (Figure 4). This last case is even more impressive than the first two because the zinc-containing boutons lose their zinc later in development, leaving the adult LGN without any zinc-containing boutons at all.

In considering these cases one is tempted to speculate that the zinc-containing boutons serve some early pioneer or induction function, laying down a skeletal substrate for later development. Experiential effects on these developmental processes have been identified (Land & Akhtar 1999; see also Dyck & Cynander

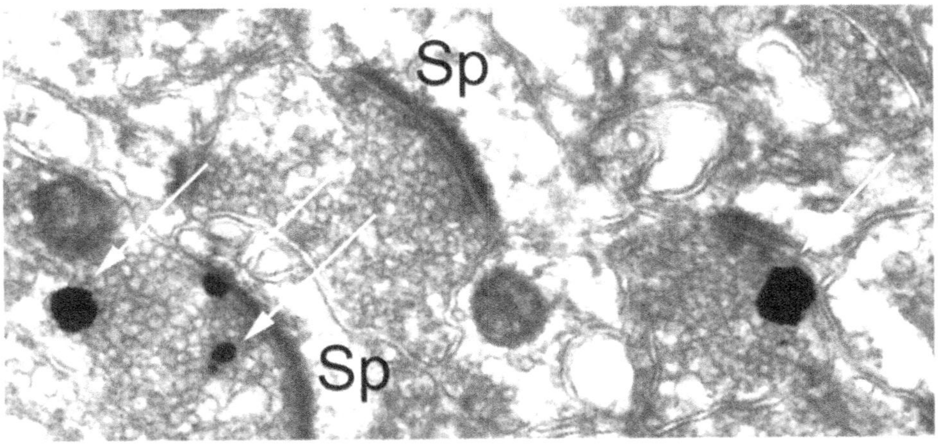

Fig. 2. Timm-Danscher method for tissue metal (which precipitates Zn^{2+} as ZnS, then coats the ZnS with silver) shows silver grains (arrows) amidst the vesicles of Type I presynaptic boutons, contacting spines (Sp). Image from the CA1 field of the rat hippocampus.

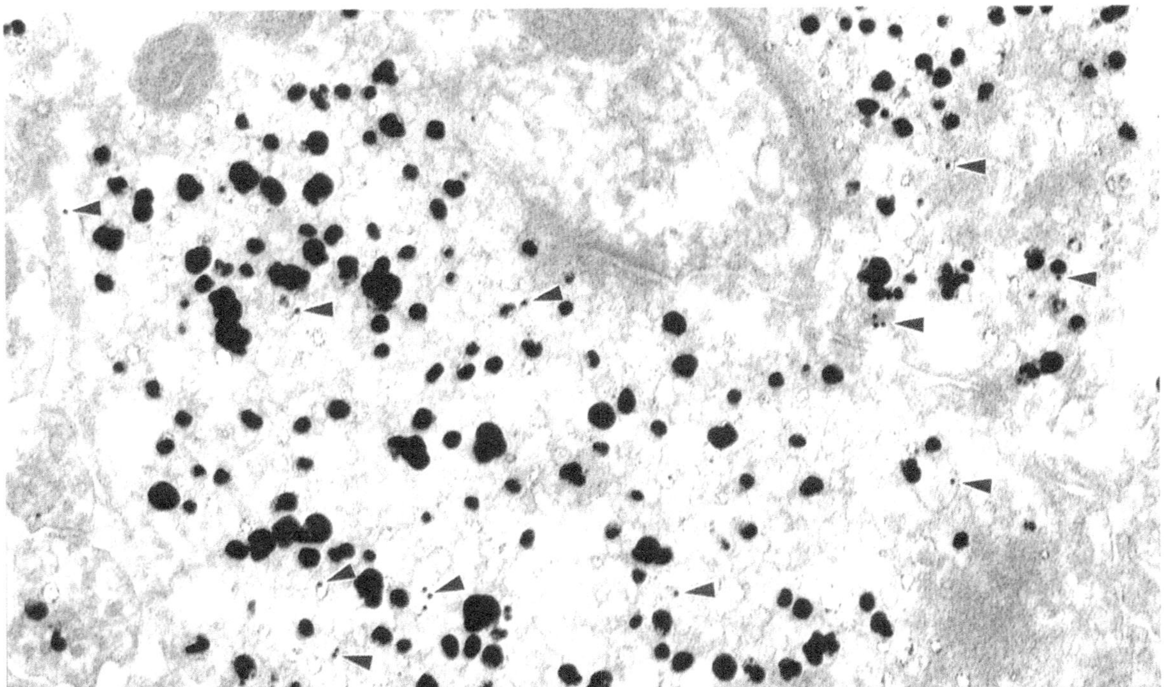

Fig. 3. Timm-Danscher silver staining for zinc shows silver grains amidst vesicles of a mossy-bouton in the rat hippocampus. Colloidal gold dots (in circles) indicate glutamate immunostaining, which co-localizes one-for-one with Timm's in the mouse hippocampus (Courtesy R. Palmiter) ~ 40,000 X.

1993). On the other hand, knockout mice congenitally lacking zinc in their vesicles develop with at least superficially normal brain organization and behavior (Cole *et al.* 2000), so whatever the developmental function of vesicular zinc might be, it would appear that developmental plasticity can compensate.

There are three kinds of Zn^{2+} signals

At the time of this writing, we can identify three distinct classes of zinc signals which we will call: Zn^{2+}-SYN, Zn^{2+}-TRANS , and Zn^{2+}-INT.

Zn^{2+}-SYN

The first of these signals, 'synaptic Zn^{2+}' (Zn^{2+} -SYN) is a conventional, transmitter-like, synaptic signal, between presynaptic bouton and postsynaptic spine of dendrite. The zinc ions for this signal are stored in presynaptic vesicles at concentrations probably reaching low millimolar levels (see calculations in Frederickson 1989; Frederickson *et al.* 2000). Upon the arrival of an action potential at the bouton, the calcium- and impulse-frequency dependent exocytosis of these zinc-filled vesicles produces a rapid 'puff' of essentially free, ionic Zn^{2+} in the extracellular fluid surrounding the boutons. These Zn^{2+} transients, or 'puffs' have rise-times of a few milliseconds, and can reach apparent concentrations of 10–30 μM Zn^{2+} in the extracellular fluid (Li *et al.* 2000, 2001; Thompson *et al.* 2001; Vogt *et al.* 2000). Zn^{2+} 'puffs' or 'flashes' comparable to those seen during synaptic release can also be seen during exocytosis of a similar zinc-rich secretory granule, the insulin-containing beta pancreatic cell (Qian *et al.* 2000), and could presumably be seen during exocytosis of Mast cell granules, salivary cell granules, and that of the dozen-odd other cell types that have zinc-filled secretory granules (Frederickson 1989; Frederickson *et al.* 1987b).

The Zn^{2+}-SYN signal reaches multiple zinc-modulated extracellular zinc-recognition sites on membrane-spanning receptors, pumps, and channels of postsynaptic neurons. The most thoroughly studied of these are the zinc-modulated amino acid receptors, with both the glutamate receptor families and the GABA receptors having diverse and potent responses to the Zn^{2+} ion signal (Smart *et al.* 1994). One of the intriguing examples of the complex and facile nature of these interactions is found at the mossy-fiber to CA3 pyramidal neuron, where GABA, zinc and glutamate are all released, with the possibility arising of differential zinc modulation of both receptors by zinc at the single synapse (Walker *et al.* 2001). The time-varying expression of receptor-ionophore splice variants which have different zinc sensitivity give the nervous system another degree of freedom with respect to the impact of zinc modulation (Chen *et al.* 1997). Ectopic sprouting of zinc-releasing fibers is another potential mechanism of zinc-mediated plasticity (Coulter 2000).

Despite the complex and protean effects of zinc upon amino acid synapses, it appears that the predominant effect of the Zn^{2+}-SYN signal in the brain is a tonic defacilitation at glutamate-gated excitatory ion channel(s). Thus, blocking Zn^{2+}-SYN by chelation tends to be proconvulsive (Mitchell & Barnes 1993), and adding modest concentrations of zinc, anticonvulsive (Morton *et al.* 1990). In ZnT-3 knockout mice (lacking stainable zinc in their vesicles) one of the surprisingly few (Cole *et al.* 2001) defects observed is that the mice have increased susceptibility to kainic-acid induced seizures (Cole *et al.* 2000). The overall electrophysiological impact of the Zn^{2+}-SYN signal on the hippocampal mossy-CA3 glutamatergic synapse was recently described in elegant detail and summarized with the statement that "... the metal ion Zn^{2+} is a neurotransmitter and ... the activity-dependent synaptic release of Zn^{2+} modulates NMDA receptor function. This modulation involves an inhibition mediated by both the high-affinity, voltage-independent binding site and the low-affinity, voltage-dependent binding site of Zn^{2+}" (Vogt *et al.* 2000).

Zn^{2+}-TRANS

The second Zn^{2+} signal that has been observed and characterized in the brain occurs roughly simultaneous with the conventional Zn^{2+}-SYN signal. This signal is analogous to the transmembrane Ca^{2+} signals, and consists of a transmembrane flux of Zn^{2+} from the extracellular milieu, through gated, zinc-permeable channels, and into the somata (Li *et al.* 2001a, b) or dendrites (Suh *et al.* 2001) of postsynaptic neurons (Figure 5). The highest conductance Zn^{2+} channel is the Ca-AK channel (Figure 6), though NMDA-gated Ca^{2+} channels will also pass zinc (Weiss & Sensi 2000; Weiss *et al.* 2000). Because the presynaptic release of Zn^{2+} is the only known source of appreciable amounts of extracellular Zn^{2+}, this signal

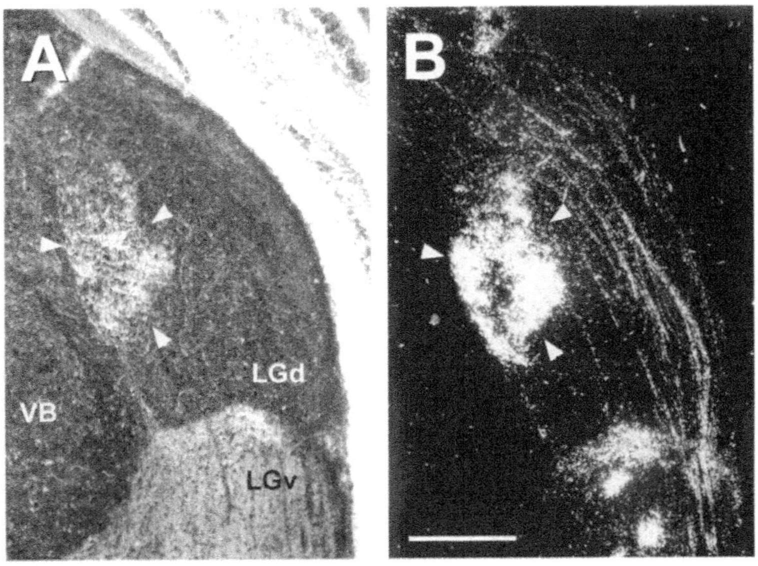

Fig. 4. Zinc staining is present briefly in uncrossed, but not crossed axons innervating the lateral geniculate nucleus of the albino rat. Compare the zinc-staining (A) with the location of the uncrossed axon terminals (B) **A**. Darkfield photomicrograph showing distribution of histochemically reactive zinc in a coronal section through the lateral geniculate nucleus of a postnatal day 15 rat. Synaptic zinc (stained by the Timm-Danscher method, white arrowheads) is localized to a discrete zone in the medial portion of the nucleus. **B**. Section adjacent to that illustrated in A showing distribution of incrossed axon terminals (white arrowheads) that were stained by injection of horseradish peroxidase into the ipsilateral eye. Note that synaptic zinc and uncrossed retinal projections are localized to an identical region of the LGd. This developmental zinc staining disappears in the adult. Calibration bar = 300 mm for A and B. Abbreviations: LGN – lateral geniculate nucleus; LGd – lateral geniculate dorsal; LGv – lateral geniculate ventral; VB – ventrobasal thalamic nucleus. Courtesy Peter Land.

Zn^{2+}-TRANS is both a transynaptic and a transcellular signal. The temporal and spatial characteristics of these Zn^{2+}-TRANS 'flashes', or 'sparks' in neuronal cytosol have not been established, but the rise times can be within a few 10's of milliseconds of the initial stimulation of the zinc-containing presynaptic fiber system (Figure 5) (Li *et al.* 2001a; Suh *et al.* 2001). Whether Zn^{2+}-TRANS signals can, in turn, induce subsequent intracellular zinc mobilization Zn^{2+}-INT in the same way that transmembrane Ca^{2+} fluxes induce mobilization of Ca^{2+} from the SER (or SR) (Melamed-Brook *et al.* 1999; Keizer & Levine 1996) is not presently known.

Zn^{2+}-TRANS signals can be blocked either in the cleft, by bath application of a sufficiently fast and high-affinity chelator, or at the postsynaptic neuron, by the use of channel blockers (e.g., CNQX) that prevent the opening of the dominant postsynaptic Zn^{2+} channel, Ca-AK). Conversely, the Zn^{2+}-TRANS signal can be mimicked by bath application of Zn^{2+} and a suitable zinc ionophore (e.g., pyrithione). John Sarvey and his colleagues have recently shown by both the blockade and the mimicry approaches that the Zn^{2+}-TRANS signal is necessary and (in the presence of glutamate) sufficient to induce LTP of mossy fiber input in CA3 pyramidal neurons (Li *et al.* 2001b).

Zn^{2+}-INT

The third Zn^{2+} signal now under scrutiny is analogous to another class of Ca^{2+} signal, namely, the intracellular Ca^{2+} signal. In the case of calcium, the sarcoplasmic or smooth endoplasmic reticulum are established storage sites for intracellular release of the ion, and the pumps and channels and ligands that mediate the Ca^{2+} movement into and out of those depots are well characterized (Keizer & Levine 1996). With Zn^{2+}, on the other hand, there is (as yet) no candidate organelle for the storage of pools for intracellular release. The neuronal vesicle has no detectable Zn^{2+} until it moves out of the Golgi apparatus and into the orthograde axoplasmic flow down axons (Frederickson & Danscher 1990), and no other zinc-filled organelle is found in healthy neurons (Frederickson *et al.* 1992; Danscher 1996; Fredrickson & Danscher 1990).

No matter from where the intracellular Zn^{2+} signals originate, they are prominent and robust in many types of cells, including neurons. Nitric oxide is one of the more potent inducers of Zn^{2+}-INT signals. In the

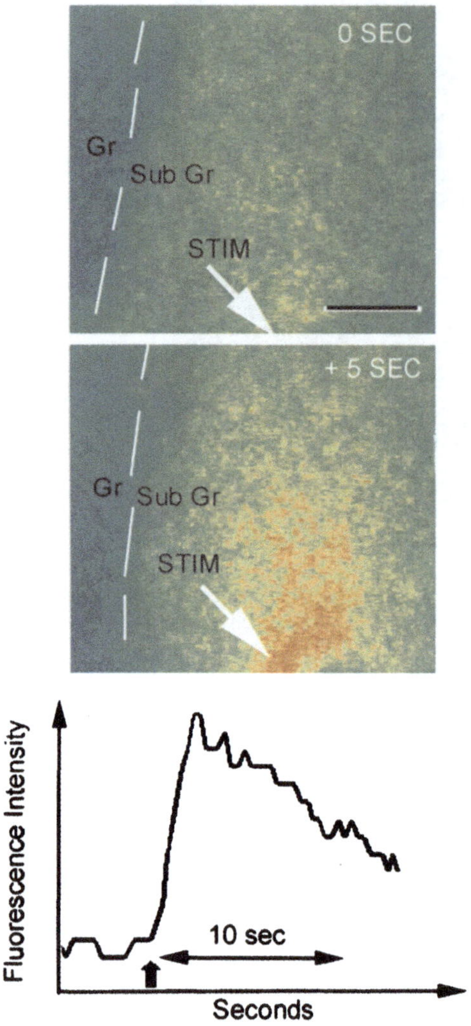

Fig. 5. Intracellular zinc 'puff' is shown in pseudo color, using the membrane-permeable ('trappable') intracellular zinc indicator, Newport green. The electrode tip (STIM) was within the hilus of the dentate gyrus with approximate location of the Granule-Sub-Granule border indicated. Note the 'puff' of zinc at +5 sec after a 5 sec, 100 Hz stimulation. Note rapid rise time of this Zn^{2+}-TRANS fluorescent signal in the quantitative figure at bottom. Quantitative figure reflects changes within pixels located at the center of the 'puff'. Courtesy Yang Li, Chris Hough and John Sarvey. See: Li *et al.* (2001a).

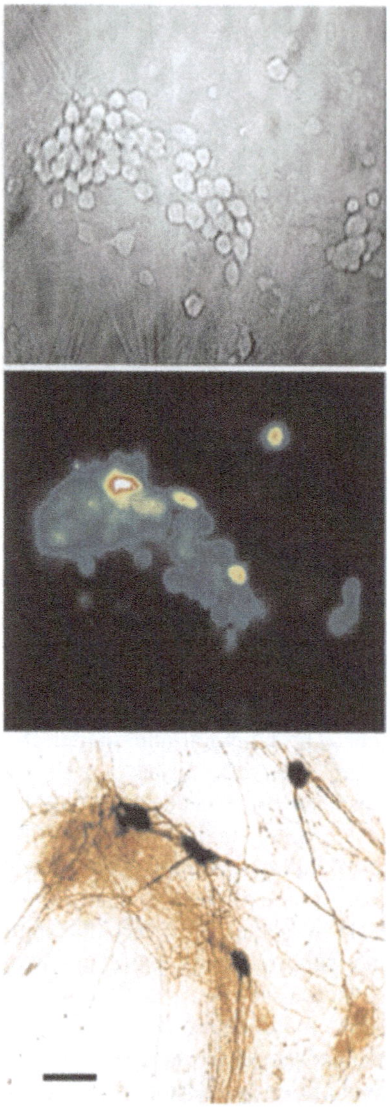

Fig. 6. Neurons that fill with Zn^{2+} during a glutamate + zinc challenge, are the same neurons that express abundant Ca-AK channels. Cultured neurons (top) respond differentially to combined stimulation with zinc and glutamate, with four somata attaining high levels of intracellular zinc (yellow-red-white in the pseudo color, middle panel), and the rest showing only marginal intracellular zinc. Double staining by the cobalt method which labels via the Ca-AK channels (bottom panel) reveals that the 4 zinc-filled neurons each had robust expression of the high-zinc-permeability Ca-AK channels (Courtesy John Weiss & Stefano Sensi; see Weiss & Sensi 2000).

otherwise intact brain, NO* stimulation causes immediate appearance of perikaryal and nuclear zinc staining (Figure 7) (Cuajungco & Lees 1998; Frederickson *et al.* 2001a). In excitotoxicity accompanying seizures, ischemia, hemorrhage, or blunt trauma, similar Zn^{2+} staining occurs within 10's of minutes of the insult (Figure 7) (Frederickson *et al.* 1988, 1989; Prough *et al.* 2001; Suh & Frederickson 2001). Because these Zn^{2+}-INT signals are seen (i) in neurons far from any

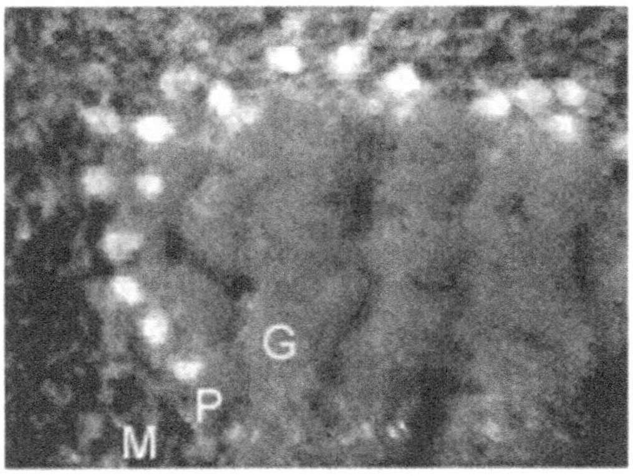

Fig. 7. The cerebellar cortex is normally completely negative for zinc staining throughout the Molecular (M), Purkinje (P) and Granule (G) strata. However, infusion of a NO* donor (spermine NO) *in vivo* causes Purkinje somata to rapidly develop intense TSQ fluorescence for zinc (white spheres).

synaptic zinc source (Figure 6; Frederickson *et al.* 2001a), and are seen (ii) in neurons of mutant mice lacking detectable synaptic zinc (Lee *et al.* 2000), and are seen (iii) in a variety of dissociated cell types in culture (Zangger *et al.* 2001 and references therein), there is little doubt that the intracellular zinc can arise from some still-undiscovered zinc depot within cells. One likely candidate for the source of the Zn^{2+}-INT signals would be zinc-sequestering proteins such as metallothionein, which can release up to 7 zinc ions per molecule upon command. The NO* stimulus has been shown to release zinc from proteinaceous storage sites (including MT) by nitrosylation of ligands (Zangger *et al.* 2001). Other redox and oxidative stimuli can cause a similar Zn^{2+}-INT zinc release (Maret 2000; Ye *et al.* 2001).

The targets and physiological functions of the Zn^{2+}-INT signals are still largely undiscovered. Because the Zn^{2+}-INT signals appear in conditions of biological stress, one focus of research has been on Zn^{2+}-INT control of mitochondrial function and of apoptotic events (Truong-Tran *et al.* 2001; Yoon *et al.* 2000; Hyun *et al.* 2000). One powerful new technology that will doubtless accelerate progress is the development of antibodies for the zinc-binding sites that will distinguish the zinc-site-occupied (zincylated ?) from the zinc-site-unoccupied (dezincylated ?) condition (Herwald *et al.* 2001). One physiological response to a Zn^{2+}-INT signal has been tentatively identified by this antibody strategy (Herwald *et al.* 2001).

Zinc versus calcium

Conventional wisdom holds that zinc and calcium have fundamentally different roles in the brain. Zinc is viewed as an enduring component of proteins and an ion that is sometimes temporarily 'free' (i.e., between proteins) whereas calcium is typically viewed as a 'free' extracellular ion, that is generally bound in the intracellular milieu, (i.e., between signaling duties) in calcium-sequestering proteins that serve to 'send' calcium signals. Thus, the term 'zinc-binding protein' generally refers to any of hundreds of metalloproteins that have zinc tightly bound into more-or-less permanent sites (e.g., zinc-finger motifs; Berg 1990) at critical folding points or active catalytic sites (Vallee & Falchuk 1993). Calcium-binding protein, in contrast generally refers to any of dozens of proteins known to scavenge, store, and release the ion, in the course of calcium signaling events (e.g., Williams 1996; Toutenhoofd & Strehler 2000).

It is now clear that the conventional wisdom about zinc is wrong, at least as concerns the brain. As described above, zinc ions are stored, released, taken up, and translocated across membranes and within cells, and there are myriad proteinaceous targets and receptors specifically tuned to these zinc signals. Furthermore, it is now appreciated that there are zinc-binding proteins that are kinetically-labile, having relatively rapid on- and off-rates (Maret 2000; Auld 1995).

If both zinc and calcium serve as signal ions, then one may ask why there are two such signals and how their differences (and similarities) could be utilized by cells. As Table 1 shows, one salient difference is the fact that extracellular free zinc (unlike extracellular calcium) is essentially absent in the extracellular fluid, with the estimated concentration being below the detection limits for most analytical methods (~ 1 pM; see Simons 1991; Powell *et al.* 1999; discussion in Frederickson 1989). Therefore, whereas transmembrane calcium signals can be initiated by merely opening a calcium channel, transmembrane zinc signals require the release of some ion into the extracellular space *and* opening of transmembrane channels. So far, zinc-containing presynaptic vesicles, with mM levels of zinc ions, are the only known source of such extracellular zinc in brain tissue. This means that transmembrane zinc signals are always 'AND' signals, with synaptic release *AND* membrane channel opening both required.

The other notable difference is that there are already dozens of calcium scavenging, sequestering,

Table 1. Comparison of zinc and calcium signal properties.

Attribute	Zinc (Zn^{2+})	Calcium (Ca^{2+})
Approximate ion concentration in deep ocean	1 pM - 1 nM	10 mM
Approximate concentration of free ion extracellularly	not established, but likely 1 pM to 1 nM	22 mM
Approximate concentration of free ion in cytosol	not established, but ~005–10 nM	~50 nM
Approximate concentration of ion in storage (releasable) pools	~3–30 mM in presynaptic vesicles	? low μM in SER ?
Magnitude of transient physiological ion signals	~1000-fold increases in cytosol	~ 1000-fold increases in cytosol
Rise-time of transient physiological signals	1–10 msec	1–10 msec
Specific membrane-bound storage organelles ?	YES: presynaptic vesicles many other secretory granules	YES: SER (& synaptic vesicles) in neurons; SR in muscle
Specific releasable pools ?	YES: presynaptic vesicles; MT	YES: SER, SR
Membrane-spanning gated channels for the ion ?	YES: voltage and ligand gated	YES: voltage and ligand gated
Ion- sequestering proteins ?	YES (few described, eg, metallothionein)	YES (dozens described, eg, calbindin, calmodulin)
Ion-sensitive ion channels ?	YES: (NMDA, AMPA KA, K^+; GABAa,b all modulated by extracellular Zn^{2+})	YES, Na^+, K^+, CA^{++} all modulated by intracellular Ca^{++}
Ion modulated enzymes	YES	YES
Cytotoxic at high intracellular levels	YES 100's nM fatal	YES 100's nM fatal
Number of ion-containing proteins	hundreds of zinc finger and other zinc-containing proteins	many calcium storage and calcium-modulated proteins
References	Li *et al.* 2001a,b; Thomson *et al.* 2000, 2001; Canzoniero *et al.* 1999; Simons 1991; Smart *et al.* 1994; Frederickson *et al.* 2000; Lobner *et al.* 2000; Powell *et al.* 1999; Frederickson 1989; Choi & Koh 1998	Melamed-Brook *et al.* 1999; ZhuGe *et al.* 2000; Keizer & Levine 1996; Brown & MacLeod 2000; Van Assche *et al.* 1996; Wilson 1996

and releasing proteins identified, as well as calcium-modulated signal cascades, whereas only metallothionein (Maret 2000) and the Zn-T zinc pumps (ZnT1, 2, 3) have been established as 'helper proteins' for storing and releasing zinc signals (Pamiter *et al.* 1996). This may be mostly due to a lack of interest in the past: the search for proteins that control the zinc signal pathways is doubtless now beginning.

Pathophysiology

Zinc dysregulation is implicated as a contributing factor in two types of neuropathology: (i) Alzheimer's disease, and (ii) the so-called 'excitotoxicity' which injures neurons after ischemia, hemorrhage, seizures, or mechanical brain traumas.

In broad outline, the evidence implicating zinc is roughly the same in both cases. Thus, in the first place, both conditions are marked by the appearance of vivid, anomalous, and emblematic staining for zinc in the histochemical centers of the disease processes in the brain (Frederickson *et al.* 1989; Suh *et al.* 2000a, b). In the second place, both conditions can be ameliorated by the simple strategy of treating the brain with a zinc chelator to lower the extracellular Zn^{2+} burden (Koh *et al.* 1996; Frederickson *et al.* 2001; Suh *et al.* 2000; Cherny *et al.* 2001). Third, both conditions can be simulated in certain *in vitro* test models by the addition of excess exogenous zinc to the brain tissue under study (Bush *et al.* 1994a, b; Weiss & Sensi 2000). Fourth, both conditions are plausibly triggered (or accelerated) by the pathological release of an excess of Zn^{2+} into the extracellular milieu from the only known store of such Zn^{2+}, the zinc-containing boutons. Fifth, both conditions tend to strike preferentially in those cerebrocortical regions (hippocampal formation, amygdala, frontal cortex) where the concentration of zinc-containing axonal boutons is highest (reviews in Suh *et al.* 2000a, b; Frederickson *et al.* 1989, 1992, 2000).

Alzheimer's Disease (AD)

One of us (AIB) has developed a model where abnormalities of the zinc homeostatic control mechanisms may explain the pathological features found in AD. The model is based on the tendency of amyloid beta (Aβ a primary constituent of AD plaques and AD angiopathy) to precipitate in the presence of sufficient zinc to occupy its low affinity (micromolar) binding site (Figure 8).

Amyloid beta peptide binds zinc at both high- and low-affinity binding sites (Bush *et al.* 1994b). Zinc concentrations above 300 nM rapidly precipitate synthetic human Aβ1-40 (Bush *et al.* 1994a, b). Interestingly, zinc preserves the α-helical conformation of Aβ1-40 and its complexation is completely reversed with chelation treatment (Huang *et al.* 1997). This result suggests that zinc-induced Aβ aggregation may possibly inhibit its intrinsic neurotoxicity. Meanwhile, rat Aβ1-40 (with substitutions of Arg$\rightarrow$Gly, Tyr$\rightarrow$Phe and His$\rightarrow$Arg at positions 5, 10 and 13, respectively) binds zinc less avidly (K_A = 3.8 μM) and is unaffected by zinc at these concentrations, perhaps that is why rats do not form Alzheimer's like plaques (Shivers *et al.* 1988). In the absence of zinc, the solubilities of the rat and the human Aβ species are indistinguishable (Bush *et al.* 1994a).

Zinc-induced Aβ precipitation at pH 7.4 is highly specific for zinc, however, copper Cu^{2+} and iron Fe^{2+} can also bind, especially at lower (6.6) pH (Atwood *et al.* 1998). While Cu^{2+} and Fe^{3+} binding to Aβ induces O_2 dependent H_2O_2 production and toxicity (Huang *et al.* 1999a, b), co-incubation with Zn^{2+} (>4:1, Zn:Cu) inhibits H_2O_2 production (Cuajungco *et al.* 2000). Zn^{2+} had previously been reported to protect cell cultures from Aβ toxicity (Lovell *et al.* 1999). Extending this observation, we found that Zn^{2+} rescued primary cortical and human embryonic kidney 293 cells that were exposed to Aβ1-42 (Cuajungco *et al.* 2000). Since plaques contain exceptionally high concentrations of Zn^{2+} ($\approx$1 mM, see below), we examined the relationship between oxidation (8-OH guanosine) levels in AD-affected tissue and histological amyloid burden, and found a highly significant negative correlation. Therefore, Zn^{2+} loading into plaques may represent an attempt at protective homeostatic response in AD, where plaques form as the result of a more robust Zn^{2+} antioxidant response to the underlying oxidative attack, and that the Aβ in the plaques has been redox-silenced by the high concentrations of Zn^{2+}, whereas the diffuse and soluble Aβ accumulations within the brain would be a source of H_2O_2 and oxidative damage (McLean *et al.* 1999).

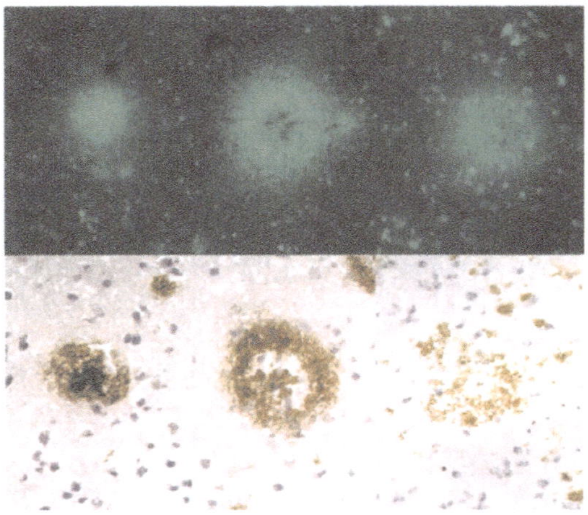

Fig. 8. Autopsy material from patients deceased of Alzheimer's disease show TSQ staining (upper panels) of amyloid plaques; shown, for comparison, stained with AB1-42 immunostaining (lower panels). The six plaques are all from different sections, and have been combined digitally for illustration. Courtesy of Math Cuajungco, Sang Won Suh, Al Rampy, and Zoran Gatalica.

Recently, we have shown that Aβ binds Cu^{2+} and Zn^{2+} through selective binding sites. When synthetic Aβ is co-incubated with excess but equal amounts of Cu^{2+} and Zn^{2+}, $\approx$1.5 equivalents of each metal ion bind to each mole of peptide. Since the affinity of the low and high affinity Cu^{2+} binding sites on Aβ ranges

Table 2. Zinc-binding and zinc-dependent proteins implicated in AD pathology.

Proteins	Zinc effects	Role in Alzheimer's	References
α_2-Macroglobulin	zinc-binding protein	zinc binding triggers Zn+Aβ + α_2-M complex formation	Du *et al.* 1997; Hughes *et al.* 1998
Nerve Growth Factor-β	removal of zinc 'zipper' releases active fragment from zymogen	neurotrophic agent - combats AD advance	Pattison & Dunn 1976a, b; Ross *et al.* 1997
S100β	zinc-containing protein, neurite extension factor	elevated amidst plaques and NFT in AD brains	Griffin *et al.* 1989; Marshak *et al.* 1991
Metallothionein	zinc-sequestering protein	altered in AD brains	Zambenedetti *et al.* 1998; Adlard *et al.* 1999
Alpha- and beta-secretases	cleave proteins at zinc-binding site	cleave APP releasing Aβ	Bush *et al.* 1994
Matrix metalloproteinases	zinc-containing enzymes	can degrade Aβ1-41 and Aβ 1-42	Backstrom *et al.* 1996; Roher *et al.* 1994
Caspase	zinc-modulated enzyme: up- or down-regulated, depending on concentration	modulates apoptosis may modulate AD apoptotic cell death	Cuajungco & Lees 1997; Choi & Koh 1998

from nM- attoM, and since the highest affinity Zn^{2+} site on Aβ is 100 nM, the finding that Cu^{2+} does not compete for all of the available metal binding sites when co-incubated with Zn^{2+} implies that Aβ possesses separate and selective Cu^{2+} and Zn^{2+} binding sites (Atwood *et al.* 2000). Importantly, mildly acidic conditions, representing physiological acidosis (e.g., pH 6.8), abolish Zn^{2+} binding to Aβ, but enhance Cu^{2+} binding to Aβ, so that when Cu^{2+} and Zn^{2+} are co-incubated with Aβ at pH 6.8, $\approx$3.0 equivalents of Cu^{2+} bind to the peptide, but virtually no Zn^{2+} (Atwood *et al.* 1998; Atwood *et al.* 2000). Physiological acidosis may therefore be one mechanism by which Aβ loses the redox-protective Zn^{2+} binding, and may then be liable for inappropriate redox activity.

In keeping with our prediction that Aβ is a zinc metalloprotein, we have now published observations that amyloid plaques in post-mortem AD have a marked enrichment of zinc (to mM levels (Lovell *et al.* 1998)) that is histochemically visible (Figure 8; Suh *et al.* 2000b). Intriguingly, Zn^{2+} is also markedly enriched in the neocortical amyloid plaques of APP transgenic mice (Lee *et al.* 1999), supporting the likelihood that abnormal Zn-Aβ interaction is responsible for plaque formation *in vivo*. The significance of Zn (and Cu) being present in the amyloid mass in the brain in AD is that it may be possible to create therapeutic drugs for AD that safely target the Aβ-Zn interaction. Zn/Cu-selective chelators reverse Zn/Cu-induced aggregation of synthetic Aβ *in vitro* (Huang *et al.* 1997; Atwood *et al.* 1998), inhibit Aβ-mediated H_2O_2 formation (Huang *et al.* 1999a, b; Bush *et al.* 1999), and solubilize Aβ from amyloid deposits in post-mortem AD-affected brain tissue (Cherny *et al.* 1999). Recently, we reported the profound inhibition of Aβ deposition in the pellet phase of brain homogenates (375 μg/g wet weight, $P = 0.0001$) in a blinded study of APP2576 transgenic mice treated orally for 9 weeks with clioquinol, an antibiotic and bioavailable Cu/Zn chelator. This was accompanied by a modest increase in soluble Aβ (1.45% of total cerebral Aβ); but APP, synaptophysin and GFAP levels were unaffected. Behavioral and body weight parameters were significantly more stable in the treated animals, and there was no evidence of systemic metal depletion (Cherny *et al.* 2001). The affinity of clioquinol for Zn is only nanomolar, so therefore while the molecule therapeutically targets the metals that induce Aβ aggregation or redox activity, unlike common chelators (e.g., EDTA, desferrioxamine) clioquinol does not appear to have sufficient affinity for Zn and Cu to disturb metal-dependent biochemistry.

Beyond the immediate interactions of zinc and amyloid, there are a number of less direct pathways by which zinc dysregulation can affect the rate and severity of the AD pathophysiology. Table 2 lists some examples of zinc-containing and zinc-sensitive protein signals and enzymes that can modify the course of the AD pathology, and would be themselves perturbed in the face of any primary disturbance of zinc homeostasis in the brain.

Excitotoxicity

In conditions of compromised cerebral blood flow and in sustained status epilepticus, the so-called 'excitotoxic' cell injury cascade is triggered in the brain. The release of copious glutamate and consequent depolarization of neurons that constitute excitotoxicity is accompanied by the appearance of very high levels of free Zn^{2+} in the somata of (and only of) the dying neurons (Frederickson *et al.* 1988, 1989; Tonder *et al.* 1990; Suh *et al.* 2000a; Suh & Frederickson 2001; Prough *et al.* 2001) (Figure 9). Because this anomalous, pathological intracellular Zn^{2+} burden can be found in neurons not surrounded by appreciable zinc-containing innervation (Frederickson *et al.* 2001a), and can be found in neurons of knockout mice congenitally-lacking detectable Zn^{2+} in their presynaptic vesicles (Lee *et al.* 2000), it seems certain that there is a Zn^{2+}-INT signal contributing to the excitotoxic zinc signal. As mentioned earlier, nitric oxide and superoxide stimuli mobilizing zinc off proteins such as metallothionein probably contribute part of this Zn^{2+}-INT signal.

At the same time, in the brains of otherwise normal animals, the massive release of glutamate during excitotoxicity is accompanied by an equally massive release of Zn^{2+} from the presynaptic boutons (Frederickson *et al.* 1988, 2001b; Suh *et al.* 2000; Sorensen *et al.* 1998). Therefore it is hard to imagine that there is not a Zn^{2+}-TRANS signal that also contributes to the zinc-loading of neurons during excitotoxic crises.

Regardless of the relative contributions of Zn^{2+}-INT and Zn^{2+}-TRANS, the Zn^{2+} apparently kills cells by entering mitochondria and disrupting function, with both a release of reactive oxidative species and induction of both apoptosis and necrosis (depending on paradigms) and the death of the cell (Weiss *et al.* 2000; Weiss & Sensi 2000; Choi & Koh 1998).

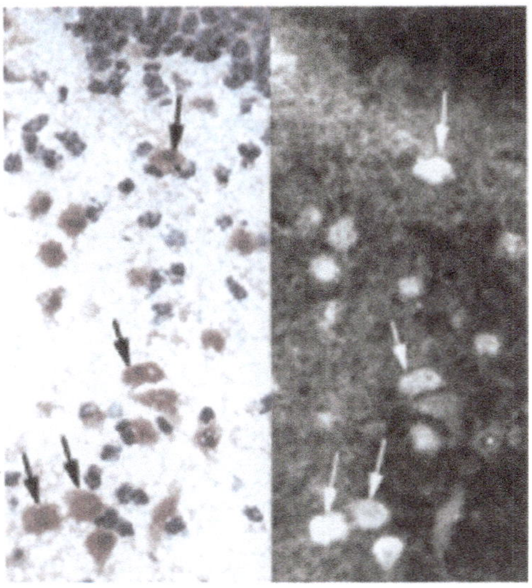

Fig. 9. In the rabbit subjected to brief global ischemia, neurons throughout the cerebrum (arrows) show TSQ staining for zinc (right panel) and (in the corresponding neurons) acidophilic changes indicative of injury/degeneration (left panel). Double-staining of the same section; 1 h post ischemia. Courtesy of Mark Zornow, Don Prough, and Sang Won Suh.

As discovered by Choi and his colleagues, therapeutic administration of an extracellular zinc-specific chelator can reduce excitotoxic cell loss by up to 80% (Koh *et al.* 1994). This has been shown in experimental animals with excitotoxicity induced by ischemia (Koh *et al.* 1994), trauma (Suh *et al.* 2000a), or trauma plus hemorrhage (Prough & Frederickson, unpublished) and (with some mixed results) in some seizure models (Lees *et al.* 1998). In all cases, the neuron sparing is judged by the reduced numbers of neurons showing acidophilic signs of injury up to 24 h after the initial insult. Interestingly, this rescue of neurons by a chelator can be done even if the zinc insult (from exogenous zinc) is applied first then terminated, and the therapeutic zinc chelation (by CaEDTA) is applied some time later. In this latter case, the zinc-staining of cells induced by a prior insult is reversed by the subsequent addition of CaEDTA to the medium (Frederickson *et al.* 2001b). In addition to conventional chelators, such as CaEDTA, the highly-specific zinc-stripped (apo) form of a zinc-containing protein (carbonic anhydrase) is also neuroprotective when administered intraventricularly in a trauma + hemorrhage rat model (Prough & Frederickson, unpublished). Because the on- and off-rates of carbonic anhydrase can be separately modified by genetic engineering (Thompson *et al.* 2000, 2001), this latter approach may prove useful for design and fabrication of intracerebral zinc buffers.

Acknowledgements

We thank Richard Thompson, John Sarvey and Chris Hough for discussion. Supported in part by NS42015, NS 46668, and NS 38585.

References

Adlard PA, West AK, Vickers JC. 1998 Increased density of metallothionein I/II immunopositive cortical glial cells in the early stages of Alzheimer's disease. *Neurobiol Dis* **5**, 349–356.

Atwood CS, Moir RD, Huang X, Scarpa RC, Bacarra NME, Romano DM, Hartshorn MA, Tanzi RE, Bush AI. 1998 Dramatic aggregation of Alzheimer Aß by Cu(II) is induced by conditions representing physiological acidosis. *J Biol Chem* **273**, 12817–12826.

Auld DS. 1995 Removal and replacement of metal ions in metallopeptidases. *Methods Enzymol* **248**, 228–242.

Backstrom JR, Miller CA, Tokes ZA. 1992 Characterization of neutral proteinases from Alzheimer-affected and control brain specimens, identification of calcium-dependent metalloproteinases from the hippocampus. *J Neurochem* **58**, 983–992.

Baranano DE, Ferris CD, Snyder SH. 2001 Atypical neural messengers. *Trends Neurosci* **24**, 99–106.

Beaulieu C, Dyck R, Cynader M. 1992 Enrichment of glutamate in zinc-containing terminals of the cat visual cortex. *Neuroreport* **10**, 861–864.

Berg JM. 1990 Zinc finger domains: hypotheses and current knowledge. *Annu Rev Biophys Biophys Chem* **19**, 405–421.

Birinyi A, Parker D, Antal M, Shupliakov O. 2001 Zinc co-localizes with GABA and glycine in synapses in the lamprey spinal cord. *J Comp Neurol* **433**, 208–221.

Brown EM, MacLeod RJ. 2001 Extracellular calcium sensing and extracellular calcium signaling. *Physiol Rev* **81**, 239–297.

Bush AI, Pettingel WH, Multhaup G, Paradis MD, Vonsattel JP, Gusella JF, Beyreuther K, Masters CL, Tanzi RE. 1994a Rapid induction of Alzheimer A amyloid formation by zinc. *Science* **265**, 1464–1467.

Bush AI, Pettingel WH, Paradis MD, Tanzi RE. 1994b. Modulation of Aß adhesiveness and a secretase site cleavage by zinc. *J Biol Chem* **269**, 12152–12158.

Bush AI, Huang X, Fairlie DP. 1999 The possible origin of free radicals from amyloid β peptides in Alzheimer's disease. *Neurobiol Aging* **20**, 335–337.

Canzoniero LMT, Turetsky DM, Choi DW. 1999 Measurement of Intracellular Free Zinc Concentrations Accompanying Zinc-Induced Neuronal Death. *J Neurosci* **19**(RC31), 1–6.

Casanovas-Aguilar C, Reblet C, Perez-Clausell J, Bueno-Lopez JL. 1998 Zinc-rich afferents to the rat neocortex, projections to the visual cortex traced with intracerebral selenite injections. *J Chem Neuroanat* **15**, 97–109.

Chen N, Moshaver A, Raymond LA. 1997 Differential sensitivity of recombinant N-methyl-D-aspartate receptor subtypes to zinc inhibition. *Mol Pharmacol* **51**, 1015–1023.

Cherny RA, Atwood CS, Xilinas X, Gray DN, Jones WD, McLean CA, Barnham KJ, Volitakis I, Fraser FW, Kim Y-S, Huang X, Goldstein LE, Moir RD, Lim JT, Zheng H, Beyreuther K, Tanzi RE, Masters CL, Bush AI. 2001 Treatment with a copper-zinc chelator markedly and rapidly inhibits β-amyloid accumulation in Alzheimer's disease transgenic mice. *Neuron* **30**, 665–676.

Choi DW, Koh JY. 1998 Zinc and brain injury. *Annu Rev Neurosci* **21**, 347–375.

Christensen MK, Frederickson CJ. 1998 Zinc containing afferent projections to the rat corticomedial amygdaloid complex, a retrograde tracing study. *J Comp Neurol* **400**, 375–390.

Cole TB, Wenzel HJ, Kafer KE, Schwartzkroin PA, Palmiter RD. 1999 Elimination of zinc from synaptic vesicles in the intact mouse brain by disruption of the ZnT3 gene. *Proc Natl Acad Sci USA* **96**, 1716–1721.

Cole TB, Martyanova A, Palmiter RD. 2001 Removing zinc from synaptic vesicles does not impair spatial learning, memory, or sensorimotor functions in the mouse. *Brain Res* **891**, 253–265.

Cole TB, Robbins CA, Wenzel HJ, Schwartzkroin PA, Palmiter RD. 2000 Seizures and neuronal damage in mice lacking vesicular zinc. *Epilepsy Res* **39**, 153–169.

Coulter DA. 2000 Mossy fiber zinc and temporal lobe epilepsy, pathological association with altered 'epileptic' gamma-aminobutyric acid A receptors in dentate granule cells. *Epilepsia* **41**(Suppl 6), S96–S99.

Cuajungco MP, Lees GJ. 1998 Nitric oxide generators produce accumulation of chelatable zinc in hippocampal neuronal perikarya. *Brain Res* **799**, 118–129.

Cuajungco MP, Lees GJ, Kydd RR, Tanzi RE, Bush AI. 2000 Zinc and Alzheimer's disease, an update. *Nutl Neurosci* **2**, 191–208.

Danscher G. 1996 The autometallographic zinc-sulfide method. A new approach *in vivo* creation of nanometer sized zinc sulfide crystal lattices in zinc-enriched synaptic and secretory vesicles. *J Histochem* **28**, 363–372.

Du Y, Ni BF, Glinn M, Dodel RC, Bales KR, Zhang Z, Hyslop PA, Paul SM. 1997 α2-macroglobulin as a β-amyloid peptide-binding plasma protein. *J Neurochem* **69**, 299–305.

Dyck RH, Cynader MS. 1993 An interdigitated columnar mosaic of cytochrome oxidase, zinc, and neurotransmitter-related molecules in cat and monkey visual cortex. *Proc Natl Acad Sci USA* **90**, 9066–9069.

Franco-Pons N, Casanovas-Aguilar C, Arroyo S, Rumia J, Perez-Clausell J, Danscher G. 2000 Zinc-rich synaptic boutons in human temporal cortex biopsies. *Neuroscience* **98**, 429–435.

Frederickson CJ. 1989 Neurobiology of zinc and zinc-containing neurons. *Int Rev Neurobiol* **31**, 145–238.

Frederickson CJ, Cuajungco MP, Suh SW. 2001 Nitric oxide causes apparent release of zinc from presynaptic boutons. *Neuroscience* (in press).

Frederickson CJ, Danscher G. 1990 Zinc-containing neurons in the hippocampus and related CNS structures. *Prog Brain Res* **83**, 71–84.

Frederickson CJ, Danscher G, Cravens KJ, Slomianka L, Sylvan LB. 1991 Staining for zinc reveals columnar patches in the hippocampus of the newborn rat. *Soc Neurosci Abs* **17**, 1131.

Frederickson CJ, Hernandez MD, Goik SA, Morton JD, McGinty JF. 1988 Loss of zinc staining from hippocampal mossy fibers during kainic acid-induced seizures, a histofluorescence study. *Brain Res* **446**, 383–386.

Frederickson CJ, Hernandez MD, McGinty JF. 1989 Translocation of zinc may contribute to seizure-induced death of neurons. *Brain Res* **480**, 317–321.

Frederickson CJ, Kasarskis EJ, Ringo D, Frederickson RE. 1987a A quinoline fluorescence method for visualizing and assaying the histochemically-reactive zinc in the brain. *J Neurosci Meth* **20**, 91–103.

Frederickson CJ, Klitenick MA, Manton WI, Kirkpatrick JB. 1983 Cytoarchitectonic distribution of zinc in the hippocampus of man and the rat. *Brain Res* **273**, 335–339.

Frederickson CJ, Moncrieff DW. 1994 Zinc-containing neurons. *Biol Signals* **3**, 127–139.

Frederickson CJ, Perez-Clausell J, Danscher G. 1987b Zinc-Containing 7S-NGF Complex, Evidence from Zinc Histochemistry for Localization in Salivary Secretory Granules. *J Histochem Cytochem* **35**, 579–583.

Frederickson CJ, Rampy BA, Reamy-Rampy S, Howell GA. 1992 Distribution of histochemically reactive zinc in the forebrain of the rat. *J Chem Neuroanat* **5**, 521–530.

Frederickson CJ, Suh SW, Cha K, Koh JY, Cuajungco MP. 2001b Depletion of Intracellular Zinc in Neurons by Use of an Extracellular Chelator *in vivo* and *in vitro*. *Neuroreport* (in press).

Frederickson CJ, Suh SW, Silva D, Frederickson CJ, Thompson RB. 2000 Importance of zinc in the central nervous system, the zinc-containing neuron. *J Nutr* **130** (Suppl 5), 14715–14835.

Griffin WST, Stanley LC, Ling C, White L, MacLeod V, Perrot LJ, White III CL, Araoz C. 1989 Brain interleukin-1 and S-100b immunoreactivity are elevated in Down syndrome and Alzheimer's disease. *Proc Natl Acad Sci USA* **86**, 7611–7615.

Haug F-MS. 1967 Electron microscopical localization of the zinc in hippocampal, mossy fiber synapses by a modified sulfide silver procedure. *Histochemie* **8**, 355–368.

Herwald H, Morgelin M, Svensson HG, Sjobring U. 2001 Zinc-dependent conformational changes in domain D5 of high molecular mass kininogen modulate contact activation. *Eur J Biochem* **268**, 396–404.

Huang X, Atwood CS, Hartshorn MA, Multhaup G, Goldstein LE, Scarpa, RC, Cuajungco MP, Gray DN, Lim J, Moir RD, Tanzi RE, Bush AI. 1999a The Aβ peptide of Alzheimer's Disease directly produces hydrogen peroxide through metal ion reduction. *Biochemistry* **38**, 7609–7616.

Huang X, Atwood CS, Moir RD, Hartshorn MA, Vonsattel JP, Tanzi RE, Bush AI. 1997 Zinc-induced Alzheimer's Abeta1-40 aggregation is mediated by conformational factors. *J Biol Chem* **272**, 26464–26470.

Huang X, Cuajungco MP, Atwood CS, Hartshorn MA, Tyndall J, Hanson GR, Stokes KC, Leopold M, Multhaup G, Goldstein LE, Scarpa, RC, Saunders AJ, Lim J, Moir RD, Glabe C, Bowden EF, Masters CL, Fairlie DP, Tanzi RE, Bush AI. 1999b Cu(II) potentiation of Alzheimer Aβ neurotoxicity, correlation with cell-free hydrogen peroxide production and metal reduction. *J Biol Chem* **274**, 37111–37116.

Hughes SR, Khorkova O, Goyal S, Knaeblein J, Heroux J, Riedel NG, Sahasrabudhe S. 1998 α2-Macroglobulin associates with β-amyloid peptide and prevents fibril formation. *Proc Natl Acad Sci USA* **95**, 3275–3280.

Hyun HJ, Sohn J, Ahn YH, Shin HC, Koh JY, Yoon YH. 2000 Depletion of intracellular zinc induces macromolecule synthesis-and caspase-dependent apoptosis of cultured retinal cells. *Brain Res* **869**, 39–48.

Keizer J, Levine L. 1996 Ryanodine receptor adaptation and Ca2+(-)induced Ca2+ release-dependent Ca2+ oscillations. *Biophys J* **71**, 3477–3487.

Koh JY, Suh SW, Gwag BJ, He YY, Hsu CY, Choi DW. 1996 The role of zinc in selective neuronal death after transient global cerebral ischemia. *Science* **272**, 1013–1016.

Kovacs KJ, Larson AA. 1997 Zn^{2+} inhibition of [3H]MK-801 binding is different in mouse brain and spinal cord, effect of glycine and glutamate. *Eur J Pharmacol* **324**, 117–123.

Land PW, Akhtar ND. 1999 Experience-dependent alteration of synaptic zinc in rat somatosensory barrel cortex. *Somatosens Mot Res* **16**, 139–150.

Lee JY, Cole TB, Palmiter RD, Koh JY. 2000 Accumulation of zinc in degenerating hippocampal neurons of ZnT3-null mice after seizures, evidence against synaptic vesicle origin. *J Neurosci* **1**(20), RC79.

Lee J-Y, Mook-Jung I, Koh J-Y. 1999 Histochemically reactive zinc in plaques of the Swedish mutant beta-amyloid precursor protein transgenic mice. *J Neurosc* **19** (RC10), 1–5.

Lees GJ, Cuajungco MP, Leong W. 1998 Effect of metal chelating agents on the direct and seizure-related neuronal death induced by zinc and kainic acid. *Brain Res* **799**, 108-117.

Li Y, Hough C, Suh SW, Sarvey J, Frederickson CJ. 2001a Evidence that Synaptically-released Zn^{2+} is translocated into postsynaptic neuons during synaptic transmission. *J Neurophysiol* (in press).

Li Y, Hough C, Frederickson CJ, Sarvey JA. 2001b Zinc release and entry into pre- and postsynaptic neurons is required for LTP in the mossy fiber-CA3 synapse. *Soc Neurosci Abs* (in press).

Lobner D, Canzoniero LM, Manzerra P, Gottron F, Ying H, Knudson M, Tian M, Dugan LL, Kerchner GA, Sheline CT, Korsmeyer SJ, Choi DW. 2000 Zinc-induced neuronal death in cortical neurons. *Cell Mol Biol* **46**, 797–806.

Long YY, Frederickson CJ. 1994 A Zinc-Containing Fiber System of Thalamic Origin. *NeuroReport* **5**, 2026–2028.

Lovell MA, Robertson JD, Teesdale WJ, Campbell JL, Markesbery WR. 1998 Copper, iron and zinc in Alzheimer's disease senile plaques. *J Neurol Sci* **158**, 47–52.

Maret W. 2000 The function of zinc metallothionein, a link between cellular zinc and redox state. *J Nutr* **130** (Suppl 5), 1455S–1458S.

Marshak DR, Pesce SA, Stanley LC, Griffin WST. 1991 Increased S-100b neurotrophic activity in Alzheimer's disease temporal lobe. *Neurobiol Aging* **13**, 1–7.

Maske H. 1955 Uber den topochemischen Nachweis von Zink in Ammonshorn verschiedener Säugetiere. *Naturwissenschaften* **42**, 424.

McLardy T. 1970 Anatomical rationale of ablative surgery for temporal lobe seizures and dyscontrol, suggested stereo-chemode chelate-blockade alternative. *Acta Neurochir* **23**, 119–124.

McLean C, Cherny R, Fraser F, Fuller S, Smith M, Beyreuther K, Bush A, Masters C. 1999 Soluble pool of Aβ amyloid as a determinant of severity of neurodegeneration in Alzheimer's. *Disease Ann Neurol* **46**, 860–966.

Melamed-Book N, Kachalsky SG, Kaiserman I, Rahamimoff R. 1999 Neuronal calcium sparks and intracellular calcium 'noise'. *Proc Natl Acad Sci USA* **96**, 15217–15221.

Mitchell C, Barnes M. 1993 Proconvulsant action of diethyldithiocarbamate in stimulation of the perforant path. *Neurotox Teratol* **15**, 165–171.

Morton JD, Howell GA, Frederickson CJ. 1990 Effects of subcutaneous injections of zinc chloride on seizures induced by noise and by kainic acid. *Epilepsia* **31**, 139–144.

Morvan J, Hunter D-F, Krause L, Flinn JM, Jones BF. 2000 Synchrotron X-ray fluorescence demonstrates increased zinc in the cortex of rats raised on drinking water containing enhanced levels of zinc. *Soc Neurosci Abs.*

Palmiter RD, Cole TB, Quaife CF, Findley SD. 1996 ZnT-3, a putative transporter of zinc into synaptic vesicles. *Proc Natl Acad Sci USA* **93**, 14934–14939.

Pattison SE, Dunn MF. 1976a On the mechanism of divalent metal ion chelator induced activation of the 7S nerve growth factor esteropeptidase, activation by 2,2′,2″-terpyridine and by 8-hydroxyquinoline-5-sulfonic acid. *Biochemistry* **15**, 3691–3696.

Pattison SE, Dunn MF. 1976b On the mechanism of divalent metal ion chelator induced activation of the 7S nerve growth factor esteropeptidase, thermodynamics and kinetics of activation. *Biochemistry* **15**, 3696–3703.

Powell JJ, Burden TJ, Greenfield SM, Taylor PD, Thompson RP. 1999 Urinary excretion of essential metals following intravenous calcium disodium edetate, an estimate of free zinc and zinc status in man. *J Inorg Biochem* **75**, 159–165.

Prough DS, Suh SW, Frederickson CJ, Li ZY, DeWitt DS. 2001 Traumatic brain injury and hemorrhagic hypotension increase cortical and hippocampal zinc translocation in rats. *Crit Care Med* **28**(76, Supplement S).

Qian WJ, Aspinwall CA, Battiste MA, Kennedy RT. 2000 Detection of secretion from single pancreatic beta-cells using extracellular fluorogenic reactions and confocal fluorescence microscopy. *Anal Chem* **72**, 711–717.

Roher AE, Kasunic TC, Woods AS, Cotter RJ, Ball MJ, Fridman R. 1994 Proteolysis of Aß peptide from Alzheimer's disease brain by gelatinase A. *Biochem Biophys Res Commun* **205**, 1755–1761.

Ross GM, Shamovsky IL, Lawrence G, Solc M, Dostaler SM, Jimmo SL, Weaver DF, Riopelle RJ. 1997 Zinc alters conformation and inhibits biological activities of nerve growth factor and related neurotrophins. *Nat Med* **8**, 872–878.

Rubio ME, Juiz JM. 1998 Chemical anatomy of excitatory endings in the dorsal cochlear nucleus of the rat, differential synaptic distribution of aspartate aminotransferase, glutamate, and vesicular zinc. *J Comp Neurol* **399**, 341–358.

Shivers BD, Hilbich C, Multhaup G, Salbaum M, Beyreuther K, Seeburg PH. 1988 Alzheimer's disease amyloidogenic glycoprotein, expression pattern in rat brain suggests role in cell contact. *EMBO J.* **7**, 1365–1370.

Simons TJ. 1991 Intracellular free zinc and zinc buffering in human red blood cells. *J Membr Biol* **123**, 63–71.

Smart TG, Xie X, Krishek BJ. 1994 Modulation of inhibitory and excitatory amino acids receptor ion channels by zinc. *Prog Neurobiol* **42**, 393–441.

Sorensen J, Mattson B, Andreasen A, Johansson B. 1998 Rapid disappearance of zinc positive terminals in focal brain ischemia. *Brain Res* **812**, 265–269.

Suh SW, Chen JW, Motamedi M, Bell B, Listiak K, Pons NF, Danscher G, Frederickson CJ. 2000a Evidence that synaptically-released zinc contributes to neuronal injury after traumatic brain injury. *Brain Res* **852**, 268–273.

Suh SW, Frederickson CJ. 2001 Loss of vesicular zinc and appearance of perikaryal zinc after seizures induced by pilocarpine. *NeuroReport* **12**, 1523–1525.

Suh SW, Jensen KB, Jensen MS, Silva DS, Kesslak PJ, Danscher G, Frederickson CJ. 2000b Histochemically-reactive zinc in amyloid plaques, angiopathy, and degenerating neurons of Alzheimer's diseased brains. *Brain Res* **852**, 274–278.

Suh SW, Thompson RB, Zeng YA, Hough C, Li Y, Sarvey J, Frederickson CJ. 2001 Zinc signals in apical dendrites after stimulation of zinc-containing (but not zinc-free) synaptic inputs. *Soc Neurosci Abs* (in press).

Thompson RB, Whetsell WO, Maliwal BP, Fierke CA, Frederickson CJ. 2000 Fluorescence Microscopy of stimulated Zn(II) release from organotypic cultures of mammalian hippocampus using a carbonic anhydrase-based biosensor system. *J Neurosci Meth* **96**, 35–45.

Thompson RB, Suh SW, Fierke C, Frederickson CJ. 2001 Ratiometric quantitation of fast Zn^{2+} signals in the brain. *SPIE* (in press).

Tonder N, Johansen FF, Frederickson CJ, Zimmer J, Diemer NH. 1990 Possible role of zinc in the selective degeneration of dentate hilar neurons after cerebral ischemia in the adult rat. *Neurosci Lett* **109**, 247–252.

Toutenhoofd SL, Strehler EE. 2000 The calmodulin multigene family as a unique case of genetic redundancy, multiple levels of regulation to provide spatial and temporal control of calmodulin pools? *Cell Calcium* **28**, 83–96.

Truong-Tran AQ, Carter J, Ruffin R, Zalewski PD. 2001 New insights into the role of zinc in the respiratory epithelium. *Immunol Cell Biol* **79**, 170–177.

Vallee BL, Falchuk KH. 1993 The biochemical basis of zinc physiology. *Physiol Rev* **73**, 79–118.

Van Assche F, van Tilborg W, Waeterschoot H. 1996 Environmental Risk Assessment for Essential Elements - Case Study Zinc. In: '*Report of the International Workshop on Risk Assessment of Metals and their Inorganic Compounds*' Ottawa: ICME; 171–180.

Vincent SR, Semba KA. 1989 A heavy metal marker of the developing striatal mosaic. *Dev Brain Res* **45**, 155–159.

Vogt K, Mellor J, Tong G, Nicoll R. 2000 The actions of synaptically released zinc at hippocampal mossy fiber synapses. *Neuron* **26**, 187–196.

Walker MC, Ruiz A, Kullmann DM. 2001 Monosynaptic GABAergic Signaling from Dentate to CA3 with a Pharmacological and Physiological Profile Typical of Mossy Fiber Synapses. *Neuron* **29**, 703–715.

Weiss JH, Sensi SL, Koh JY. 2000 Zn(2+), a novel ionic mediator of neural injury in brain disease. *Trends Pharmacol Sci* **10**, 395–401.

Weiss JH, Sensi SL. 2000 Ca^{2+}-Zn^{2+} permeable AMPA or kainate receptors, possible key factors in selective neurodegeneration. *Trends Neurosci* **20**, 365–371.

Williams, RJP. 1996 Calcium binding proteins in normal and transformed cells. *Cell Calcium* **20**, 87–93.

Ye B, Maret W, Vallee BL. 2001 Zinc metallothionein imported into liver mitochondria modulates respiration. *Proc Natl Acad Sci USA* **98**, 2317–2322.

Yoon YH, Jung KH, Sadun AA, Shin HC, Koh JY, Park JA, Lee JY, Sato TA, Koh JY. 2000 Co-induction of p75NTR and p75NTR-associated death executor in neurons after zinc exposure in cortical culture or transient ischemia in the rat. *Neurosci* **20**, 9096–9103.

Zambenedetti P, Giordano R, Zatta, P. 1998 Metallothioneins are highly expressed in astrocytes and microcapillaries in Alzheimer's disease. *J Chem Neuroanat* **15**, 21–26.

Zangger K, Oz G, Haslinger E, Kunert O, Armitage IM. 2001 Nitric oxide selectively releases metals from the N-terminal domain of metallothioneins, potential role at inflammatory sites. *FASEB J* **15**, 1303–1305.

ZhuGe R, Fogarty KE, Tuft RA, Lifshitz LM, Sayar K, Walsh JV Jr. 2000 Dynamics of signaling between Ca(2+) sparks and Ca(2+)- activated K(+) channels studied with a novel image-based method for direct intracellular measurement of ryanodine receptor Ca(2+) current. *J Gen Physiol* **16**, 845–864.

BioMetals 14: 367–383, 2001.

Review

Extracellular and immunological actions of zinc

Lothar Rink* & Philip Gabriel
*Institute of Immunology and Transfusion Medicine, University of Lübeck School of Medicine, Ratzeburger Allee 160, D-23538 Lübeck, Germany; *Author for correspondence (Tel: +49-451-500 3694; Fax: +49-451-500 3069; E-mail: rink@immu.mu-luebeck.de)*

Received 27 December 2000; accepted 15 March 2001

Key words: cell biology, immunology, review, trace elements, zinc

Abstract

Zinc is an essential trace element for the immune system, but also very important in other organ systems. Every highly proliferating cell system is dependent on sufficient availability of zinc. During the last decades the influence of zinc on various cell systems have been investigated. Multiple effects of exogenously added zinc have been described in *in vitro* culture systems and in *in vivo* systems. However, most of these effects are so far poorly understood, and the dosages used in the *in vitro* systems are not comparable and sometimes unphysiologically high. Especially in the immune system a number of effects were described and over the last ten years we have come to understand some molecular mechanisms of zinc in this cell system. A zinc deficiency is accompanied by an immunodeficiency, resulting in an increased number of infections. However, the immune function is delicately regulated by zinc, since both increased and decreased zinc levels result in a disturbed immune function. Therefore, zinc supplementation must be accurately supervised. In this review, we discuss the activity of extracellular zinc in four sections. 1. The effect of zinc on different *in vitro* cell systems, including keratinocytes, osteocytes and leukocytes, and the concentrations of zinc needed for a specific cell response. 2. The modulation of the innate immune system *in vitro* and *in vivo*. 3. The role of zinc in the B cell response and antibody production. 4. Effects of zinc on the development and function of T cells.

Abbreviations: AIDS – acquired immune deficiency syndrome; BSA – bovine serum albumin; CD – cluster of differentiation; FCS – fetal calf serum; FIV – feline immune deficiency virus; HLA – human leukocyte antigen; IFN – interferon; Ig – immunoglobulin; IL – interleukin; IRAK – interleukin 1 receptor associated kinase; KIR – killer cell inhibitory receptor; LPS – lipopolysaccharide; MHC – major histocompatibility complex; MLC – mixed lymphocyte culture; MLR – mixed lymphocyte reaction; MT – metallothionein; NDV – Newcastle disease virus; NK – natural killer; PBMC – peripheral blood mononuclear cells; PMA – phorbol myristate acetate; PMN – polymorphonuclear neutrophils; ROS – reactive oxygen species; SF – serum free; SOD – superoxide dismutase; STZ – serum-treated zymosan; TCR – T cell receptor; TH – T helper; TNF – tumor necrosis factor; ZIP – zinc regulated metal transporter (ZRT) iron regulated metal transporter (IRT) like protein; ZnT – zinc transporter.

Introduction

Zinc is an essential trace element for all organisms (Raulin 1869; Todd *et al.* 1934). In mammals, a zinc deficiency is primarily observed by its effects on highly proliferating cell systems like the skin and the immune system. Prasad *et al.* (1963) described a zinc deficiency syndrome in children from Persia practicing geophagia, which was characterized by anaemia, hypogonadism, hepatosplenomegaly, skin alterations, growth and mental retardation. With the discovery of acrodermatitis enteropathica (a rare autosomal recessive inheritable disease) it was clearly shown that these symptoms are dependent on zinc deficiency due to a zinc-specific malabsorption syndrome (Neldner & Hambidge 1975). This disease shows a number of immunological alterations like thymic atrophy and a high frequency of bacterial, viral and fungal infections. Without treatment, this disease leads to death within a few years, whereas pharmacological zinc supplementation can reverse all symptoms (Neldner & Hambidge 1975). Since these early observations there is no doubt about the importance of zinc for the integrity of the immune system. During the last two decades, a number of reviews have reflected these issues. However, the groups focused on different topics other than the *in vivo* mouse model (King *et al.* 1995), *in vitro* systems (Bach 1981; Wellinghausen *et al.* 1997a; Wellinghausen & Rink 1998; Rink & Kirchner 2000), clinical trials (Prasad 2000) or nutritional aspects of zinc and immunology (Rink & Gabriel 2000).

The major problem in zinc biology is that there is no specialized zinc storage system in the body. Therefore we have to reach a steady state of zinc intake and excretion. The bioavailability of zinc depends on the composition of the diet and is influenced by a number of different factors, as reviewed elsewhere (Valberg *et al.* 1984; Favier & Favier 1990; Rink & Gabriel 2000). Besides the composition of the diet, the constitution (Weiss *et al.* 1998; Klainman *et al.* 1981; Yuzbasiyan-Gurkan *et al.* 1989) and age (Cakman *et al.* 1996; Rink *et al.* 1998) of the consumer is important for zinc resorption, leading to a number of contradictory recommendations according to the daily intake of zinc (Rink & Gabriel 2000).

Due to these problems, clinical trials are somewhat problematic. The total body content of zinc in humans is 2–4g, but zinc is called a trace element since its plasma concentration is only 12–16 μM (definetively normal) and with ranges from 10.1–16.8 μM in women and 10.6–17.9 μM in men. However, the plasma pool is the smallest zinc pool in the body, but a highly mobile and immunologically important one (Mills 1989; Favier & Favier 1990). In the serum, zinc is predominantly bound to albumin (60%, low-affinity), α_2-macroglobulin (30%, high-affinity) and transferrin (10%) (Scott & Bradwell 1983). These distributions and affinities are also important for *in vitro* culture systems.

Zinc supplementation in *in vitro* systems

The effect of extracellularly added zinc ions was investigated in different cell systems. However, the effective zinc dosages are difficult to compare due to the fact that different culture media and zinc sources were used. Generally, higher zinc dosages are needed, if the culture medium contains serum. Therefore, the percentage of serum as well as the source of the serum is important, since some zinc binding proteins are enhanced in fetal serum and the total protein content varies between different species. Furthermore, some zinc effects, like the IFN-γ induction in T cells (Driessen *et al.* 1994; Wellinghausen *et al.* 1997b), is only observed in the presence of serum. This problem indicates that serum- or protein-free media are not an advantage every time, but the zinc effect is more clear and the amounts of zinc to be used are strongly reduced in comparison to media containing serum. However, the composition of serum-free media is not normally published by the manufacturer. Since most serum-free and all protein-free media (to the authors' knowledge) contain zinc themselves, the real zinc concentration is questionable for the investigator and the reader. The same is true for some conventional cell culture media, which vary in their zinc content from minute amounts up to 3 μM zinc, which is all below the stimulatory level. In serum-free media, albumin, transferrin and insulin are the normal protein supplements. All three have a zinc-binding capacity and influence the zinc-dependent response of the investigated cells (Wellinghausen *et al.* 1996b). Table 1 summarizes the effects of zinc in different *in vitro* cell systems like: keratinocytes, monocytes, T cells, thymocytes, neutrophils, neuroblastoma cells, pheochromocytoma cells, hepatocytes, fibroblasts, spermatozoa, astrocytes, osteocytes, osteoblasts, osteoclasts, epithelial cells and pancreatic islet cells.

Zinc has various effects on completely different cell systems, but the mechanism of zinc influx was controversial for a long period of time. Zinc added to a cell culture enters the cells within minutes (Wellinghausen *et al.* 1996b; Reyes 1996). Recently, Gaither & Eide (2000) described a human zinc transporter (hZIP) for zinc uptake from the environment, whereas so far only some zinc-specific transporters (ZnT) which avoid the efflux of zinc from intracellular pools were described. The ZnTs seem to be involved in intracellular redistribution of zinc and were first described in the nervous system (Palmiter & Findley 1995; Palmiter *et al.* 1996a,b; Tsuda *et al.* 1997). There is no report as to whether or not the ZnTs are associated with zinc uptake, whereas the transferrin receptor (CD71) and calcium ion channels were discussed in terms of unspecific transport of zinc in addition to facilitated

Table 1. Effect of extracellular zinc in *in vitro* cell systems. The table gives examples for *in vitro* effects of zinc on different cell types. Interestingly, the effective zinc content showed extreme variation. The examples are listed in increasing zinc amounts in the experimental system. Furthermore the culture conditions are indicated, since the protein amounts influences the free zinc content as discussed in the text.

Zinc [μM]	Medium*	Cell type	Effect	Reference
10^{-8}–100	BSA	rat osteoclasts	zinc is a highly potent inhibitor of osteoclastic bone resorption	Moonga & Dempster 1995
2–8	SF	human keratinocytes	zinc gluconate induces the expression of Vα, α3, α2 & α6 - integrins	Tenaud *et al.* 1999
2–20	SF	human keratinocytes	zinc gluconate reduces the very late antigen(VLA)-3 expression induced by nickel gluconate	Sainte-Marie *et al.* 1998
25	FCS	human neuro-blastoma BE(2)-cells	zinc sulfate decreases the level of apoptosis in neuronal cells exposed to toxin	Ho *et al.* 2000
50	FCS	rat astrocytes	zinc chloride protects diethyldithiocarbamate-mediated toxicity associated with an increase of MT concentration	Wilson & Trombetta 1999
15–100	SF	D10 N T-cell line	100 μM zinc sulfate inhibits the IL-1 type I receptor-associated kinase (IRAK)	Wellinghausen *et al.* 1997
100	FCS	HeLa human Hepa mouse	zinc sulfate facilitates activation of the DNA binding activity of recombinant MTF-1	Bittel *et al.* 1998
10–100	FCS	Osteoblast-like cells MC3T3-E1	zinc sulfate inhibits mineralization during tissue formation	Togari *et al.* 1993
100	FCS	PBMC of HIV positive patients	zinc chloride decreases the percentage of apoptotic cells compared with cells treated only with PHA	Neves *et al.* 1998
100	FCS	mouse pancreatic islets	zinc sulfate induces MT to protect islets against toxicity mediated by reactive oxygen species	Ohly & Gleichmann 1995
100	BSA	human monocytes	zinc aspartate moderately activates monocytes	Herold *et al.* 1995
50–150	FCS	human keratinocytes	100 μM zinc chloride induces cell proliferation	Parat *et al.* 1999
25–200	BSA	chicken osteocytes	zinc sulfate alters bone resorptive rates	Chen *et al.* 1998
80–200	FCS	mouse thymocytes	zinc sulfate induces apoptosis in CD4+CD8+$\alpha\beta$ TCR^{lo}CD3ε^{lo} thymocytes	Telford & Fraker 1995
12–250	FCS	intestinal epithelial cell line IEC-6	zinc sulfate promotes intestinal epithelial wound healing by enhancement of epithelial cell restitution	Cario *et al.* 2000
30–250	FCS	human PBMC	zinc sulfate induces IL-1, 6, TNF-γ, sIL-2R & IFN-γ	Wellinghausen *et al.* 1996a,b, Driessen *et al.* 1994

250	FCS or SF	monocytes, Mono-Mac cell line	zinc sulfate induces IL-1β in both systems and SF or FCS containing medium	Driessen *et al.* 1994
3–300	FCS	rat pheo-chromocytoma PC12 cells	zinc potentiates the dopamine release evoked by ATP	Koizumi *et al.* 1995
30–300	FCS	human PBMC	zinc chloride reduces the frequency of cell division and induces blast formation	Santra *et al.* 2000
100–300	FCS	swiss 3T3 mouse fibroblasts	zinc chloride mimics the action of growth factors on intracellular MAP kinase activation and protein tyrosine phosphorylation	Hansson 1996
25–500	FCS	mouse thymocytes	zinc sulfate inhibits glucocorticoid induced apoptosis in mouse thymocytes. Zinc concentrations lower than 25 μM had to be combined with ionophores. Concentrations between 80–200 μM induce apoptosis.	Telford & Fraker 1995
740	FCS	neonatal mouse skin cells	zinc chloride protects against UV-induced genotoxicity	Record *et al.* 1996
500–1000	BSA	human neutrophils	zinc chloride attracts leukocytes by inducing and promoting the chemotactic response	Hujanen *et al.* 1995
1000	SF	human spermatozoa	zinc chloride elicits an inhibition of superoxide anion production and SOD-like activity	Gavella *et al.* 1999

*Culture conditions without serum (SF), with serum (FCS) or bovine serum albumin (BSA)

diffusion through amino acids and anionic exchange (Bentley 1992; Hogstrand *et al.* 1996). However, there are also contradictory reports for these mechanisms (Wellinghausen *et al.* 1996b). Since the exogenously added zinc increases the free intracellular zinc about 70% (measured by zinquin), but the total zinc uptake is about 300–600%, depending on the cell system (measured by atomic absorption spectroscopy), there must be a fast binding process to intracellular proteins (Wellinghausen *et al.* 1996b; Fischer *et al.*, manuscript in preparation). Both, the free and the total zinc uptake shows a fast increase within the first minutes and a saturation at the described maxima after 30–60 min (Wellinghausen *et al.* 1996b; Fischer *et al.*, manuscript in preparation). The new described human ZIP seem to be the main way of zinc influx into human cells (Gaither & Eide 2000), but their distribution in leukocyte subsets is so far not investigated.

General effects on eukaryotic cells influencing immune functions

Zinc is a cofactor for more than three hundred enzymes out of all six classes of enzymes (Coleman 1992a,b; Vallee & Falchuk 1993) as it is important for the structural integrity or enzymatic activity of the enzymes (reviewed in Rink & Gabriel 2000). Furthermore, zinc modulates the activity of a number of enzymes. Factors interacting with DNA or RNA, like transcription and replication factors, contain a zinc finger motif (reviewed in Rink & Gabriel 2000). Therefore, a variety of general cell functions are influenced by the zinc concentration. For this reason, cell proliferation is strictly zinc-dependent and, without zinc, highly proliferating cell systems, like the immune system, the skin and the reproductive system, show diverse dysfunctions. The dysfunction reflects two aspects, the ageing of the cells with functional deficits and the missing regeneration of the system by the production of new completely functional cells. Furthermore, different factors important for signal transduction need zinc for a regular function (reviewed by Beyersmann & Haase in this issue and Rink & Gabriel 2000).

Apoptosis, the physiological method of programmed cell death, is very important in the development and differentiation of complex organisms. The apoptosis is regulated by zinc (reviewed by Truong-Tran *et al.* in this issue). Especially in the immune system regulation and normal function are strictly dependent on apoptosis to exclude autoimmune T cells and B cells and to kill infected or tumorous cells by cytotoxic T cells or NK cells without side effects (reviewed by Wellinghausen & Rink 1998; Rink & Gabriel 2000).

These different zinc effects are very important but not restricted to any cell or organ system, as shown by the *in vitro* systems above. However, a slightly decreased zinc status may first influence the immune system, due to an increased number of infections. Despite these general consequences, there are also some direct effects of zinc on the immune system.

Modulation of immunological functions by zinc

The immune system can be divided into different parts. The first line of defense is the innate immune system with granulocytes, monocytes and natural killer (NK) cells. These cells are completely differentiated in the peripheral blood and do not need further education for their function. Therefore, the response is very fast but lacks a memory. In contrast, the specific immune system, with the two parts, humoral (B cells) and cellular immunity (cytotoxic T cells) are produced as precursors and educated to recognize their specific antigen in the thymus (T cells) or bone marrow (B cells). The resulting naive (before antigen contact) lymphocytes differentiate after antigen contact into effector cells and memory cells. The memory cells are the basis for the immunological memory and the stronger reaction to a known antigen as a secondary response. Nowadays there is no doubt that zinc is an essential trace element for the immune system. The effects of zinc are multi-faceted and influence the innate as well as the specific part of the immune system. Furthermore, not only proliferation of the immune system depends on zinc but also the proliferation of the pathogens, thus decreasing zinc in the plasma is one acute phase response in infection. However, cellular and molecular mechanisms of zinc within the immune system were discovered only during the last 10 years.

Innate immunity

The earliest step of an immune response is the recruitement of leukocytes from the blood stream to the infected tissue via chemotaxis, adhesion and diapedesis of the leukocytes. Zinc induces adhesion of myelomonocytic cells to the endothelium, whereas zinc chelation diminishes cell recruitment (Chavakis *et al.* 1999). The chemotaxis of neutrophils, the step before the adhesion, is decreased under zinc deficiency *in vivo*. *In vitro*, zinc itself showed a chemotactic activity on neutrophil granulocytes (PMN) (Hujanen *et al.* 1995). However, more important for the PMN is the general effect of zinc on cell proliferation, since PMN are produced and released by the bone marrow at a rate of 60 million cells per minute. Furthermore, the main functions of the cells of the innate immune system are impaired under zinc deficiency: natural killer (NK) cell activity, phagocytosis of macrophages and neutrophils, and generation of the oxidative burst (Keen & Gershwin 1990; Allen *et al.* 1983). Neutrophils do not respond with cytokine production to zinc, but seem to have an influence on the viability of these short-living cells (unpublished data). This may be due to the fact that PMN contains a high concentration of zinc binding proteins. Release of the S-100 Ca^{2+} binding protein calprotectin during degradation of neutrophils inhibits reproduction of bacteria and *Candida albicans* by zinc chelation (Murthy *et al.* 1993; Clohessy & Golden 1995, Sohnle *et al.* 1991). Effects on neutrophil granulocytes are summarized in Figure 1.

The influence of zinc on NK cells could be partially explained on the molecular level, since zinc is required for the interaction of the p58 killer cell inhibitory receptor (KIR) on NK cells with MHC class I molecules (mainly HLA-C) on target cells (Rajagopalan *et al.* 1995). In contrast to the influence on the killer inhibitory signal, the positive signals did not require zinc (Rajagopalan *et al.* 1995). This may result in unspecific killing and functional loss of NK cells during zinc deficiency. In healthy elderly persons

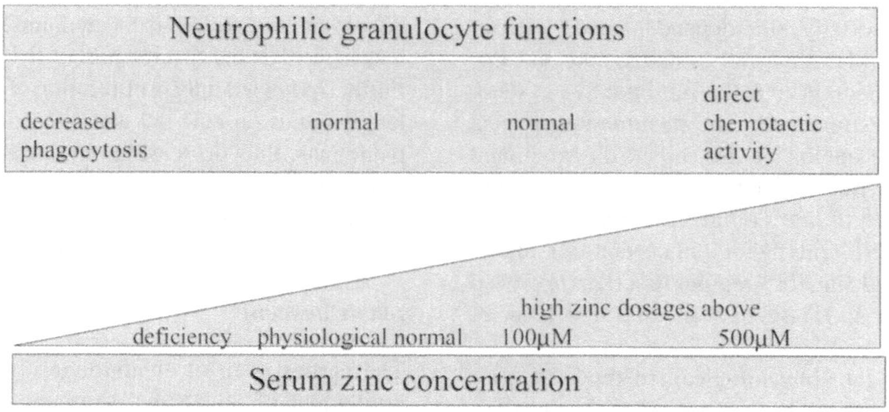

Fig. 1. Influence of zinc on the function of neutrophil granulocytes. Neutrophil increase their main immune functions with increasing zinc concentrations.

(SENIEUR-elderly), a group with decreased serum zinc without malnutrition, the number of NK cells is increased, but the killing activity is decreased (Rink *et al.* 1998; Rink & Seyfarth 1997). This effect on NK cells is also observed under experimental conditions of zinc deficiency *in vivo* and *in vitro* (Prasad 1998, 2000). However, *in vitro* zinc showed no effect on purified NK cells (Crea *et al.* 1990). Effects on NK cells are summarized in Figure 2.

A number of effects of zinc on monocytes were described *in vitro*. Zinc induced activation (Herold *et al.* 1993) and cytokine production in isolated monocytes as well as in monocytic cell lines (Driessen *et al.* 1994; Wellinghausen *et al.* 1997b). Furthermore, a number of cytokines induced in peripheral blood mononuclear cells (PBMC) could be related to being produced by the monocyte fraction, such as IL-1, IL-6 and TNF-α since these cytokines are produced in the absence of T cells as well (Driessen et al. 1994; Wellinghausen *et al.* 1996, 1997). At least for TNF-α it was shown that zinc induced a de novo synthesis of the mRNA (Wellinghausen 1996a). Monocyte activation by zinc is specifically enhanced by insulin and transferrin in the culture medium, whereas high serum content of the medium prevents the stimulation (Crea *et al.* 1990; Phillips & Azari 1974; Driessen *et al.* 1995; Wellinghausen *et al.* 1996b). This synergism is not mediated by the specific receptors (Wellinghausen *et al.* 1996b). Under serum-free conditions 50–100 μM zinc are sufficient for cytokine induction in monocytes, whereas under serum supplementation 250 μM are necessary (Driessen *et al.* 1994, 1995; Wellinghausen *et al.* 1996b). How monocytes are directly activated by zinc is unresolved, but protein tyrosine kinases as well as cAMP- and cGMP-dependent protein kinases are clearly involved (Wellinghausen *et al.* 1996). Monocytes showed a higher tolerance to exogenous zinc than lymphocytes, but there is no difference in the zinc uptake (Goode *et al.* 1989; Bulgarini *et al.* 1989; Wellinghausen *et al.* 1996b, 1997b). Interestingly, monocytes from zinc-deficient elderly persons showed a higher proinflammatory cytokine response to lipopolysaccharide and phorbol ester stimulation and have a preactivation of monocytes (Rink *et al.* 1998; Fagiolo *et al.* 1993, and unpublished data). On the other hand, *in vitro* zinc supplementation could restore the defective IFN-α production of PBMC (note that monocytes and dendritic cells are the main IFN-α producers) from zinc deficient elderly (Cakman *et al.* 1997). The effects on monocytes are summarized in Figure 3. In conclusion, the innate immune system needs zinc for the generation of the great number of cells and for the function on a molecular level. The effects on the innate immune system are summarized in Tables 2a–d.

B cells

Although B cells are the producers of antibodies and therefore the most important cells of the humoral immunity, there is little knowledge about these cells with regard to zinc. Zinc itself seems to have no direct influence on the activity of B cells (Crea *et al.* 1990). However, zinc deficient patients, like elderly and hemodialysis patients, showed a reduced response to vaccination (Fraker *et al.* 1986; Lighart *et al.* 1984; Bonomini *et al.* 1993; Sandstead *et al.* 1982; Cakman *et al.* 1996). For hemodialysis patients, at least we were able to correlate the response to the serum zinc concentration (Kreft *et al.* 2000). However, vari-

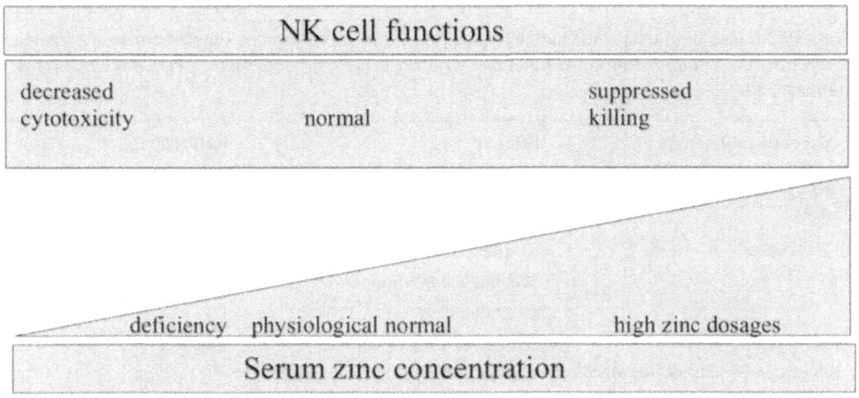

Fig. 2. Influence of zinc on the function of natural killer (NK) cells. Only zinc levels within the normal range can guarantee effective NK cell function.

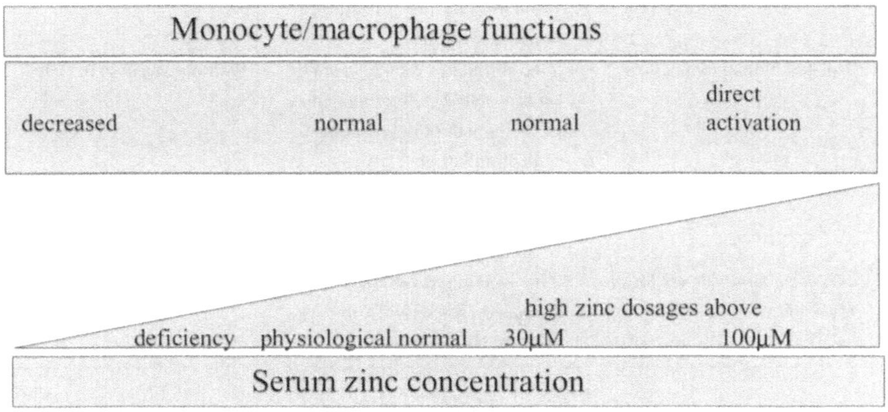

Fig. 3. Influence of zinc on the function of monocytes and macrophages. Monocytes are the only cell population that can be directly induced by zinc ions.

Table 2a. Innate immunity: zinc deficiency *in vivo*. Cells of the innate immune system showed impaired *in vitro* functions after *in vivo* zinc deficiency.

Experimental system	Effect	Reference
NK cells	NK cell lytic activity decreases after 20 weeks of deficiency	Prasad 2000
Elderly subjects	Reduced IFN-α production after stimulation with NDV	Cakman et al. 1997
NK cells	NK cell lytic activity is decreased in zinc deficiency	Prasad 1998
Zinc deficient diet for 3 weeks (Rat model)	Decreased NK and LPS activated NK cell activity is associated with zinc deficiency	Ozturk *et al.* 1994
Human granulocytes	Zinc deficiency showed *in vivo* the reduction of the oxidative burst	Keen & Gershwin 1990 Allen *et al.* 1983
Human monocytes	Zinc deficiency impairs phagocytosis	Allen *et al.* 1983 Keen & Gershwin 1990

Table 2b. Innate immunity: zinc supplementation *in vitro*. The *in vitro* supplementation of zinc can reverse or rarely improve immune functions with regard to cytoprotection or specific capabilities of immune cells.

Experimental system	Effect	Reference
AK-5 cells, NK cells	Pretreatment of AK-5 with zinc sulfate resulted in complete inhibition of antibody-dependent NK-induced DNA fragmentation	Bright *et al.* 1995
Human granulocytes: phagocytosis and killing of *S. aureus* and *S. epidermidis*	Cytoprotection of zinc against staphylococcal toxins	Sunzel *et al.* 1995
Rat granulocytes isolated from peritoneal cavity	1 mM zinc chloride attracts leukocytes by inducing and promoting the chemotactic response	Hujanen *et al.* 1995
Isolated human monocytes	Zinc stimulates monocytes, no other isolated cell component of the human blood responds with stimulation	Wellinghausen *et al.* 1997
Septic rat monocytes	Zinc inhibits the superoxide production after stimulation of both PMA and STZ	Srinivas *et al.* 1989
NK cells, target clones HLA-Cw4 and 8	Zinc is required for HLA-C mediated protection from lysis by NK cells	Rajagopalan 1995

Table 2c. Innate immunity: zinc supplementation *in vivo*. The *in vivo* zinc supplementation can modify and reverse immune dysfunctions caused by mild or severe zinc deficiency.

Experimental system	Effect	Reference
Human monocytes from patients with leukemia	Orally administered zinc aspartate increases the capacity of monocytes to release of ROS after *in vitro* stimulation	Herold 1993
Plasma of cervical carcinoma patients	Zinc supplementation increases IL-2 production of PBMC and restores thymulin production and NK cytotoxicity	Mocchegiani *et al.* 1999
Septic rat monocytes	Increased superoxide production after PMA or STZ stimulation	Srinivas *et al.* 1989
NK cells	NK cell lytic activity returns to normal range	Prasad 2000

Table 2d. Innate immunity: therapeutic zinc application. Four main examples regarding the therapeutic use of zinc as a modulator of the immune system. The positive effect of using orally applicated zinc solutions is supported by these examples.

Disease	Possible effect of zinc	Reference
Common cold	Zinc gluconate stabilizes the cell membrane against viral penetration and increases IFN-α	Mossad *et al.* 1996
Acrodermatitis enteropathica	Increases NK cell activity with regard to IL-2 induction and increases the phagocytic activity of phagocytes	Prasad *et al.* 1995
Rheumatoid arthritis	impairment of PMN phagocytosis	Zoli *et al.* 1998
Herpes simplex infection	Increased IFN-α production	Varadinova *et al.* 1993

Table 3a. In vitro effects of zinc deficiency on B cell functions (mouse model). Specific cell experiments support the hypothesis that B cell maturation depends on zinc.

Cell type	Effect	Reference
precursor B cells ($CD45^+$ $CD43^-IgM^-$)	zinc deficiency induces apoptosis and reduces cell count 50–70%	Fraker *et al.* 2000
Immature B cells ($CD45^+IgM^+IgD^-$)	zinc deficiency induces apoptosis and reduces cell count 50–70%	Fraker *et al.* 2000
Pro-B cells (CD $45^+CD43^+6C3^+$)	high bcl-2 level protects against apoptosis caused by zinc deficiency	Fraker *et al.* 2000
Mature B cells (IgM^+IgD^+)	high bcl-2 level protects against apoptosis caused by zinc deficiency	Fraker *et al.* 2000

Table 3b. In vivo effects of zinc deficiencies on B cells. *In vivo* zinc deficiency experiments support the findings made in *in vitro* models.

Cell type	Effect	Reference
B cells	91% decrease in severely deficient mice; 43% decrease in moderately deficient mice	Fraker *et al.* 1995
$CD45^+$ IgM^-	56–96% decrease	Fraker *et al.* 1995
Immature B cells $CD45^+$ IgM^+	35–80% decrease	Fraker *et al.* 1995
Mature B cells $CD45^+IgM^+IgD^+$	5–70% decrease	Fraker *et al.* 1995
B cells	IgM, IgG & IgA levels are increased	Rink & Seyfarth 1997

Table 3c. In vitro effects of zinc supplementation on B cells. *In vitro* supplementation with zinc reverses the dysfunctions induced by the zinc deficiency.

Cell type	Effect	Reference
70Z/3 murine pre-B leukemia cell line	zinc induces IL-4 associated CD5 downregulation	Jyonouchi *et al.* 1991

Table 3d. In vivo effects of zinc supplementation on B cells. Zinc supplementation *in vivo* benefits the treated subjects according to the immune function.

Subject	Effect	Reference
6–35-month-old infants	lower respiratory infections were reduced by 10 mg/d	Fraker 2000
Elderly	improved IgG antibody response to tetanus vaccine	Duchateau *et al.* 1981b

ous vaccination studies were done with additional zinc supplementation, but in most cases, there was no increase of the antibody titer against the vaccine (Rawer *et al.* 1987; Grekas *et al.* 1992; Brodersen *et al.* 1995; Turk *et al.* 1998; Provinciali *et al.* 1998). The major problem in all these studies was, that the zinc uptake was not controlled and that the amount of zinc applied to the probands was not comparable and sometimes definitely too high (400 mg/day), since different groups reported a suppression of immune functions at high zinc dosages like 100 mg/day (Porter *et al.* 1977; Chandra 1984; Patterson *et al.* 1985; Provinciali *et al.* 1998; Rheinhold *et al.* 1999; Rink & Kirchner 1999). But if zinc supplementation is done in the right way, IgG response to vaccination could be improved (Duchateau *et al.* 1981b). This may be related to the induction of apoptosis in immature B cells and B cell precursors by zinc deficiency (Fraker *et al.* 2000). Since mature B cells due to a high Bcl-2 level are more resistant to zinc deficiency, B cell memory is less affected than a primary response, like initial vaccination (Fraker *et al.* 2000). Other possible mechanisms are the increase of IFN-α production by zinc (Cakman *et al.* 1997) or the restoration of impaired T cell help (Sandstead *et al.* 1982; Mocchegiani *et al.* 1995a). Both these explanations could also explain the failure of studies with high zinc dosages, since these inhibit IFN-α production as well as T cell functions (Cakman *et al.* 1997; Wellinghausen *et al.* 1997b). Effects of zinc on B cells and B cell functions are summarized in Figure 4 and Tables 3a–d.

T cells

One of the first *in vivo* observations regarding zinc was thymic atrophy, which resulted in an impaired T cell development and decreased T cell counts (Osati-ashtiani *et al.* 1998; Fraker *et al.* 1995). Essential steps in thymic function are dependent on the thymic hormone thymulin (a nonapeptide), which is only active after binding of zinc as a cofactor (Bach 1981, 1983). Thymulin is secreted by thymic epithelial cells and induces markers of differentiation in immature T cells (Saha *et al.* 1995). Besides these intrathymic functions on thymocytes and immature T cells, thymulin also acts on mature peripheral T cells. It modulates the cytokine release by PBMC and induces proliferation of CD8 T cells in combination with IL-2 (Coto *et al.* 1992; Safie-Garabedian *et al.* 1993). Therefore, zinc influences immature and mature T cells through the activation of thymulin. As expected, substitution of zinc can reverse the zinc deficiency-induced changes in the thymus and on peripheral cells (Mocchegiani *et al.* 1995). This effect is also observed in AIDS patients (Mocchegiani *et al.* 1995). In contrast to other lymphocyte populations, a direct effect of zinc on T cells was observed. Thirty years ago it was first described that zinc induced blast transformation in human lymphocytes (Berger & Skinner 1974; Sood *et al.* 1999; Kirchner & Rühl 1970; Rühl *et al.* 1971). Furthermore, zinc induced the expression of the high affinity receptor for IL-2 (Tanaka *et al.* 1989), one effect resulting in decreased proliferation of T cells in zinc deficiency (Crea *et al.* 1990; Dowd *et al.*

Table 4a. In vitro effects of zinc deficiency on T cell function. Zinc is essential for the effectiveness of T cells according to their immune functions.

Cell type	Effect	Reference
HUT-78 (Th_0 T cell line) precultured in zinc deficient medium	IL-2 gene expression, IL-2R and NF-κB is reduced	Prasad 2000

Table 4b. In vivo effects of zinc deficiency on T cells. *In vivo* experiments point out the predominant role of zinc according to T cell maturation.

Experimental system	Effect	Reference
Dietary induction of zinc deficiency	within 8 weeks reduced lymphocyte, granulocyte, and platelet counts	Prasad 2000
Dietary induction of zinc deficiency	reduced thymulin activity in serum. Imbalance of TH_1 and TH_2. Decrease in the percentage of $CD_8^+CD_{73}^+$ T cells (cytotoxic T cell precursor)	Prasad 1998
Elderly subjects	Reduced T cell counts	Cakman *et al.* 1997
Th_1-T cells of zinc deficient subjects	Lower IL-2 and IFN-γ production	Prasad 2000

1986). IL-2 itself, as well as the soluble IL-2 receptor (sIL-2R) and interferon (IFN)-γ (all mainly T cell products) are induced by zinc in human PBMC (Salas & Kirchner 1987; Scuderi 1990; Driessen *et al.* 1994). However, at least the induction of IFN-γ is dependent on the presence of monocytes (Salas & Kirchner 1987; Driessen *et al.* 1994; Rühl & Kirchner 1978; Wellinghausen *et al.* 1997b). The secretion of IFN-γ by T cells depends on the induction of IL-1 in monocytes, since anti-IL-1 could inhibit the T cell activation (Driessen *et al.* 1994). However, zinc concentrations over 100 μM in serum-free culture medium stimulate monocytes but inhibit T cell functions, since T cells have a lower intracellular zinc concentration and are more susceptible to increasing zinc levels than monocytes (Goode *et al.* 1989; Bulgarini *et al.* 1989; Wellinghausen *et al.* 1997b). Since the increase of intracellular free zinc in monocytes and T cells is equal after exogenous addition of zinc (Wellinghausen *et al.* 1996b, 1997b), the lower tolerance leads to a T cell blockade. Therefore, stimulation of monocytes and T cells by zinc is dependent on the amount of free zinc ions as a counterpart to the protein composition of culture media, as discussed above.

While the zinc-induced activation of T cells is IL-1-dependent, the molecular mechanism is the inhibition of the IL-1 type I receptor associated kinase IRAK by zinc (Driessen *et al.* 1994; Wellinghausen *et al.* 1997b). This mechanism is also the basis for the inhibition of the IL-1-dependent growth of the murine IL-1 indicator cell line D10 (Wellinghausen *et al.* 1997b). Whereas for the IL-1 blockade amounts of 100 μM are necessary, the alloreactivity of T cells in the mixed lymphocyte reaction/culture (MLR or MLC) could already be inhibited by amounts over 50 μM (Campo *et al.* 2001).

In contrast to T cell stimulation, T cell inhibition by an excess of zinc could also be observed *in vivo* (Chandra 1984; Duchateau *et al.* 1981a), but these effects are similar to those observed in zinc deficiency. This means, that the T cell activity is critically regulated by the zinc concentration. This may be the reason why some autoimmune diseases with a T cell pathology, like rheumatoid arthritis, are associated with moderate zinc deficiency (Simkin 1976). In some clinical trials, at least a zinc supplementation reduced the pain score in rheumatoid arthritis (Simkin 1976). This led to the presumption that zinc deficiency increased allo- or autoreactivity, whereas it is inhibited by high zinc dosages. The observation that decreased plasma zinc levels in preganancy are associated with an increased risk of preterm delivery and abortion fits in

Table 4c. In vivo effects of zinc supplementation on T cells. Zinc supplementation can reverse the T cell dysfunctions caused by zinc deficiency.

Zinc supplementation	Effect	Reference
25 mg zinc sulfate for the treatment of residents	Zinc increases the number of $CD_4^+DR^+$T cells and cytotoxic T-lymphocytes	Fortes *et al.* 1998
2 mg/kg/d zinc acetate	Weight gain and recovery from marasmus	Castillo-Duran *et al.* 1987
Zinc supplementation	Zinc restores the decreased thymulin activity	Prasad 1998
Oral zinc supplementation in old mice for 1 month	Full recovery of thymic functions after zinc supplementation	Mocchegiani *et al.* 1995
Elderly subjects	Restores T-cell help	Cakman *et al.* 1996
Low weight infants	Doubled responders to DTH after zinc supplementation of 2 mg/kg/d	Fraker *et al.* 2000

Table 4d. In vitro effects of zinc on T cells. Zinc effects on T cells seem to be contradictory because they stimulate T cell functions and decrease alloreactivity and apoptosis.

Cell type	Effect	Reference
PBMC with regard to T cells	Zinc induces an increased IFN-γ production	Driessen *et al.* 1995
Enriched human T cells and D10 mouse T cells	Higher zinc concentrations directly inhibit the IL-1β dependent T cell stimulation	Wellinghausen *et al.* 1997
Mixed lymphocyte culture (MLC)	Zinc suppresses alloreactivity in the MLC	Campo *et al.* 2001
Interleukin-2-dependent feline T-lymphocyte cell line inoculated with NCSU-1 (FIV) were supplemented with 1 mM zinc chloride	Zinc decreases the percentage of cells undergoing apoptosis and prevented the loss of CD_4^+ lymphocytes	Johnson *et al.* 1996

Table 4e. Zinc therapy with regard to T cells. The predominant effects caused by zinc deficiency in regard to T cells are based on a zinc dependent T cell maturation. So these defects can be reversed by zinc supplementation.

Disease	Possible zinc effect	Reference
Sickle cell disease	Reconstitution of thymocyte function decreases hospitalization and vaseoocclusive pain	Prasad *et al.* 1999
Acrodermatitis enteropathica	Zinc reconstitutes thymocyte functions and reverses all skin and systemic symptoms	Prasad 1995
AIDS	Zinc inhibits T cell apoptosis and increases thymocyte proliferation	Mocchegiani *et al.* 1995
Rheumatoid arthritis	T cell suppression and blocking IL-1 signal transduction	Zoli *et al.* 1998
Down syndrome	Zinc directly influences leukocytes and thymus hormones and reverses the haematological symptoms	Trubiani *et al.* 1996
Crohn's disease	Zinc suppresses T-cells and reconstitutes the thymus function. This results in an alleviation of skin lesions and improvement of visual acuity	Brignola *et al.* 1993

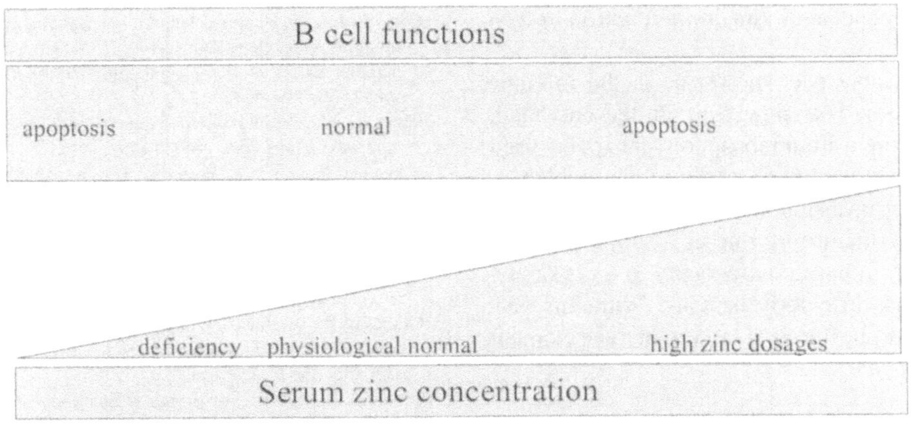

Fig. 4. Influence of zinc on the function of B cells. The normal range of zinc concentration is obligatory for the correct B cell function. B cells sensitively respond to zinc level changes.

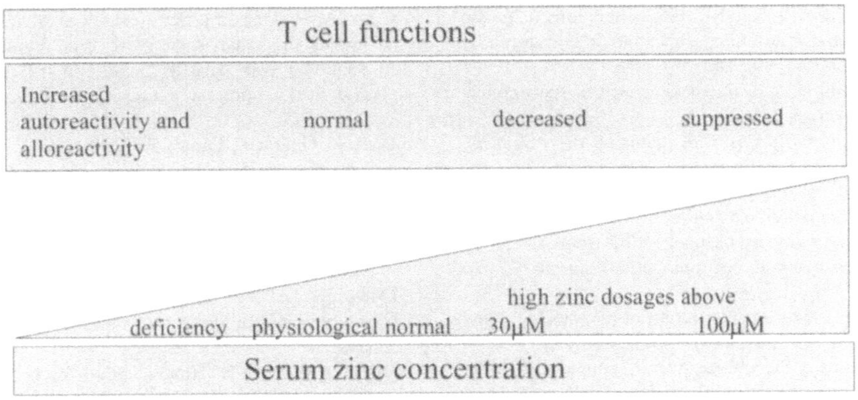

Fig. 5. Influence of zinc on the function of T cells. T cell functions are delicatly regulated by the serum zinc level. Low zinc increases abnormal functions whereas high zinc amounts unspecifically suppresses T cells.

with this model (Bedwal & Bahuguna 1994; Jameson 1993; Favier 1992).

In conclusion, the T cell activity is regulated by zinc and the normal physiological value seems to be slightly below the optimal concentration of T cell functions. A further reduction of this value leads to a dysfunction with autoreactivity, whereas at concentrations above 30 μM the T cell inhibitory effects of zinc take place. These inhibitory effects of zinc might be a new tool for zinc therapies. The influence of zinc on T cells *in vivo* and *in vitro* is summarized in Figure 5 and Tables 4a–e.

Perspective

The reviewed data clearly indicate that zinc is essential for an intact immune system. However, we need more information on a molecular basis to understand the role of zinc on the different cell subsets. Furthermore, investigations regarding the dose response of zinc have to be done in detail, since we still do not know the best effective dose of zinc *in vitro* and, more importantly, *in vivo*. Controlled clinical studies, with zinc supplementation and longitudinal measurement of the zinc content in the serum and cell compartments are missing. Since all experiments investigating the immune system need specific stimulants for the leukocyte subsets to be investigated, we also need further information regarding the influence of zinc on the stimulants themselves (reviewed by Wellinghausen & Rink 1998; Rink & Gabriel 2000). Since cytokine functions and detection are also influenced by zinc, i.e., through zinc-activated α_2-macroglobulin, some results have to be reevaluated or the experimental design has to be changed (James 1990). In conclusion, it is difficult to find an immunostimulant and test system completely independent of zinc in its function *in vivo* or *in vitro*.

However, the perspective of zinc in the immune system seems to be very powerful. On the one hand, zinc may be a new immunosuppressant to be used without massive side effects, a role of zinc which so far has not been investigated in detail. More important seems to be the use of zinc as a supplement for different patient groups and especially to increase the immune response in elderly persons. With this perspective, zinc may be the most important trace element in public health.

References

Allen JI, Perri RT, McClain CJ, Kay NE. 1983 Alterations in human natural killer cell activity and monocyte cytotoxicity induced by zinc deficiency. *J Lab Clin Med* **102**, 577–589.

Bach JF. 1981 The multi-faceted zinc dependency of the immune system. *Immunol Today* **4**, 225–227.

Bach JF. 1983 Thymulin (FTS-Zn). *Clin Immunol Allerg* **3**, 133–150.

Bedwal RS, Bahuguna A. 1994 Zinc, copper and selenium in reproduction. *Experientia* **50**, 626–640.

Bentley PJ. 1992 Influx of zinc by channel catfish (Ictalurus punctatus): uptake from external environmental solutions. *Comp Biochem Physiol C* **101**, 215–217.

Berger NA, Skinner M. 1974 Characterization of lymphocyte transformation induced by zinc ions. *J Cell Biol* **61**, 45–55.

Bittel D, Dalton T, Samson SLA, Gedamu L, Andrews GK. 1998 The DNA binding activity of metal response element-binding transcription factor-1 is activated *in vivo* and *in vitro* by zinc, but not by other transition metals. *J Biol Chem* **273**, 7127–7133.

Bonomini M, Di Paolo B, De Risio F, Niri L, Klinkmann H, Ivanowich P, Albertazzi A. 1993 Effects of zinc supplementation in chronic haemodialysis patients. *Nephrol Dial Transplant* **8**, 1166–1168.

Brignola C, Belloli C, De Simone G, Evangelisti A, Parente R, Mancini R, Innan P, Mocchegiani E, Fabris N, Morini MC. 1993 Zinc supplementation restores plasma concentrations of zinc and thymulin in patients with Crohn's disease. *Aliment Pharmacol Ther* **7**, 75–280.

Brodersen HP, Holtkamp W, Larbig D, Beckers B, Thiery J, Lautenschlager J, Probst HJ, Ropertz S, Yavari A. 1995 Zinc supplementation and hepatitis B vaccination in chronic haemodialysis patients: a multicentre study. *Nephrol Dial Transplant* **10**, 1780.

Bulgarini D, Habetswallner D, Boccoli G, Montesoro E, Camagna A, Mastroberardino G, Rosania C, Testa U, Peschle C. 1989 Zinc modulates the mitogenic activation of human peripheral blood lymphocytes. *Ann Ist Super Sanita* **25**, 463–470.

Cakman I, Kirchner H, Rink L. 1997 Zinc supplementation reconstitutes the production of interferon-α by leukocytes from elderly persons. *J Interferon Cytokine Res* **17**, 469–472.

Cakman I, Rohwer J, Schütz RM, Kirchner H, Rink L. 1996 Dysregulation between TH1 and TH2 T cell subpopulations in the elderly. *Mech Ageing and Dev* **87**, 197–209.

Campo CA, Wellinghausen N, Faber C, Fischer A, Rink L. 2001 Zinc inhibits the mixed lymphocyte culture. *Biol Trace Elem Res* **79**, 15–22.

Cario E, Jung S, Harder d'Heureuse J, Schulte C, Sturm A, Wiedenmann B, Goebell H, Dignass AU. 2000 Effects of exogenous zinc supplementation on intestinal epithelial repair *in vitro*. *Eur J Clin Invest* **30**, 419–428.

Chandra RK. 1984 Excessive intake of zinc impairs immune responses. *JAMA* **252**, 1443–1446.

Chavakis T, May AE, Preissner KT, Kanse SM. 1999 Molecular mechanisms of zinc-dependent leukocyte adhesion involving the urokinase receptor and β2-integrins. *Blood* **93**, 2976–2983.

Chen D, Waite LC, Pierce jr. WM. 1998 *In vitro* bone resorption is dependent on physiological concentrations of zinc. *Biol Trace Elem Res* **61**, 9–18.

Clohessy PA, Golden BE. 1995 Calprotectin-mediated zinc chelation as a biostatic mechanism in host defense. *Scand J Immunol* **42**, 551–556.

Coleman JE. 1992a Zinc proteins: Enzymes, storage proteins, transcription factors and replication proteins. *Annu Rev Biochem* **16**, 897–946.

Coleman JE. 1992b Structure and mechanism of alkaline phosphatase. *Annu Rev Biophys Biomol Struct* **21**, 441–83.

Coto JA, Hadden EM, Sauro M, Zorn N, Hadden JW. 1992 Interleukin 1 regulates secretion of zinc-thymulin by human thymic epithelial cells and its action on T-lymphocyte proliferation and nuclear protein kinase *C. Proc Natl Acad Sci USA* **89**, 7752–7756.

Crea A, Guérin V, Ortega F, Hartemann P. 1990 Zinc et système immunitaire. *Ann Med Interne* **141**, 447–451.

Castillo-Duran C, Heresi G, Fisberg M, Uauy R. 1987 Controlled trial of zinc supplementation during recovery from malnutrition: effects on growth and immune function. *Am J Clin Nutr* **45**, 602–608.

Dowd PS, Kelleher J, Guillou PJ. 1986 T-lymphocyte subsets and interleukin-2 production in zinc-deficient rats. *Br J Nutr* **55**, 59–69.

Driessen C, Hirv K, Rink L, Kirchner H. 1994 Induction of cytokines by zinc ions in human peripheral blood mononuclear cells and separated monocytes. *Lymphokine Cytokine Res* **13**, 15–20.

Driessen C, Hirv K, Wellinghausen N, Kirchner H, Rink L. 1995 Influence of serum on zinc, toxic shock syndrome toxin-1, and lipopolysaccharide-induced production of IFN-γ and IL-1β by human mononuclear cells. *J Leukoc Biol* **57**, 904–908.

Duchateau J, Delespesse G, Vereecke P. 1981a Influence of oral zinc supplementation on the lymphocyte response to mitogens of normal subjects. *Am J Clin Nutr* **34**, 88–93.

Duchateau J, Delespesse G, Vrijens R, Collet H. 1981b Beneficial effects of oral zinc supplementation on the immune response of old people. *Am J Med* **70**, 1001–1004.

Fagiolo U, Cossarizza A, Scala E, Fanales-Belasio E, Ortolani C, Cozzi E, Monti D, Franceschi C, Paganelli R. 1993 Increased cytokine production in mononuclear cells of healthy elderly people. *Eur J Immunol* **23**, 2375–2378.

Favier A, Favier M. 1990 Consequences des deficits en zinc durant la grossesse pour la mère et le nouveau-né. *Rev Fr Gynecol Obstet* **85**, 13–27.

Favier AE. 1992 The role of zinc in reproduction. Hormonal mechanisms. *Biol Trace Elem Res* **32**, 363–382.

Fortes C, Forastiere F, Agabiti N, Fano V, Pacifici R, Virgili F, Piras G, Guidi L, Bartoloni C, Tricerri A, Zuccaro P, Ebraim S, Perucci CA. 1998 The effect of zinc and vitamin A supplementation on immune response in an older population. *J Am Geriatr Soc* **46**, 19–26.

Fraker PJ, Gershwin ME, Good RA, Prasad A. 1986 Interrelationships between zinc and immune functions. *Fed Proc* **45**, 1474–1479.

Fraker PJ, Osati-ashtiani F, Wagner MA, King LE. 1995 Possible roles for glucocorticoids and apoptosis in the suppression of lymphopoiesis during zinc deficiency: a review. *J Am Coll Nutr* **14**, 11–17.

Fraker PJ, King LE, Laakko T, Vollmer TL. 2000 The dynamic link between the integrity of the immune system and zinc status. *J Nutr* **130**, 1399S–1406S.

Gaither LA, Eide DJ. 2000 Functional expression of the human hZIP2 zinc transporter. *J Biol Chem* **275**, 5560–5564.

Gavella M, Lipovac V, Vucic M, Sverko V. 1999 *In vitro* inhibition of superoxide anion production and superoxide dismutase activity by zinc in human spermatozoa. *Int J Androl* **22**, 266–274.

Goode HF, Kelleher J, Walker BE. 1989 Zinc concentrations in pure populations of peripheral blood neutrophils, lymphocytes and monocytes. *Ann Clin Biochem* **26**, 89–95.

Grekas D, Alivanis P, Kotzadamis N, Kiriazopoulos M, Tourkantonis A. 1992 Influenza vaccination in chronic hemodialysis patients. The effect of zinc supplementation. *Ren Fail* **14**, 575–578.

Hadden JW. 1995 The treatment of zinc is an immunotherapy. *Int J Immunopharmacol* **17**, 697–701.

Hansson A. 1996 Extracellular zinc ions induces mitogen-activated protein kinase activity and protein tyrosine phosphorylation in bombesin-sensitive Swiss 3T3 fibroblasts. *Arch Biochem Biophys* **328**, 233–238.

Herold A, Bucurenci N, Mazilu E, Szegli G, Sidenco L, Baican I. 1993 Zinc aspartate *in vivo* and *in vitro* modulation of reactive oxygen species production by human neutrophils and monocytes. *Rom Arch Microbiol Immunol* **52**, 101–108.

Ho LH, Ratnaike RN, Zalewski PD. 2000 Involvement of intracellular labile zinc in suppression of DEVD-caspase acvtivity in human neuroblastoma cells. *Biochem Biophys Res Commun* **268**, 148–154.

Hogstrand C, Verbost PM, Bonga SE, Wood CM. 1996 Mechanisms of zinc uptake in gills of freshwater rainbow trout interplay with calcium transport. *Am J Physiol* **270**, R1141–1147.

Hujanen ES, Seppä ST, Virtanen K. 1995 Polymorphnuclear leukocyte chemotaxis induced by zinc, copper and nickel *in vitro*. *Biochim Biophys Acta* **1245**, 145–152.

James K. 1990 Interaction between cytokines and α_2-macroglobulin. *Immunol Today* **11**, 163–166.

Jameson S. 1993 Zinc status in pregnancy: the effect of zinc therapy on perinatal mortality, prematurity, and placental ablation. *Ann NY Acad Sci* **678**, 178–192.

Johnson CM, Benson NA, Papadi GP. 1996 Apoptosis and CD4+ lymphocyte depletion following feline immunodeficiency virus infection of a T-lymphocyte cell line. *Vet Pathol* **33**, 195–203.

Jyonouchi H, Voss RM, Good RA. 1991 Zinc depletion modifies CD5 expression by 70Z/3 murine pre-B leukemia cell line. *Proc Soc Exp Biol Med* **198**, 818–825.

Keen CL, Gershwin ME. 1990 Zinc deficiency and immune function. *Ann Rev Nutr* **10**, 415–431.

King LE, Osati-Ashtiani F, Fraker PJ. 1995 Depletion of cells of the B lineage in the bone marrow of zinc-deficient mice. *Immunology* **85**, 69–73.

Kirchner H, Rühl H. 1970 Stimulation of human peripheral lymphocytes by Zn^{2+} *in vitro*. *Exp Cell Res* **61**, 229–230.

Klaiman AP, Victery W, Kluger MJ, Vander AJ. 1981 Urinary excretion of zinc and iron following acute injection of dead bacteria in dogs. *Proc Soc Exp Biol Med* **167**, 165–171.

Koizumi S, Ikeda M, Inoue K, Nakazawa K, Inoue K. 1995 Enhancement by zinc of ATP-evoked dopamine release from rat pheochromocytoma PC12 cells. *Brain Res* **673**, 75–82.

Kreft B, Fischer A, Krüger S, Sack K, Kirchner H, Rink L. 2000 The impaired immune response to diphtheria vaccination in elderly chronic hemodialysis patients is related to zinc deficiency. *Biogerontology* **1**, 61–66.

Lighart GJ, Coberand JX, Fournier C, Galanaud P, Hijmans W, Kennes B, Müller-Hermelink HK, Steinmann GG. 1984 Admission criteria for immunogerontological studies in man: The Senieur Protocol. *Mech Ageing Dev* **28**, 47–55.

Mills CF. 1989 *Zinc in Human Biology*. London: Springer.

Mocchegiani E, Santarelli L, Muzzioli M, Fabris N. 1995a Reversibility of the thymic involution and of age-related peripheral immune dysfunction by zinc supplementation in old mice. *Int J Immunopharmacol* **17**, 703–718.

Mocchegiani E, Veccia S, Ancarani F, Scalise G, Fabris N. 1995b Benefit of oral zinc supplementation as an adjunct to zidovudine (AZT) therapy against opportunistic infections in AIDS. *Int J Immunopharmacol* **17**, 719–727.

Mocchegiani E, Ciavattini A, Santarelli L, Tibaldi A, Muzzioli M, Bonazzi P, Giacconi R, Fabris N, Garzetti GG. 1999 Role of zinc and alpha2 macroglobulin on thymic endocrine activity and on peripheral immune efficiency (natural killer activity and interleukin 2) in cervical carcinoma. *Br J Cancer* **79**, 244–250.

Moonga BS, Dempster DW. 1995 Zinc is a potent inhibitor of osteoclastic bone resorption *in vitro*. *J Bone Miner Res* **10**, 453–457.

Mossad SB, Macknin ML, Medendorp SV, Mason P. 1996 Zinc gluconate lozenges for treating the common cold. *Ann Inter Med* **125**, 81–88.

Murthy ARK, Lehrer RI, Harwig SSL, Miyasaki KT. 1993 *In vitro* candidastatic properties of the human neutrophil calprotectin complex. *J Immunol* **151**, 6291–6301.

Neldner KH, Hambidge KM. 1975 Zinc therapy in acrodermatitis enteropathica. *N Engl J Med* **292**, 879–882.

Neves jr I, Bertho AL, Veloso VG, Nascimento DV, Campos-Mello DLA, Morgado MG. 1998 Improvement of the lymphoproliferative immune response and apoptosis inhibition *in vitro* treatment with zinc of peripheral blood mononuclear cells (PBMC) from HIV+ individuals. *Clin Exp Immunol* **11**, 264–268.

Ohly P, Gleichmann H. 1995 Metallothionein: *in vitro* induction with zinc and streptozotocin in pancreatic islet of mice. *Exp Clin Endocrinol* **103**, 79–82.

Osatiashtiani F, King LE, Fraker PJ. 1998 Variance in the resistance of murine early bone marrow B cells to a deficiency in zinc. *Immunology* **94**, 94–100.

Ozturk G, Erbas D, Imir T, Bor NM. 1994 Decreased natural killer (NK) cell activity in zinc-deficient rats. *Gen Pharmacol* **25**, 1499–503.

Palmiter RD, Findley SD. 1995 Cloning and functional characterization of a mammalian zinc transporter that confers resistance to zinc. *EMBO J* **14**, 639–649.

Palmiter RD, Cole TB, Findley SD. 1996a ZnT-2, a mammalian protein that confers resistance to zinc by facilitating vesicular sequestration. *EMBO J* **15**, 1784–1791.

Palmiter RD, Cole TB, Quaife CJ, Findley SD. 1996b ZnT-3, a putative transporter of zinc into synaptic vesicles. *Proc Natl Acad Sci USA* **93**, 14934–14939.

Parat MO, Richard MJ, Meplan C, Favier A, Beani JC. 1999 Impairment of cultured cell proliferation and metallothionein expression by metal chelator NNN′N′-tetrakis-(2-pyridylmethyl)ethylene diamine. *Biol Trace Elem Res* **70**, 51–68.

Patterson WP, Winkelmann M, Perry MC. 1985 Zinc-induced copper deficiency: megamineral sideroblastic anemia. *Ann Inter Med* **103**, 385–386.

Phillips JL, Azari P. 1974 Zinc transferrin: Enhancement of nucleic acid synthesis in phytohemagglutinin-stimulated human lymphocytes. *Cell Immunol* **10**, 31–37.

Porter KG, McMaster D, Elmes ME, Love AH. 1977 Anaemia and low serum-copper during zinc therapy. *Lancet* **2**, 774.

Prasad AS, Miaie Ajr, Farid Z, Schulert A, Sandstead HH. 1963 Zinc metabolism in patients with the syndrome of iron deficiency, hypgonadism and dwarfism. *J Lab Clin Med* **83**, 537–549.

Prasad AS. 1995 Zinc: an overview. *Nutrition* **11**, 93–99.

Prasad AS. 1996 Zinc deficiency in women, infants and children. *J Am Coll Nutr* **15**, 113–120.

Prasad AS. 1998 Zinc and immunity. *Mol Cell Biochem* **188**, 63–69.

Prasad AS, Beck FW, Kaplan J, Chandrasekar PH, Ortega J, Fitzgerald JT, Swerdlow P. 1999 Effect of zinc supplementation on incidence of infections and hospital admission in sickle cell disease (SCD). *Am J Hematol* **61**, 194–202.

Prasad AS. 2000 Effects of zinc deficiency on immune functions. *J Trace Elem Exp Med* **13**, 1–20.

Provinciali M, Montenovo A, Di-Stefano G, Colombo M, Daghetta L, Cairati M, Veroni C, Cassino R, Della-Torre F, Fabris N. 1998 Effect of zinc or zinc plus arginine supplementation on antibody titre and lymphocyte subsets after influenza vaccination in elderly subjects: a randomized controlled trial. *Age Ageing* **27**, 715–722.

Rajagopalan S, Winter CC, Wagtmann N, Long EO. 1995 The Ig-related killer cell inhibitory receptor binds zinc and requires zinc for recognition of HLA-C on target cells. *J Immunol* **155**, 4143–4146.

Raulin J. 1869 Etudes chimique sur la vegetation. *Ann Sci Natl Botan Biol Veget* **11**, 293–299.

Rawer P, Willems WR, Breidenbach T, Guttmann W, Pabst W, Schutterle G. 1987 Seroconversion rate, hepatitis B vaccination, hemodialysis, and zinc supplementation. *Kidney Int S* **22**, 149–152.

Record IR, Jannes M, Dreosti IE. 1995 Protection by zinc against UVA- and UVB-induced cellular and genomic damage *in vivo* and *in vitro*. *Biol Trace Elem Res* **53**, 19–25.

Recommendation of the German Society of Nutrition 1995 Ausschuß Nahrungsbedarf der DGE Zufuhrempfehlungen und Nährstoffbedarf. Teil II: Vergleich der Vorschläge von SCF/EC mit den Empfehlungen der DGE. *Ernährungs-Umschau* **42**, 4–10.

Reinhold D, Ansorge S, Grüngreiff K. 1999 Immunobiology of zinc and zinc therapy. *Immunol Today* **20**, 102.

Reyes JG. 1996 Zinc transport in mammalian cells. *Am J Physiol* **270**, C401–C410.

Rink L, Cakman I, Kirchner H. 1998 Altered cytokine production in the elderly. *Mech Ageing Dev* **102**, 199–210.

Rink L, Gabriel P. 2000 Zinc and the Immune System. *Proc Nutr Soc* **59**, 541–552.

Rink L, Kirchner H. 1999 Reply to Reinhold *et al.*. *Immunol Today* **20**, 102–103.

Rink L, Kirchner H. 2000 Zinc-Altered Imune Function and Cytokine Production. *J Nutr* **130**, 1407S–1411S.

Rink L, Seyfarth M. 1997 Besonderheiten immunologischer Untersuchungsergebnisse im Alter. *Z Gerontol Geriat* **30**, 220–225.

Rühl H, Kirchner H. 1978 Monocyte-dependent stimulation of human T cells by zinc. *Clin Exp Immunol* **32**, 484–488.

Rühl H, Kirchner H, Borchert G. 1971 Kinetics of the Zn^{2+}-stimulation of human peripheral lymphocytes *in vitro*. *Proc Soc Exp Biol Med* **137**, 1089-1092.

Safie-Garabedian B, Ahmed K, Khamashta MA, Taub NA, Hughes GRV. 1993 Thymulin modulates cytokine release by peripheral blood mononuclear cells: a comparison between healthy volunteers and patients with systemic lupus erythematodes. *Int Arch Allergy Immunol* **101**, 126–131.

Saha AR, Hadden EM, Hadden JW. 1995 Zinc induces thymulin secretion from human thymic epithelial cells *in vitro* and augments splenocyte and thymocyte responses *in vivo*. *Inter J Immunopharmacol* **17**, 729–733.

Sainte-Marie I, Jumbou O, Tenaud I, Dreno B. 1998 Comparative study of the *in vitro* inflammatory activity of three nickel salts on keratinocytes. *Acta Dermatol Venerol* **78**, 169–172.

Salas M, Kirchner H. 1987 Induction of interferon-γ in human leukocyte cultures stimulated by Zn^{2+}. *Clin Immunol Immunopathol* **45**, 139–142.

Sandstead HH, Henriksen LK, Greger JL, Prasad AS, Good RA. 1982 Zinc nutriture in the elderly in relation to taste acuity, immune response, and wound healing. *Am J Clin Nutr* **36**, 1046–1059.

Santra M, Talukder G, Sharma A. 2000 Clastogenic effects of zinc chloride on human peripheral blood leukocytes *in vitro*. *Cytobios* **102**, 55–62.

Scott BJ, Bradwell AR. 1983 Identification of the serum binding proteins for iron, zinc, cadmium, nickel and calcium. *Clin Chem* **29**, 629–633.

Simkin PA. 1976 Oral zinc sulphate in rheumatoid arthritis. *Lancet* **2**, 539–542.

Sohnle PG, Collins-Lech C, Wiessner JH. 1991 The zinc-reversible antimicrobial activity of neutrophil lysates and abscess fluid supernatants. *J Infect Dis* **164**, 137–142.

Sood SM, Wu MX, Hill KA, Slattery CW. 1999 Characterization of zinc-depleted alanyl-tRNA synthetase from *Escherichia coli*: role of zinc. *Arch Biochem Biophys* **368**, 380–384.

Scuderi P. 1990 Differential effects of copper and zinc on human peripherial blood monocyte cytokine secretion. *Cell Immunol* **126**, 391–405.

Srinivas U, Jeppsson B, Braconier JH. 1989 Superoxide production of peritoneal macrophages in experimental gram-negative sepsis; influence of *in vitro* and *in vivo* supplements of zinc. *APMIS* **97**, 682–688.

Sunzel B, Holm S, Reuterving CO, Soderberg T, Hallman G, Hanstrom L. 1995 The effect of zinc on bacterial phagocytosis, killing and cytoprotection in human polymorphnuclear leucocytes. *APMIS* **103**, 635–644.

Tanaka Y, Shiozawa S, Morimoto I, Fujita T. 1989 Zinc inhibits pokeweed mitogen-induced development of immunoglobulin-secreting cells through augmentation of both CD4 and CD8 cells. *Inter J Immunopharmacol* **11**, 673–679.

Telford WG, Fraker PJ. 1995 Preferential induction of apoptosis in mouse $CD4{+}CD8{+}\alpha\beta TCR_{lo}CD3\epsilon_{lo}$ thymocytes by zinc. *J Cell Physiol* **164**, 259–270.

Tenaud I, Sainte-Marie I, Jumbouo, Litoux P, Dreno B. 1999 *In vitro* modulation of keratinocyte wound healing integrins by zinc, copper and manganese. *Br J Dermatol* **140**, 26–34.

Todd WK, Elvelym A, Hart EB. 1934 Zinc in the nutrition of the rat. *Am J Physiol* **107**, 146–156.

Togari A, Arakawa S, Arai M, Matsumoto S. 1993 Alterations of *in vitro* bone metabolism and tooth formation by zinc. *Gen Pharmac* **24**, 1133–1140.

Trubiani O, Antonucci A, Palka G, Di-Primio R. 1996 Programmed cell death of peripheral myeloid precursor cells in Down patients: effect of zinc therapy. *Ultrastruct Pathol* **20**, 457–462.

Tsuda M, Imaizumi K, Katayama T, Kitagawa K, Wanaka A, Tohyama M, Takagi T. 1997 Expression of zinc transporter gene, ZnT-1, is induced after transient forebrain ischemia in the gerbil. *J Neurosci* **17**, 6678–6684.

Turk S, Bozfakioglu S, Ecder ST, Kahraman T, Gurel N, Erkoc, Aysuna N, Turkmen A, Bekiroglu N, Ark E. 1998 Effects of zinc supplementation on the immune system and on antibody response to multivalent influenza vaccine in hemodialysis patients. *Inter J Artif Organs* **21**, 274–278.

Valberg LS, Flanagan PR, Chamberlain MJ. 1984 Effect of iron, tin and copper on zinc absorption in humans. *Am J Clin Nutr* **40**, 536–541.

Vallee BL, Falchuk KH. 1993 The biochemical basis of zinc physiology. *Physiol Rev* **73**, 79–118.

Varadinova TL, Bontchev PR, Nachev CK, Shishkov SA, Strachilov D, Paskalev Z, Toutekova A, Panteva M. 1993 Mode of action of Zn-complexes on herpes simplex virus type 1 infection *in vitro*. *J Chemother* **5**, 3–9.

Weiss G, Widner B, Zoller H, Fuchs D. 1998 The immunobiology of zinc and the kidney. *Immunol Today* **19**, 193.

Wellinghausen N, Driessen C, Rink L. 1996a Stimulation of human peripheral blood mononuclear cells by zinc and related cations. *Cytokine* **18**, 767–771.

Wellinghausen N, Fischer A, Kirchner H, Rink L. 1996b Interaction of zinc ions with human peripheral blood mononuclear cells. *Cell Immunol* **171**, 255–261.

Wellinghausen N, Kirchner H, Rink L. 1997a The immunobiology of zinc. *Immunol Today* **18**, 519–521.

Wellinghausen N, Martin M, Rink L. 1997b Zinc inhibits IL-1 dependent T cell stimulation. *Eur J Immunol* **27**, 2529–2535.

Wellinghausen N, Rink L. 1998 The significance of zinc for leukocyte biology. *J Leuk Biol* **64**, 571–577.

Wilson A, Trombetta LD. 1999 The protective effects of zinc on diethyldithiocarbamate cytotoxicity in rat astrocytes *in vitro*. *Toxicol Lett* **105**, 129–140.

Yuzbasiyan-Gurkan VA, Brewer GJ, Vander AJ, Guenter MJ, Prasad AS. 1989 Net renal tubular reabsorption of zinc in healthy man and impaired handling in sickle cell anemia. *Am J Hematol* **31**, 87–90.

Zoli A, Altomonte L, Caricchio R, Gaossi A, Mirone L, Ruffini MP, Magaro M. 1998 Serum zinc and copper in active rheumatoid arthritis: correlation with interleukin 1 beta and tumour necrosis factor alpha. *Clin Rheumatol* **17**, 378–382.

BioMetals 14: 385–395, 2001.

Review

Zinc physiology and biochemistry in oocytes and embryos

Kenneth H. Falchuk* & Marcelo Montorzi
*Center for Biochemical and Biophysical Sciences and Medicine, and Department of Medicine, Brigham and Women's Hospital, Harvard Medical School, Boston, Massachusetts, 02115, USA; *Author for correspondence (Tel: (617) 432 - 1367; Fax: (617) 566-3137; E-mail: kenneth_falchuk@hms.harvard.edu)*

Received 2 February 2001; accepted 28 May 2001

Key words: eggs, embryos, oocytes, transcription factors, transport proteins, yolk platelets, storage, zinc

Abstract

The essential role of zinc in embryogenesis was identified through studies of its presence in eggs and embryos, the effects of its deficiency and its role in metallo proteins required for organ development and formation. The *Xenopus laevis* oocyte zinc content varies during oogenesis. It increases from 3 to 70 ng zinc/oocyte as it progresses from stage I to VI. The oocyte zinc is derived from the maternal liver as part of a metallo-complex with vitellogenin. The latter transports the metal in plasma and into the oocyte. Once internalized, most of the zinc is stored within yolk platelets bound to lipovitellin, one of the processed products of vitellogenin. About 90% of the total zinc is associated with the yolk platelet lipovitellin while the remaining 10% is in a compartment associated with hitherto unknown molecule(s) or organelle(s) of the cytoplasm. The bi-compartmental distribution remains constant throughout embryogenesis since the embryo behaves as a closed system for zinc after fertilization. The yolk platelet zinc is used after the tadpole is hatched while we proposed that the 10% of the zinc in the non-yolk platelet pool is the one used for embryogenesis. It provides zinc to newly synthesized molecules responsible for the development of zinc-dependent organ genesis. Interference with the availability of this zinc by the chelating agent 1,10-phenanthroline results in the development of embryos that lack dorsal organs, including brain, eyes and spinal cord. The extensive teratology is proposed to be due to altered or absent zinc distribution between the cytosolic pool and zinc-transcription factors. The data identify the components of a zinc transport, storage and distribution system in a vertebrate organism.

Introduction

Zinc is a constituent of many molecules involved in protein, lipid and carbohydrate metabolism. It also participates in the synthesis of viral, prokaryotic and eukaryotic nucleic acids. It is present in all living cells. Its function at the cellular level has depended on a large body of phenomenological information available on the effects of its deficiency (Vallee & Falchuk 1993). Thus, zinc deficiency induces proliferative arrest in many cell types, suppresses growth of plants and animals and causes congenital malformations in offspring of zinc-deprived animals. The teratology in vertebrate embryos is striking. In the mouse, the plasma zinc pool is responsible for delivering zinc to the growing fetus (Dreosti *et al.* 1968). Within 24 h of maternal zinc deprivation in the diet, the amount of available zinc is depleted and the embryo is deprived of the metal. The consequences include high mortality or malformations that involve nearly all organ systems. The most pervasive malformations are those of dorsal organs such as the head structures, neural tissues such as brain and spinal cord, as well as musculo-skeletal abnormalities (Keen & Hurley 1989). The most sensitive period for teratology appears to be in the early developmental stages, in humans corresponding to the first trimester.

There is limited knowledge regarding the basis for its role in cellular proliferation, differentiation and growth, in general, and for the teratology arising from its deficiency, in particular. Similarly, there is sparse information on zinc uptake and distribution within cells. Finally, the foundation for understanding the metabolism of zinc in the oocyte and embryo is at its inception despite the acceptance that the embryo is a sensitive target of zinc deficiency (Falchuk 1998). To provide the basis for that understanding, we will review the information that is available on the content and distribution of zinc during oogenesis and embryogenesis as well as the underlying molecular events that are dependent on the metal and are crucial for embryogenesis.

Egg metal content

Eggs from oviparous animals behave as a closed systems while those from mammals behave as open systems. Thus, once eggs from e.g. sea urchin or frog are fertilized, they are deposited into a sea, river, lake or placed on the undersurface of leaves or other sites to develop into embryos. In those environments, these fertilized eggs must have all of the necessary nutrients or constituents to form a complete embryo without depending on an external and variable supply (Davidson 1990; Nomizu *et al.* 1993). Therefore, during maturation in the maternal ovary, the egg must acquire all essential chemical substances, including metals, for use after fertilization. In this sense, it is a closed system. On the other hand, mammalian eggs take up zinc, and likely other metals and nutrients, from the fallopian fluid once they enter the tube and begin dividing as fertilized eggs (Hurley & Shrader 1975; Gallaher & Hurley 1980). Moreover, once implantation has taken place, and the maternal blood supply is linked to that of the placenta, exchange of nutrients occurs between mother and fetus and, therefore, the latter is open to and dependent on a maternal source of metals, vitamins, etc.

The composition and biochemistry of individual eggs from many oviparous animals have been studied for years. Over the past decades, it also has become possible to obtain single eggs from mammals. Therefore, sea urchin, frog and mouse eggs can be individually collected and prepared for metal analysis by obviating spurious metal contamination using buffers, water, glass and plastic ware rendered metal-free. The collection of eggs, their preparation and the operational conditions for analyses have been reported (Nomizu *et al.* 1993; Vallee & Falchuk 1993). Briefly, the frogs are stimulated to ovulate by injection with chorionic gonadotropin. Hundreds of eggs are readily collected as they are spawned. Sea urchins are injected with KCl and they immediately release eggs, also in the hundreds. Mice are induced to ovulate with chorionic gonadotropin and ten or more eggs and embryos are collected within the fallopian tubes.

Table 1. Zinc content of a white blood cell and an individual mature egg.

Species	Zinc content, ng/egg or cell	Relative amounts
Human leukocyte	0.01	1
Mouse	1	100
Sea urchin	20	2000
Xenopus laevis frog	70	7000

The zinc contents of single eggs from mice, sea urchin and frog are shown in Table 1. The zinc content calculated for a single leukocyte (a polymorphonuclear white blood cell) is shown for comparison as an example of a fully differentiated adult cell. All three eggs have higher zinc content than the white blood cell. This suggests a higher requirement for zinc and/or the existence of a storage site, quite distinct from that of adult cells. Moreover, the two closed systems, sea urchin and frog eggs, have a much higher zinc content than the open system, the murine egg.

The information shown in Table 1 has been used to select a suitable system to study zinc metabolism during oogenesis and embryogenesis. The mature *Xenopus laevis* egg offers the advantage over the other two egg types that its zinc content is higher. Another advantage is that large numbers of oocytes at each stage of oogenesis and embryos at different points of development can be readily obtained and used for study of metal metabolism.

As part of that study, the contents of other metals in *Xenopus laevis* eggs also has been determined (Table 2).

The amounts of each metal in the egg differ. There are two metals, magnesium and calcium that are higher than zinc while four are less. Two are not detected. The metal contents can be arranged in the following order: Mg>Ca>Zn>Fe>Mn>Cu>Ni.

Zinc during oogenesis

The frog ovaries are located within the abdomen forming the most immediate visible structure on entering

Table 2. Metal content of oocytes.

Metal	Content, ng/oocyte*
Mg	329 ± 32
Ca	161 ± 15
Zn	70 ± 3
Fe	33 ± 4
Mn	10 ± 1
Cu	2 ± 0.1
Ni	0.2 ± 0.03
Co	ND
Cd	ND

*Average value ± standard deviation; $n = 10$; ND = not detected.

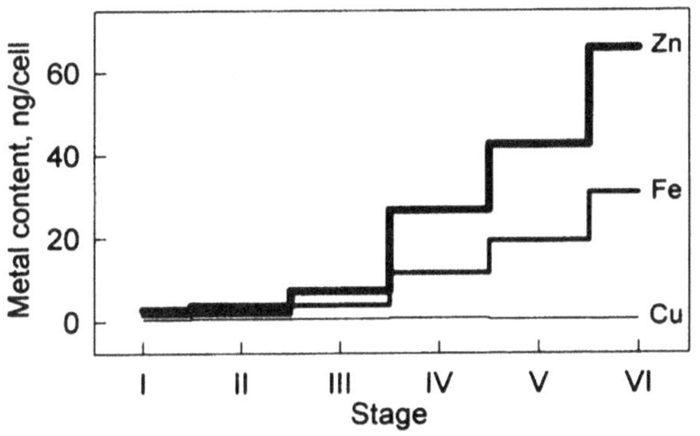

Fig. 1. Changes in metal content during oogenesis. As oocytes mature, they take up metals from maternal plasma. Zinc and iron increase during the entire oogenic period while copper attains its maximal value by stage I.

the cavity. It is composed of six lobes each with hundreds of oocytes at all stages of development. Their removal does not injure the oocytes and they are easily exposed by dissection of the fibrous membrane encasing the oocytes.

Oocytes can be separated on the basis of the six stages of their maturation since their size and morphological appearance differ from 50 to 1300 μm in stages I and VI, respectively (Hansen & Riebessell 1991). Their color changes from clear and transparent in the earliest stage to green/black pigment in the last stage. The zinc content of oocytes varies as a function of the stage of maturation. During the initial stages of maturation (stages I to III), the zinc content increases from 2 to 7 ng/oocyte (Figure 1). However, from stage III to VI it increases to approximately 70 ng/oocyte, a 35-fold increment from its original value. The contents of iron and copper are shown for comparison. Similar increase in iron is observed during oogenesis though the total amount is always less than zinc at all stages. The maximum copper content is attained during stage I and remains constant throughout oogenesis.

The zinc content of stage VI oocytes does not vary in eggs of any given frog though it varies in those of different frogs. The range of values in eggs of different frogs is about 65 to 133 ng/egg, a two-fold variation (Nomizu *et al.* 1993). Assuming an average egg volume of 1 μl (Hansen & Riebessell 1991) and a zinc content of 70 ng the concentration of zinc in the oocyte is approximately 1 mM.

The final zinc content of stage VI oocytes is achieved over a period of about 3 years, the amount of time that is required to terminate oogenesis. During the first three stages of oogenesis, the increases are quantitatively less than during the last three stages (Figure 1). This behavior corresponds exactly to the pre-vitellogenic and vitellogenic phases of oocyte development. These phases are descriptive of the rate of uptake of the protein vitellogenin by the oocyte during oogenesis. The concurrent uptake of zinc and vitellogenin by the oocyte is due to the relationship between the two processes.

Zinc is transported in plasma by, and is taken into the oocyte bound to, vitellogenin. Estrogen induces the liver to synthesize the phospho-glyco-lipo-metallo-protein vitellogenin. Within two weeks of the hormonal stimulation, the frog liver up-regulates its synthesis of the protein and secretes it into the plasma in large quantities. The protein is purified from the serum by chromatography on a Mono Q column and identified on the basis of its molecular weight and amino acid composition, specifically its high serine content with about 30% phospho-serine (Montorzi *et al.* 1995). Vitellogenin is a metallo protein that contains one g/at of zinc per 220 kDa monomer but no other group IIB or transition metal and 1.5 mol of calcium per monomer (Table 3). Since vitellogenin is a dimer, its total metal content, therefore, is 5 mol of metal, 2 mol of zinc and 3 mol of calcium per molecule. These data demonstrate that zinc is transported in plasma bound to vitellogenin.

The zinc protein is taken up from the plasma by oocytes through receptor-mediated endocytosis (Wallace 1978; Wallace *et al.* 1983; Wallace & Jared 1968; Banaszak *et al.* 1991; Hansen & Riebessell 1991). The receptor is a 115-kDa membrane protein (Stifani *et al.* 1990). The receptor – vitellogenin complex is internalized in vesicles by endocytosis. These fuse with other vesicles and lysosomes to form multivesicular bodies and process the protein into lipovitellin and

Table 3. Metal content of vitellogenin, lipovitellin and phosvitin.

Metal	Vitellogenin (mol/220kDa)	Lipovitellin (mol/141kDa)	Phosvitin (mol/30kDa)
Zn	1.02	1.06	0.20
Ca	1.50	ND*	2.10
Mg	0.15	0.10	3.0
Cd	ND	ND	ND
Mn	0.06	ND	0.05
Fe	0.15	0.10	0.5
Co	ND	ND	ND
Ni	ND	ND	ND
Cu	0.09	0.03	ND

* ND = Not detected

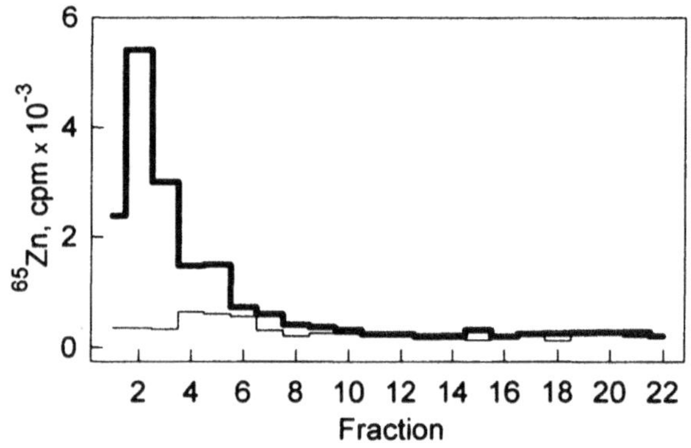

Fig. 2. Stage II oocyte ^{65}Zn vitellogenin uptake. Oocytes were incubated for 30 min with either ^{65}Zn vitellogenin (bold line) or free ^{65}Zn (thin line). Oocytes were homogenized and the constituents separated in a sucrose gradient. Fractions 0–5 contain cytosol, ribonucleoprotein particles and other small organelles. Free zinc does not enter the oocyte. In contrast, when zinc is bound to vitellogenin, it enters and is distributed only in the low density fractions.

phosvitin. The latter fuse further into yolk platelets of progressively larger size and weight. The vesicles, multivesicular bodies and yolk platelets can be separated in a sucrose gradient on the basis of their differing densities. Yolk platelets localize in sucrose densities between 1.20 and 1.24 g/ml. They are distributed bi-modally with the lighter ones concentrating at 1.21 g/ml and the heavier ones at 1.23 g/ml. As the oocytes mature, more of the yolk platelets sediment to the higher density. One of the resultant effects of this organelle formation is the entry of zinc into the oocyte and its appearance in yolk platelets.

This conclusion has been confirmed experimentally. The chelating agent 1,10-phenanthroline (OP) removes the zinc from vitellogenin in a concentration dependent manner. When dialyzed against 10^{-6} M OP, the metal protein complex remains intact while at 10^{-3} M all of the zinc is removed (pK for removal is 4.8). The metal removal by OP is very rapid and the apoprotein is formed within minutes of incubation with the chelating agent. A ^{65}Zn species has been generated from the apo vitellogenin and used to study the protein's role in zinc uptake by oocytes *in vitro* (Falchuk *et al.* 1995). Neither oocytes in the previtellogenic (stage II) or in the vitellogenic phase (stage IV) take up free ^{65}Zn during an incubation period of 24 h (Figure 2). However, both stage II and IV oocytes take up a zinc-vitellogenin complex but their distribution of zinc differs. Within 30 min of incubation, ^{65}Zn-bound vitellogenin enters both the stage II and IV oocytes. In stage II oocytes, all of the ^{65}Zn is localized entirely to the cytosolic fluid and remains in that space throughout the entire period of observation (Figure 2). In contrast, the distribution of ^{65}Zn within the more developed stage IV oocytes mirrors the time course and the formation of the vesicles, multivesicular bodies and yolk platelets. It changes in the first 300 min of incubation (Figure 3). After 30 min, the major peak of zinc is found in the 1.10–1.12 g/ml gradient fraction, the region of multivesicular bodies. After 90 min, a second zinc peak appears in the 1.20–1.24 g/ml region, where light and heavy yolk platelets are found. In the course of the next 300 min, over 90% of the zinc localizes to the yolk platelet region. This distribution remains unchanged for the next 20 h of the experimental period. The zinc content of heavy yolk platelets, 13.8 fg/platelet, is about 3 times greater than that of light platelets, 3.2 fg/platelet (Figure 4). Thus, within hours of uptake, zinc is transferred from the multivesicular bodies into a yolk platelet pool during the vitellogenic phase of oogenesis. At the mature

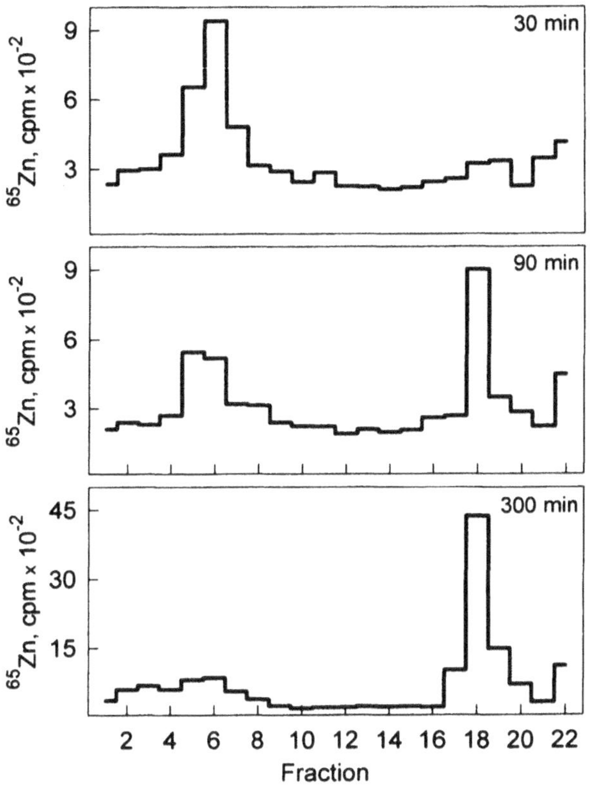

Fig. 3. Stage IV oocyte ^{65}Zn vitellogenin uptake. Stage IV oocytes were incubated and the components separated as described in Figure 2. Free zinc does not enter the oocyte. Within 30 min, ^{65}Zn vitellogenin has entered and is found within the fractions containing multivesicular bodies. During the next 300 min nearly all of the ^{65}Zn vitellogenin is distributed to the region of yolk platelets (fractions 17–20).

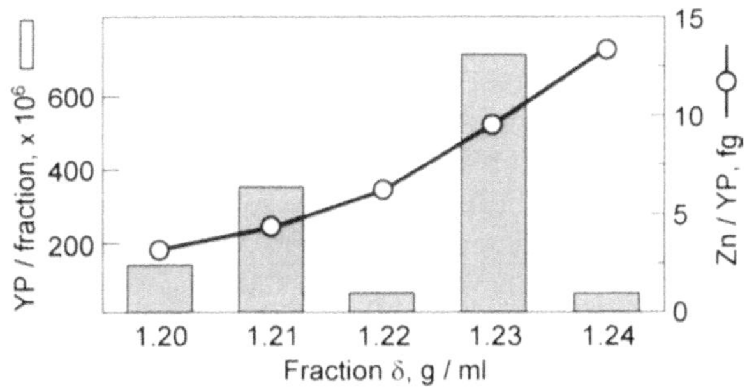

Fig. 4. Yolk platelet zinc content. As yolk platelets mature their density increases. This results in two populations. The light platelets distribute to a density of about 1.21 g/ml sucrose while the heavy organelles are concentrated around 1.23 g/ml. The zinc content of platelets increases progressively as the platelets mature.

stage, about 10% of the zinc is present in a low density fraction (pool I) while 90% is found within yolk platelets (pool II).

Processing of zinc-vitellogenin in the oocyte

The generation of both zinc pools described above from the same process of vitellogenin uptake implies that in stage II oocytes, vitellogenin does not undergo the same processing as in stage IV, and, therefore, is not transferred into yolk platelets in the early compared with latter stages. It is not known if stage II oocytes take up vitellogenin through receptor-mediated endocytosis, as in stage IV, but rather an alternate, non specific pathway, as shown with albumin and other proteins that enter the oocyte (Opresko 1991).

Alternatively, zinc could be released from vitellogenin in the cytosol following entry and be transferred to other molecules, analogous to the behavior of iron as it is exchanged between transferrin and ferritin. A number of molecules in the stage II oocytes are potential candidates as zinc acceptors in the cytosolic pool. The 7S and 42S ribonucleoprotein particles are distributed in the pertinent sucrose gradient regions (Denis & Le Maire 1972) where zinc is located in the stage II oocytes. Both particles contain zinc and one of the proteins, TFI IIA, of the 7S one, is a zinc protein (Hanas *et al.* 1983; Miller *et al.* 1985). Moreover, TFI-IIA is synthesized during stages I and II and zinc must be available in the cytosol at that time to fully load the apoprotein. Metallothionein is another zinc acceptor that can distribute zinc to other proteins and it has to be given consideration in the embryo.

The different behavior of vitellogenin taken up by stage IV oocytes leads to its storage in oocyte special organelles. At this stage, vitellogenin does enter the oocyte through the receptor-mediated endocytosis as already described. In the multivesicular body, vitellogenin dissociates from its receptor and it is cleaved into two proteins, lipovitellin and phosvitin. This cleavage is a necessary step for further fusion of the multivesicular bodies to form first light and then heavy yolk platelets that contain condensed and crystalized lipovitellin and phosvitin complexes (Wallace & Jared 1968; Wallace & Opresko 1983; Opresko & Karf 1987). Lipovitellin and phosvitin are readily differentiated on the basis of their solubility properties, electrophoretic mobilities, amino acid composition and sequences, high phosphoserine and lipid contents (Banaszak *et al.* 1991; Montorzi *et al.* 1995; Falchuk *et al.* 1995). Both yolk platelet proteins are solubilized by 1 M NaCl and can be separated from each other by differential precipitation with 66% ammonium sulfate. Lipovitellin precipitates by this treatment while phosvitin remains in the supernatant. Lipovitellin can then be resolubilized and obtained in pure form following chromatography on Sepharose 6B (Montorzi *et al.* 1995). The yolk platelet zinc is associated entirely with lipovitellin. In fact, the only metal that is bound to lipovitellin in stochiometric amounts is zinc, 1 mol/141 kDa (Table 3). The other major metal of vitellogenin is calcium and it is in the domain that is cleaved into phosvitin. At some point during the uptake and processing of vitellogenin by the oocyte and its assembly in the yolk platelets, phosvitin acquires magnesium (Table 3). The zinc is tightly bound to lipovitellin since it survives the purification procedure that includes extensive dialysis. Furthermore, to remove the zinc requires use of 10^{-3} M OP (other chelating agents will also work) or exposure to acid at pH below 5. The removal is completed within minutes. X-ray absorption fine structure analysis (XFAS) identified the zinc coordination sites in both vitellogenin and lipovitellin (Auld *et al.* 1999). The amino acid ligands for vitellogenin and lipovitellin have been shown to be two histidines and two other N/O ligands. Recently, lipovitellin from the chicken has also been shown to be a zinc protein (Groche *et al.* 2000).

Distribution of zinc and other metals in the embryo

While it requires three years for the oocyte to mature and complete its uptake and storage of zinc into cytosolic and yolk platelet pools, its utilization of some of that zinc takes place over a period of days. Once the egg is fertilized, the two pools remain constant during the entire cleavage and gastrulation periods as well as until the embryo has hatched, a period of less than 48 h depending on the temperature. Beyond that stage, the zinc is progressively transferred out from the yolk platelets. We have proposed that the pool that is used to deliver zinc to any newly formed apoprotein is the cytosolic pool. Any transfer of zinc from the yolk platelet pool would appear in the cytosolic fraction and would have been readily detected. Based on our finding that the oocyte zinc content is about 60–70 ng, the low density pool I contains 6–7 ng while the yolk platelet pool II has the remainder. The zinc content of a typical fully differentiated cell contains 0.01 ng (Table 1). Therefore, if the differentiated phenotype is associated with this quantity of zinc, there is sufficient zinc in the cytosolic pool alone for about 7 million cells.

The distribution of zinc in two pools, i.e., the cytosol and the oocyte/egg yolk platelet, differs from the distribution of other metals (Figure 5). Iron is mostly distributed in the region of the mitochondria and vesicular bodies. Copper is mostly in the low density with a second smaller fraction in the region of the heavy yolk platelets/nuclei. Calcium, magnesium and manganese are in both a low density fraction and the yolk platelet one. Manganese is also found in the region of mitochondria.

Role of zinc in nucleus and transcription underlying development

The zinc in the embryo must be distributed to newly formed apoproteins that function in the generation of the phenotype. When there are insufficient quantities of zinc in the embryo, the phenotype is totally disturbed and in some organism over 80% of all organs are either malformed or are not made at all. The molecular basis for these phenomena is only now emerging and will be reviewed here with the intent of setting the stage to understanding how the zinc and zinc proteins of the oocyte described above are ultimately used to direct the formation of organs by the embryo. A major role of zinc is to regulate chromatin structure and function (Vallee & Falchuk 1981, 1993). At the level of the nucleus, zinc content determines the types and amounts of chromatin binding proteins and their effect on its organization and capability to be transcribed. When zinc is present in normal amounts, a full complement of histones is associated with nuclear DNA. When zinc is reduced in the cell, new histones are not formed and instead basic polypeptides are synthesized (Stankiewicz *et al.* 1983; Mazus *et al.* 1984; Czupryn *et al.* 1987). The consequence is chromatin condensation and an overall reduction in total mRNA formation. Moreover, different proteins are transcribed leading to a different phenotype. These findings led to the formulation of a hypothesis that zinc regulates the expression of a defined set of genes (Vallee & Falchuk 1981). One of the types of proteins requiring zinc for function was proposed to be gene regulatory molecules acting at the level of transcription. Nuclear zinc interacts with these regulatory proteins to activate (or in some cases repress) transcription of particular genes, which in turn, determine the types of proteins formed. The

Table 4. Zinc transcription regulatory proteins.

Protein	Zinc, g at mole	Reference
X. laevis TFIIIA	2	Hanas *et al.* 1983
X. laevis TFIIIA	7–11	Miller *et al.* 1985
Glucocorticoid Receptor (407–556) Fragment	2	Freedman *et al.* 1988
Estrogen Receptor (185–250) Fragment	2	Schwabe *et al.* 1990
Yeast GAL4 (1–147) Fragment	2	Johnston 1987; Pan *et al.* 1990
HIV tat protein	2	Frankel *et al.* 1988
A. nidulans ALCR (7–58) Fragment	2	Sequeval *et al.* 1994
Yeast CYP1(HAP1) (49–126) Fragment	2	Timmerman *et al.* 1994
Yeast PPRI (1–118) Fragment	2	Ball *et al.* 1995
LIM domain (lin-11, RBTN, ISI-1)	2	Li *et al.* 1991; Archer *et al.* 1994
Hela cell SP1 (614–778) Fragment	3	Kuwahara *et al.* 1990
K. lactis LAC9 (1–128) Fragment	2	Halvorsen *et al.* 1990

model described has been amply confirmed. At least 10 transcription factors have now been identified by direct analyses of zinc to require the metal for function (Table 4). The first of these, TFIIIA, has a primary structure with highly conserved sequences comprised of two cysteine and two histidine residues separated by variable numbers of amino acids in 9 repeat units of about 30 amino acids. The Cys and His residues in each of these conserved repeat units could serve as zinc ligands forming tetrahedral coordination complexes with one zinc atom (Miller *et al.* 1985). In the case of TFIIIA, the presence of zinc generates an intervening compact loop structure containing the DNA binding domain of the protein in the sequence intervening between the pair of Cys and His residues. This is the 'zinc finger' DNA binding motif. The number of zinc atoms needed for DNA binding is 2–3 since a peptide containing 'fingers 1–3' binds almost as tightly to specific DNA fragments as the nine 'finger' molecule (Liao *et al.* 1992). While the actual number needed in the intact molecule has yet to be resolved experimentally, it is noteworthy that 2 or 3 zinc atoms is the most frequent number required for function by all of the other transcription factors examined to date (Table 4).

Following the insight on zinc binding motifs, hundreds of transcription factors have been identified with homologous sequences comprised of different combinations of Cys and His residues. These are presumed to be zinc binding sites and the proteins to require zinc for function. None have been isolated to homogeneity and submitted for metal analysis since they are present in exceedingly small quantities in the cell. They must be considered, therefore, putative zinc proteins. Many of the ones listed in Table 4 have been cloned as fragments containing the DNA binding domain of the entire molecule. These fragments bind specific DNA sequences only when zinc is present and associated with the fragment.

In the absence of zinc, a number of regulatory proteins either might not be formed at all or, if formed, might remain as apomolecules that would lack function. On the other hand, other genes might be activated and their transcription products result in inhibitory polypeptides. Together, such effects on synthesis of functional proteins and gene repression and/or activation could produce the phenotype of zinc deficient cells and organisms. Thus, failure to express or to generate active factors by providing zinc to apoproteins

Table 5. Putative zinc transcription factors involved in development.

Gene	Tissues expressing gene	Effect of absence or mutation	Reference
Scratch	Neural precursor cells	Decreased eye photoreceptors and neural loss	Roark *et al.* 1995
Castor	Delaminated CNS neuroblasts, ventral midline glial precursors	Reduction in CNS axonal network	Mellerick *et al.* 1992
Spelt	Terminal pattern elements	Absence of terminal elements	Kuhnlein *et al.* 1994
Krox20	Hindbrain	Loss of rhombomers 3 and 5, fusion of trigeminal, facial and vestibular ganglia	Swiatek *et al.* 1993
Kitz-1	Olfactory epithelial cells	? Malformation	Bernard *et al.* 1994
SKr2	Schwan cells, cephalic and neural crest derived tissues	? Malformation	Schutz *et al.* 1994
Zic	Early embryonic stage neural tube and granule cells of developing cerebellum	Neural tube defects	Nagai *et al.* 1997
Zfh-4	Midbrain	? Malformation	Kostich *et al.* 1995
Ovo	Female gametes	? Altered gametogenesis	Mevel-Ninio *et al.* 1991
Kr	Abdominal segments	Absence of thoracic and abdominal segments	Redeman *et al.* 1988
MZF1	Hematopoietic cells	Altered hematopoiesis	Perrotti *et al.* 1995
Egr-1	Hematopoietic cells	Altered hematopoiesis	Krishnaraju *et al.* 1995
Snail Escargot, Worniu	Central Nervous System	Defective neuroblast asymmetric divisions	Cai *et al.* 2001
Forkheak genes, Homeodomain box A5, Gli, Pod1		Defective lung morphogenesis	Costa *et al.* 2001

at critical junctures in the process of organogenesis could lead to failure to produce tissues or organs. The identification of the specific zinc transcription factors that could be altered and are responsible for the phenotype characteristic of zinc deficiency has yet to be carried out. A number of transcription factors categorized as containing one form of zinc binding domains or another have been recognized (Table 4). While they belong to the large group that has not been certified to contain zinc by direct analysis, their functions in development and the effects of their absence or mutation are of particular interest. Specifically, the effects are those known to be targets of zinc deficiency, namely formation of tissues of the nervous, reproductive, musculoskeletal and hematopoietic systems. These genes link a functional alteration in a putative zinc transcription factor and a specific developmental abnormality. Hence, they provide the experimental tools to study

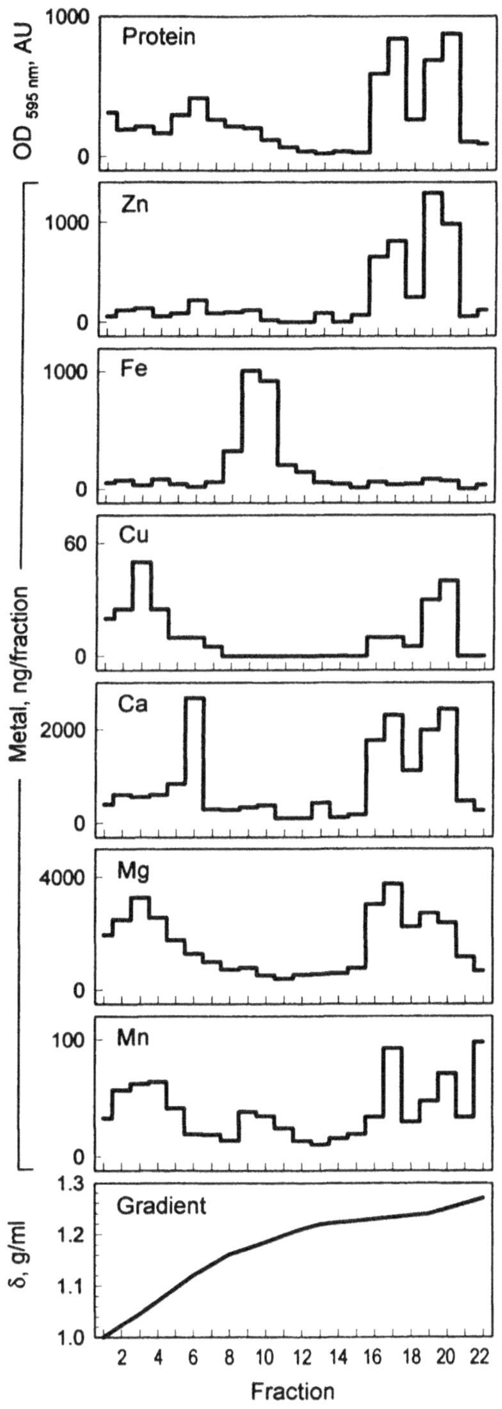

Fig. 5. Distribution of metals in oocyte compartments. Homogenates were centrifuged in a sucrose gradient from 1.0 to 1.25 g/ml (lower panel). Protein was measured by the Bio Rad total protein method. Metals were analyzed by atomic absorption spectrometry (Nomizu *et al.* 1993). The majority (80%) of the protein is in the yolk platelets localized to fractions 17–22. Nearly 90% of the zinc is located in those organelles. Other metals also found in the platelets include copper, calcium, magnesium and manganese. All of these metals are also distributed in less dense fractions containing cytosol, mitochondria and multivesicular bodies.

the distribution of zinc in the embryo and the molecular events that result when zinc is not incorporated into crucial proteins.

The first steps toward this objective have been taken with *Xenopus laevis*. Curtailment of zinc available from the putative cytosolic stores has been accomplished with incubation of the embryos in solutions containing the chelating agent OP. These studies provide support for the proposal that the cytosolic pool, in fact, is the one that distributes zinc to apoproteins during embryogenesis. An OP concentration of 10^{-3} M is required to remove zinc from lipovitellin, as described above. Incubating embryos with OP at concentrations of 100-fold lower does not remove zinc from lipovitellin, yet results in embryos that lack formation of head structures, including the brain and eyes among other classic teratology of zinc deficiency (Jornvall *et al.* 1993; Montorzi *et al.* 2000). About 74% of the embryos hatch. The embryos are smaller, manifest craniofacial malformations, including microcephaly. They do not form head structures; the brain and eyes are absent and there is extrusion of ocular bud regions. The spine and tail regions evidence blebs. Somites and heart are absent. The malformed embryos survive for about 24–48 h after hatching. The stages of embryogenesis that are most sensitive to the teratogenic effects of the chelating agent are 7–15, the period when migration of germ cell lines and their organization into future organs is achieved.

Additional information is emerging relating transcription factors directly involved in development that is considered to be zinc dependent. These would be molecules that could be affected in zinc deficiency. Thus, as described above, zinc deficiency results in pattern of activation and/or repression of a set of genes that is distinct from that of zinc sufficient cells. This manifests in the formation of particular gene products together with the synthesis of others. These products function as necessary components of proliferation and are presumed to be necessary for development as well. Failure to express them at a critical juncture in the process of organogenesis or to supply and incorporate zinc into these macromolecules could result in abnormal phenotypes.

The identification of the specific zinc transcription factors that could be affected and which could be responsible for the pathology of zinc deficiency has yet to be carried out. A number of transcription factors, expressed in cells and tissues of developing embryos, are examples of the many candidates that continue to emerge in the literature (Table 5). They have been categorized as belonging to the class of zinc dependent transcription factors because of the existence in their primary amino acid sequences of 'zinc binding motifs' that are homologous to those of the known zinc transcription factors (Table 4). While these have not been isolated or characterized as actual zinc proteins, their functions in developmental processes are of particular interest. The ones shown (Table 5) are selected on the basis of their apparent involvement in the formation of tissues, such as the nervous, reproductive, musculoskeletal and hematological systems, all known to be targets of zinc deficiency (Keen & Hurley 1989). In some instances, gene mutation or knock out experiments have been carried out and resulted in lack of expression of normal gene products accompanied by either abnormal or even absent anatomical structures and/or organs.

Concluding remarks

The gene products listed in Table 5 serve to generate an initial list of specific examples with which to study the biochemical functions and characteristics of their transcription products in both zinc sufficient and deficient organisms. Towards this end, the information on the content, uptake and distribution of zinc in *Xenopus laevis* together with the effects of chelating agents on its development provides a suitable biological system to study the relationship between zinc, transcription factors, differentiation and organogenesis.

Acknowledgements

This work was supported, in part, by the Endowment for Research in Human Biology, Inc. and the Bert and Natalie Vallee Foundation, Boston, Massachusetts.

References

Archer VE, Breton J, Sanchez-Garcia I, Osada H, Forester A, Thompson AJ, Rabbitts TH. 1994 Cysteine-rich LIM domains of LIM–homeodomain and LIM- only proteins contain zinc but not iron. *Proc Natl Acad Sci USA* **91**, 316–320.

Auld DS, Falchuk KH, Zhang K, Montorzi M, Vallee BL. 1996 X-ray absorption fine structure as a monitor of zinc coordination sites during oogenesis of *Xenopus laevis*. *Proc Natl Acad Sci USA* **93**, 3227–3231.

Ball LJ, Diakun GP, Gadhavi PL, Young NA, Armstrong EM, Garner CD, Laue ED. 1995 Zinc coordination in the DNA binding domain of the yeast transcription activator PPR1. *FEBS Lett* **358**, 278–282.

Banaszak L, Sharrock W, Timmins P. 1991 Structure and function of a lipoprotein: lipovitellin. *Annu Rev Biophys Chem* **20**, 221–246.

Bernard O, Ganiatsas S, Kannourakis G, Dringen R. 1994 Kizl, a protein with LIM zinc finger and kinase domains, is expressed mainly in neurons *Cell Growth Diff.* **5**, 159–1171.

Cai Y, Chia W, Yang X. 2001 A family of snail-related zinc finger proteins regulates two distinct and parallel mechanisms that mediate *Drosophila* neuroblast asymmetric divisions. *EMBO J* **20**, 1704–1714.

Costa RH, Kalinichenko VV, Lim L. 2001 Transcription factors in mouse lung development and function. *Am J Physiol* **280**, L823–L838.

Czupryn M, Falchuk KH, Vallee BL. 1987 Zinc deficiency and metabolism of histones and nonhistone proteins in *Euglena gracilis*. *Biochemistry* **26**, 8263–8269.

Davidson EH. 1990 How embryos work: A comparative view of diverse modes of cell specification. *Development* **108**, 365–389.

Denis H, Le Maire M. 1983 Thesaurisomes, a novel kind of nucleoprotein particle. *Subcell Biochem* **9**, 263–297.

Dreosti IE, Tao S, Hurley LS. 1968 Plasma zinc and leukocyte changes in weanling and pregnant rats during zinc deficiency. *Proc Soc Exp Biol Med* **127**, 169–174.

Falchuk KH. 1998 The molecular basis for the role of zinc in developmental biology. *Mol Cell Biochem* **188**, 41–48.

Falchuk KH, Montorzi M, Vallee BL. 1995 Zinc uptake and distribution in *Xenopus Laevis* oocytes and embryos. *Biochemistry* **34**, 16524–16531.

Frankel AD, Chen L, Cotter RJ, Pabo CO. 1988 tat protein from human immunodeficiency virus forms a metal linked dimer. *Science* **240**, 70–73.

Freedman LP, Luisi BF, Korszun ZR, Basavappa R, Sigler PB, Yamamoto KR. 1988 The function and structure of the metal coordination sites within the glucocorticoid receptor DNA binding domain. *Nature (London)* **334**, 543–546.

Gallaher D, Hurley LS. 1980 Low zinc concentration in rat uterine fluid after 4 days of dietary deficiency. *J Nutr* **110**, 591–593.

Groche D, Rashkovetsky LG, Falchuk KH, Auld DS. 2000 Subunit composition of the zinc proteins alpha- and beta-lipovitellin from chicken. *J Protein Chem* **19**, 379–387.

Hanas JS, Hazuda D, Bogenhagen DF, Wu FY-H, Wu C-W. 1983 *Xenopus* transcription factor A requires zinc for binding to 5S gene. *J Biol Chem* **258**, 14120–14125.

Hansen P, Riebessell M. 1991 *The early development of Xenopus Laevis*. Berlin: Springer-Verlag: 1–18.

Halvorsen YC, Nandabaln K, Dickson RC. 1990 LAC 9 DNA-binding domain coordinates two zinc atoms per monomer and contacts DNA as a dimer. *J Biol Chem* **265**, 13283–13289.

Hurley LS, Shrader RE. 1975 Abnormal development of preimplantation rat eggs after three days of maternal dietary zinc deficiency. *Nature (London)* **254**, 427–429.

Jornvall H, Falchuk KH, Geraci G, Vallee BL. 1993 1,10-phenanthroline and *Xenopus laevis* teratology. *Biochem Biophys Res Comm* **200**, 1398–1406.

Johnston M. 1987 Genetic evidence that zinc is an essential cofactor in the DNA binding domain of GAL4 protein. *Nature (London)* **328**, 353–355.

Keen CL, Hurley LS. 1989 Zinc and reproduction: Effects of deficiency on foetal and postnatal development. In: Mills CF. ed. *Zinc in Human Biology*. London: Springer-Verlag: 183–220.

Kostich WA, Sanes JR. 1995 Expression of zfh-4, a new member of the zinc finger-homeodomain family, in developing brain and muscles. *Dev Dyn* **202**, 145–152.

Krishnaraju K, Nguyen HQ, Liebermann DA, Hoffman B. 1995 The zinc finger transcription factor Egr-I potentiates macrophage differentiation of hematopoietic cells. *Mol Cell Biol* **15**, 5499–5507.

Kuwahara J, Coleman JE. 1990 Role of zinc (II) ions in the structure of the three finger DNA binding domain and the SP1 transcription factor. *Biochemistry* **29**, 8628–8631.

Kuhnlein RP, Frommer G, Friedrich M, Gonzalez-Gaitain M, Weber A, Wagner-Bernholz JF, Gehring WJ, Jackle H, Schuh R. 1994 spalt encodes an evolutionary conserved zinc finger protein of novel structure which provides homeotic gene function in the head and tail region of the *Drosophila* embryo. *EMBO J* **13**, 168–179.

Li PM, Reichter J, Freyd G, Horvitz HR, Walsh CT. 1991 The LIM region of a presumptive *Caenorhabditis elegans* transcription factor is an iron-sulfur and zinc containing metallodomain. *Proc Natl Acad Sci USA* **88**, 9210–9213.

Liao X, Clemens KR, Tennant PE, Wright JM, Gottesfeld M. 1992 Specific interaction of the first three zinc fingers of TFIIIA with the internal control region of the *Xenopus* 5S RNA gene. *J Mol Biol* **223**, 857–871.

Mazus B, Falchuk KH, Vallee BL. 1984 Histone formation, gene expression and zinc deficiency in *Euglena gracilis*. *Biochemistry* **23**, 42–44.

Mellerick DM, Kassis JA, Zhang SD, Odenwald WF. 1992 castor encodes a novel zinc finger protein required for the development of a subset of CNS neurons in *Drosophila*. *Neuron* **9**, 789–803.

Mevel-Ninio M, Terracol R, Kafatos FC. 1991 The ovo gene of *Drosphila* encodes a zinc finger protein required for female germ line development. *EMBO J* **10**, 2259–2266.

Miller J, McLachlan AD, Klug A. 1985 Repetitive zinc binding domains in the protein transcription factor IIIA from *Xenopus oocytes*. *EMBO J* **4**, 1609–1614.

Montorzi M, Falchuk KH, Vallee BL. 1995 Vitellogenin and Lipovitellin: zinc proteins of *Xenopus Laevis* oocytes. *Biochemistry* **34**, 10851–10858.

Montorzi M, Burgos MH, Falchuk KH. 2000 *Xenopus laevis* embryo development: arrest of epidermal cell differentiation by the chelating agent 1,10 phenanthroline. *Mol Reprod Dev* **55**, 75–82.

Nagai T, Aruga J, Takada S, Gunther T, Sporle R, Schugart K, Mikoshiba, K. 1997 The expressions of the mouse Zicl, Zic2 and Zic3 gene suggest an essential role for zic genes in body pattern formation. *Dev Biol* **182**, 299–313.

Nomizu T, Falchuk KH, Vallee BL. 1993 Zinc, iron and copper contents of *Xenopus laevis* oocytes and embryos. *Mol Reprod Dev* **36**, 419.

Opresko LK. 1991 *Xenopus Laevis*: Practical Uses in Cell and Molecular Biology. *Methods Cell Biol* **36**, 117–132.

Opresko L, Karpf RA. 1987 Specific proteolysis regulates fusion between endocytic compartments in *Xenopus* oocytes. *Cell* **51**, 557–568.

Pan T, Coleman JE. 1990 GAL4 transcription factor is not a 'zinc finger' but forms a Zn(II)2Cys6 binuclear cluster. *Proc Natl Acad Sci USA* **87**, 2077–2081.

Perrotti D, Melotti P, Skowedki T, Casella I, Peschle C, Calebretta B. 1995 Over-expression of the zinc protein MZI inhibits hematopoietic development from embryonic stem cells. Correlation with negative regulation of CD34 and cmyc promoter activity. *Mol Cell Biol* **15**, 6075–6087.

Redemann N, Gaul U, Jackle H. 1988 Disruption of a putative cys-zinc interaction eliminates the biological activity of the Kruppel finger protein. *Nature (London)* **332**, 90–92.

Roark M, Sturtevant MA, Emery J, Vaessin H, Grell E, Brier E. 1995 scratch, a pan-neural gene encoding a zinc finger protein related to snail, promotes neuronal development. *Genes Dev* **9**, 2384–2398.

Schutz B, Niessing J. 1994 Cloning and structure of a chicken zinc finger cDNA: Restricted expression in developing neural crest cells. *Gene* **148**, 227–236.

Schwabe JWR, Neuhaus D, Rhodes D. 1990 Solution structure of DNA binding domain of the oestrogen receptor. *Nature (London)* **348**, 458–460.

Sequeval D, Felenbok B. 1994 Relationship between zinc content and DNA binding activity of the DNA- binding motif of the transcription factor ALCR in *Aspergillus nidulans*. *Mol Gen Genet* **242**, 33–39.

Stankiewicz A, Falchuk KH, Vallee BL. 1983 Composition and structure of zinc deficient *Euglena gracilis* chromatin. *Biochemistry* **22**, 5150–5156.

Stifani S, Nimpf J, Schneider WJ. 1990 Vitellogenesis in *Xenopus laevis* and chicken: cognate ligands and oocyte receptors. The binding site for vitellogenin is located on lipovitellin I. *J Biol Chem* **265**, 882–888.

Swiatek PJ, Gridely T. 1983 Perinatal lethality and defects in hindbrain development in mice homozygous for a targeted mutation of the zinc finger gene krox20. *Genes Dev* **7**, 2071–2084.

Timmerman JE, Guiard B, Schechter E, Delsuc MA, Lallemand JY, Gervais M. 1994 The DNA-binding domain of the yeast Saccharomyces cerevisiae CYP1 (HAP1) transcription factor possesses two zinc ions which are complexed in a zinc cluster. *Eur Biochem* **225**, 593–599.

Vallee BL, Falchuk KH. 1981 Zinc and gene expression. *Philos Trans R Soc Lond B Biol Sci* **294**, 185–197.

Vallee BL, Falchuk KH. 1993 The biochemical basis of zinc physiology. *Physiol Rev* **73**, 79–111.

Wallace RA. 1978 *The Vertebrate Ovary*. In: Jones RE. ed. New York: Plenum: 469–502.

Wallace RA, Jared DW. 1968 Estrogen induces lipophosphoprotein in serum of male *Xenopus laevis*. *Science* **160**, 91–92.

Wallace RA, Nickol JM, Ho T, Jared OW. 1972 Studies on amphibian yolk. X. The relative roles of autosynthetic and heterosynthetic processes during yolk protein assembly by isolated oocytes. *Dev Biol* **29**, 225–272.

Wallace RA, Opresko L, Wiley HS, Selman K. 1983 The oocyte as an endocytic cell. *Ciba Found Symp* **98**, 228–248.

BioMetals **14:** 397–412, 2001.

Review

Zinc metabolism and homeostasis: The application of tracer techniques to human zinc physiology

Nancy F. Krebs* & K. Michael Hambidge
*Section of Nutrition, Department of Pediatrics, University of Colorado School of Medicine, Denver, Colorado, USA; *Author for correspondence (Tel: (303) 315-7037; Fax: (303) 315-3273; E-mail: Nancy.Krebs@UCHSC.edu)*

Received 15 February 2001; accepted 17 May 2001

Key words: absorption, compartmental modeling, endogenous zinc excretion, stable isotopes, zinc homeostasis

Abstract

Tracer kinetic techniques based on zinc stable isotopes have a vital role in advancing knowledge of human zinc physiology and homeostasis. These techniques have demonstrated the complexity of zinc metabolism, and have been critical to estimating the size and interrelationships of those pools of zinc that exchange rapidly with zinc in plasma and which are likely to be especially important for zinc dependent biology. This paper presents findings from recent research linking a steady state compartmental model with non-steady state post-prandial sampling from the intestine, utilizing a combination of intestinal intubation/perfusion and stable isotope tracer kinetic techniques. The gastrointestinal tract has a central role in maintaining whole body zinc homeostasis. While the fractional absorption of zinc from a meal depends on the quantity of exogenous zinc and on such dietary factors as phytic acid, the fractional absorption does not appear to be dependent on the size of the rapidly exchanging pool of the host. In contrast, the quantity of endogenous zinc excreted via the intestine is positively correlated with both the amount of absorbed zinc and the zinc 'status' of the host, and thus this process has an equally critical role in maintaining zinc homeostasis. The observed alterations in zinc metabolism in some disease states can be understood in the context of known homeostatic processes. In other conditions, however, such alterations as inflammation-associated hyperzincuria and zinc redistribution, the links between homeostatic perturbation and cellular biology are yet to be explained. Thus the challenge remains for research at the whole body level to carefully characterize zinc distribution and exchange under diverse circumstances, while research at the cellular level must elucidate the regulatory processes and the factors to which they respond.

Abbreviations: EZP – exchangeable zinc pool; DS – Down Syndrome.

Introduction

Parallel with the exceptional recent advances in our appreciation of the diversity, versatility and extraordinary importance of the cellular biology of zinc (Cousins 1998; McMahon & Cousins 1998), there has been notable progress in understanding human zinc physiology and homeostasis (Hambidge *et al.* 1998; King *et al.* 2000). Furthermore there has been a great expansion in recognition of the clinical and public health importance of this essential micronutrient (Bhutta *et al.* 1999). The purpose of this paper is to synthesize current concepts of human zinc physiology and homeostasis, based largely on isotopic tracer techniques. To complete the process of integration as far as is possible, the paper will conclude with a review of some clinical conditions and the observed changes in zinc homeostasis, for in these circumstances there may well be clues to both normal and abnormal physiology. This also serves as a challenging reminder that there remains a great deal of essential research ahead before knowledge of the cellular biology of this metal can be adequately integrated with the clinical and public health sequelae of zinc deficiency.

Tracer techniques

Both radioisotopes and stable isotopes of zinc have been utilized in investigations of human zinc physiology and homeostasis. In experienced hands and with sensitive equipment, radio-tracer techniques have made important contributions to knowledge of human zinc physiology over a period of more than half a century. Prior to the 1980s, tracer studies of human zinc physiology depended on radioisotope techniques and these have been employed effectively to develop detailed compartmental models of zinc metabolism (Wastney *et al.* 1986) and to study zinc absorption and bioavailability (Sandstrom & Lonnerdal 1989).

Over approximately the past twenty years, there has been a steadily growing body of experience and expertise in the application of zinc stable isotope techniques to investigate whole body human zinc homeostasis and physiology. This has been facilitated by advances in analytical instrumentation, especially in the development and application of inductively coupled plasma mass spectrometry (ICPMS). State of the art ICPMS instrumentation is capable of relatively rapid, precise, and accurate measurements of zinc stable isotope ratios.

Of greater fundamental importance, zinc has three stable isotopes for which the natural abundance is sufficiently low to allow their utilization as 'tracers'. These are ^{67}Zn (natural abundance 4.1%), ^{68}Zn (18.8%) and ^{70}Zn (0.6%). The availability of these 3 stable isotopes of zinc in low natural abundance concentration makes it possible to administer all three tracers essentially simultaneously via different routes. One example of the application of multi-tracer techniques is in the development of our compartmental model of zinc metabolism to be described later. The development of this model utilized kinetic data derived from administration of different zinc stable isotopes intravenously, orally in the post-absorptive state and orally with all meals on the same day. A second example is the facilitation of the comparison of the effects of different diets and different chemical forms of zinc on zinc bioavailability and homeostasis. Even with improved analytical sensitivity, however, caution is required to ensure that quantities of isotope administered are not themselves of sufficient magnitude to perturb the very physiology that is being examined.

The safety of the stable isotopes is unquestioned, and it is thus possible to utilize them in studies in women during the reproductive cycle and in the growing child, population groups for which zinc nutriture is of special interest. These techniques can be applied in studies of populations far removed from major research centers, especially in the developing world. This provides unique opportunities to learn from populations whose habitual diets are low in zinc (Lei *et al.* 1996) or are high in factors, especially phytic acid, that impair the bioavailability of this micronutrient (Manary *et al.* 2000). The intricacies of appropriate study design and reliable analyses, however, ideally dictate the involvement of experienced investigators and laboratories.

Tracer techniques open the door to probing key variables of zinc homeostasis. The refinement and application of this methodology represent a major advance in our ability to explore human zinc physiology, to understand the complexities of human zinc homeostasis and to gain new insights into why and when zinc deficiency occurs. Although the ability to perform complex whole body kinetic studies is clearly of importance, simpler applications are also frequently of special value. For example, arguably the most important information that has been derived from their application relates to the central role of the gastrointestinal tract in maintaining whole body zinc homeostasis. This can be achieved quantitatively and is dependent on only simple algebraic equations for data processing. In contrast to traditional metabolic balance methodology, application of these tracer techniques allows separation of individual variables of zinc homeostasis while providing greater accuracy and precision of measurements.

Whole body zinc physiology: Kinetic studies and compartmental analysis

Kinetic and other data derived from a combination of human zinc tracer and metabolic studies quickly become very complex. In these circumstances, model-based compartmental analysis is not only of immediate practical value in data analysis but also of heuristic value in exploring and better understanding the complexities of mammalian zinc metabolism. These models can be extended effectively to assist in linking the cellular biology of zinc to whole body zinc metabolism (Dunn & Cousins 1989).

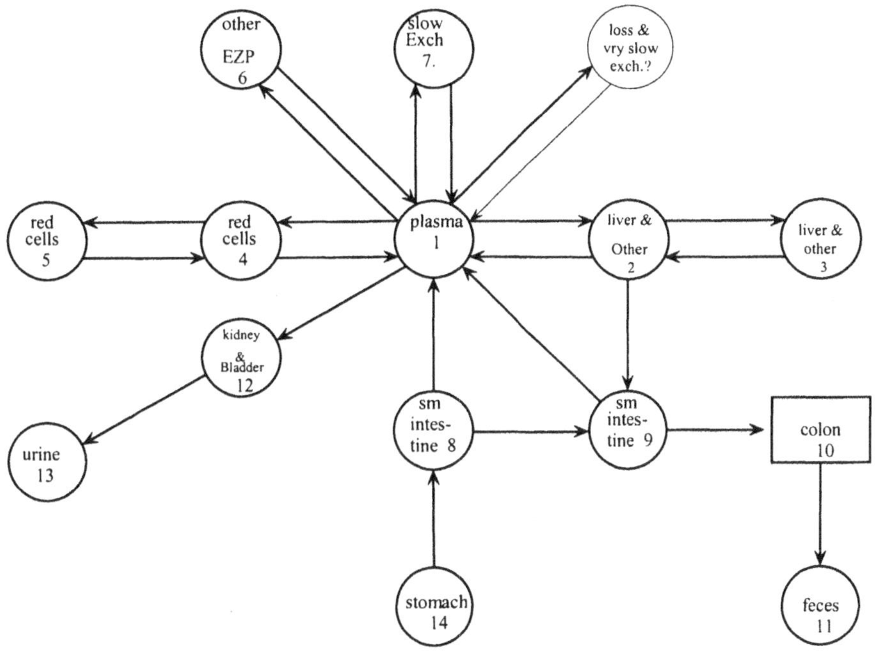

Figure 1. Structure of compartmental model developed to fit zinc stable isotope tracer kinetic data from 5 subjects. The circles represent compartments and are labeled with physiologic/anatomic or kinetic designation. The rectangle indicates the non-mixing delay compartment. (Figure adapted from compartmental model as described in reference Miller *et al.* 2000).

The complexity of human zinc physiology is apparent with the administration of a zinc tracer intravenously even when sampling is limited to blood (plasma and erythrocytes) and excreta. Adequate analysis of such data requires more than a sum of exponential analysis and investigators have turned increasingly to model-based compartmental analyses. Typically, these models have also incorporated additional data derived from the oral administration of tracers, and, in some instances, from regional scanning (radio-tracers only) and a range of steady state data including that derived from simple algebraic equations. Several such models have been reported (Lowe *et al.* 1997; Miller *et al.* 1998, 2000; Wastney, 1989; Wastney *et al.* 1986, 1996, 1991), the complexity of which varies according to the amount of data that is required to fit and other factors. We have recently published a model-based compartmental analysis of the steady state kinetic data obtained from studies in normal adults who received oral (fasting and with meals) and intravenous stable zinc isotopes. The extended multiple studies analysis (EMSA) program of SAAM/CONSAM (Miller *et al.* 1998, 2000) was applied to the individuals' data to derive a composite model (Figure 1). This is in contrast to other reports for which population parameter values were derived from the arithmetic means of individual parameters (Lowe *et al.* 1997; Wastney *et al.* 1986). In our experience, this is the simplest compartmental model that provides an adequate fit for data derived primarily from measurements of enrichment in plasma, erythrocytes, urine and feces over approximately a two week period following administration of zinc stable isotopes orally and intravenously. Both the structural identifiability and validity of this model were thoroughly documented. It includes fourteen compartments and twenty-five kinetic parameters that were not measured directly. Although more slowly exchanging pools were not completely identifiable with the limited duration of these sample collections, this model serves as one illustration of the complexity of human zinc physiology (Miller *et al.* 2000).

Both a strength and a weakness of these compartmental analyses is that, while they help to link kinetic data to some specific organs, the compartments in these models do not, in general, correspond precisely to a specific organ, except for those tissues in which tracer is measured directly. Regional scanning after administration of ^{65}Zn demonstrated, for example, the central role of the liver in zinc metabolism (Wastney *et al.* 1986). It is the principal organ that accounts for the second and third plasma exponential decay curves for a zinc tracer administered intravenously, and more detailed investigations with an animal model have identified a third rapidly exchanging liver compartment, attributable to hepatic metallothionein (Dunn & Cousins 1989; Lowe *et al.* 1991).

Other rapidly exchanging compartments are less well defined anatomically, but are known to represent zinc in multiple organs. These have been identified to some extent by the use of animal models, in which tracer and tracee have been analyzed in selected individual tissues. For example, zinc in kidney and spleen has been specifically shown to be part of the rapidly exchanging system (House & Wastney 1997). These authors speculated that other components of the immune system contribute to this compartment, and other investigators have demonstrated that the zinc tracer in bone marrow is rapidly exchanging (Dunn & Cousins 1989). Such animal studies have been useful in several ways. For example, the demonstration that even those tissues which account for the great part of the slowly exchanging zinc (e.g., bone) also contain more rapidly exchanging tracer (House & Wastney 1997), suggesting different carriers and/or transporters in tissue subtypes. Even with these animal models, however, there is a notable lack of analytical data for some organs that are of special interest with respect to zinc metabolism, including the central nervous system (Frederickson *et al.* 2000) and the pancreas (Andrews *et al.* 1990; Dalton *et al.* 1996; De Lisle *et al.* 1996; Kelly *et al.* 1996; Onosaka *et al.* 1988; Rofe *et al.* 1999). The importance of these rapidly exchanging pools, defined by their kinetic parameters, will be discussed later in relation to key processes of zinc homeostasis.

Despite the limitations of these models, their full potential has likely not yet been tapped. They have been used only to a limited extent to compare zinc physiology in different populations, for example the elderly (Wastney *et al.* 1992), in conditions of varying zinc intake (Wastney *et al.* 1986), or in disease states in which zinc homeostasis is likely to be perturbed (Lowe *et al.* 1995; Wastney *et al.* 1996, 1999).

The next several paragraphs are devoted to a description of unpublished data by Krebs *et al.* which is included to illustrate the compartmental modeling of a more complex zinc kinetic study that yielded both steady state and non-steady state data and required more detailed modeling of the gastrointestinal tract and intestinal – systemic interchange (Krebs *et al.* 1999). The development of the recently published 'composite' steady state model (Miller *et al.* 2000) (Figure 1) provided the framework to which intestinal perfusion and aspiration data for a four hour non-steady state period after a test meal have been incorporated. The study design included passage of a multilumen intestinal tube into the proximal jejunum, which had duodenal and jejunal perfusion ports for perfusion of nonabsorbable marker to be used to calculate flow rates, and aspiration ports in the duodenum and jejunum for aspiration of intestinal contents after the test meal. The particular subject to whom reference will be made in this text ingested a liquid test meal containing ^{67}Zn as an extrinsic label. Intravenous infusion of ^{70}Zn preceded the test meal by one hour, and was followed by frequent blood sampling to provide kinetic data. To evenly label all of the exchangeable pools, including sources of intestinal endogenous zinc, by the time of the intestinal intubation, ^{68}Zn was infused intravenously 10 days prior. The use of 3 tracers thus allowed us to model separately the movement of exogenous and endogenous zinc.

The non-steady state model of the intestinal aspiration data, including 3 sampling ports in the proximal small bowel, 3 isotopes (tracers) and natural zinc (tracee) (Krebs *et al.* 1999; Krebs *et al.* 1998b), is presented for this subject in Figures 2 and 3. To simplify presentation of these complex data, the models illustrating flow of exogenous zinc and endogenous zinc are shown separately. Additionally, although the steady state model (Figure 1) is not shown in these figures because of space limitations, it is critical to note that the data and models shown below are 'linked' to the steady state system, so the models fit both systems (steady state and non-steady state), and reflect exchange of all 3 tracers.

Figure 2 indicates the total flow of exogenous zinc (labelled with ^{67}Zn) over the $\sim$4 h after the test meal, which contained a total of 5.48 mg Zn. The numbers beside arrows going from the intestinal compartments into the plasma indicate zinc absorbed into the system. The figures between the intestinal compartments indicate amounts (mg) of zinc flowing 'down' the intestine. The amount of exogenous zinc exiting the system via each of the aspiration ports is also shown (circles at bottom of figure). The maximal absorption (0.69 mg) is seen from the most proximal compartment (duodenum), which represents $\sim$12.5% of intake from meal and dose. At the most distal port, 4 mg of exogenous zinc has flowed past, either to be absorbed more distally or be excreted in the feces. The fractional absorption (FAZ) calculated by the model, based on addition of amounts transferred

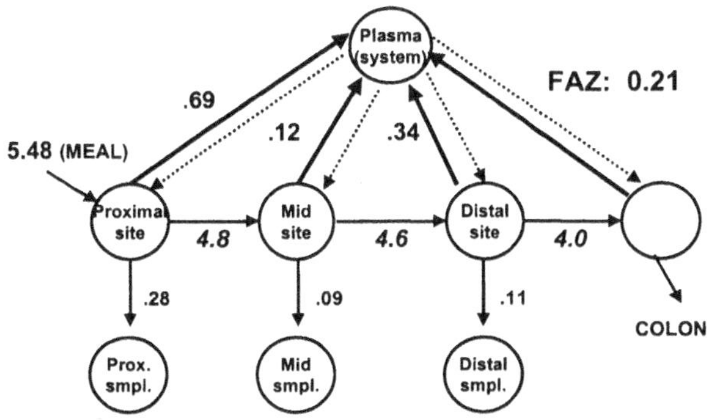

Figure 2. Total flow of exogenous zinc (mg) after test meal (0–4.3 h), based on flow of oral isotope (^{67}Zn). Full non-steady state model is based on intestinal aspiration data, including 3 sampling ports, 3 isotopes (tracers) and natural zinc (tracee), and is linked to steady state model with same tracers. The flow of endogenous zinc for same subject and over same time period is shown in Figure 3. The outflow compartments indicate exogenous zinc exiting the system via sampling from aspiration ports. FAZ = fractional absorption of zinc.

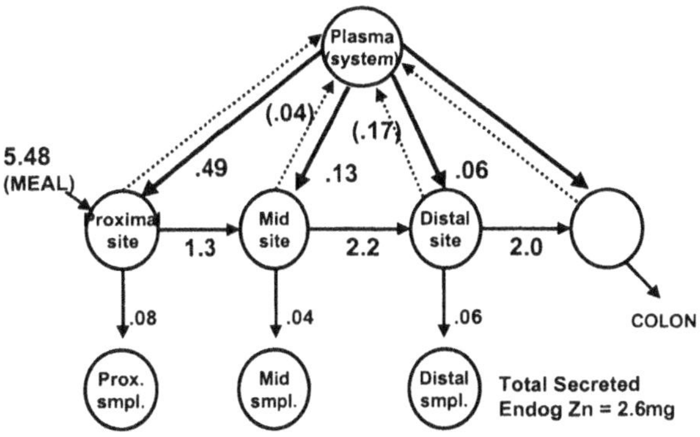

Figure 3. Total flow of endogenous zinc (mg) after test meal (0–4.3 h), based on flow of two intravenously administered isotopes (^{68}Zn and ^{70}Zn). The amounts of zinc flowing between the aspiration sites (proximal site, mid site, and distal site) reflect endogenous zinc present at each of the sites prior to sampling plus amount secreted into the gut from the system after the test meal. Full non-steady state model is based on intestinal aspiration data, including 3 sampling ports, 3 isotopes (tracers) and natural zinc (tracee), and is linked to steady state model with same tracers. The flow of exogenous zinc for same subject and over same time period is shown in Figure 2. The outflow compartments indicate endogenous zinc exiting the system via sampling from aspiration ports.

into the system = 0.21, is in good agreement with algebraic flow data from the aspirations, and with calculations of fractional absorption from urine isotope ratios (= 0.23). The plasma appearance of the oral tracer remained high through ~9 h after ingestion of the test meal, whereas the data from intestinal aspirates indicated no further disappearance from the intestinal lumen after ~ 3 h (Data not shown). Our tentative interpretation of this is that release of zinc from the enterocytes into the portal circulation or from hepatocytes into the peripheral circulation occurred over the longer time frame.

Figure 3 indicates flow of endogenous zinc after the test meal, based on flow of both of the intravenously administered isotopes (^{68}Zn & ^{70}Zn). The model projects not only the amount of each tracer, but the amount/flow of natural zinc, based on isotope 'dilution.' In this portion of the model, the exchange between the plasma and small intestine is bidirectional, indicating some apparent reabsorption of endogenously secreted zinc. The reabsorption figures are in parentheses between intestinal compartments and the plasma/system. It should be noted that the 'plasma' compartment as the source of the endogenous zinc is viewed as preliminary. Subsequent modeling from additional subjects suggests that the model may be better fitted to the data by having endogenous zinc originate from the liver/other compartment (see Figure 1).

The model predicts that the maximum secretion of endogenous zinc after a meal occurs in the proximal compartments, consistent with the hypothesis that a major part of the endogenous zinc secretion comes from the pancreaticobiliary secretions. The total amount secreted with the meal is projected to equal ~ 2.6 mg. The majority of reabsorption occurs from the distal compartment (arrow indicating 0.17 mg into plasma). Relating this to the anatomic location of the aspiration ports, this would be mid-jejunum and possibly extending into ileum. The model predicts 2.0 mg

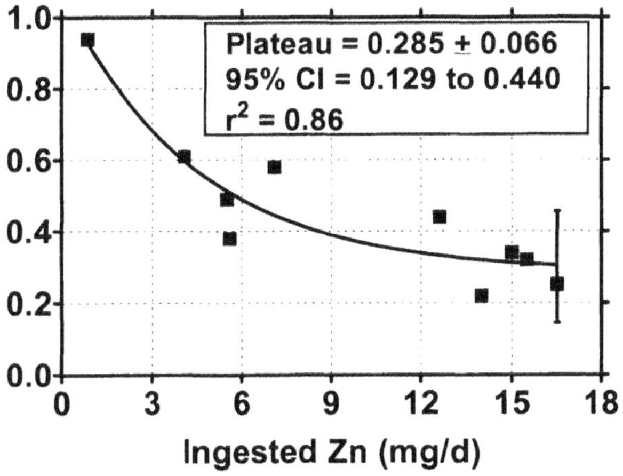

Figure 4. Inverse relationship between amount of ingested zinc and fractional absorption of zinc. Data points represent mean fractional absorption measurements based on stable isotope methods from several studies in healthy adult men. References to specific studies are provided in the text.

endogenous zinc passing between the distal site (jejunum) and the colon during the post-prandial period alone, which can either be reabsorbed more distally or excreted in the feces. Mean daily endogenous fecal zinc over 4 days for this subject was 4.5 mg. We thus predict that some of the endogenous zinc secreted in conjunction with this single test meal is likely to have been reabsorbed in the distal small bowel.

In summary, these data from the intestinal aspiration studies, using multiple tracers, and combining steady state with non-steady state kinetic data, represent an example of very complex application of compartmental modeling. In fact, the compartmental analysis is essentially the only way that all of the data can be analyzed simultaneously to characterize the exchange of tracer and tracee between the gut and the rest of the body. The model supports and extends calculations from data obtained by direct aspiration from the intestinal lumen: absorption of exogenous zinc likely primarily occurs in the proximal small bowel, i.e., between mid-distal duodenum and proximal jejunum, and disappearance from the intestinal lumen is apparently complete within 3 h of intake. If this is correct, there are implications for the anatomic distribution of the cellular absorptive transport mechanism. Secondly, the model suggests that the majority of endogenously secreted zinc enters in the proximal small bowel, consistent with a major source being the pancreaticobiliary secretions, although not necessarily the exclusive source. The model also supports the concepts that substantial amounts of zinc are secreted with meals (Matseshe *et al.* 1980), that maintenance of normal zinc homeostasis will be dependent on some reabsorption of the endogenous zinc, and that this most likely occurs in more distal small bowel, e.g., jejunum and possibly ileum. The role of the gastrointestinal tract in maintaining whole body zinc homeostasis will be considered further in the next major section.

Apart from their contribution to better understanding of the physiology of zinc and of its homeostasis, tracer techniques have potential to provide useful information about zinc nutritional 'status'. This has appeal for at least two reasons. First, despite intensive efforts, no sensitive biomarker of zinc status has yet been identified. Second, a number of studies have demonstrated that even modest depletion of critical pools of zinc result in functional compromise in zinc dependent processes, such as growth and immune function. There is thus considerable attractiveness to measurement of changes in these critical pools, especially rapidly exchanging pools, which may provide useful insights into zinc status (Lei *et al.* 1996; Lowe *et al.* 1995; Miller *et al.* 1994).

To give one example, the compartmental model has allowed us to evaluate the accuracy of estimates of the quantity of rapidly exchanging zinc, the exchangeable pool (EZP), which we define as zinc that exchanges/intermixes with zinc in plasma within three days and which accounts for only approximately 10% of the total body zinc content. In Figure 1, the EZP is comprised of the sum of the masses of compartments 1, 2, 3, 4 and 6.

Sum of exponential analyses after administration of a zinc tracer into the plasma of the systemic circulation indicates that four exponential decay terms are required to fit the tracer disappearance data from the plasma over the first 24 h. Extrapolation of the linear regression line fitting the log-transformed intravenous tracer enrichment data between 3 and 10 days after tracer administration to the y-axis (time 0) provides an estimate of the exchangeable zinc pool (EZP) (Miller *et al.* 1994; Miller *et al.* 1997). This is an estimate of the EZP that can be derived from urine as an alternative to plasma kinetic data and can be obtained under field conditions for adults (Lei *et al.* 1996), children (Manary *et al.* 2000), and infants (Krebs *et al.* 2000a). As will be discussed in the section addressing interrelationships between variables of zinc homeostasis, our experience is that EZP determinations can provide useful insights into zinc homeostasis and to differences in zinc status of individuals.

Zinc homeostasis and the gastrointestinal system

The gastrointestinal tract has the principal role in maintaining whole body zinc homeostasis. This is accomplished by modulation of the quantity of exogenous dietary zinc absorbed and of the quantity of endogenous zinc excreted. In no other organ system is it more necessary to amalgamate advances in knowledge of the cellular biology of zinc with parallel advances in our understanding of zinc physiology.

Fractional absorption of zinc

The fraction of dietary zinc absorbed is affected first by other dietary factors, especially those that reduce the fraction of zinc that is available for absorption by the intestinal mucosa. In general, the efficiency of absorption of zinc ingested with meals of any composition is less than that of zinc ingested as a simple salt in solution. Human milk has commonly been regarded as promoting the absorption of zinc. Fractional absorption of zinc in some breastfed infants is as high as 0.80, although average fractional absorption is ~0.6 (Krebs *et al.* 1996); (Krebs, unpublished data). Comparison of these figures for absorption of zinc from human milk with that from an aqueous solution (Lei *et al.* 1993), suggests that rather than promoting absorption, zinc absorption is not inhibited to a discernible extent by the sum of other factors in human milk.

The dietary factor that has received most recognition as a major inhibitor of zinc bioavailability is inositol hexaphosphate, or phytic acid (Sandstrom 1997; Sandstrom & Lonnerdal 1989). Phytic acid is present in all seeds, especially grains and legumes and is considered to be a major etiologic factor in human zinc deficiency globally (Gibson 1994). It is present in especially high concentration in cereal grains and legumes which provide the major food staples for many populations in the developing world. The inhibitory effects of phytic acid may be especially noteworthy at times of high requirement (Manary *et al.* 2000). Lumenal factors affecting zinc bioavailability, while important in determining dietary zinc requirements (WHO 1996), are not particularly relevant to bridging the whole body physiology and the cellular biology of zinc. Accordingly, bioavailability per se will not be a focus of this paper.

With the consumption of diets of relatively high zinc bioavailability, there is an inverse relationship between the quantity of zinc ingested and the fractional absorption of that zinc. This is illustrated in Figure 4, which is derived from the mean data for stable isotope studies of young, healthy adult men (Hunt *et al.* 1992; Jackson *et al.* 1984; Lee *et al.* 1993; Taylor *et al.* 1991; Turnlund *et al.* 1986, 1984; Wada *et al.* 1985). This relationship has a major impact on the absolute quantity of zinc absorbed. The decline in fractional absorption with increasing dietary zinc is an outstanding factor in maintaining zinc homeostasis when intake is excessive (Lowe & Jackson 2000; Wastney *et al.* 1986; Weigand 1983). Although fractional absorption increases with dietary zinc restriction (King *et al.* 2000; Lee *et al.* 1993; Taylor *et al.* 1991; Wada *et al.* 1985), there is uncertainty about how effectively this increase is maintained over periods of many months (Lee *et al.* 1993). Despite the inverse relationship between fractional absorption and ingested zinc, the quantity of zinc absorbed each day varies directly with the quantity of ingested zinc over a wide range of intake (Food and Nutrition Board 2001, pre-print; Lei *et al.* 1996). This implies that the changes in fractional absorption in response to changes in the quantity of ingested zinc are alone inadequate to maintain zinc homeostasis, especially with restricted levels of intake.

Whether fractional absorption of zinc is regulated in response to changes in zinc 'status' is not entirely clear. Typically, in zinc depletion studies, the quantity of zinc in tracer-labeled test meals has corresponded to that in the experimental low zinc diet rather than to the quantity of zinc in the baseline 'normal zinc' test meals. It is not possible to determine from such studies if observed increases in fractional absorption of zinc are related to changes in the zinc status of the host rather than attributable to the smaller quantity of absorbable zinc present at the brush border of enterocytes involved in zinc absorption.

There are, to date, only a few observations that are consistent with and most readily explained by regulation of fractional absorption of zinc in response to changes in the physiologic state of the host. The best documented of these is lactation, itself a very special physiologic state. In a rural population in northeast China that has an habitually low dietary zinc intake (Lei *et al.* 1996), fractional absorption of zinc was strikingly higher at six weeks' lactation: 0.53, compared to 0.33 in nonlactating women from the same area on similar diets (Lei *et al.* 2000). This and other reports of increased fractional absorption during lactation (Fung *et al.* 1997; Jackson *et al.* 1988; Moser-Veillon *et al.* 1996) strongly suggest that fractional absorption is indeed responsive to changes in host physiologic condition. The basis of the enhanced absorption during human lactation, which has not been consistently observed during pregnancy (Fung *et al.* 1997), is not known. The hormonal milieu of lactation may affect the intestinal absorptive surface, the transporters involved in zinc absorption, gastrointestinal motility, or other factors (Davies & Williams 1977).

The impact of other physiologic conditions in the host on fractional absorption is less clear. For example, when the typical high phytic acid content of the Malawian diet is reduced during recovery from both malnutrition and infection in young children, fractional absorption of zinc increased significantly (Manary *et al.* 2000). In contrast, an increase in fractional absorption with identical phytate reduction was not observed in relatively well Malawian children whose baseline fractional absorption was similar to that of the malnourished children. This difference in response may most readily be explained by host differences in physiologic requirements. Because the requirements were relatively low in the well children, it was hypothesized that an inhibitory effect of phytic acid on efficiency of utilization was not detectable, in contrast to the recovering malnourished children who likely needed a higher fractional absorption to meet requirements. This higher absorption could be achieved only when the inhibitory effect of high dietary phytate was removed. Caution is, however, required in interpreting these data as the subject numbers were small and the study was not designed prospectively to address this question.

There are other data from studies of the inhibitory effects of phytic acid on zinc absorption that can be plausibly explained by an effect of zinc 'status' on fractional absorption of zinc. Specifically, when healthy subjects whose habitual diets contain relatively little phytic acid are fed a high phytic acid test meal (Sandstrom & Sandberg 1992) or high phytic acid meals for a single day (Adams *et al.* 2001), fractional zinc absorption is relatively low. Recent observations in Malawi (Manary *et al.* 2000), (Manary, unpublished data) in subjects whose habitual diet is high in phytic acid suggest, however, that humans may be able to up-regulate absorption over time.

Some evidence also suggests that fractional absorption of zinc is not affected by zinc status. For example, we have observed that three weeks on a moderately zinc restricted diet was not associated with an increase in fractional absorption when an identical test meal was given for the two periods (Krebs *et al.* 2001). The lack of correlation between the size of EZP and fractional absorption in studies in both adults and infants also argues against a specific effect of 'status' on absorption (Krebs *et al.* 2000a; Lei *et al.* 1996).

Prospective human tracer studies carefully designed to address the factors affecting absorption are needed. Specific issues to be clarified include the effects of host factors, such as zinc status and physiologic state, vs. intralumenal factors, such as the amounts of zinc, phytic acid, and other dietary components. Clearly, there may also be interactions among these factors. It is apparent that parallel progress at a sub-cellular/molecular level and a human physiology level will be mutually invaluable in attaining this goal.

Total absorbed exogenous zinc

Although fractional absorption has been given considerable attention, especially since it is typically the variable that is actually measured with extrinsic labelling tracer techniques, it is the quantity of zinc absorbed per day, rather than the fractional absorption that seems to be of most practical importance. The total absorbed zinc (fractional absorption × zinc intake) is the variable that is directly impacted by changes in intake of available zinc. Beyond a certain level of zinc intake (which may correspond to approximate dietary requirements), increases in absorption of zinc are limited. Homeostatic mechanisms, however, do not prevent a small but progressive increase in absorption with increasing intake (Hambidge & Krebs 2001; Weigand 1983). At low intakes, total absorption progressively and relatively rapidly declines directly with the severity of zinc restriction. This is despite progressive increases in fractional absorption of the available zinc, which appear inadequate to maintain homeostasis alone even with relatively mild degrees of dietary zinc restriction. The direct relationship between daily zinc absorption and ingested zinc has been depicted in both animal models (Weigand 1983) and in humans (Food and Nutrition Board 2001, pre-print; Hambidge & Krebs 2001).

Excretion of endogenous zinc

Endogenous zinc is excreted via several routes, including the intestine, kidneys, integument, and semen. The intestine is not only the major route, but also the only one that is clearly subject to regulation at typical as well as at extreme levels of intake. Endogenous zinc excreted in the feces is typically at least twice that excreted via all other routes and can be several-fold higher. The quantity of endogenous zinc excreted via the intestine, i.e. in the feces, depends on both recent (Jackson *et al.* 1984; Johnson *et al.* 1993; Taylor *et al.* 1991) and long-term (Lee *et al.* 1993; Lei *et al.* 1996) zinc intake over a wide range of ingested zinc (Hambidge & Krebs 2001). The quantity can vary by an order of magnitude depending on zinc intake.

In contrast to fractional absorption, excretion of intestinal endogenous zinc is apparently regulated in response to changes in the zinc 'status' of the host over a wide range of typical dietary intake. Regulation appears to be quite rapidly responsive to changes in zinc 'status' and may be sensitive to minor changes. Adjustments in excretion of intestinal endogenous zinc to changes in zinc intake are maintained over prolonged periods (Lee *et al.* 1993; Lei *et al.* 1996). These characteristics identify endogenous intestinal zinc as a variable of zinc homeostasis that is of cardinal importance.

The quantity of endogenous zinc excreted in the feces is the difference between that secreted and that reabsorbed. With the anticipated rapid exchange of most zinc ligands (Williams 1989), it is difficult, although not impossible, to hypothesize differential reabsorption of endogenous zinc compared to absorption of exogenous zinc from the intestinal lumen. Therefore, if fractional absorption of dietary zinc is not regulated by zinc 'status', this is likely to be equally true for reabsorption of endogenous zinc. If this is so, the effects of zinc 'status' on the regulation of intestinal excretion of endogenous zinc should then be directed to the quantity of endogenous zinc secreted.

Relatively small amounts of endogenous zinc are secreted into the gastrointestinal tract in saliva, gastric juices and bile (Finley *et al.* 1994; Sullivan *et al.* 1965), with more possibly secreted through the small intestinal mucosal cells (Sturniolo *et al.* 1999), although the documentation in the human of the latter is quite limited. There is evidence, particularly from animal studies, to support the pancreatic secretions as a major source of endogenous zinc in the intestinal lumen (Adler *et al.* 1980; Birnstingle *et al.* 1956; Dijkstra *et al.* 1991; Lee *et al.* 1990; Van Wouwe & Uijlenbroek 1994). Results of intestinal aspiration/perfusion studies in humans are compatible with this conclusion, although none have distinguished pancreatic from biliary secretion (Krebs *et al.* 1999, 1998b; Lee *et al.* 1990; Sullivan *et al.* 1965). As noted earlier, both intestinal aspiration of labeled endogenous zinc and compartmental modeling suggest that the majority of endogenous secretion is quite proximal (Figure 3), i.e., consistent with pancreaticobiliary secretions into the duodenum (Krebs *et al.* 1999, 1998b). It is tempting to speculate that the rapid induction of metallothionein in response to zinc administration, including metallothionein in the pancreas, may have a role in the regulation of zinc secretion (Andrews *et al.* 1990; Dalton *et al.* 1996; De Lisle *et al.* 1996; Kelly *et al.* 1996). Recent detection of metallothionein in pancreaticobiliary secretions in the human duodenal lumen is consistent with such a hypothesis (Krebs, unpublished data).

The quantity of endogenous zinc secreted with a test meal appears to be substantial relative to the daily fecal excretion of endogenous zinc (Krebs *et al.* 1999, 1998b; Matseshe *et al.* 1980). To maintain normal zinc homeostasis, it thus seems probable that reabsorption of some endogenous zinc is essential. Based on the calculated net endogenous zinc flow at the most distal aspiration site (proximal jejunum) in our intestinal perfusion/aspiration studies and the resultant compartmental model (Figure 3), a large portion of the reabsorption is likely to occur in the more distal small bowel (Krebs *et al.* 1999, 1998b). The role of jejunum and ileum in reabsorption of endogenous zinc is further supported by the strong positive correlation observed between endogenous fecal zinc and fecal fat in infants with pancreatic insufficiency due to cystic fibrosis and with consequent fat malabsorption (Krebs *et al.* 2000b). This observation is discussed in more detail in a later section.

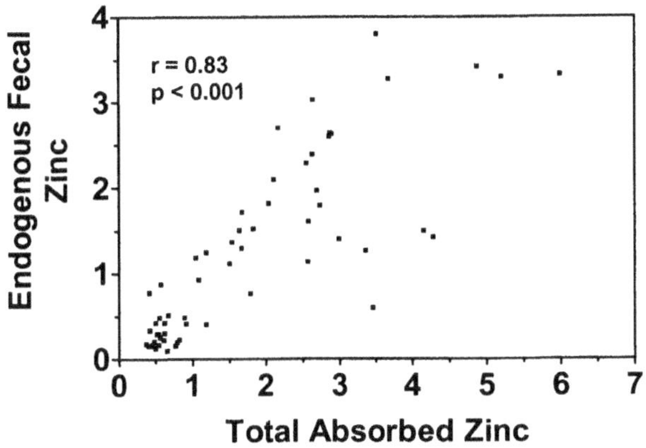

Figure 5. Correlation between total absorbed zinc and endogenous fecal zinc. Data points represent individual subjects from zinc stable isotope studies in infants and adults. Specific references are provided in the text.

Despite recent advances in characterization of zinc transporters and possible interaction with metallothionein, the molecular mechanisms and cellular processes in the gastrointestinal tract that are responsible for regulation of absorption of exogenous zinc and secretion and reabsorption of endogenous zinc await further clarification. Meanwhile, studies of the central role of the gastrointestinal tract in maintaining human zinc homeostasis at a whole body level are giving perspective to the quantitative importance of the key processes, their specific roles, and their gross anatomic localization.

Interrelationships between key variables of zinc homeostasis

As an increasing body of knowledge accumulates from the application of zinc stable isotope techniques, it is becoming apparent that there are some consistent and predictable interrelationships between key variables of zinc homeostasis under normal circumstances. Notable among these is the positive correlation between endogenous fecal zinc and total absorbed zinc, illustrated across a range of absorbed zinc in infants and adult subjects in Figure 5 (Krebs & Westcott 2001, in press; Lei *et al.* 1996). Of considerable practical importance is the growing yet incomplete evidence that this relationship holds not only when absorption exceeds physiologic requirement, but also at very low levels of absorption (Hambidge & Krebs 2001). The direct relationship between these two variables is clearly of central importance to zinc homeostasis and the achievement of zinc balance. It demands close attention in any factorial approach to calculating dietary zinc requirements (Food and Nutrition Board 2001, pre-print).

This association suggests that the quantity of endogenous zinc excreted in the feces is responsive to recent and habitual absorbed zinc (the quantity of, not the fraction of). If this is mediated through the effect of recent zinc absorption on zinc 'status', the effect on endogenous secretion is rapid enough (Jackson *et al.* 1984) that it is likely to be triggered by an increase in a component of the rapidly exchanging zinc pools (Miller *et al.* 1994). It was concluded by Chesters many years ago that the effects of zinc deprivation on feeding patterns and growth in mammalian models were so rapid that they must result from subtle, but physiologically important changes in the quantity of zinc in one or more rapidly exchangeable pools. Moreover, the quantity of zinc in this pool(s) must be very sensitive to dietary zinc (Chesters 1982). An observation that fits with this hypothesis is the positive correlation that has been observed between dietary zinc (Miller *et al.* 1994), and especially, total absorbed zinc and the size of the EZP (Krebs *et al.* 2000a; Lei *et al.* 1996).

The size of the EZP is also normally positively correlated with the quantity of endogenous zinc in the feces, consistent with the conclusion that it is some component of this rapidly exchanging system that is responsible for the regulation of the quantity of endogenous zinc secreted into and eventually excreted via the intestine. Incidentally, parallel correlations with plasma zinc have not been a consistent observation and there is some evidence that homeostatic mechanisms may maintain plasma zinc in circumstances that are associated with reduction in the size of the EZP (Lei *et al.* 1996).

Alterations in zinc metabolism during disease states

In general, understanding is limited for the specific metabolic processes underlying the observed changes in zinc transport and distribution in the setting of various pathologic conditions. Furthermore, the interplay between disease and dietary zinc deficiency has not been well characterized, particularly not in the human.

As described by Beisel, infection-induced malnutrition, the most common form of cytokine-induced malnutrition, occurs from the actions of proinflammatory cytokines, which initiate the acute phase response (APR) (Beisel 1995). In addition to the systemic symptoms (fever, malaise, myalgia etc.), a number of metabolic-nutritional responses also result from the APR, including protein catabolism, stimulation of metallothionein synthesis and sequestration of zinc, and many endocrinologic changes (Beisel 1995; Gabay & Kushner 1999). Extensive experimental work has demonstrated that hepatic metallothionein is involved in the response to stress. This can be induced by infusion of dexamethasone or other glucocorticoids, endotoxin and cytokines, as well as to a number of hormonal stimuli, including glucagon and epinephrine (Bremner & Beattie 1990; Cousins *et al.* 1986; Hernandez *et al.* 1996; Lehman-McKeeman *et al.* 1988; McCormick *et al.* 1981; Prasad 1993; Quinones & Cousins 1984; Schroeder & Cousins 1990). In hepatic tissue, the increase in metallothionein mRNA and metallothionein itself are also strongly correlated with increased hepatic zinc, and corresponding reduction of circulating zinc. In experiments with rats, the effects of stress and/or endotoxin on hepatic metallothionein synthesis were found to be significantly and synergistically enhanced by pretreatment with zinc, whether administered parenterally or enterally (Hernandez *et al.* 1996), emphasizing the ability of zinc itself to induce metallothionein synthesis. What is not by any means clear either from subcellular or whole animal, and certainly not from human research, is what role these profound changes in zinc metabolism have in combating stress and infection. Nor is it apparent what the end result of these changes in zinc metabolism is on zinc homeostasis and 'status'. It is unknown for example, if endogenous zinc losses are decreased, increased or unchanged.

The impact of disease state on zinc homeostasis can be considered under the broad headings of excessive losses (e.g. gastrointestinal tract and kidney), increased requirements (e.g. rapid growth, 'catch-up', tissue repair, immune stimulation), and redistribution (e.g. inflammation, closed head injury, Down syndrome, possibly Alzheimer disease). In general, tracer techniques have not been applied to systematically and comprehensively study zinc homeostasis under these clinical conditions.

Excessive losses

The dominant role of the gastrointestinal tract in normal zinc homeostasis has been described. It is thus not surprising that involvement of this organ system can result in significant perturbation of zinc homeostasis. A circular relationship of zinc deficiency and diarrhea is well recognized: severe zinc deficiency causes diarrhea and diarrhea may cause zinc deficiency. Proposed mechanisms for the diarrhea associated with zinc deficiency have included induction of certain proteins that result in increased fluid and possibly zinc secretion into the gastrointestinal tract. Examples include uroguanylin, cholecystokinin, and inducible nitric oxide synthase, all of which have increased expression during zinc deficiency (Abou-Mohamed *et al.* 1998; Blanchard & Cousins 1997; Wapnir 2000). Zinc deficiency is also associated with immune dysfunction. Impairment of the extensive immune system in the gastrointestinal tract may predispose to invasion by microorganisms as well as alter systemic immune responses (Scott & Koski 2000). Diarrhea from non-nutritional causes may cause excessive zinc losses and predispose to zinc deficiency by altering transit and/or the absorptive surface and thus impacting both absorption and reabsorption of exogenous and endogenous zinc. Despite the limitations in understanding of the complexities of zinc physiology in the setting of diarrheal disease, the results of a recent meta-analysis emphasize the remarkable benefit of zinc supplementation in the treatment and prevention of diarrhea in developing countries (Bhutta *et al.* 1999).

Cystic fibrosis represents a specific example of a disease with perturbed zinc homeostasis. Although pathological changes are discernible throughout the gastrointestinal tract, the outstanding pathophysiologic feature in the gastrointestinal system of this autosomal recessively inherited disease is pancreatic insufficiency. Effects on zinc metabolism, even in young infants at early stages of disruption of exocrine pancreatic function, include impairment of absorption of exogenous dietary zinc and excessive intestinal excretion of endogenous zinc (Easley *et al.* 1998; Krebs *et al.* 2000b). The quantity of the endogenous zinc excreted in the feces is positively correlated with fecal fat, which is typically excessive in this disease due to lipase deficiency. Since fat is absorbed primarily in the ileum, these findings suggest that this region of the intestine normally has a substantial role in the reabsorption of endogenous zinc that is secreted post-prandially. This observation serves as a further reminder of the need to evaluate all regions of the intestine, especially the small intestine, in investigating the mechanism(s) responsible for zinc absorption and reabsorption. The fat malabsorption associated with pancreatic insufficiency, as in cystic fibrosis, and the accompanying excessive losses of endogenous zinc, would certainly predispose to zinc deficiency if persistent. Indeed, we have reported that in infants identified by newborn screening to have cystic fibrosis, approximately one third have hypozincemia, most likely representing zinc deficiency (Krebs *et al.* 1998a).

A number of conditions are associated with hyperzincuria, but the underlying mechanism has not been characterized, nor is it clear whether there may be more than one mechanism. Hyperzincuria is associated with many chronic inflammatory states, including especially liver disease (Hambidge *et al.* 1987; Narkewicz *et al.* 1999; Sullivan & Lankford 1965), but also inflammatory bowel disease (Fleming *et al.* 1981), closed head injury (McClain 1990), skeletal trauma (Askari *et al.* 1982), cancer (Melichar *et al.* 1994), and diabetes (Chausmer, 1998). Whether there is any relationship between metallothionein in the kidney and hyperzincuria in these clinical conditions has not been reported. It is also tempting to speculate that one of the recently characterized zinc transporters, such as ZnT-1, which has been suggested to have a zinc exporting function, may be induced by the inflammatory response (Palmiter & Findley 1995). The relatively rapid normalization of the hyperzincuria observed after liver transplant in patients with chronic liver disease also suggests that there may be systemic signals, such as cytokines, that drive the hyperzincuria (Narkewicz *et al.* 1999).

Increased requirements

Several clinical conditions are characterized by tissue proliferation and by relatively high zinc requirements. Early infancy and childhood, adolescence, and the reproductive cycle are obvious times during the normal life cycle when zinc requirements are increased (Food and Nutrition Board 2001, pre-print; King 2000; King & Turnlund 1989; Krebs & Hambidge 1986). Zinc deficiency has been documented in all of these conditions (Brown *et al.* 1998; Caulfield *et al.* 1998, 1999; Goldenberg *et al.* 1995; Hambidge *et al.* 1972; Prasad *et al.* 1961; Walravens & Hambidge 1976).

Infants born prematurely have a particularly high requirement for zinc absorption and retention to achieve intrauterine accretion rates. Stable isotope methodology, including compartmental analysis, has been applied to this population to characterize variables of zinc homeostasis (Ehrenkranz *et al.* 1989; Friel *et al.* 1996; Jalla *et al.* 1997; Wastney *et al.* 1996, 1999). Results of these studies have generally concluded that healthy growing premature infants can achieve *in utero* zinc accretion rates (Ehrenkranz *et al.* 1989; Jalla *et al.* 1997; Wastney *et al.* 1999). Further, we found a significant positive correlation between average daily rate of weight gain and net absorbed zinc, emphasizing the importance of optimizing zinc retention (Jalla, unpublished data). To date, the rigorous demands of the application of tracer methods has limited their use to relatively stable preterm infants. Given the critical role of zinc in normal growth and development, such techniques are likely to offer important insights into potential differences in zinc homeostasis between normal and growth retarded neonates.

Other less well characterized clinical circumstances in which zinc requirements are exceptionally high are traumatic and surgical wound healing, conditions which are often complicated by considerable inflammatory response and concurrent increased zinc losses (Agren 1990; Iwata *et al.* 1999). Activation of the immune response is associated with an increase in the need for zinc, due to its involvement with cell replication and lymphocyte clonal expansion, as well as with lymphocyte activation (Fraker *et al.* 2000; Shankar & Prasad 1998). There is ample documentation of the detrimental effect of zinc deficiency on the immune response, but little is known about the impact of immune stimulation on whole body zinc homeostasis, that is, in changes in distribution and exchange rates between tissues, in uptake and excretion. Zinc concentration in the circulation has been proposed to be especially important, with both low and high levels impacting leukocyte responsiveness (Rink & Kirchner 2000). Clearly the complexity of the immune system presents significant challenges to the application of tracer techniques, but likewise, whole body studies may provide important complementary insight to in vitro studies of small components of this system.

Redistribution

The shifts in zinc distribution that occur in inflammation and the development of the acute phase response have been described above. Patients with Down Syndrome (DS, Trisomy 21) have been repeatedly found to have, on average, low plasma zinc levels despite dietary zinc intakes that are unremarkable (Chiricolo *et al.* 1993, 1994a; Licastro *et al.* 1994b; Napolitano *et al.* 1990; Stabile *et al.* 1991; Sustrova & Strbak 1994). Zinc supplementation has been undertaken in several trials, with positive effects on thyroid function (Napolitano *et al.* 1990; Sustrova & Strbak 1994), growth (Napolitano *et al.* 1990), humoral and cellular immune function, and apoptosis in peripheral lymphocytes (Antonucci *et al.* 1997). Altered zinc metabolism has also been proposed to be at least part of the basis of the accelerated aging in the DS population (Licastro *et al.* 1994b). No studies have been undertaken to utilize tracer techniques to study variables of zinc homeostasis or pool sizes. Thus it is not known whether there are differences in uptake and retention of exogenous zinc or whether the apparent zinc deficit is the result of differences in the exchangeable pool sizes or in total body zinc. Although there are significant challenges to applying stable isotope techniques to this population, the information that could be gleaned from such studies is potentially invaluable to advance understanding of zinc metabolism in this specific population and in general.

Future directions

The potential rewards of synergy in zinc research between cellular biology and human physiology and nutrition are becoming increasingly apparent as progress in each of these areas accelerates. Those of us involved in human zinc research are dependent on parallel advances in research directed to the cellular biology of zinc.

Such advances at the subcellular level are essential to achieve an adequate understanding of zinc homeostatic mechanisms, their interrelationships and limitations. This, in turn, is necessary if, for example, we are to really understand dietary zinc requirements and the limitations of homeostasis beyond which zinc deficiency or toxicity will occur. The considerations in this paper highlight the gastrointestinal tract and its associated organs which have a central role in the maintenance of human zinc homeostasis. Acceleration in zinc tracer research, supplemented by special techniques such as intestinal intubation/perfusion or by regional scanning of the distribution of radio-zinc, is now at least starting to provide clearer insights into the regulation of major variables of zinc homeostasis and into the interrelationships between these variables. Temporal and anatomic aspects of homeostasis are recognized, although not yet totally clarified, especially the regulation of endogenous zinc excretion. Future progress in these areas can assist in guiding the direction of cellular and molecular research on the mechanisms and regulation of zinc absorption and excretion.

Advances in the cellular biology of zinc alert the human nutrition researcher to the remarkable scope and diversity of zinc-dependent biology and metabolism. These range from more generalized functions, including those related to transcription, cellular growth, and the diverse roles of metallothionein, to highly specific functions such as the role of zinc in synaptic signaling in the central nervous system. These advances also hold out hope of new biomarkers of zinc status, for which there is a real need. Simultaneously, progress in our understanding not only of the clinical but also of the global public health importance of human zinc deficiency, highlights those directions in which advances in cellular biology are likely to have special immediate relevance to human health. These include, for example: cellular growth and differentiation, the biological roles of zinc in the immune system, and other aspects of host defense mechanisms and the role of zinc in cognitive function. Finally, despite recent advances, wide gaps remain between recent advances in our understanding of the cellular biology of zinc and specific links with the clinical features of zinc deficiency.

Acknowledgements

This work was supported by the following grants from the National Institutes of Health, General Clinical Research Centers RR-00069 and RR00051; K08-DK-02240; Clinical Nutrition Research Unit, P30-DK-48520. The authors also acknowledge the critical input from Leland V. Miller, B.S. and Jamie E. Westcott, M.S., and other members of our research team for the data presented in this manuscript.

References

Abou-Mohamed G, Papapetropoulos A, Catravas JD, Caldwell RW. 1998 Zn^{2+} inhibits nitric oxide formation in response to lipopolysaccharides: implication in its anti-inflammatory activity. *Eur J Pharmacol* **341**, 265–272.

Adams CL, Hambridge KM, Raboy V, Dorsch JA, Sian L, Westcott JE, Krebs NF. 2001 Zinc absorption from a low phytic acid maize. *Am J Clin Nutr*, in press.

Adler M, Robberecht P, Mestdagh M, Cremer M, Delcourt A, Christophe J. 1980 The pancreatic secretion of zinc in man and rat. *Gastroenterol Clin Biol* **4**, 441–449.

Agren MS. 1990 Studies on zinc in wound healing. *Acta Derm Venereol Suppl* **154**, 1–36.

Andrews GK, Kage K, Palmiter-Thomas P, Sarras MP, Jr. 1990 Metal ions induce expression of metallothionein in pancreatic exocrine and endocrine cells. *Pancreas* **5**, 548–554.

Antonucci A, Di Baldassarre A, Di Giacomo F, Stuppia L, Palka G. 1997 Detection of apoptosis in peripheral blood cells of 31 subjects affected by Down syndrome before and after zinc therapy. *Ultrastruct Pathol* **21**, 449–452.

Askari A, Long CL, Blakemore WS. 1982 Net metabolic changes of zinc, copper, nitrogen, and potassium balances in skeletal trauma patients. *Metabolism* **31**, 1185–1193.

Beisel WR. 1995 Herman Award Lecture, 1995: Infection-induced malnutrition–from cholera to cytokines. *Am J Clin Nutr* **62**, 813–819.

Bhutta ZA, Black RE, Brown KH, Gardner JM, Gore S, Hidayat A, Khatun F, Martorell R, Ninh NX, Penny ME, Rosado JL, Roy SK, Ruel M, Sazawal S, Shankar A. 1999 Prevention of diarrhea and pneumonia by zinc supplementation in children in developing countries: pooled analysis of randomized controlled trials. Zinc Investigators' Collaborative Group. *J Pediatr* **135**, 689–697.

Birnstingle M, Stone B, Richards V. 1956 Excretion of radioactive zinc (65Zn) in bile, pancratic and duodenal secretions in the dog. *Am J Physiol* **186**, 377–379.

Blanchard RK, Cousins RJ. 1997 Upregulation of rat intestinal uroguanylin mRNA by dietary zinc restriction. *Am J Physiol* **272**, G972–G978.

Bremner I, Beattie JH. 1990 Metallothionein and the trace minerals. *Annu Rev Nutr* **10**, 63–83.

Brown KH, Peerson JM, Allen LH. 1998 Effect of zinc supplementation on children's growth: a meta-analysis of intervention trials. *Bibl Nutr Dieta* **54**, 76–83.

Caulfield LE, Zavaleta N, Figueroa A. 1999 Adding zinc to prenatal iron and folate supplements improves maternal and neonatal zinc status in a Peruvian population. *Am J Clin Nutr* **69**, 1257–1263.

Caulfield LE, Zavaleta N, Shankar AH, Merialdi M. 1998 Potential contribution of maternal zinc supplementation during pregnancy to maternal and child survival. *Am J Clin Nutr* **68**, 499S–508S.

Chausmer AB. 1998 Zinc, insulin and diabetes. *J Am Coll Nutr* **17**, 109–115.

Chesters JK. 1982 Metabolism and biochemistry of zinc. In: Prasad AS. ed. *Clinical, Biochemical, and Nutritional Aspects of Trace Elements*. Vol. 6. Current Topics in Nutrition and Disease. New York: Alan R. Liss; 221–238.

Chiricolo M, Musa AR, Monti D, Zannotti M, Franceschi C. 1993 Enhanced DNA repair in lymphocytes of Down syndrome patients: the influence of zinc nutritional supplementation. *Mutat Res* **295**, 105–111.

Cousins RJ. 1998 A role of zinc in the regulation of gene expression. *Proc Nutr Soc* **57**, 307–311.

Cousins RJ, Dunn MA, Leinart AS, Yedinak KC, DiSilvestro RA 1986 Coordinate regulation of zinc metabolism and metallothionein gene expression in rats. *Am J Physiol* **251**, E688–E694.

Dalton T, Fu K, Palmiter RD, Andrews GK. 1996 Transgenic mice that overexpress metallothionein-I resist dietary zinc deficiency. *J Nutr* **126**, 825–833.

Davies NT, Williams RB. 1977 The effect of pregnancy and lactation on the absorption of zinc and lysine by the rat duodenum *in situ*. *Br J Nutr* **38**, 417–423.

De Lisle RC, Sarras MP, Jr., Hidalgo J, Andrews GK. 1996 Metallothionein is a component of exocrine pancreas secretion: implications for zinc homeostasis. *Am J Physiol* **271**, C1103–C1110.

Dijkstra M, Kuipers F, Smit EP, de Vries JJ, Havinga R, Vonk, RJ. 1991 Biliary secretion of trace elements and minerals in the rat. Effects of bile flow variation and diurnal rhythms. *J Hepatol* **13**, 112–119.

Dunn MA, Cousins RJ. 1989 Kinetics of zinc metabolism in the rat: effect of dibutyryl cAMP. *Am J Physiol* **256**, E420–E430.

Easley D, Krebs N, Jefferson M, Miller L, Erskine J, Accurso F, Hambidge KM. 1998 Effect of pancreatic enzymes on zinc absorption in cystic fibrosis. *J Pediatr Gastroenterol Nutr* **26**, 136–139.

Ehrenkranz RA, Gettner PA, Nelli CM, Sherwonit EA, Williams JE, Ting BT, Janghorbani M. 1989 Zinc and copper nutritional studies in very low birth weight infants: comparison of stable isotopic extrinsic tag and chemical balance methods. *Pediatr Res* **26**, 298–307.

Finley JW, Johnson PE, Reeves PG, Vanderpool RA, Briske-Anderson M. 1994 Effect of bile/pancreatic secretions on absorption of radioactive or stable zinc. *In vivo* and *in vitro* studies. *Biol Trace Elem Res* **42**, 81–96.

Fleming CR, Huizenga KA, McCall JT, Gildea J, Dennis R 1981 Zinc nutrition in Crohn's disease. *Dig Dis Sci* **26**, 865–870.

Food and Nutrition Board, Institute of Medicine 2001 (pre-print). *Dietary Reference Intakes for Vitamin A, Vitamin K, Boron, Chromium, Copper, Iodine, Iron, Manganese, Molybdenum, Nickel, Silicon, Vanadium and Zinc*. Washington, DC: National Academy Press.

Fraker PJ, King LE, Laakko T, Vollmer TL. 2000 The dynamic link between the integrity of the immune system and zinc status. *J Nutr* **130**, 1399S–1406S.

Frederickson CJ, Suh SW, Silva D, Thompson RB. 2000 Importance of zinc in the central nervous system: the zinc-containing neuron. *J Nutr* **130**, 1471S–1483S.

Friel JK, Andrews WL, Simmons BS, Miller LV, Longerich HP 1996 Zinc absorption in premature infants: comparison of two isotopic methods. *Am J Clin Nutr* **63**, 342–347.

Fung EB, Ritchie LD, Woodhouse LR, Roehl R, King JC. 1997 Zinc absorption in women during pregnancy and lactation: a longitudinal study. *Am J Clin Nutr* **66**, 80–88.

Gabay C, Kushner I. 1999 Acute-phase proteins and other systemic responses to inflammation. *N Engl J Med* **340**, 448–454.

Gibson RS. 1994 Zinc nutrition in developing countries. *Nutr Res Rev* **7**, 151–173.

Goldenberg RL, Tamura T, Neggers Y, Copper RL, Johnston KE, DuBard MB, Hauth JC. 1995 The effect of zinc supplementation on pregnancy outcome. *JAMA* **274**, 463–468.

Hambidge KM, Hambidge C, Jacobs M, Baum JD. 1972 Low levels of zinc in hair, anorexia, poor growth, and hypogeusia in children. *Pediatr Res* **6**, 868–874.

Hambidge KM, Krebs NF, Lilly JR, Zerbe GO. 1987 Plasma and urine zinc in infants and children with extrahepatic biliary atresia. *J Pediatr Gastroenterol Nutr* **6**, 872–877.

Hambidge KM, Krebs NF, Miller L. 1998 Evaluation of zinc metabolism with use of stable-isotope techniques: implications for the assessment of zinc status. *Am J Clin Nutr* **68**, 410S–413S.

Hambidge M, Krebs NF. 2001. Interrelationships of key variables of human zinc homeostasis: relevance to dietary zinc requirements. *Annu Rev Nutr* **21**, 429–452.

Hernandez J, Giralt M, Belloso E, Rebollo DV, Romero B, Hidalgo J. 1996. Interactions between metallothionein inducers in rat liver and primary cultures of rat hepatocytes. *Chem Biol Interact* **100**, 27–40.

House WA, Wastney ME. 1997 Compartmental analysis of zinc kinetics in mature male rats. *Am J Physiol* **273**, R1117–R1125.

Hunt JR, Mullen LK, Lykken GI. 1992 Zinc retention from an experimental diet based on the US FDA Total Diet Study. *Nutr Res* **127**, 1335–1344.

Iwata M, Takebayashi T, Ohta H, Alcalde RE, Itano Y, Matsumura T. 1999 Zinc accumulation and metallothionein gene expression in the proliferating epidermis during wound healing in mouse skin. *Histochem Cell Biol* **112**, 283–290.

Jackson MJ, Giugliano R, Giugliano LG, Oliveira EF, Shrimpton R, Swainbank IG. 1988 Stable isotope metabolic studies of zinc nutrition in slum-dwelling lactating women in the Amazon valley. *Br J Nutr* **59**, 193–203.

Jackson MJ, Jones DA, Edwards RH, Swainbank IG, Coleman ML 1984 Zinc homeostasis in man: studies using a new stable isotope-dilution technique. *Br J Nutr* **51**, 199–208.

Jalla S, Krebs NF, Rodden IDJ, Miller LV. 1997 Zinc homeostasis in very low birth weight infants – a comparison of human milk and formulas. *Pediat Res* **41**, 233A.

Johnson PE, Hunt CD, Milne DB, Mullen LK. 1993 Homeostatic control of zinc metabolism in men: zinc excretion and balance in men fed diets low in zinc. *Am J Clin Nutr* **57**, 557–565.

Kelly EJ, Quaife CJ, Froelick GJ, Palmiter RD. 1996 Metallothionein I and II protect against zinc deficiency and zinc toxicity in mice. *J Nutr* **126**, 1782–1790.

King J. 2000 Determinants of maternal zinc status during pregnancy. *Am J Clin Nutr* **71(suppl)**, 1334S–1343S.

King J, Turnlund J. 1989 Human Zinc Requirements. In: Mills C. ed. *Zinc in Human Biology*. London: Human Nutrition Reviews. Springer-Verlag; 335–350.

King JC, Shames DM, Woodhouse LR. 2000 Zinc homeostasis in humans. *J Nutr* **130**, 1360S–1366S.

Krebs N, Westcott J, Miller L, Herrmann T, Hambidge K. 2000a Exchangeable zinc pool (EZP) in normal infants: correlates with parameters of zinc homeostasis. *FASEB J* **14**, A205.

Krebs NF, Hambidge KM. 1986 Zinc requirements and zinc intakes of breast-fed infants. *Am J Clin Nutr* **43**, 288–292.

Krebs NF, Reidinger CJ, Miller LV, Hambidge KM. 1996 Zinc homeostasis in breast-fed infants. *Pediatr Res* **39**, 661–665.

Krebs NF, Sontag M, Accurso FJ, Hambidge KM. 1998a Low plasma zinc concentrations in young infants with cystic fibrosis. *J Pediatr* **133**, 761–764.

Krebs NF, Westcott J. 2001 Zinc and breastfed infants: If and when is there a risk of deficiency? In: *Proceedings of 10th International Conference, International Society for Research in Human Milk and Lactation*, Vol. in press. Plenum Press.

Krebs NF, Westcott J, Miller LV. 1999 Localization of secretion and reabsorption of endogenous zinc by compartmental modeling of intestinal data. *FASEB J* **13**, A214.

Krebs NF, Westcott JE, Arnold TD, Kluger BM, Accurso FJ, Miller LV, Hambidge KM. 2000b Abnormalities in zinc homeostasis in young infants with cystic fibrosis. *Pediatr Res* **48**, 256–261.

Krebs NF, Westcott JE, Huffer JW, Miller LV. 1998b Absorption of exogenous zinc (Zn) and secretion of endogenous Zn in the human small intestine. *FASEB J* **12**, A345.

Krebs NF, Westcott JE, Sian L, Miller LV. 2001 Effect of dietary zinc (Zn) intake restriction on net secretion of intestinal endogenous zinc. *FASEB J* **14**, A402.

Lee DY, Prasad AS, Hydrick-Adair C, Brewer G, Johnson PE 1993 Homeostasis of zinc in marginal human zinc deficiency: role of absorption and endogenous excretion of zinc. *J Lab Clin Med* **122**, 549–556.

Lee HH, Hill GM, Sikha VK, Brewer GJ, Prasad AS, Owyang C 1990 Pancreaticobiliary secretion of zinc and copper in normal persons and patients with Wilson's disease. *J Lab Clin Med* **116**, 283–288.

Lehman-McKeeman LD, Andrews GK, Klaassen CD. 1988 Induction of hepatic metallothioneins determined at isoprotein and messenger RNA levels in glucocorticoid-treated rats. *Biochem J* **249**, 429–433.

Lei S, Hambidge KM, Westcott JL, Miller LV, Fennessey PV 1993 Influence of a meal and incremental doses of zinc on changes in zinc absorption. *Am J Clin Nutr* **58**, 533–536.

Lei S, Krebs NF, Westcott JE, Miller LV, Hambidge KM. 2001 Zinc homeostasis during lactation in a population with low zinc intakes. *Am J Clin Nutr*, in press.

Lei S, Mingyan X, Miller LV, Tong L, Krebs NF, Hambidge KM 1996 Zinc absorption and intestinal losses of endogenous zinc in young Chinese women with marginal zinc intakes. *Am J Clin Nutr* **63**, 348–353.

Licastro F, Chiricolo M, Mocchegiani E, Fabris N, Zannoti M, Beltrandi E, Mancini R, Parente R, Arena G, Masi M. 1994a Oral zinc supplementation in Down's syndrome subjects decreased infections and normalized some humoral and cellular immune parameters. *J Intellect Disabil Res* **38**, 149–162.

Licastro F, Morini MC, Davis LJ. 1994b Neuroendocrine immune modulation induced by zinc in a progeroid disease–Down's syndrome. *Ann N Y Acad Sci* **717**, 299–306.

Lowe NM, Bremner I, Jackson MJ. 1991 Plasma $_{65}$Zn kinetics in the rat. *Br J Nutr* **65**, 445–455.

Lowe NM, Hall EJ, Anderson RS, Batt RM, Jackson MJ. 1995 A stable isotope study of zinc kinetics in Irish setters with gluten-sensitive enteropathy. *Br J Nutr* **74**, 69–76.

Lowe NM, Jackson MJ. 2000 Kinetic studies of whole-body trace-element metabolism. In: Lowe NM, Jackson MJ. eds. *Advances in Isotope Methods for the Analysis of Trace Elements in Man.* London: CRC Press; 81–91.

Lowe NM, Shames DM, Woodhouse LR, Matel JS, Roehl R, Saccomani MP, Toffolo G, Cobelli C, King JC. 1997 A compartmental model of zinc metabolism in healthy women using oral and intravenous stable isotope tracers. *Am J Clin Nutr* **65**, 1810–1819.

Manary MJ, Hotz C, Krebs NF, Gibson RS, Westcott JE, Arnold T, Broadhead RL, Hambidge KM. 2000 Dietary phytate reduction improves zinc absorption in Malawian children recovering from tuberculosis but not in well children. *J Nutr* **130**, 2959–2964.

Matseshe JW, Phillips SF, Malagelada JR, McCall JT. 1980 Recovery of dietary iron and zinc from the proximal intestine of healthy man: studies of different meals and supplements. *Am J Clin Nutr* **33**, 1946–1953.

McClain CJ. 1990 The pancreas and zinc homeostasis [editorial]. *J Lab Clin Med* **116**, 275–276.

McCormick CC, Menard MP, Cousins RJ. 1981 Induction of hepatic metallothionein by feeding zinc to rats of depleted zinc status. *Am J Physiol* **240**, E414–E421.

McMahon RJ, Cousins RJ. 1998 Mammalian zinc transporters. *J Nutr* **128**, 667–670.

Melichar B, Jandik P, Tichy M, Malir F, Mergancova J, Voboril Z. 1994 Urinary zinc excretion and acute phase response in cancer patients. *Clin Investig* **72**, 1012–1014.

Miller LV, Hambidge KM, Naake VL, Hong Z, Westcott JL, Fennessey PV. 1994 Size of the zinc pools that exchange rapidly with plasma zinc in humans: alternative techniques for measuring and relation to dietary zinc intake. *J Nutr* **124**, 268–276.

Miller LV, Krebs NF, Hambidge KM. 1998 Human zinc metabolism: advances in the modeling of stable isotope data. *Adv Exp Med Biol* **445**, 253–269.

Miller LV, Krebs NF, Hambidge KM. 2000 Development of a compartmental model of human zinc metabolism: identifiability and multiple studies analyses. *Am J Physiol* **279**, R1671–R1684.

Miller LV, Krebs NF, Jefferson M, Easley D, Hambidge KM 1997 Compartmental modeling of human zinc metabolism: evaluation of a method for estimating the size of the rapidly exchanging pool of zinc. In: *Proceedings of Trace Element Metabolism in Man and Animals-9.* Ottawa: NRC Research Press; 144–145.

Moser-Veillon PB, Patterson KY, Veillon C. 1996 Zinc absorption is enhanced during lactation. *FASEB J* **10**, A729.

Napolitano G, Palka G, Grimaldi S, Giuliani C, Laglia G, Calabrese G, Satta MA, Neri G, Monaco F. 1990 Growth delay in Down syndrome and zinc sulphate supplementation. *Am J Med Genet Suppl* **7**, 63–65.

Narkewicz MR, Krebs N, Karrer F, Orban-Eller K, Sokol RJ 1999 Correction of hypozincemia following liver transplantation in children is associated with reduced urinary zinc loss. *Hepatology* **29**, 830–833.

Onosaka S, Min KS, Fujita Y, Tanaka K, Iguchi S, Okada Y 1988 High concentration of pancreatic metallothionein in normal mice. *Toxicology* **50**, 27–35.

Palmiter RD, Findley SD. 1995 Cloning and functional characterization of a mammalian zinc transporter that confers resistance to zinc. *EMBO J* **14**, 639–649.

Prasad A. 1993 Biochemistry of Metallothionein. In: *Biochemistry of Zinc* Biochemistry of the Elements. New York and London: Plenum Press; 77–92.

Prasad AS, Halsted JA, Nadimi M. 1961 Syndrome of iron deficiency anemia, hepatosplenomegaly, hypogonadism, dwarfism and geophagia. *Am J Med* **31**, 532–546.

Quinones SR, Cousins RJ. 1984 Augmentation of dexamethasone induction of rat liver metallothionein by zinc. *Biochem J* **219**, 959–963.

Rink L, Kirchner H. 2000 Zinc-altered immune function and cytokine production. *J Nutr* **130**, 1407S–1411S.

Rofe AM, Winters N, Hinskens B, Philcox JC, Coyle P. 1999 The role of the pancreas in intestinal zinc secretion in metallothionein-null mice. *Pancreas* **19**, 69–75.

Sandstrom B. 1997 Bioavailability of zinc. *Eur J Clin Nutr* **51** **(Suppl 1)**, S17–S19.

Sandstrom B, Lonnerdal B. 1989 Promotors and Antagonists of Zinc Absorption. In: Mills CF. ed. *Zinc in Human Biology.* London: ILSI Human Nutrition Reviews. Springer-Verlag; 57–78.

Sandstrom B, Sandberg AS. 1992 Inhibitory effects of isolated inositol phosphates on zinc absorption in humans. *J Trace Elem Electrolytes Health Dis* **6**, 99–103.

Schroeder JJ, Cousins RJ. 1990 Interleukin 6 regulates metallothionein gene expression and zinc metabolism in hepatocyte monolayer cultures. *Proc Natl Acad Sci USA* **87**, 3137–3141.

Scott ME, Koski KG. 2000 Zinc deficiency impairs immune responses against parasitic nematode infections at intestinal and systemic sites. *J Nutr* **130**, 1412S–1420S.

Shankar AH, Prasad AS. 1998 Zinc and immune function: the biological basis of altered resistance to infection. *Am J Clin Nutr* **68**, 447S–463S.

Stabile A, Pesaresi MA, Stabile AM, Pastore M, Sopo SM, Ricci R, Celestini E, Segni G. 1991 Immunodeficiency and plasma zinc levels in children with Down's syndrome: a long-term follow-up of oral zinc supplementation. *Clin Immunol Immunopathol* **58**, 207–216.

Sturniolo GC, Mestriner C, Irato P, Albergoni V, Longo G, D'Inca R. 1999 Zinc therapy increases duodenal concentrations of metallothionein and iron in Wilson's disease patients. *Am J Gastroenterol* **94**, 334–338.

Sullivan JF, Lankford HG. 1965 Zinc metabolism and chronic alcoholism. *Am J Clin Nutr* **17**, 57–63.

Sullivan JF, O'Grady J, Lankford HG. 1965 The zinc content of pancreatic secretion. *Gastroenterology* **48**, 438–443.

Sustrova M, Strbak V. 1994 Thyroid function and plasma immunoglobulins in subjects with Down's syndrome (DS) during ontogenesis and zinc therapy. *J Endocrinol Invest* **17**, 385–390.

Taylor CM, Bacon JR, Aggett PJ, Bremner I. 1991 Homeostatic regulation of zinc absorption and endogenous losses in zinc-deprived men. *Am J Clin Nutr* **53**, 755–763.

Turnlund JR, Durkin N, Costa F, Margen S. 1986 Stable isotope studies of zinc absorption and retention in young and elderly men. *J Nutr* **116**, 1239–1247.

Turnlund JR, King JC, Keyes WR, Gong B, Michel MC. 1984 A stable isotope study of zinc absorption in young men: effects of phytate and alpha-cellulose. *Am J Clin Nutr* **40**, 1071–1077.

Van Wouwe JP, Uijlenbroek JJ. 1994 The role of the pancreas in the regulation of zinc status. *Biol Trace Elem Res* **42**, 143–149.

Wada L, Turnlund JR, King JC. 1985 Zinc utilization in young men fed adequate and low zinc intakes. *J Nutr* **115**, 1345–1354.

Walravens PA, Hambidge KM. 1976 Growth of infants fed a zinc supplemented formula. *Am J Clin Nutr* **29**, 1114–1121.

Wapnir RA. 2000 Zinc deficiency, malnutrition and the gastrointestinal tract. *J Nutr* **130**, 1388S–1392S.

Wastney ME. 1989 Zinc absorption in humans determined using *in vivo* tracer studies and kinetic analysis. *Adv Exp Med Biol* **249**, 13–25.

Wastney ME, Aamodt RL, Rumble WF, Henkin RI. 1986 Kinetic analysis of zinc metabolism and its regulation in normal humans. *Am J Physiol* **251**, R398–R408.

Wastney ME, Ahmed S, Henkin RI. 1992 Changes in regulation of human zinc metabolism with age. *Am J Physiol* **263**, R1162–R1168.

Wastney ME, Angelus P, Barnes RM, Subramanian KN. 1996 Zinc kinetics in preterm infants: a compartmental model based on stable isotope data. *Am J Physiol* **271**, R1452–R1459.

Wastney ME, Angelus PA, Barnes RM, Subramanian KN. 1999 Zinc absorption, distribution, excretion, and retention by healthy preterm infants. *Pediatr Res* **45**, 191–196.

Wastney ME, Gokmen IG, Aamodt RL, Rumble WF, Gordon GE, Henkin RI. 1991 Kinetic analysis of zinc metabolism in humans after simultaneous administration of $_{65}$Zn and $_{70}$Zn. *Am J Physiol* **260**, R134–R141.

Weigand E. 1983 Absorption of trace elements: zinc. *Int J Vitam Nutr Res Suppl* **25**, 67–81.

WHO. 1996 *Trace Elements in Human Nutrition and Health.*: Geneva.

Williams RJP. 1989 An introduction to the biochemistry of zinc. In: Mills CF. ed. *Zinc in Human Biology*, Vol. 4. London: Springer-Verlag; 15–31.

Subject Index to Volume 14 Numbers 3-4

Author Index to Volume 14

Instructions to Authors

BioMetals

EDITOR-IN-CHIEF
G. Winkelmann
Microbiology & Biotechnology
Universität Tübingen
Auf der Morgenstelle 28
72076 Tübingen, Germany

Tel: (+49) 7071 297 304; Fax: (+49) 7071 295002

AIMS AND SCOPE

BioMetals is an international, multi-disciplinary journal devoted to the rapid publication of fundamental advances in both basic and applied research into the role of metal ions in biology, biochemistry and medicine.

The Journal aims to provide a forum for new research and clinical results on the structure and function of metal ions, metal chelates, siderophores, metal-containing proteins and biominerals in all biosystems. It is expected that *BioMetals* will stimulate cross-fertilization between medicine, biochemistry, pharmacology, toxicology, microbiology, cell biology, chemistry and plant physiology.

While the emphasis is on speed of publication, the Editorial Board will rigorously maintain refereeing standards to ensure that only the highest quality research is selected for publication.

MANUSCRIPT SUBMISSION

Kluwer-Academic Publishers prefer the submission of manuscripts and figures in electronic form in addition to a hard-copy printout. The preferred storage medium for your electronic manuscript is a 31/2 inch diskette. Please label your diskette properly, giving exact details on the name(s) of the file(s), the operating system and software used. Always save your electronic manuscript in the word processor format that you use: conversions to other formats and versions tend to be imperfect. In general, use as few formatting codes as possible. For safety's sake, you should always retain a backup copy of your file(s). **After acceptance**, please make absolutely sure that you send the latest (i.e., revised) version of your manuscript, both as hard-copy printout and on diskette.

Kluwer Academic Publishers prefer articles submitted in word processing packages such as MS Word. WordPerfect, etc. running under operating systems MS DOS. Windows and Apple Macintosh, or in the file format LaTeX. Articles submitted in other software programs, as well as articles for conventional type-setting, can also be accepted.

For submission in LaTeX. Kluwer Academic Publishers have developed a Kluwer LaTeX class file, which can be downloaded from www.kap.nl/kaphtml.htm/IFAHOME. Use of this class file is highly recommended. Do not use versions downloaded from other sites. Technical support is available at: texhelp@wkap.nl. If you are not familiar with TeX/LaTeX, the class file will be of no use to you. In that case, submit your article in a common word processor format.

For the purpose of reviewing, articles for publication may be submitted as hard-copy printout (4-fold) and on diskette to: For the Americas:

Prof. C.J. Carrano
Department of Chemistry
Southwest Texas State University
San Marcos, Texas 78666
USA
Tel: +1-516-245-3117
Fax: +1-516-245-2374
e-mail: cc05@swt.edu

For Japan:

Dr. Ichiro Okura
Dept. of Bioengineering
Tokyo Institute of Technology
Nagasutu Midori-ku
Yokohama 226-8501
Japan
Tel: +81-924-5752
Fax: +81-924-5778
e-mail: iokura@bio.titech.ac.jp

For Australia and Asia:

Prof. John Webb OAM
Dept. of Chemistry
Murdoch University
Perth. Western Australia 6150
Tel: +61-8-9360-2547
Fax: +61-8-9310-1711
e-mail: Johnwebb@central.murdoch.edu.au

For Europe and the rest of the world: Prof. G. Winkelmann. Editor-in-Chief.

Authors of reviews are requested to contact the Editor-in-Chief to discuss suitability before submitting their manuscript.

MANUSCRIPT PRESENTATION

The journal's language is English, British English or American English spelling and terminology may be used, but either one should be followed consistently throughout the article. Manuscripts should be printed or typewritten on A4 or US Letter bond paper, one side only, leaving adequate margins on all sides to allow reviewers' remarks. Please double-space all material, including notes and references. Quotations of more than 40 words should be set off clearly, either by indenting the left-hand margin or by using a smaller typeface. Use double quotation marks for direct quotations and single quotation marks for quotations within quotations and for words or phrases used in a special sense.

Number the pages consecutively with the first page containing:

- **running head (shortened tittle)**
- **article type**
- **title**
- **author(s)**
- **affiliation(s)**
- **full address for correspondence, including telephone and fax number and e-mail address**

Abstract Please provide a short abstract of 100 to 250 words. The abstract should not contain any undefined abbreviations or unspecified references.

Key words Please provide 5 to 10 key words or short phrases in alphabetical order.

Abbreviations Abbreviations and their explanations should be collected in a list.

Figures and Tables

Submission of electronic figures

In addition to hard-copy printouts of figures, authors are encouraged to supply the electronic versions of figures in either Encapsulated PostScript (EPS) or TIFF format. Many other formats, e.g., Microsoft Postscript. PiCT (Macintosh) and WMF (Windows), cannot be used and the hard copy will be scanned instead.

Figures should be saved in separate files *without* their captions, which should be included with the text of the article. Files should be named according to DOS conventions, e.g., 'figure1.eps'. For vector graphics, EPS is the preferred format. Lines should not be thinner than 0.25pts and in-fill patterns and screens should have a density of at least 10%. Font-related problems can be avoided by using standard fonts such as Times Roman and Helvetica. For bitmapped graphics. TIFF is the preferred format but EPS is also acceptable. The following resolutions are optimal: black-and-white line figures −600−1200 dpi: line figures with some grey or coloured lines – 600 dpi: photographs – 300 dpi; screen dumps – leave as is. Higher resolutions will not improve output quality but will only increase file size, which may cause problems with printing: lower resolutions may compromise output quality. Please try to provide artwork that approximately fits within the typeset area of the journal. Especially screened originals, i.e. originals with grey areas, may suffer badly from reduction by more than 10–15%.

AVOIDING PROBLEMS WITH EPS GRAPHICS

Please always check whether the figures print correctly to a PostScript printer in a reasonable amount of time. If they do not, simplify your figures or use a different graphics program.

If EPS export does not produce acceptable output, try to create an EPS file with the printer driver (see below). This option is unavailable with the Microsoft driver for Windows NT, so if you run Windows NT, get the Adobe driver from the Adobe site (www.adobe.com).

If EPS export is not an option, e.g., because you rely on OLE and cannot create separate files for your graphics, it may help us if you simply provide a PostScript dump of the entire document.

HOW TO SET UP FOR EPS AND POSTSCRIPT DUMPS UNDER WINDOWS

Create a printer entry specifically for this purpose: install the printer 'Apple Laserwriter Plus' and specify 'FILE': as printer port. Each time you send something to the 'printer' you will be asked for a filename. This file will be the EPS file or PostScript dump that we can use.

The EPS export option can be found under the PostScript tab. EPS export should be used only for single-page documents. For printing a document of several pages, select 'Optimise for portability' instead. The option 'Download header with each job' should be checked.

Submission of hard-copy figures

If no electronic versions of figures are available, submit only high-quality artwork that can be reproduced as is, i.e., without any part having to be redrawn or re-typeset. The letter size of any text in the figures must be large enough to allow for reduction. Photographs should be in black-and-white on glossy paper. If a figure contains colour, make absolutely clear whether it should be printed in black-and-white or in colour. Figures that are to be printed in black-and-white should not be submitted in colour. Authors will be charged for reproducing figures in colour.

Each figure and table should be numbered and mentioned in the text. The approximate position of figures and tables should be indicated in the margin of the manuscript. On the reverse side of each figure, the name of the (first) author and the figure number should be written in pencil: the top of the figure should be clearly indicated. Figures and tables should be placed at the end of the manuscript following the Reference section. Each figure and table should be accompanied by an explanatory legend. The figure legends should be grouped and placed on a separate page. Figures are not returned to the author unless specifically requested.

In tables, footnotes are preferable to long explanatory material in either the heading or body of the table. Such explanatory footnotes, identified by superscript letters, should be placed immediately below the table.

Section Headings

First-, second-, third, and fourth-order headings should be clearly distinguishable but not numbered.

Appendices Supplementary material should be collected in an Appendix and placed before the Notes and Reference sections.

Notes

Please use endnotes rather than footnotes. Notes should be indicated by consecutive superscript numbers in the text. A source reference not should be indicated by means of an asterisk after the title. The note should be placed at the bottom of the first page.

Cross-Referencing

In the text, a reference identified by means of an author's name should be followed by the date of the reference in parentheses and page number(s) where appropriate. When there are more than two authors, only the first author's name should be mentioned, followed by 'et al.'. In the event that an author cited has had two or more works published during the same year, the reference, both in the text and in the reference list, should be identified by a lower case letter like 'a' and 'b' after the date to distinguish the works.

Examples:

Winograd (1986, p. 204)
(Winograd 1986a, b)
(Winograd 1986; Flores et al. 1988)
(Bullen & Bennett 1990)

Acknowledgements

Acknowledgements of people, grants, funds, etc. should be placed in a separate section before the References.

References

References to books, journal articles, articles in collections and conference or workshop proceedings, and technical reports should be listed at the end of the article in order. Articles in preparation or articles submitted for publication, unpublished observations, personal communications, etc. should not be included in the reference list but should only be mentioned in the article text (e.g. T. Moore, personal communication).

References to books should include the author's name: year of publication: title: page numbers where appropriate: publisher: place of publication, in the order given in the example below.

Perrin DD, Armarego WLF. 1988 *Purification of Laboratory Chemicals*, 3rd edn. Oxford: Pergamon Press.

References to articles in an edited collection should include the author's name: year of publication: editor's name: title of collection:

first and last page numbers: publisher: place of publication in the order given in the example below.

Dionis JB, Jenny H-B, Peter HH. 1991 Therapeutically Useful Iron Chelators. In: Winkelmann G. ed. *CRC Handbook of Microbial Iron Chelates*. Boca Raton: CRC Press: 339–356.

References to articles in conference proceedings should include the author's name: year of publication: editor's name (if any): title of proceedings: first and last page numbers: place and date of conference: publisher and/or organization from which the proceedings can be obtained: place of publication, in the order given in the example below.

van Camp W, Bowler C, Villarroch R. et al. 1990 Characterisation of iron superoxide dismutase cDNAs from plants obtained by genetic complementarities in *Escherichia coli Proc Natl Acad Sci* USA **87**(24), 9903–9907

References to articles in periodicals should include the author's name: year of publication: article title: abbreviated title of periodical: volume number (issue number where appropriate): first and last page numbers, in the order given in the example below.

Kabsch W, Sander S. 1983 Dictionary of protein secondary structure: Pattern recognition of hydrogen-bonded and geometric features. *Biopolymers* **22**, 2577–2637.

References to technical reports or doctoral dissertations should include the author's name: year of publication: title of report or dissertation: institution: location of institution, in the order given in the example below.

Mukherjee AB. 1989 *The Release of Cadmium and Mercury into the Finnish Environment*. Report 64. Ministry of the Environment. Environmental Protection Department.

PROOFS
Proofs will be sent to the corresponding author. One corrected proof, together with the original, edited manuscript, should be returned to the Publisher within three days of receipt by mail (airmail overseas).

OFFPRINTS
25 offprints of each article will be provided free of charge. Additional offprints can be ordered by means of an offprint order form supplied with the proofs.

PAGE CHARGES AND COLOUR FIGURES
No page charges are levied on authors or their institutions. Colour figures are published at the author's expense only.

COPYRIGHT
Authors will be asked, upon acceptance of an article, to transfer copyright of the article to the Publisher. This will ensure the wildest possible dissemination of information under copyright laws.

PERMISSIONS
It is the responsibility of the author to obtain written permission for a quotation from unpublished material, or for all quotations in excess of 250 words in one extract or 500 words in total from any work still in copyright, and for the reprinting of figures, tables or poems from unpublished or copyrighted material.

ADDITIONAL INFORMATION
Additional information can be obtained from:
Jan Willem Wijnen, Kluwer Academic Publishers, P.O. Box 17, 3300 AA Dordrecht: telephone 078-6392155: fax 078-6392254: e-mail janwillem.wijnen@wkap.nl

GPSR Compliance

The European Union's (EU) General Product Safety Regulation (GPSR) is a set of rules that requires consumer products to be safe and our obligations to ensure this.

If you have any concerns about our products, you can contact us on ProductSafety@springernature.com

In case Publisher is established outside the EU, the EU authorized representative is:

Springer Nature Customer Service Center GmbH
Europaplatz 3
69115 Heidelberg, Germany

Batch number: 10824518

Printed by Printforce, the Netherlands